Strength of materials and structures

Strength of materials and structures

An introduction to the mechanics of solids and structures

(SI units)

JOHN CASE

M.A., F.R.Ae.S. *Formerly Head of Department of Applied Mechanics, Royal Naval Engineering College, Plymouth*

A. H. CHILVER

Vice Chancellor, Cranfield Institute of Technology. Formerly Chadwick Professor of Civil Engineering, University College, London

Edward Arnold

First Published 1959 as *Strength of Materials*
by Edward Arnold (Publishers) Limited
25 Hill Street
London WIX 8LL
Reprinted 1961 and 1964
Second edition 1971
Reprinted 1972

ISBN 07131 3243 4 Boards
07131 3244 2 Paper

Text set in 10/12 pt. Monotype Times, printed by photolithography,
and bound in Great Britain at The Pitman Press, Bath

Preface

The first edition of this book was published in 1959, and since then the book has had a number of reprintings. All these reprintings have shown that the book is widely used as an introductory text to the field of the strength of materials and structures, and it is hoped that this new edition will ensure the book's continuing usefulness. The first edition was published under the title of *Strength of Materials;* the book is used in fact as a general introduction to both the strength of materials and structures, and in the second edition this broader title has been chosen. As an elementary text, the book of course gives an introduction to the application of basic ideas in solid and structural mechanics to engineering problems.

The content covers most of the requirements of an engineering undergraduate in his first and second years, and in some cases for the whole of his course. For more advanced studies, the authors' Advanced Strength of Materials will cover the requirements for final honours degree courses and for post-graduate studies.

The book begins with a simple discussion of stresses and strains in materials and structural components, and the forms these take in tension, compression and shear; in Chapter 5 some simple general properties of stress and strain are first introduced. These basic properties are then applied to a wide range of problems, including shells, beams and shafts; plastic as well as elastic problems are treated. In Chapter 17 a simple introduction is given to the important principle of virtual work, and two special forms of this—leading to strain energy and complementary energy—are dealt with in Chapter 18. The final chapters are devoted, respectively, to buckling, vibrations and impact stresses.

Both worked examples and unsolved problems are given in the text, and all these are treated in SI units. Some of the examples, and many of the additional problems, are based on questions set by various examining bodies; the sources of these questions are shown in the text.

This new edition was begun by both authors, but because of Mr John Case's death in 1969, the new edition was completed by Dr Chilver. During the life of the first edition, many useful comments and corrections were suggested by readers; corrections and amendments based on these have been incorporated in this second edition; but readers' comments will still be most welcome.

Principal notation

a length
b breadth
c wave velocity, distance
d diameter
e eccentricity
h depth
j number of joints
l length
m mass, modular ratio, number of members
n frequency, load factor, distance
p pressure
q shearing force per unit length
r radius
s distance
t thickness
u displacement
v displacement, velocity
w displacement, load intensity, force
x coordinate
y coordinate
z coordinate

α coefficient of linear expansion
γ shearing strain
δ deflection
ϵ direct strain
η efficiency
θ temperature, angle of twist
ν Poisson's ratio

A area
C complementary energy
D diameter
E Young's modulus
F shearing force
G shearing modulus
H force
I second moment of area
J torsion constant
K bulk modulus
L length
M bending moment
P force
Q force
R force, radius
S force
T torque
U strain energy
V force, volume, velocity
W work done, force
X force
Y force
Z section modulus

ρ density
σ direct stress
τ shearing stress
ω angular velocity
Δ deflection
Φ step-function

Note on SI units

The units used throughout the book are those of the Système Internationale d'Unités; this is usually referred to as the SI system. In the field of the strength of materials and structures we are concerned with the following basic units of the SI system:

length	metre (m)
mass	kilogramme (kg)
time	second (s)
temperature	kelvin (K)

There are two further basic units of the SI system—electric current and luminous intensity—which we need not consider for our present purposes, since these do not enter the field of the strength of materials and structures. For temperatures we shall use conventional degrees centigrade (° C), since we shall be concerned with temperature changes rather than absolute temperatures. The units which we derive from the basic SI units, and which are relevant to our field of study, are:

force	newton (N)	$kg.m.s^{-2}$
work, energy	joule (J)	$kg.m^2.s^{-2} = Nm$
power	watt (W)	$kg.m^2.s^{-3} = Js^{-1}$
frequency	hertz (Hz)	cycle per second

The acceleration due to gravity is taken as:

$$g = 9{\cdot}81 \ ms^{-2}$$

Linear distances are expressed in metres and multiples or divisions of 10^3 of metres, i.e.

kilometre (km)	10^3 m
metre (m)	1 m
millimetre (mm)	10^{-3} m

In many problems of stress analysis these are not convenient units, and others, such as the centimetre (cm), which is 10^{-2} m, are more appropriate.

The unit of force, the newton (N), is the force required to give unit acceleration (ms^{-2}) to unit mass (kg). In terms of newtons the common force units in the foot-pound-second system (with $g = 9{\cdot}81 \ ms^{-2}$) are

$$1 \text{ lb.wt} = 4{\cdot}45 \text{ newtons (N)}$$

$$1 \text{ ton.wt} = 9{\cdot}96 \times 10^3 \text{ newtons (N)}$$

In general, decimal multiples in the SI system are taken in units of 10^3. The prefixes we make most use of are:

kilo	k	10^3
mega	M	10^6
giga	G	10^9

Thus:

$$1 \text{ ton.wt} = 9{\cdot}96 \text{ kN}$$

The unit of force, the newton (N), is used for external loads and internal forces, such as shearing forces. Torques and bending moments are expressed in newton-metres (Nm).

An important unit in the strength of materials and structures is stress. In the foot-pound-second system, stresses are commonly expressed in lb.wt/in^2, and tons/in^2. In the SI system these take the values:

$$1 \text{ lb.wt/in}^2 = 6{\cdot}89 \times 10^3 \text{ N/m}^2 = 6{\cdot}89 \text{ kN/m}^2$$

$$1 \text{ ton.wt/in}^2 = 15{\cdot}42 \times 10^6 \text{ N/m}^2 = 15{\cdot}42 \text{ MN/m}^2$$

Yield stresses of the common metallic materials are in the range:

$$200 \text{ MN/m}^2 \quad \text{to} \quad 750 \text{ MN/m}^2$$

Again, Young's modulus for steel becomes:

$$E_{\text{steel}} = 30 \times 10^6 \text{ lb.wt/in}^2 = 207 \text{ GN/m}^2$$

Thus, working and yield stresses will be expressed in MN/m^2 units, while Young's modulus will be given in GN/m^2 units.

Contents

1 Tension and compression: direct stresses

1.1 Introduction

The strength of a material, whatever its nature, is defined largely by the internal stresses, or intensities of force, in the material. A knowledge of these stresses is essential to the safe design of a machine, aircraft, or any type of structure. Most practical structures consist of complex arrangements of many component members; an aircraft fuselage, for example, is an elaborate system of interconnected sheeting, longitudinal stringers, and transverse rings. The detailed stress analysis of such a structure is a difficult task, even when the loading conditions are simple. The problem is complicated further because the loads experienced by a structure are variable and sometimes unpredictable. We shall be concerned mainly with stresses in materials under relatively simple loading conditions; we begin with a discussion of the behaviour of a stretched wire, and introduce the concepts of direct stress and strain.

1.2 Stretching of a steel wire

One of the simplest loading conditions of a material is that of *tension*, in which the fibres of the material are stretched. Consider, for example, a long steel wire held rigidly at its upper end, Fig. 1.1, and loaded by a mass hung from the lower end. If vertical movements of the lower end are observed during loading it will be found that the wire is

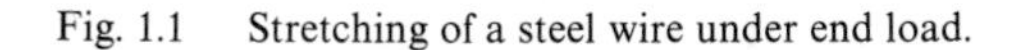

Fig. 1.1 Stretching of a steel wire under end load.

stretched by a small, but measurable, amount from its original unloaded length. The material of the wire is composed of a large number of small crystals which are only visible under microscopic study; these crystals have irregularly shaped boundaries, and largely random orientations with respect to each other; as loads are applied to the wire, the crystal structure of the metal is distorted.

For small loads it is found that the extension of the wire is roughly proportional to the applied load, Fig. 1.2. This linear relationship between load and extension was discovered by Robert Hooke in 1678; a material showing this characteristic is said to obey *Hooke's law.*

As the tensile load in the wire is increased, a stage is reached where the material ceases to show this linear characteristic; the corresponding point on the load-extension curve of Fig. 1.2 is known as the *limit of proportionality.* If the wire is made of a high-strength steel then the load-extension curve up to the *breaking point* has the form shown in Fig. 1.2. Beyond the limit of proportionality the extension of the wire increases non-linearly up to the breaking point.

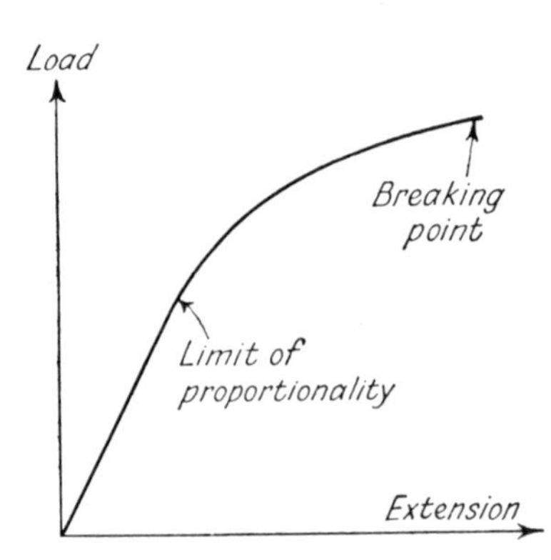

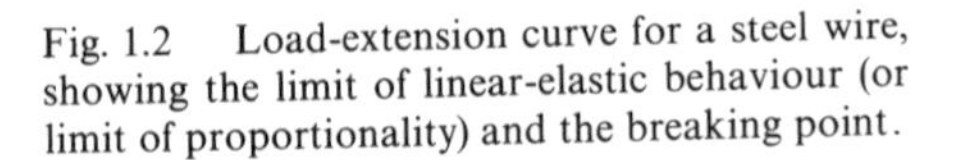
Fig. 1.2 Load-extension curve for a steel wire, showing the limit of linear-elastic behaviour (or limit of proportionality) and the breaking point.

The limit of proportionality is important because it divides the load-extension curve into two regions. For loads up to the limit of proportionality the wire returns to its original unstretched length on removal of the loads; this property of a material to recover its original form on removal of the loads is known as *elasticity*; the steel wire behaves, in fact, as a stiff elastic spring. When loads are applied above the limit of proportionality, and are then removed, it is found that the wire recovers only part of its extension and is stretched permanently; in this condition the wire is said to have undergone an *inelastic*, or *plastic*, extension.

In the case of elastic extensions, work performed in stretching the wire is stored as *strain energy* in the material; this energy is recovered when the loads are removed. During inelastic extensions work is performed in making permanent changes in the internal structure of the material; not all the work performed during an inelastic extension is recoverable on removal of the loads; this energy reappears in other forms, mainly as heat.

The load-extension curve of Fig. 1.2 is not typical of all materials; it is reasonably typical, however, of the behaviour of *brittle* materials, which are discussed more fully in §1.5. An important feature of most engineering materials is that they behave elastically up to the limit of proportionality, that is, all extensions are recoverable for loads up to this limit. The concepts of linearity and elasticity* form the basis of the theory of small deformations in stressed materials.

1.3 Tensile and compressive stresses

The wire of Fig. 1.1 was pulled by the action of a mass attached to the lower end; in this condition the wire is in *tension*. Consider a cylindrical bar *ab*, Fig. 1.3, which has

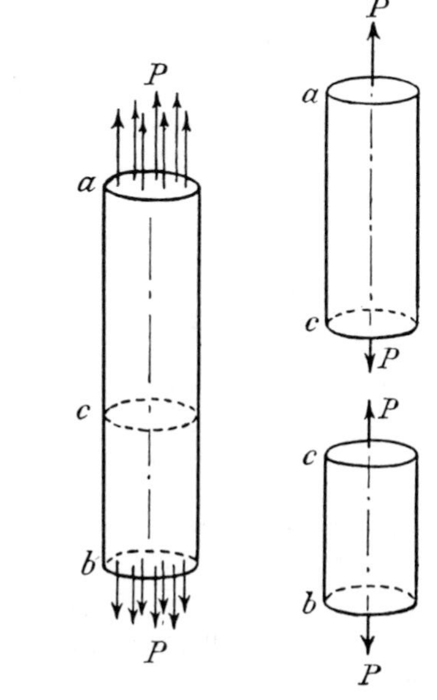

Fig. 1.3 Cylindrical bar under uniform tensile stress; there is a similar state of tensile stress over any imaginary normal cross-section.

a uniform cross-section throughout its length. Suppose that at each end of the bar the cross-section is divided into small elements of equal area; the cross-sections are taken normal to the longitudinal axis of the bar. To each of these elemental areas an equal tensile load is applied normal to the cross-section and parallel to the longitudinal axis of the bar. The bar is then uniformly stressed in tension.

Suppose the total load on the end cross-sections is P; if an imaginary break is made perpendicular to the axis of the bar at the section c, Fig. 1.3, then equal forces P are required at the section c to maintain equilibrium of the lengths ac and cb. This is equally true for any section across the bar, and hence on any imaginary section perpendicular to the axis of the bar there is a total force P.

* The definition of elasticity requires only that the extensions are recoverable on removal of the loads; this does not preclude the possibility of a non-linear relation between load and extension, although no such non-linear elastic relationships are known for materials in common use in engineering.

When tensile tests are carried out on steel wires of the same material, but of different cross-sectional areas, the breaking loads are found to be proportional approximately to the respective areas of the wires. This is so because the tensile strength is governed by the *intensity* of force on a normal cross-section of a wire, and not by the total force. This intensity of force is known as *stress*; in Fig. 1.3 the *tensile stress* σ at any normal cross-section of the bar is

$$\sigma = \frac{P}{A} \tag{1.1}$$

where P is the total force on a cross-section, and A is the area of the cross-section.

In Fig. 1.3 uniform stressing of the bar was ensured by applying equal loads to equal small areas at the ends of the bar. In general we are not dealing with equal force intensities of this type, and a more precise definition of stress is required. Suppose δA is an element of area of the cross-section of the bar, Fig. 1.4; if the normal force acting

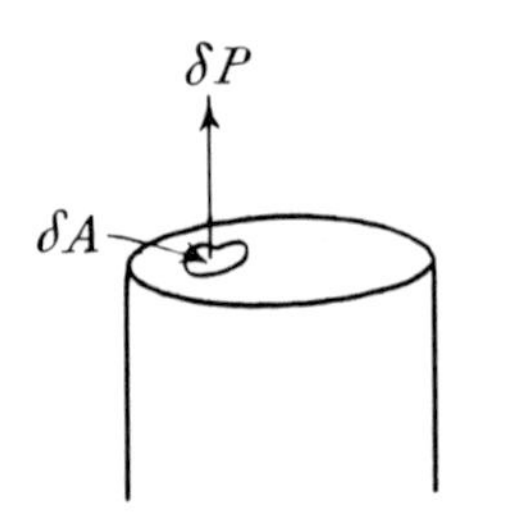

Fig. 1.4 Normal load on an element of area of the cross-section.

on this element is δP, then the tensile stress at this point of the cross-section is defined as the limiting value of the ratio $(\delta P/\delta A)$ as δA becomes infinitesimally small. Thus

$$\sigma = \underset{\delta A \to 0}{\text{Limit}} \frac{\delta P}{\delta A} = \frac{dP}{dA} \tag{1.2}$$

This definition of stress is used in studying problems of non-uniform stress distribution in materials.

When the forces P in Fig. 1.3 are reversed in direction at each end of the bar they tend to *compress* the bar; the loads then give rise to *compressive stresses.* Tensile and compressive stresses are together referred to as *direct stresses.*

Problem 1.1: A steel bar of rectangular cross-section, 3 cm by 2 cm, carries an axial load of 30 kN. Estimate the average tensile stress over a normal cross-section of the bar.

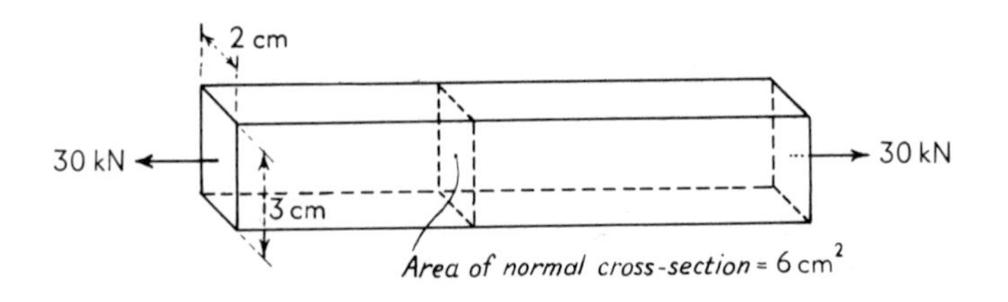

Solution

The area of a normal cross-section of the bar is

$$A = 0{\cdot}03 \times 0{\cdot}02 = 0{\cdot}6 \times 10^{-3}\ \text{m}^2$$

The average tensile stress over this cross-section is then

$$\sigma = \frac{P}{A} = \frac{30 \times 10^3}{0{\cdot}6 \times 10^{-3}} = 50\ \text{MN/m}^2$$

Problem 1.2: A steel bolt, 2·50 cm in diameter, carries a tensile load of 40 kN. Estimate the average tensile stress at the section a and at the screwed section b, where the diameter at the root of the thread is 2·10 cm.

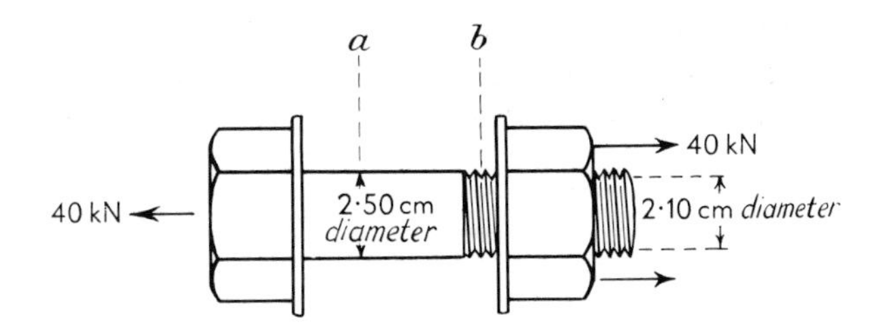

Solution

The cross-sectional area of the bolt at the section a is

$$A_a = \frac{\pi}{4}(0{\cdot}025)^2 = 0{\cdot}491 \times 10^{-3}\ \text{m}^2$$

The average tensile stress at A is then

$$\sigma_a = \frac{P}{A_a} = \frac{40 \times 10^3}{0{\cdot}491 \times 10^{-3}} = 81{\cdot}4\ \text{MN/m}^2$$

The cross-sectional area at the root of the thread, section b, is

$$A_b = \frac{\pi}{4}(0{\cdot}021)^2 = 0{\cdot}346 \times 10^{-3}\ \text{m}^2$$

The average tensile stress over this section is

$$\sigma_b = \frac{P}{A_b} = \frac{40 \times 10^3}{0{\cdot}346 \times 10^{-3}} = 115{\cdot}6\ \text{MN/m}^2$$

1.4 Tensile and compressive strains

In the steel wire experiment of Fig. 1.1 we discussed the extension of the whole wire. If we measure the extension of, say, the lowest quarter-length of the wire we find that for a given load it is equal to a quarter of the extension of the whole wire. In general we find that, at a given load, the ratio of the extension of any length to that length is constant for all parts of the wire; this ratio is known as the *tensile strain*.

Suppose the initial unstrained length of the wire is L_0, and that e is the extension due to straining; the tensile strain ϵ is defined as

$$\epsilon = \frac{e}{L_0} \tag{1.3}$$

This definition of strain is useful only for small distortions, in which the extension e is small compared with the original length L_0; this definition is adequate for the study of most engineering problems, where we are concerned with values of ϵ of the order 0·001, or so.

If a material is compressed the resulting strain is defined in a similar way, except that e is the contraction of a length.

We note that strain is a *non-dimensional* quantity, being the ratio of the extension, or contraction, of a bar to its original length.

Problem 1.3: A cylindrical block of concrete is 30 cm long and has a circular cross-section 10 cm diameter. It carries a total compressive load of 70 kN, and under this load contracts 0·02 cm. Estimate the average compressive stress over a normal cross-section and the compressive strain.

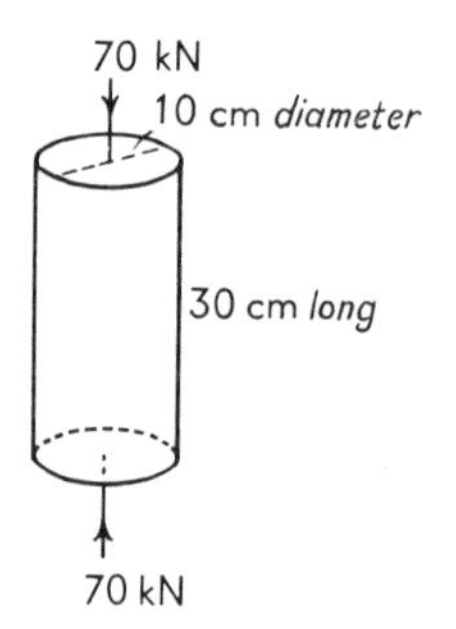

Solution

The area of a normal cross-section is

$$A = \frac{\pi}{4}(0{\cdot}10)^2 = 7{\cdot}85 \times 10^{-3}\ \text{m}^2$$

The average compressive stress over this cross-section is then

$$\sigma = \frac{P}{A} = \frac{70 \times 10^3}{7{\cdot}85 \times 10^{-3}} = 8{\cdot}92\ \text{MN/m}^2$$

The average compressive strain over the length of the cylinder is

$$\epsilon = \frac{0{\cdot}02 \times 10^{-2}}{30 \times 10} = 0{\cdot}67 \times 10^{-3}$$

1.5 Stress-strain curves for brittle materials

Many of the characteristics of a material can be deduced from the tensile test. In the experiment of Fig. 1.1 we measured the extensions of the wire for increasing loads;

it is more convenient to compare materials in terms of stresses and strains, rather than loads and extensions of a particular specimen of a material.

The tensile *stress-strain curve* for a high-strength steel has the form shown in Fig. 1.5. The stress at any stage is the ratio of the load to the *original* cross-sectional area of the test specimen; the strain is the elongation of a unit length of the test specimen. For stresses up to about 750 MN/m^2 the stress-strain curve is linear, showing that the material obeys Hooke's law in this range; the material is also elastic in this range, and no permanent extensions remain after removal of the stresses. The ratio of stress to strain for this linear region is usually about 200 GN/m^2 for steels; this ratio is known as *Young's modulus* and is denoted by E. The strain at the limit of proportionality is of the order 0·003, and is small compared with strains of the order 0·100 at fracture.

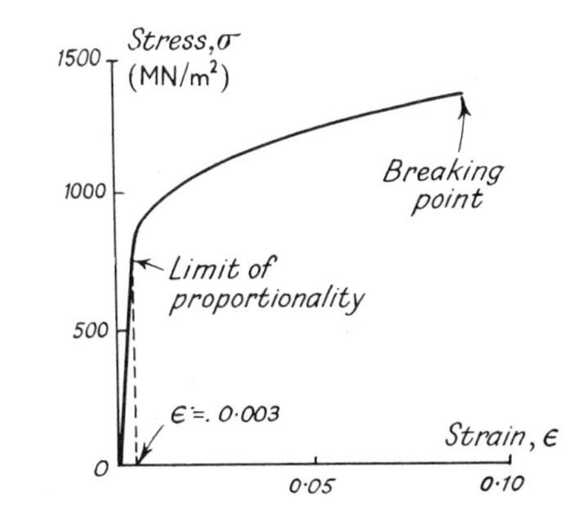

Fig. 1.5 Tensile stress-strain curve for a high-strength steel.

We note that Young's modulus has the units of a stress; the value of E defines the constant in the linear relation between stress and strain in the elastic range of the material. We have

$$E = \frac{\sigma}{\epsilon} \tag{1.4}$$

for the linear-elastic range. If P is the total tensile load in a bar, A its cross-sectional area, and L_0 its length, then

$$E = \frac{\sigma}{\epsilon} = \frac{P/A}{e/L_0} \tag{1.5}$$

where e is the extension of the length L_0. Thus the extension is given by

$$e = \frac{PL_0}{EA} \tag{1.6}$$

If the material is stressed beyond the linear-elastic range the limit of proportionality is exceeded, and the strains increase non-linearly with the stresses. Moreover, removal of the stress leaves the material with some permanent extension; this range is then both non-linear and inelastic. The maximum stress attained may be of the order of 1500 MN/m^2, and the total extension, or *elongation*, at this stage may be of the order of 10%.

The curve of Fig. 1.5 is typical of the behaviour of *brittle* materials—as, for example, high-strength steels, cast iron, high-strength light alloys, and concrete. These materials

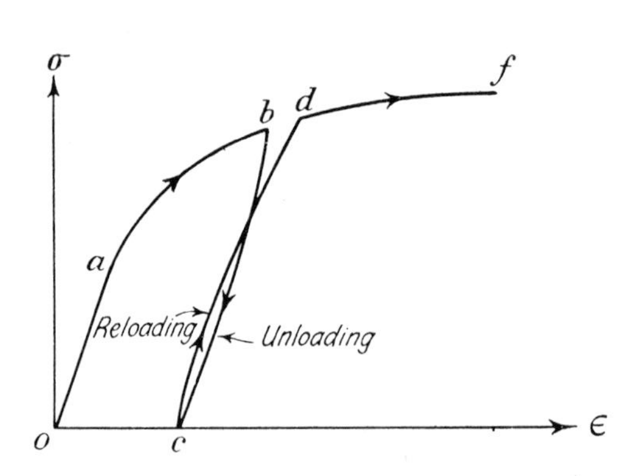

Fig. 1.6 Unloading and reloading of a material in the inelastic range; the paths *bc* and *cd* are approximately parallel to the linear-elastic line *oa*.

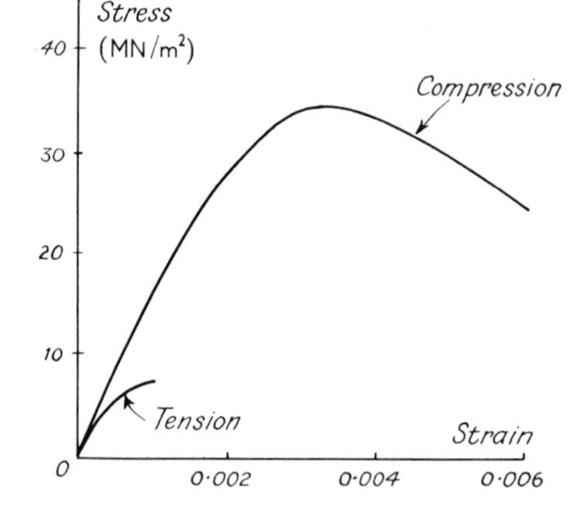

Fig. 1.7 Typical compressive and tensile stress-strain curves for concrete, showing the comparative weakness of concrete in tension.

are characterized by small permanent elongation at the breaking point; in the case of metals this is usually 10%, or less.

When a material is stressed beyond the limit of proportionality and is then unloaded, permanent deformations of the material take place. Suppose the tensile test-specimen of Fig. 1.5 is stressed beyond the limit of proportionality, (point *a* in Fig. 1.6), to a point *b* on the stress-strain diagram. If the stress is now removed, the stress-strain relation follows the curve *bc*; when the stress is completely removed there is a residual strain given by the intercept *Oc* on the ϵ-axis. If the stress is applied again, the stress-strain relation follows the curve *cd* initially, and finally the curve *df* to the breaking point. Both the unloading curve *bc* and the reloading curve *cd* are approximately parallel to the elastic line *Oa*; they are curved slightly in opposite directions. The process of unloading and reloading, *bcd*, has little or no effect on the stress at the breaking point, the stress-strain curve being interrupted by only a small amount *bd*, Fig. 1.6.

The stress-strain curves of brittle materials for tension and compression are usually similar in form, although the stresses at the limit of proportionality and at fracture may be very different for the two loading conditions. Typical tensile and compressive stress-strain curves for concrete are shown in Fig. 1.7; the maximum stress attainable in tension is only about one-tenth of that in compression, although the slopes of the stress-strain curves in the region of zero stress are nearly equal.

1.6 Ductile materials

A brittle material is one showing relatively little elongation at fracture in the tensile test; by contrast some materials, such as mild steel, copper, and synthetic polymers, may be stretched appreciably before breaking. These latter materials are *ductile* in character.

If tensile and compressive tests are made on a mild steel, the resulting stress-strain curves are different in form from those of a brittle material, such as a high-strength steel. If a tensile test specimen of mild steel is loaded axially, the stress-strain curve is linear and elastic up to a point *a*, Fig. 1.8; the small strain region of Fig. 1.8 is reproduced

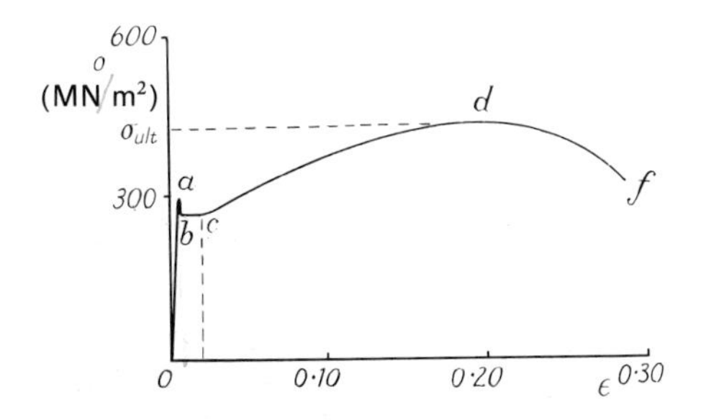

Fig. 1.8 Tensile stress-strain curve for an annealed mild steel, showing the drop in stress at yielding from the upper yield point *a* to the lower yield point *b*.

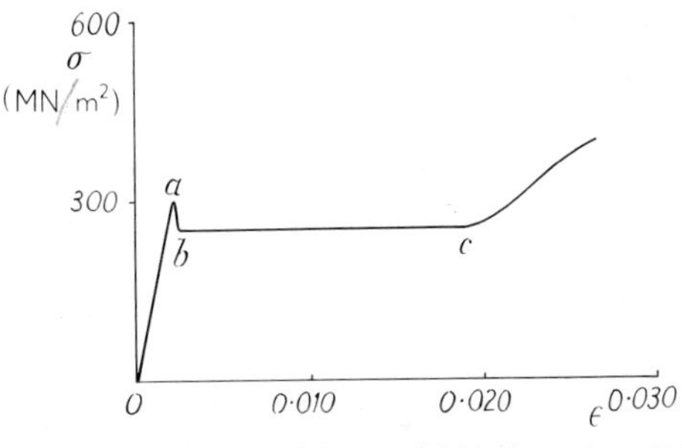

Fig. 1.9 Upper and lower yield points of a mild steel.

to a larger scale in Fig. 1.9. The ratio of stress to strain, or Young's modulus, for the linear portion *Oa* is usually about 200 GN/m², i.e. 200×10^9 N/m². The tensile stress at the point *a* is of order 300 MN/m², i.e., 300×10^6 N/m². If the test specimen is strained beyond the point *a*, Figs. 1.8 and 1.9, the stress must be reduced almost immediately to maintain equilibrium; the reduction of stress, *ab*, takes place rapidly, and the form of the curve *ab* is difficult to define precisely. Continued straining proceeds at a roughly constant stress along *bc*. In the range of strains from *a* to *c* the material is said to *yield*; *a* is the *upper yield point*, and *b* the *lower yield point*. Yielding at constant stress along *bc* proceeds usually to a strain about ten times greater than that at *a*; beyond the point *c* the material *strain-hardens*, and stress again increases with strain. The stress for a tensile specimen attains a maximum value at *d* if the stress is evaluated on the basis of the original cross-sectional area of the bar; the stress corresponding to the point *d* is known as the *ultimate stress*, σ_{ult}, of the material. From *d* to *f* there is a reduction in the nominal stress until fracture occurs at *f*. The ultimate stress in tension is attained at a stage when *necking* begins; this is a reduction of area at a relatively weak cross-section of the test-specimen. It is usual to measure the diameter of the neck

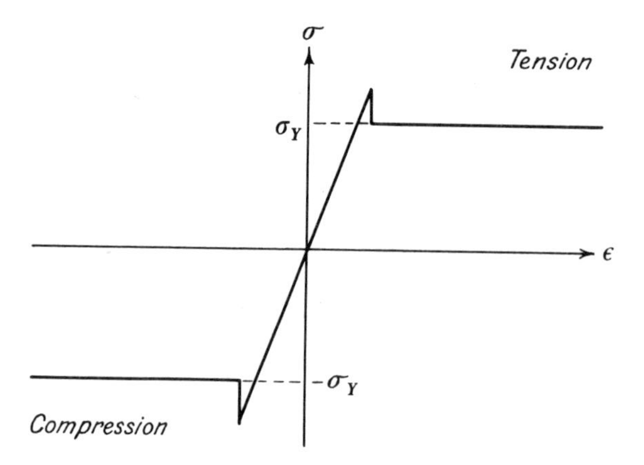

Fig. 1.10 Tensile and compressive stress-strain curves for an annealed mild steel; in the annealed condition the yield stresses in tension and compression are approximately equal.

after fracture, and to evaluate a true stress at fracture, based on the breaking load and the reduced cross-sectional area at the neck. Necking and considerable elongation before fracture are characteristics of ductile materials; there is little or no necking at fracture for brittle materials.

Compressive tests of mild steel give stress-strain curves similar to those for tension. If we consider tensile stresses and strains as positive, and compressive stresses and strains as negative, we can plot the tensile and compressive stress-strain curves on the same diagram; Fig. 1.10 shows the stress-strain curves for an annealed mild steel. In determining the stress-strain curves experimentally, it is important to ensure that the bar is loaded axially; with even small eccentricities of loading the stress distribution over any cross-section of the bar is non-uniform, and the upper yield point stress is not attained in all fibres of the material simultaneously. For this reason the lower yield point stress is taken usually as a more realistic definition of yielding of the material.

Some ductile materials show no clearly defined upper yield stress; for these materials the limit of proportionality may be lower than the stress for continuous yielding. The term *yield stress* refers usually to the stress for continuous yielding of a material; this implies the lower yield stress for a material in which an upper yield point exists; the yield stress is denoted by σ_Y.

Tensile failures of some steel bars are shown in Fig. 1.11; specimen (ii) is a brittle material, showing little or no necking at the fractured section; specimens (i) and (iii) are ductile steels showing a characteristic necking at the fractured sections. The tensile specimens of Fig. 1.12 show the forms of failure in a ductile steel and a ductile light-alloy material; the steel specimen (i) fails at a necked section in the form of a 'cup and cone'; in the case of the light-alloy bar, two 'cups' are formed. The compressive failure of a brittle cast-iron is shown in Fig. 1.13. In the case of a mild steel, failure in compression occurs in a 'barrel-like' fashion, as shown in Fig. 1.14.

The stress-strain curves discussed in the preceding paragraphs refer to static tests carried out at negligible speed. When stresses are applied rapidly the yield stress and ultimate stresses of metallic materials are usually raised. At a strain rate of 100 per second the yield stress of a mild-steel may be twice that at negligible speed.

Fig. 1.11 Tensile failures in steel specimens showing necking in mild-steel, (i) and (iii), and brittle failure in high-strength steel, (ii).

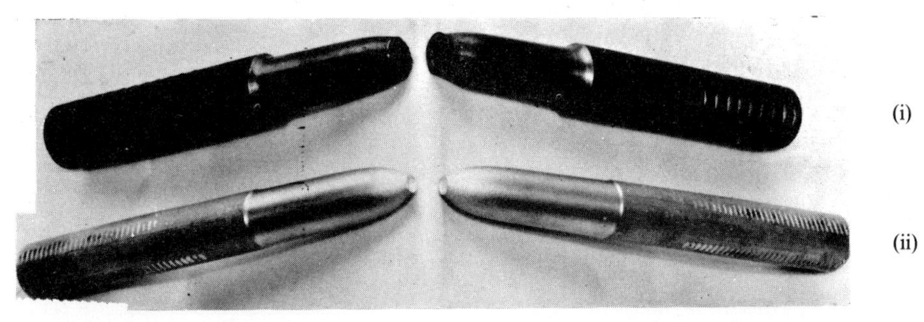

Fig. 1.12 Necking in tensile failures of ductile materials. (i) Mild-steel specimen showing 'cup and cone' at the broken section. (ii) Aluminium-alloy specimen showing double 'cup' type of failure.

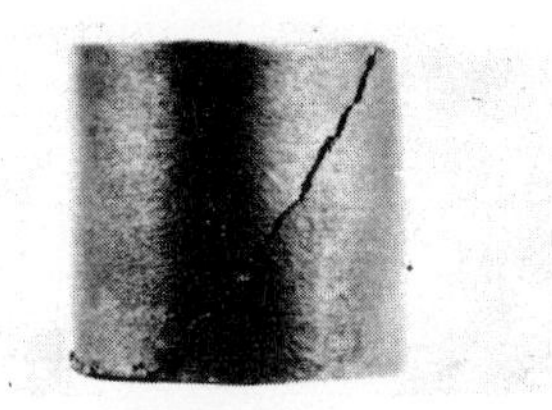

Fig. 1.13 Failure in compression of a circular specimen of cast-iron, showing fracture on a diagonal plane.

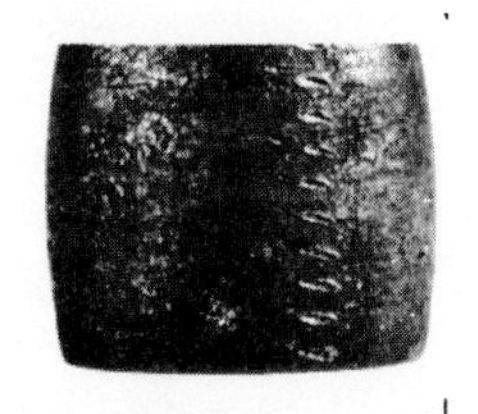

Fig. 1.14 Barrel-like failure in a compressed specimen of mild-steel.

Problem 1.4: A tensile test is carried out on a bar of mild steel of diameter 2 cm. The bar yields under a load of 80 kN. It reaches a maximum load of 150 kN, and breaks finally at a load of 70 kN. Estimate

(i) the tensile stress at the yield point,
(ii) the ultimate tensile stress,
(iii) the average stress at the breaking point, if the diameter of the fractured neck is 1 cm.

Solution

The original cross-section of the bar is

$$A_0 = \frac{\pi}{4}(0{\cdot}020)^2 = 0{\cdot}314 \times 10^{-3}\ \text{m}^2$$

(i) The average tensile stress at yielding is then

$$\sigma_Y = \frac{P_Y}{A_0} = \frac{80 \times 10^3}{0{\cdot}314 \times 10^{-3}} = 254\ \text{MN/m}^2$$

(ii) The ultimate stress is the nominal stress at the maximum load, i.e.

$$\sigma_{\text{ult}} = \frac{P_{\text{max}}}{A_0} = \frac{150 \times 10^3}{0{\cdot}314 \times 10^{-3}} = 477\ \text{MN/m}^2$$

(iii) The cross-sectional area in the fractured neck is

$$A_f = \frac{\pi}{4}(0{\cdot}010)^2 = 0{\cdot}0785 \times 10^{-3}\ \text{m}^2$$

The average stress at the breaking point is then

$$\sigma_f = \frac{P_f}{A_f} = \frac{70 \times 10^3}{0{\cdot}0785 \times 10^{-3}} = 892\ \text{MN/m}^2$$

Problem 1.5: A circular bar of diameter 2·50 cm is subjected to an axial tension of 20 kN. If the material is elastic with a Young's modulus $E = 70\ \text{GN/m}^2$, estimate the percentage elongation.

Solution

The cross-sectional area of the bar is

$$A = \frac{\pi}{4}(0{\cdot}025)^2 = 0{\cdot}491 \times 10^{-3}\ \text{m}^2$$

The average tensile stress is then

$$\sigma = \frac{P}{A} = \frac{20 \times 10^3}{0{\cdot}491 \times 10^{-3}} = 40{\cdot}7\ \text{MN/m}^2$$

The longitudinal tensile strain will therefore be

$$\epsilon = \frac{\sigma}{E} = \frac{40{\cdot}7 \times 10^6}{70 \times 10^9} = 0{\cdot}582 \times 10^{-3}$$

The percentage elongation will therefore be

$$(0{\cdot}582 \times 10^{-3})100 = 0{\cdot}058\%$$

Problem 1.6: The piston of a hydraulic ram is 40 cm diameter, and the piston rod 6 cm diameter. The water pressure is 1 MN/m^2. Estimate the stress in the piston rod and the elongation of a length of 1 m of the rod when the piston is under pressure from the piston-rod side. Take Young's modulus as $E = 200\ GN/m^2$.

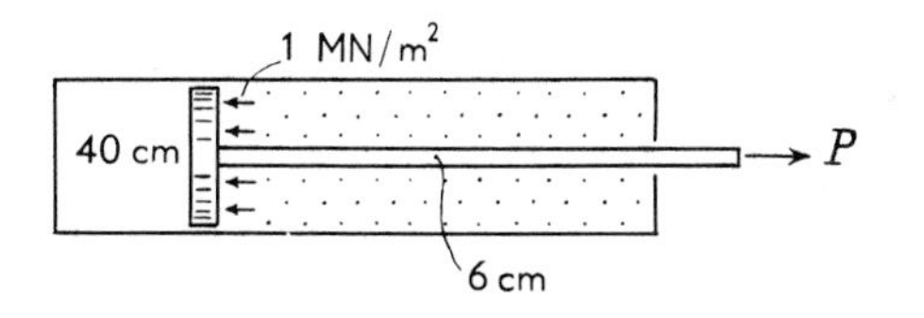

Solution

The pressure on the back of the piston acts on a net area

$$\frac{\pi}{4}[(0{\cdot}40)^2 - (0{\cdot}06)^2] = \frac{\pi}{4}(0{\cdot}46)(0{\cdot}34) = 0{\cdot}123\ \text{m}^2$$

The load on the piston is then

$$P = (1)(0{\cdot}123) = 0{\cdot}123\ \text{MN}$$

Area of the piston rod is

$$A = \frac{\pi}{4}(0{\cdot}060)^2 = 0{\cdot}283 \times 10^{-3}\ \text{m}^2$$

The average tensile stress in the rod is then

$$\sigma = \frac{P}{A} = \frac{0{\cdot}123 \times 10^6}{0{\cdot}283 \times 10^{-3}} = 43{\cdot}5\ \text{MN/m}^2$$

From equation (1.6), the elongation of a length $L = 1$ m is

$$e = \frac{PL}{EA} = \frac{P}{A}\left(\frac{L}{E}\right) = \frac{\sigma L}{E}$$

$$= \frac{(43{\cdot}5 \times 10^6)(1)}{200 \times 10^9}$$

$$= 0{\cdot}218 \times 10^{-3}\ \text{m}$$

$$= 0{\cdot}0218\ \text{cm}$$

Problem 1.7: The steel wire working a signal is 750 m long and 0·5 cm diameter. Assuming a pull on the wire of 1·5 kN, find the movement which must be given to the signal-box end of the wire if the movement at the signal end is to be 17·5 cm. Take Young's modulus as 200 GN/m^2.

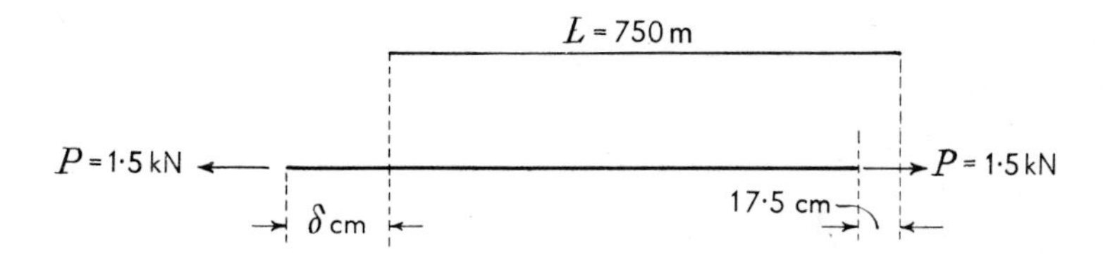

Solution

If δ(cm) is the movement at the signal-box end, the actual stretch of the wire is

$$e = (\delta - 17{\cdot}5) \text{ cm}$$

The longitudinal strain is then

$$\epsilon = \frac{(\delta - 17{\cdot}5)10^{-2}}{750}$$

Now the cross-sectional area of the wire is

$$A = \frac{\pi}{4}(0{\cdot}005)^2 = 0{\cdot}0196 \times 10^{-3} \text{ m}^2$$

The longitudinal strain can also be defined in terms of the tensile load, namely,

$$\epsilon = \frac{e}{L} = \frac{P}{EA} = \frac{1{\cdot}5 \times 10^3}{(200 \times 10^9)(0{\cdot}0196 \times 10^{-3})}$$

$$= 0{\cdot}383 \times 10^{-3}$$

On equating these two values of ϵ,

$$\frac{(\delta - 17{\cdot}5)10^{-2}}{750} = 0{\cdot}383 \times 10^{-3}$$

This equation gives

$$\delta = 46{\cdot}2 \text{ cm}$$

Problem 1.8: A circular, metal rod of diameter 1 cm is loaded in tension. When the tensile load is 5 kN, the extension of a 25 cm length is measured accurately and found to be 0·0227 cm. Estimate the value of Young's modulus, E, of the metal.

Solution

The cross-sectional area is

$$A = \frac{\pi}{4}(0{\cdot}01)^2 = 0{\cdot}0785 \times 10^{-3} \text{ m}^2$$

The tensile stress is then

$$\sigma = \frac{P}{A} = \frac{5 \times 10^3}{0{\cdot}0785 \times 10^{-3}} = 63{\cdot}7 \text{ MN/m}^2$$

The measured tensile strain is

$$\epsilon = \frac{e}{L} = \frac{0{\cdot}0227 \times 10^{-2}}{25 \times 10^{-2}} = 0{\cdot}910 \times 10^{-3}$$

Then Young's modulus is defined by

$$E = \frac{\sigma}{\epsilon} = \frac{63{\cdot}7 \times 10^6}{0{\cdot}91 \times 10^{-3}} = 70 \text{ GN/m}^2$$

Problem 1.9: A straight, uniform rod of length L rotates at uniform angular speed ω about an axis through one end and perpendicular to its length. Estimate the maximum tensile stress generated in the rod and the elongation of the rod at this speed. The density of the material is ρ and Young's modulus is E.

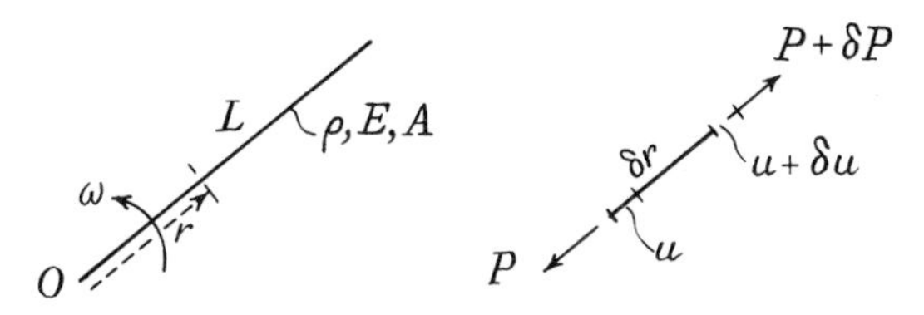

Solution

Suppose the radial displacement of any point a distance r from the axis of rotation is u. The radial displacement a distance $(r + \delta r)$ from O is then $(u + \delta u)$, and the elemental length δr of the rod is stretched therefore an amount δu. The longitudinal strain of this element is therefore

$$\epsilon = \text{Limit}_{\delta r \to 0} \frac{\delta u}{\delta r} = \frac{du}{dr}$$

The longitudinal stress in the elemental length is then

$$\sigma = E\epsilon = E\frac{du}{dr}$$

If A is the cross-sectional area of the rod, the longitudinal load at any radius r is then

$$P = \sigma A = EA\frac{du}{dr}$$

The centrifugal force acting on the elemental length δr is

$$(\rho A\, \delta r)\omega^2 r$$

Then, for radial equilibrium of the elemental length,

$$\delta P + \rho A\omega^2 r\, \delta r = 0$$

This gives

$$\frac{dP}{dr} = -\rho A\omega^2 r$$

On integrating, we have

$$P = -\tfrac{1}{2}\rho A\omega^2 r^2 + C$$

where C is an arbitrary constant; if $P = 0$ at the remote end, $r = L$, of the rod, then

$$C = \tfrac{1}{2}\rho A\omega^2 L^2$$

and

$$P = \tfrac{1}{2}\rho A\omega^2 L^2\left(1 - \frac{r^2}{L^2}\right)$$

The tensile stress at any radius is then

$$\sigma = \frac{P}{A} = \tfrac{1}{2}\rho\omega^2 L^2\left(1 - \frac{r^2}{L^2}\right)$$

This is greatest at the axis of rotation, $r = 0$, so that

$$\sigma_{max} = \tfrac{1}{2}\rho\omega^2L^2$$

The longitudinal stress, σ, is defined by

$$\sigma = E\frac{du}{dr}$$

so

$$\frac{du}{dr} = \frac{\sigma}{E} = \frac{\rho\omega^2L^2}{2E}\left(1 - \frac{r^2}{L^2}\right)$$

On integrating,

$$u = \frac{\rho\omega^2L^2}{2E}\left(r - \frac{r^3}{3L^2} + D\right)$$

where D is an arbitrary constant; if there is no radial movement at 0, then $u = 0$ at $r = 0$, and we have $D = 0$. Thus,

$$u = \frac{\rho\omega^2L^2}{2E}\left[r\left(1 - \frac{r^2}{3L^2}\right)\right]$$

At the remote end, $r = L$,

$$u_L = \frac{\rho\omega^2L^2}{2E}\,[L(\tfrac{2}{3})] = \frac{\rho\omega^2L^3}{3E}$$

1.7 Proof stresses

Many materials show no well-defined yield stresses when tested in tension or compression. A typical stress-strain curve for an aluminium alloy is shown in Fig. 1.15; the limit of proportionality is in the region of 300 MN/m^2, but the exact position of this limit is difficult to determine experimentally. To overcome this problem a *proof stress* is defined; the 0·1 % proof stress is the stress required to produce a permanent strain of 0·001 on removal of the stress. Suppose we draw a line from the point 0·001

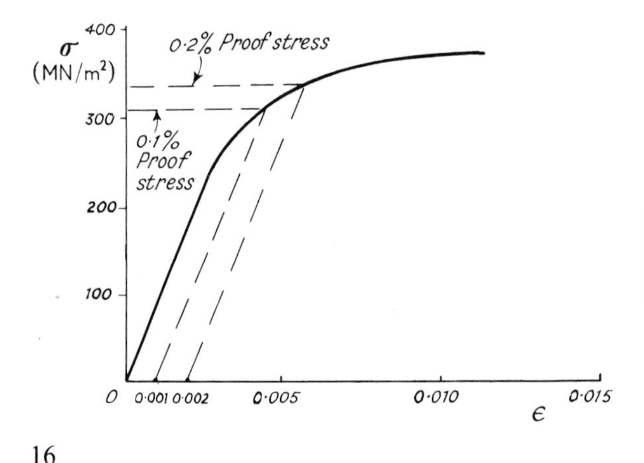

Fig. 1.15 Proof stresses of an aluminium-alloy material; the proof stress is found by drawing the line parallel to the linear-elastic line at the appropriate proof strain.

on the strain axis, Fig. 1.15, parallel to the elastic line of the material; the point where this line cuts the stress-strain curve defines the proof stress. The 0·2% proof stress is defined in a similar way.

1.8 Working stresses

In many engineering problems the loads sustained by a component of a machine or structure are reasonably well-defined; for example, the lower stanchions of a tall building support the weight of material forming the upper storeys. The stresses which are present in a component, under normal working conditions, are called the *working stresses*; the ratio of the yield stress, σ_Y, of a material to the largest working stress, σ_w, in the component is the *stress factor* against yielding. The stress factor on yielding is then

$$\frac{\sigma_Y}{\sigma_w} \tag{1.7}$$

If the material has no well-defined yield point, it is more convenient to use the *proof stress*, σ_p; the stress factor on proof stress is then

$$\frac{\sigma_p}{\sigma_w} \tag{1.8}$$

Some writers refer to the stress factor defined above as a 'safety factor'. It is preferable, however, to avoid any reference to 'safe' stresses, since the degree of safety in any practical problem is difficult to define. The present writers prefer the term 'stress factor' since this defines more precisely that the working stress is compared with the yield, or proof, stress of the material. Another reason for using 'stress factor' will become more evident after the reader has studied §1.9.

1.9 Load factors

The *stress factor* in a component gives an indication of the working stresses in relation to the yield, or proof, stress of the material. In practical problems working stresses can only be estimated approximately in stress calculations. For this reason the stress factor may give little indication of the degree of safety of a component.

A more realistic estimate of safety can be made by finding the extent to which the working loads on a component may be increased before collapse or fracture occurs. Consider, for example, the continuous beam in Fig. 1.16, resting on three supports. Under working conditions the beam carries lateral loads P_1, P_2 and P_3, Fig. 1.16(i). If all these loads can be increased simultaneously by a factor n before collapse occurs, the load factor against collapse is n. In some complex structural systems, as for example continuous beams, the collapse loads, such as nP_1, nP_2 and nP_3, can be estimated

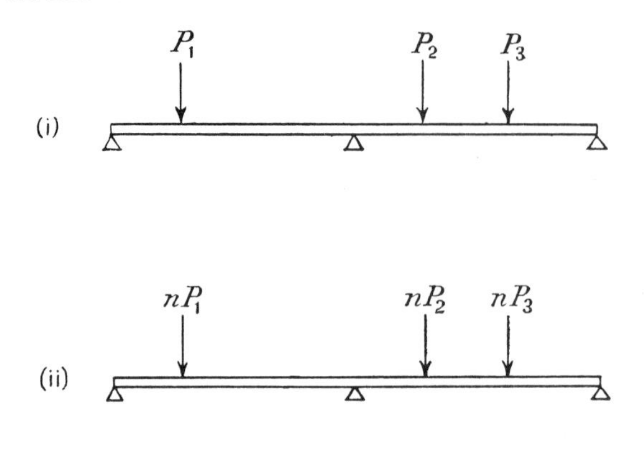

Fig. 1.16 Factored loads on a continuous beam. (i) Working loads. (ii) Factored working loads leading to collapse.

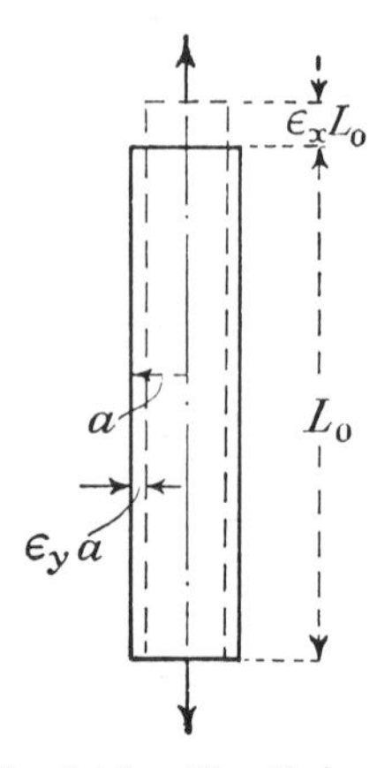

Fig. 1.17 The Poisson ratio effect leading to lateral contraction of a bar in tension.

reasonably accurately; the value of the load factor can then be deduced to give working loads P_1, P_2 and P_3.

1.10 Lateral strains due to direct stresses

When a bar of a material is stretched longitudinally—as in a tensile test—the bar extends in the direction of the applied load. This longitudinal extension is accompanied by a lateral contraction of the bar, as shown in Fig. 1.17. In the linear-elastic range of a material the lateral strain is proportional to the longitudinal strain; if ϵ_x is the longitudinal strain of the bar, then the lateral strain is

$$\epsilon_y = \nu\epsilon_x \tag{1.9}$$

The constant ν in this relation is known as *Poisson's ratio*, and for most metals has a value of about 0·3 in the linear-elastic range. If the longitudinal strain is tensile, the lateral strain is a contraction; for a compressed bar there is a lateral expansion.

With a knowledge of the lateral contraction of a stretched bar it is possible to calculate the change in volume due to straining. The bar of Fig. 1.17 is assumed to have a square cross-section of side a; L_0 is the unstrained length of the bar. When strained longitudinally an amount ϵ_x, the corresponding lateral strain of contraction is ϵ_y. The bar extends therefore an amount $\epsilon_x L_0$, and each side of the cross-section contracts an amount $\epsilon_y a$. The volume of the bar before stretching is

$$V_0 = a^2 L_0$$

After straining the volume is

$$V = (a - \epsilon_y a)^2 (L_0 + \epsilon_x L_0)$$

which may be written

$$V = a^2L_0(1 - \epsilon_y)^2(1 + \epsilon_x) = V_0(1 - \epsilon_y)^2(1 + \epsilon_x)$$

If ϵ_x and ϵ_y are small quantities compared to unity, we may write

$$(1 - \epsilon_y)^2(1 + \epsilon_x) \doteq (1 - 2\epsilon_y)(1 + \epsilon_x) \doteq 1 + \epsilon_x - 2\epsilon_y$$

ignoring squares and products of ϵ_x and ϵ_y. The volume after straining is then

$$V = V_0(1 + \epsilon_x - 2\epsilon_y)$$

The *volumetric strain* is defined as the ratio of the change of volume to the original volume, and is therefore

$$\frac{V - V_0}{V_0} = \epsilon_x - 2\epsilon_y$$

If $\epsilon_y = \nu\epsilon_x$, then the volumetric strain is

$$\epsilon_x(1 - 2\nu) \qquad (1.10)$$

Problem 1.10: A bar of steel, having a rectangular cross-section 7·5 cm by 2·5 cm, carries an axial tensile load of 180 kN. Estimate the decrease in the length of the sides of the cross-section if Young's modulus, E, is 200 GN/m^2 and Poisson's ratio, ν, is 0·3.

Solution

The cross-sectional area is

$$A = (0{\cdot}075)(0{\cdot}025) = 1{\cdot}875 \times 10^{-3}\ \text{m}^2$$

The average longitudinal tensile stress is

$$\sigma = \frac{P}{A} = \frac{180 \times 10^3}{1{\cdot}875 \times 10^{-3}} = 96{\cdot}0\ \text{MN/m}^2$$

The longitudinal tensile strain is therefore

$$\epsilon = \frac{\sigma}{E} = \frac{96{\cdot}0 \times 10^6}{200 \times 10^9} = 0{\cdot}48 \times 10^{-3}$$

The lateral strain is therefore

$$\nu\epsilon = 0{\cdot}3(0{\cdot}48 \times 10^{-3}) = 0{\cdot}144 \times 10^{-3}$$

The 7·5 cm side contracts then by an amount

$$(0{\cdot}075)(0{\cdot}144 \times 10^{-3}) = 0{\cdot}0108 \times 10^{-3}\ \text{m}$$
$$= 0{\cdot}00108\ \text{cm}$$

The 2·5 cm side contracts an amount

$$(0{\cdot}025)(0{\cdot}144 \times 10^{-3}) = 0{\cdot}0036 \times 10^{-3}\ \text{m}$$
$$= 0{\cdot}00036\ \text{cm}$$

Table 1.1 APPROXIMATE STRENGTH PROPERTIES OF SOME ENGINEERING MATERIALS

Material	Limit of proportionality (MN/m²)	Ultimate stress σ_{ult} (MN/m²)	Elongation at tensile fracture (as a fraction of the original length)	Young's modulus E (GN/m²)	Density ρ (kg/m³)	$\frac{\sigma_{ult}}{\rho}$ (m/s)²	$\frac{E}{\rho}$ (m/s)²	Coefficient of linear expansion α (per °C)
Medium-strength mild steel	280	370	0·30	200	7840	47×10^3	25×10^6	$1{\cdot}2 \times 10^{-5}$
High-strength steel	770	1550	0·10	200	7840	198×10^3	25×10^6	$1{\cdot}3 \times 10^{-5}$
Medium-strength aluminium alloy	230	430	0·10	70	2800	154×10^3	25×10^6	$2{\cdot}3 \times 10^{-5}$
Titanium alloy	385	690	0·15	120	4500	153×10^3	27×10^6	$0{\cdot}9 \times 10^{-5}$
Magnesium alloy	155	280	0·08	45	1800	156×10^3	25×10^6	$2{\cdot}7 \times 10^{-5}$
Wrought iron	185	310	—	190	7670	40×10^3	25×10^6	$1{\cdot}2 \times 10^{-5}$
Cast iron (tension)	—	155	—	140	7200	—	20×10^6	$1{\cdot}1 \times 10^{-5}$
Cast iron (compression)	—	700	—	140	7200	$97^* \times 10^3$	20×10^6	$1{\cdot}1 \times 10^{-5}$
Concrete (tension)	—	3·0	—	14	2410	—	6×10^6	$1{\cdot}2 \times 10^{-5}$
Concrete (compression)	—	30·0	—	14	2410	$12^* \times 10^3$	6×10^6	$1{\cdot}2 \times 10^{-5}$
Nylon (polyamide)	77	90	1·00	2	1140	79×10^3	$1{\cdot}8 \times 10^6$	10×10^{-5}
Polystyrene	46	60	0·03	3·5	1050	57×10^3	$3{\cdot}3 \times 10^6$	10×10^{-5}
Fluon (tetrafluoroethylene)	8	15	2·00	0·4	2220	7×10^3	$0{\cdot}2 \times 10^6$	11×10^{-5}
Polythene (ethylene)	6	12	5·00	0·2	915	13×10^3	$0{\cdot}2 \times 10^6$	28×10^{-5}
High-strength glass-fibre composite	—	1600	—	60	2000	800×10^3	30×10^6	—
Carbon-fibre composite	—	1400	—	170	1600	875×10^3	105×10^6	—
Boron composite	—	1300	—	270	2000	650×10^3	135×10^6	—

* Evaluated on the compressive value of σ_{ult}.

1.11 Strength properties of some engineering materials

The mechanical properties of some engineering materials are given in Table 1.1. Most of the materials are in common engineering use, although a number of potentially important materials—namely, glass-fibre composites, carbon-fibre composites and boron composites—are shown for comparison with established materials. In the case of some brittle materials, such as cast iron and concrete, the ultimate stress in tension is considerably smaller than in compression.

1.12 Weight and stiffness economy of materials

In some machine components and structures it is important that the weight of material should be as small as possible. This is particularly true of aircraft, for example, in which less structure weight leads to a larger pay-load. If σ_{ult} is the ultimate stress of a material in tension and ρ is its density, then a measure of the strength economy is the ratio

$$\frac{\sigma_{ult}}{\rho}$$

The materials shown in Table 1.1 are compared on the basis of strength economy in Table 1.2, from which it is clear that the modern fibre-reinforced composites offer distinct savings in weight over the more common materials in engineering use.

In some engineering applications, stiffness rather than strength is required of materials; this is so in structures and components governed by deflection limitations. A measure of the stiffness economy of a material is the ratio

$$\frac{E}{\rho}$$

some values of which are shown in Table 1.3. Boron composites and carbon-fibre composites show outstanding stiffness properties, whereas glass-fibre composites fall more into line with the best materials already in common use.

1.13 Strain energy and work done in the tensile test

As a tensile specimen extends under load, the forces applied to the ends of the test specimen move through small distances. These forces perform work in stretching the bar. If, at a tensile load P, the bar is stretched a small additional amount δe, Fig. 1.18, then the work done on the bar is approximately

$$P\,\delta e$$

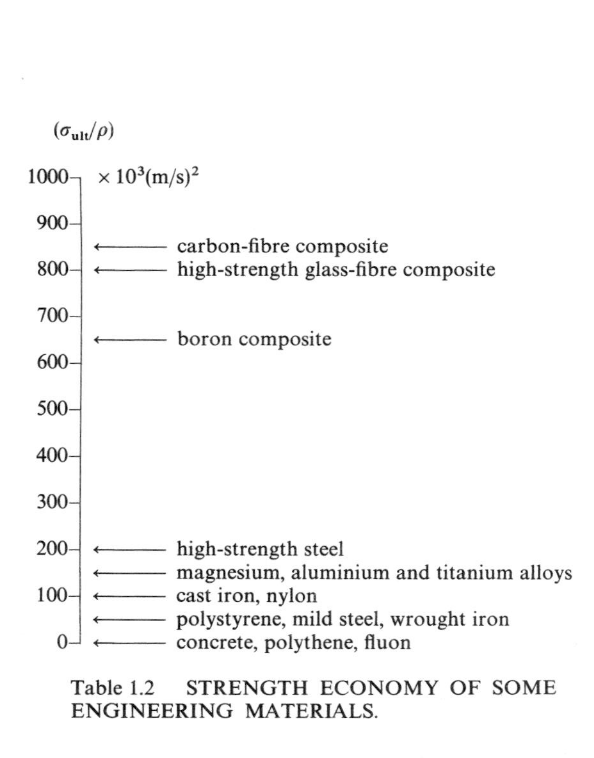

Table 1.2 STRENGTH ECONOMY OF SOME ENGINEERING MATERIALS.

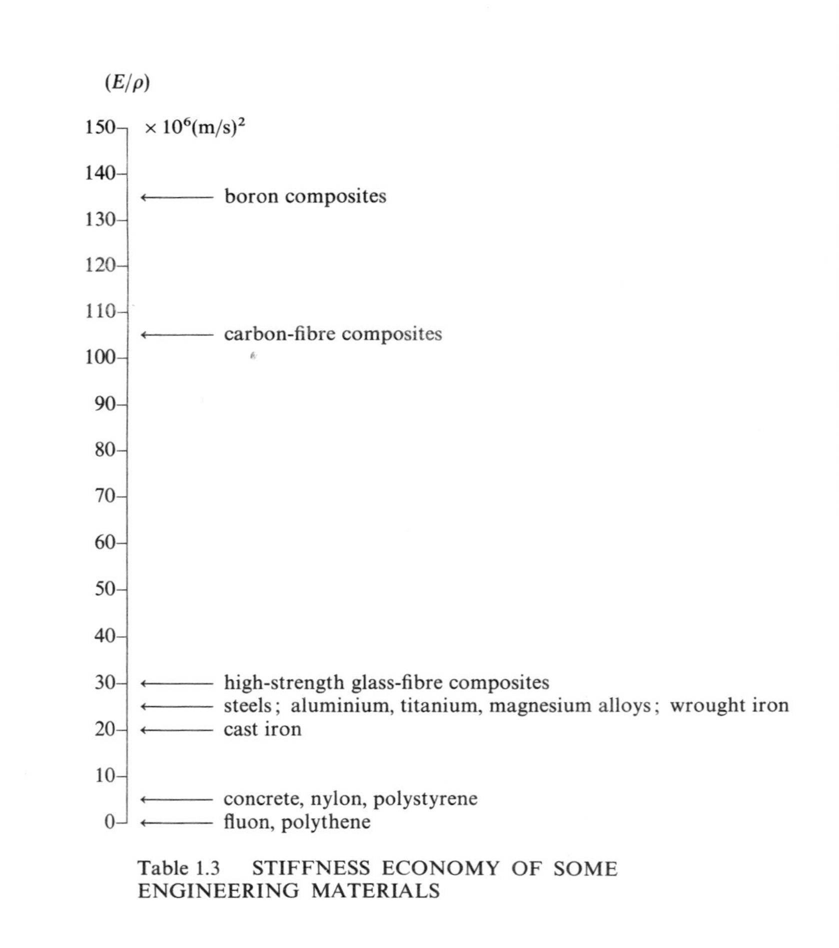

Table 1.3 STIFFNESS ECONOMY OF SOME ENGINEERING MATERIALS

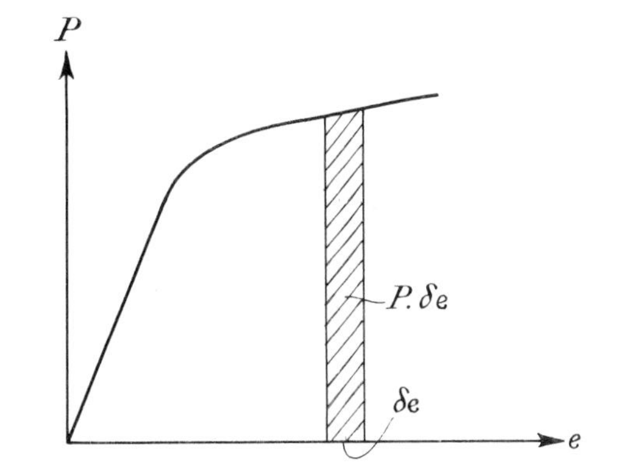

Fig. 1.18 Work done in stretching a bar through a small extension δe.

The total work done in extending the bar to the extension e is then

$$W = \int_0^e P\,de \tag{1.11}$$

which is the area under the $P - e$ curve up to the stretched condition. If the limit of proportionality is not exceeded, the work done in extending the bar is stored as *strain energy*, which is directly recoverable on removal of the load. For this case, the strain energy, U, is

$$U = W = \int_0^e P\,de \tag{1.12}$$

But in the linear-elastic range of the material, we have from equation (1.6) that

$$e = \frac{PL_0}{EA}$$

where L_0 is the initial length of the bar, A is its cross-sectional area and E is Young's modulus. Then equation (1.12) becomes

$$U = \int_0^e \frac{EA}{L_0}\,e\,de = \frac{EA}{2L_0}(e^2) \tag{1.13}$$

In terms of P,

$$U = \frac{EA}{2L_0}(e^2) = \frac{L_0}{2EA}(P^2) \tag{1.14}$$

Now (P/A) is the tensile stress σ in the bar, and so we may write

$$U = \frac{AL_0}{2E}(\sigma^2) \tag{1.15}$$

Moreover, AL_0 is the original volume of the bar, and so the strain energy per unit volume is

$$\frac{\sigma^2}{2E} \tag{1.16}$$

When the limit of proportionality of a material is exceeded, the work done in extending the bar is still given by equation (1.11); however, not all this work is stored as strain energy; some of the work done is used in producing permanent distortions in the material, the work reappearing largely in the form of heat. Suppose a mild-steel bar is stressed beyond the yield point, Fig. 1.19, and up to the point where strain-hardening

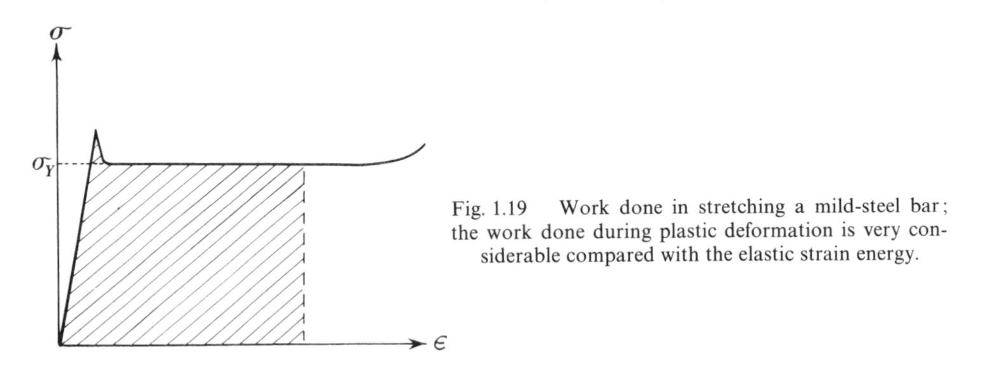

Fig. 1.19 Work done in stretching a mild-steel bar; the work done during plastic deformation is very considerable compared with the elastic strain energy.

begins; the strain at the limit of proportionality is small compared with this large inelastic strain; the work done per unit volume in producing a strain ϵ is approximately

$$W = \sigma_Y \epsilon \tag{1.17}$$

in which σ_Y is the yield stress of the material. This work is considerably greater than that required to reach the limit of proportionality. A ductile material of this type is useful in absorbing relatively large amounts of work before breaking.

1.14 Initial stresses

It frequently happens that, before any load is applied to some part of a machine or structure, it is already in a state of stress. In other words, the component is *initially stressed* before external forces are applied. Bolted joints and connections, for example, involve bolts which are pre-tensioned; subsequent loading may, or may not, affect the tension in a bolt. Most forms of welded connections introduce initial stresses around the welds, unless the whole connection is stress-relieved by a suitable heat treatment; in such cases, the initial stresses are not usually known with any real accuracy. Initial stresses can also be used to considerable effect in strengthening certain materials; for example, concrete can be made a more effective material by pre-compression in the form of pre-stressed concrete.

Problem 1.11: A 2·5 cm diameter steel bolt passes through a steel tube 5 cm internal diameter, 6·25 cm external diameter, and 40 cm long. The bolt is then tightened up onto the tube through rigid end blocks until the tensile force in the bolt is 40 kN. The distance between the head of the bolt and the nut is 50 cm. If an external force of 30 kN is applied to the end blocks, tending to pull them apart, estimate the resulting tensile force in the bolt.

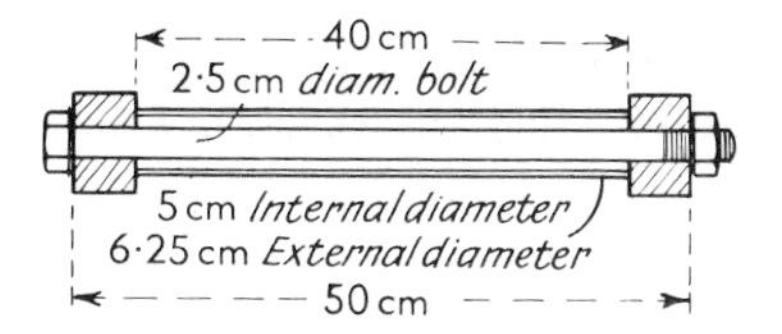

Solution

The cross-sectional area of the bolt is

$$\frac{\pi}{4}(0{\cdot}025)^2 = 0{\cdot}491 \times 10^{-3}\ \text{m}^2$$

The cross-sectional area of the tube is

$$\frac{\pi}{4}[(0{\cdot}0625)^2 - (0{\cdot}050)^2] = \frac{\pi}{4}(0{\cdot}1125)(0{\cdot}0125) = 0{\cdot}110 \times 10^{-3}\ \text{m}^2$$

Before the external load of 30 kN is applied, the bolt and tube carry internal loads of 40 kN. When the external load of 30 kN is applied, suppose the tube and bolt are each stretched amounts δ; suppose further

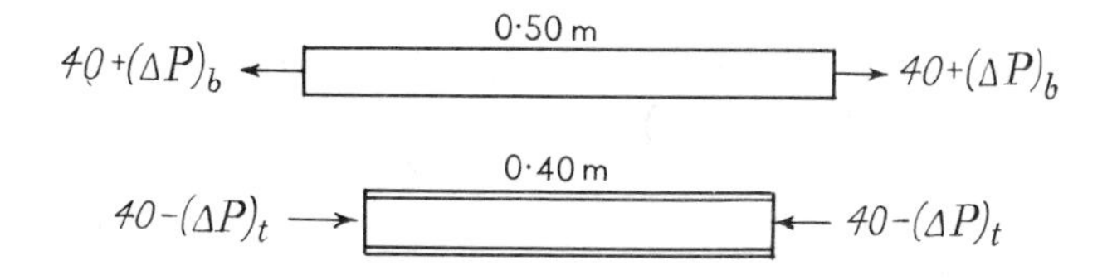

that the *change* of load in the bolt is $(\Delta P)_b$, tensile, and the *change* of load in the tube is $(\Delta P)_t$, tensile. Then the elastic stretch of each component due to the additional external load of 30 kN is

$$\delta = \frac{(\Delta P)_b(0{\cdot}50)}{(0{\cdot}491 \times 10^{-3})E} = \frac{(\Delta P)_t(0{\cdot}40)}{(0{\cdot}110 \times 10^{-3})E}$$

where E is Young's modulus. Then

$$(\Delta P)_b = 3{\cdot}58(\Delta P)_t$$

But for equilibrium of internal and external forces,

$$(\Delta P)_b + (\Delta P)_t = 30\ \text{kN}$$

These two equations give

$$(\Delta P)_b = 23{\cdot}44\ \text{kN}, \quad (\Delta P)_t = 6{\cdot}56\ \text{kN}$$

The resulting tensile force in the bolt is

$$40 + (\Delta P)_b = 63{\cdot}4\ \text{kN}$$

1.15 Composite bars in tension or compression

A composite bar is one made of two materials, such as steel rods embedded in concrete. The construction of the bar is such that constituent components extend or contract equally under load. To illustrate the behaviour of such bars consider a rod made of two materials, 1 and 2, Fig. 1.20; A_1, A_2 are the cross-sectional areas of the bars, and E_1, E_2 are the values of Young's modulus. We imagine the bars to be rigidly connected

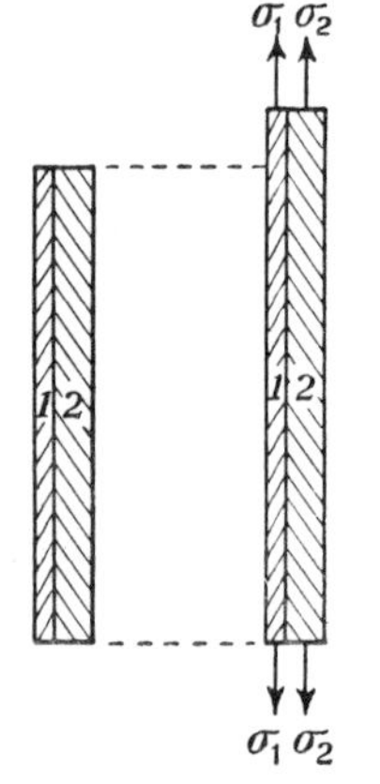

Fig. 1.20 Composite bar in tension; if the bars are connected rigidly at their ends, then they suffer the same extensions.

together at the ends; then for the longitudinal strains to be the same when the composite bar is stretched we must have

$$\epsilon = \frac{\sigma_1}{E_1} = \frac{\sigma_2}{E_2} \tag{1.18}$$

where σ_1 and σ_2 are the stresses in the two bars. But the total tensile load is

$$P = \sigma_1 A_1 + \sigma_2 A_2. \tag{1.19}$$

Equations (1.18) and (1.19) give

$$\sigma_1 = \frac{PE_1}{A_1 E_1 + A_2 E_2}, \qquad \sigma_2 = \frac{PE_2}{A_1 E_1 + A_2 E_2} \tag{1.20}$$

Problem 1.12: A concrete column, 50 cm square, is reinforced with four steel rods, each 2·5 cm in diameter, embedded in the concrete near the corners of the square. If Young's modulus for steel is 200 GN/m^2 and that for concrete is 14 GN/m^2, estimate the compressive stresses in the steel and concrete when the total thrust on the column is 1 MN.

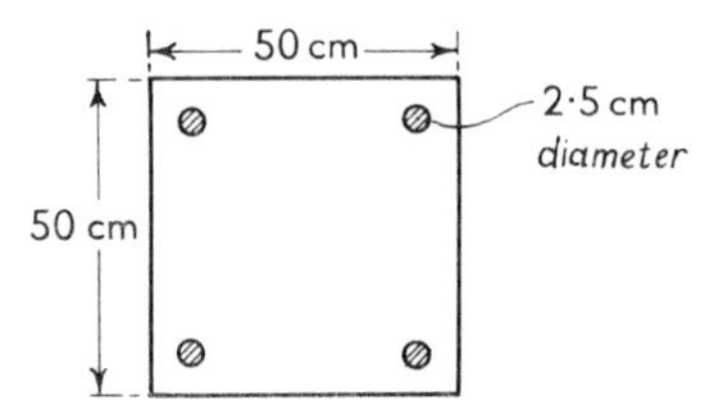

Solution

Suppose subscripts c and s refer to concrete and steel, respectively. The cross-sectional area of steel is

$$A_s = 4\left[\frac{\pi}{4}(0{\cdot}025)^2\right] = 1{\cdot}96 \times 10^{-3}\ \text{m}^2$$

and the cross-sectional area of concrete is

$$A_c = (0{\cdot}50)^2 - A_s = 0{\cdot}248\ \text{m}^2$$

Equations (1.20) then give

$$\sigma_c = \frac{10^6}{(0{\cdot}248) + (1{\cdot}96 \times 10^{-3})(200/14)} = 3{\cdot}62\ \text{MN/m}^2$$

$$\sigma_s = \frac{10^6}{(0{\cdot}248)(14/200) + (1{\cdot}96 \times 10^{-3})} = 51{\cdot}5\ \text{MN/m}^2$$

Problem 1.13: A uniform beam weighing 500 N is held in a horizontal position by three vertical wires, one attached to each end of the beam, and one at the mid-length. The outer wires are brass of diameter 0·125 cm, and the central wire is of steel of diameter 0·0625 cm. If the beam is rigid and the wires are of the same length, and unstressed before the beam is attached, estimate the stresses in the wires. Young's modulus for brass is 85 GN/m² and for steel is 200 GN/m².

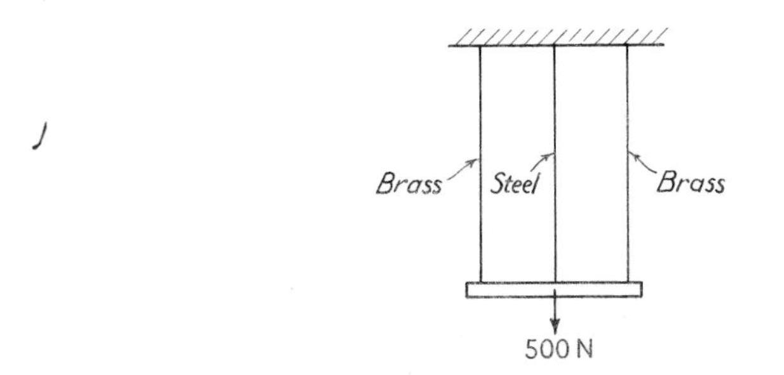

Solution

On considering the two outer brass wires together, we may take the system as a composite one consisting of a single brass member and a steel member. The area of the steel member is

$$A_s = \frac{\pi}{4}(0{\cdot}625 \times 10^{-3})^2 = 0{\cdot}306 \times 10^{-6}\ \text{m}^2$$

The total area of the two brass members is

$$A_b = 2\left[\frac{\pi}{4}(1{\cdot}25 \times 10^{-3})^2\right] = 2{\cdot}45 \times 10^{-6}\ \text{m}^2$$

Equations (1.20) then give, for the steel wire

$$\sigma_s = \frac{500}{(0{\cdot}306 \times 10^{-6}) + (2{\cdot}45 \times 10^{-6})(85/200)} = 370\ \text{MN/m}^2$$

and for the brass wires

$$\sigma_b = \frac{500}{(0{\cdot}306 \times 10^{-6})(200/85) + (2{\cdot}45 \times 10^{-6})} = 158\ \text{MN/m}^2$$

1.16 Temperature stresses

When the temperature of a body is raised, or lowered, the material expands, or contracts. If this expansion or contraction is wholly or partially resisted, stresses are set up in the body. Consider a long bar of a material; suppose L_0 is the length of the bar at a temperature θ_0, and that α is the coefficient of linear expansion of the material. The bar is now subjected to an increase θ in temperature. If the bar is completely free to expand, its length increases by $\alpha L_0 \theta$, and the length becomes $L_0(1 + \alpha\theta)$. If this expansion is prevented, it is as if a bar of length $L_0(1 + \alpha\theta)$ were compressed to a length L_0; in this case the compressive strain is

$$\epsilon = \frac{\alpha L_0 \theta}{L_0(1 + \alpha\theta)} \doteq \alpha\theta$$

since $\alpha\theta$ is small compared with unity; the corresponding stress is

$$\sigma = E\epsilon = \alpha\theta E \tag{1.21}$$

By a similar argument the tensile stress set up in a constrained bar by a fall θ in temperature is $\alpha\theta E$. It is assumed that the material remains elastic.

In the case of steel $\alpha = 1{\cdot}3 \times 10^{-5}$ per °C; the product αE is approximately 2·6 MN/m^2 per °C, so that a change in temperature of 4 °C produces a stress of approximately 10 MN/m^2 if the bar is completely restrained.

1.17 Temperature stresses in composite bars

In a component or structure made wholly of one material, temperature stresses arise only if external restraints prevent thermal expansion or contraction. In composite bars made of materials with different rates of thermal expansion, internal stresses can be set up by temperature changes; these stresses occur independently of those due to external restraints.

Consider, for example, a simple composite bar consisting of two members—a solid circular bar, 1, contained inside a circular tube, 2, Fig. 1.21. The materials of the bar and tube have different coefficients of linear expansion, α_1 and α_2, respectively. If the ends of the bar and tube are attached rigidly to each other, longitudinal stresses are set up by a change of temperature. Suppose firstly, however, that the bar and tube are quite free of each other; if L_0 is the original length of each bar, Fig. 1.21, the extensions due to a temperature increase θ are $\alpha_1\theta L_0$ and $\alpha_2\theta L_0$, Fig. 1.21(ii). The difference in lengths of the two members is $(\alpha_1 - \alpha_2)\theta L_0$; this is now eliminated by compressing the inner bar with a force P, and pulling the outer tube with an equal force P, Fig. 1.21(iii). If A_1 and E_1 are the cross-sectional area and Young's modulus, respectively, of the inner bar, and A_2 and E_2 refer to the outer tube, then the contraction of the inner bar due to P is

$$e_1 = \frac{PL_0}{E_1 A_1}$$

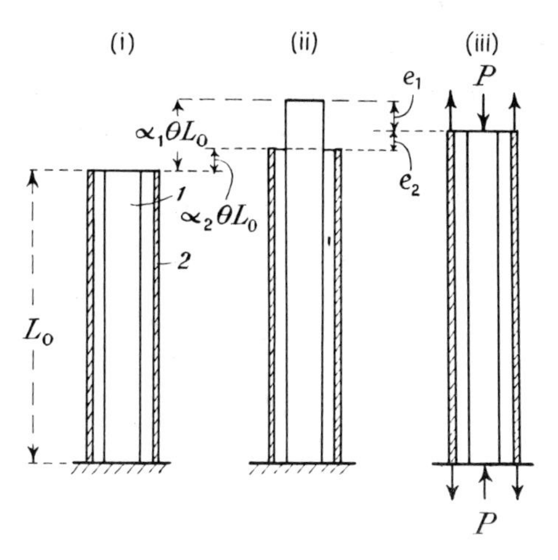

Fig. 1.21 Temperature stress in a composite bar.

and the extension of the outer tube due to P is

$$e_2 = \frac{PL_0}{E_2A_2}$$

The difference in lengths, $(\alpha_1 - \alpha_2)\theta L$, is eliminated completely when

$$(\alpha_1 - \alpha_2)\theta L_0 = e_1 + e_2$$

On substituting for e_1 and e_2, we have

$$(\alpha_1 - \alpha_2)\theta L_0 = PL_0\left(\frac{1}{E_1A_1} + \frac{1}{E_2A_2}\right) \tag{1.22}$$

The force P is induced by the temperature change θ if the ends of the two members are attached rigidly to each other; from equation (1.22), P has the value

$$P = \frac{(\alpha_1 - \alpha_2)\theta}{\left(\frac{1}{E_1A_1} + \frac{1}{E_2A_2}\right)} \tag{1.23}$$

An internal load is only set up if α_1 is different from α_2.

Problem 1.14: An aluminium rod 2·2 cm diameter is screwed at the ends, and passes through a steel tube 2·5 cm internal diameter and 0·3 cm thick. Both are heated to a temperature of 140° C, when the nuts on the rod are screwed lightly on to the ends of the tube. Estimate the stress in the rod when the common temperature has fallen 20 °C. For steel, $E = 200$ GN/m^2 and $\alpha = 1{\cdot}2 \times 10^{-5}$ per °C, while for aluminium, $E = 70$ GN/m^2 and $\alpha = 2{\cdot}3 \times 10^{-5}$ per ° C, where E is Young's modulus and α is the coefficient of linear expansion.

Solution

Let subscript a refer to the aluminium rod and subscript s to the steel tube. The problem is similar to the one discussed in §1.17, except that the composite rod has its temperature lowered, in this case from 140° C to 20° C. From equation (1.23), the common force between the two components is

$$P = \frac{(\alpha_a - \alpha_s)\theta}{\dfrac{1}{(EA)_a} + \dfrac{1}{(EA)_s}}$$

The stress in the rod is therefore

$$\frac{P}{A_a} = \frac{(\alpha_a - \alpha_s)\theta}{\dfrac{1}{E_a} + \dfrac{A_a}{E_s A_s}} = \frac{(\alpha_a - \alpha_s)E_a\theta}{1 + \dfrac{E_a A_a}{E_s A_s}}$$

Now

$$(EA)_a = (70 \times 10^9)\left[\frac{\pi}{4}(0{\cdot}022)^2\right] = 26{\cdot}6 \text{ MN}$$

Again,

$$(EA)_s = (200 \times 10^9)[\pi(0{\cdot}025)(0{\cdot}003)] = 54{\cdot}8 \text{ MN}$$

Then

$$\frac{P}{A_a} = \frac{[(2{\cdot}3 - 1{\cdot}2)10^{-5}](70 \times 10^9)(120)}{1 + (26{\cdot}6/54{\cdot}8)} = 62{\cdot}2 \text{ MN/m}^2$$

1.18 Circular ring under radial pressure

When a thin circular ring is loaded radially, a circumferential force is set up in the ring; this force extends the circumference of the ring, which in turn leads to an increase in the radius of the ring. Consider a thin ring of mean radius r, Fig. 1.22(i), acted upon by an internal radial force of intensity p per unit length of the boundary. If the ring is cut across a diameter, Fig. 1.22(ii), circumferential forces P are required at the cut sections of the ring to maintain equilibrium of the half-ring. For equilibrium

$$2P = 2pr$$

so that

$$P = pr \tag{1.24}$$

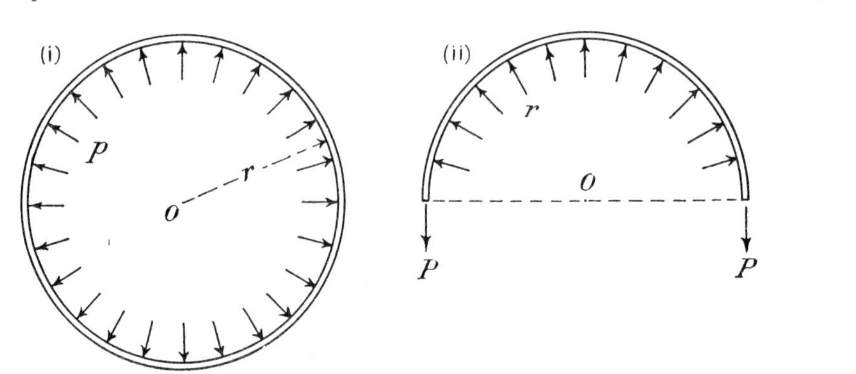

Fig. 1.22 Thin circular ring under uniform radial loading, leading to a uniform circumferential tension.

A section may be taken across any diameter, leading to the same result; we conclude, therefore, that P is the circumferential tension in all parts of the ring.

If A is the cross-sectional area of the ring at any point of the circumference, then the tensile circumferential stress in the ring is

$$\sigma = \frac{P}{A} = \frac{pr}{A} \tag{1.25}$$

If the cross-section is a rectangle of breadth b, (normal to the plane of Fig. 1.22), and thickness t, (in the plane of Fig. 1.22), then

$$\sigma = \frac{pr}{bt} \tag{1.26}$$

Circumferential stresses of a similar type are set up in a circular ring rotating about an axis through its centre. We suppose the ring is a uniform circular one, having a cross-sectional area A at any point, and that it is rotating about its central axis at uniform angular velocity ω. If ρ is the density of the material of the ring, then the centrifugal force on a unit length of the circumference is

$$\rho A \omega^2 r$$

In equation (1.25) we put this equal to p; thus, the circumferential tensile stress in the ring is

$$\sigma = \frac{pr}{A} = \rho \omega^2 r^2 \tag{1.27}$$

which we see is independent of the actual cross-sectional area. Now, ωr is the circumferential velocity, V (say), of the ring, so

$$\sigma = \rho V^2 \tag{1.28}$$

For steel we have $\rho = 7840\ \text{kg/m}^3$; to produce a tensile stress of $10\ \text{MN/m}^2$, the circumferential velocity must be

$$V = \sqrt{\sigma/\rho} = \sqrt{(10 \times 10^6)/7840} = 35{\cdot}7\ \text{m/sec}$$

Problem 1.15: A circular cylinder, containing oil, has an internal bore of 30 cm diameter. The cylinder is 1·25 cm thick. If the tensile stress in the cylinder must not exceed 75 MN/m^2, estimate the maximum load W which may be supported on a piston sliding in the cylinder.

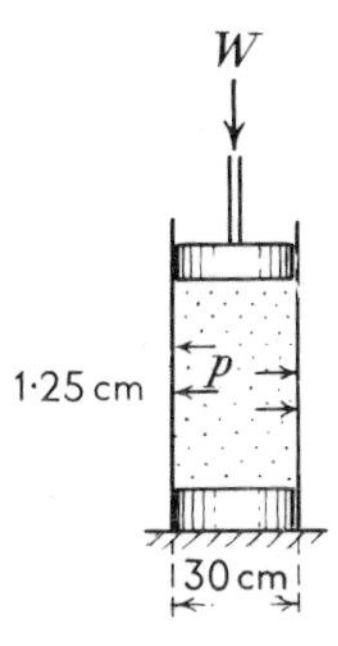

Solution

A load W on the piston generates an internal pressure p given by

$$W = \pi r^2 p$$

where r is the radius of the cylinder. In this case

$$p = \frac{W}{\pi r^2} = \frac{W}{\pi(0{\cdot}150)^2}$$

A unit length of the cylinder is equivalent to a circular ring subjected to an internal load of p per unit length of circumference. The circumferential load set up by p in this ring is, from equation (1.24),

$$P = pr = p(0{\cdot}150)$$

The circumferential stress is, therefore,

$$\sigma = \frac{P}{1 \times t} = \frac{P}{0{\cdot}0125} = 80P$$

where t is the thickness of the wall of the cylinder. If σ is limited to 75 MN/m^2, then

$$80P = 75 \times 10^6$$

But

$$80P = 80[p(0{\cdot}150)] = 12p = \frac{12W}{\pi(0{\cdot}150)^2}$$

Then

$$\frac{12W}{\pi(0{\cdot}150)^2} = 75 \times 10^6$$

giving

$$W = 441\,\text{kN}$$

Problem 1.16: An aluminium-alloy cylinder of internal diameter 10·000 cm and wall thickness 0·50 cm is shrunk onto a steel cylinder of external diameter 10·004 cm and wall thickness 0·50 cm. If the values of Young's modulus for the alloy and the steel are 70 GN/m^2 and 200 GN/m^2, respectively, estimate the circumferential stresses in the cylinders and the radial pressure between the cylinders.

Solution

We take unit lengths of the cylinders as behaving like thin circular rings. After the shrinking operation, we suppose p is the radial pressure between the cylinders. The mean radius of the steel tube is

$$\tfrac{1}{2}[10{\cdot}004 - 0{\cdot}50] = 4{\cdot}75\ \text{cm}$$

The compressive circumferential stress in the steel tube is then

$$\sigma_s = \frac{pr}{t} = \frac{p(0{\cdot}0475)}{0{\cdot}0050} = 9{\cdot}50p$$

The circumferential strain in the steel tube is then

$$\epsilon_s = \frac{\sigma_s}{E_s} = \frac{9{\cdot}50p}{200 \times 10^9}$$

The mean radius of the alloy tube is

$$\tfrac{1}{2}[10{\cdot}000 + 0{\cdot}50] = 5{\cdot}25\ \text{cm}$$

The tensile circumferential stress in the alloy tube is then

$$\sigma_a = \frac{pr}{t} = \frac{p(0{\cdot}0525)}{(0{\cdot}0050)} = 10{\cdot}5p$$

The circumferential strain in the alloy tube is then

$$\epsilon_a = \frac{\sigma_a}{E_a} = \frac{10{\cdot}5p}{70 \times 10^9}$$

The circumferential expansion of the alloy tube is

$$2\pi r\epsilon_a$$

so the mean radius increases effectively an amount

$$\delta_a = r\epsilon_a = 0{\cdot}0525\epsilon_a$$

Similarly, the mean radius of the steel tube contracts an amount

$$\delta_s = r\epsilon_s = 0{\cdot}0475\epsilon_s$$

For the shrinking operation to be carried out we must have that the initial lack of fit, δ, is given by

$$\delta = \delta_a + \delta_s$$

Then

$$\delta_a + \delta_s = 0{\cdot}002 \times 10^{-2}$$

On substituting for δ_a and δ_s, we have

$$0{\cdot}0525\left[\frac{10{\cdot}5p}{70 \times 10^9}\right] + 0{\cdot}0475\left[\frac{9{\cdot}50p}{200 \times 10^9}\right] = 0{\cdot}002 \times 10^{-2}$$

This gives

$$p = 1{\cdot}97 \text{ MN/m}^2$$

The compressive circumferential stress in the steel cylinder is then

$$\sigma_s = 9{\cdot}50p = 18{\cdot}7 \text{ MN/m}^2$$

The tensile circumferential stress in the alloy cylinder is

$$\sigma_a = 10{\cdot}5p = 20{\cdot}7 \text{ MN/m}^2$$

1.19 Creep of materials under sustained stresses

At ordinary laboratory temperatures most metals will sustain stresses below the limit of proportionality for long periods without showing additional measurable strains. At these temperatures metals deform continuously when stressed above the elastic range. This process of continuous inelastic strain is called *creep*. At high temperatures metals lose some of their elastic properties, and creep under constant stress takes place more rapidly.

When a tensile specimen of a metal is tested at a high temperature under a constant load, the strain assumes instantaneously some value ϵ_0, Fig. 1.23. If the initial strain is in the inelastic range of the material then creep takes place under constant stress. At first the creep rate is fairly rapid, but diminishes until a point a is reached on the strain-time curve, Fig. 1.23; the point a is a point of inflection in this curve, and continued application of the load increases the creep rate until fracture of the specimen occurs at b, Fig. 1.23.

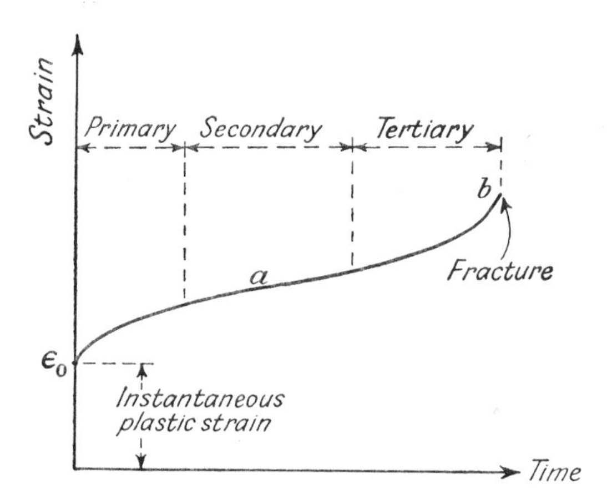

Fig. 1.23 Creep curve for a material in the inelastic range; ϵ_0 is the instantaneous plastic strain.

At ordinary atmospheric temperatures concrete shows creep properties; these may be important in pre-stressed members, where some of the initial stresses in the concrete may be lost after a long period due to creep.

1.20 Fatigue under repeated stresses

When a material is subjected to repeated tensile stresses within the elastic range, it is found that the material 'tires' and fractures rather suddenly after a large, but finite, number of repetitions of stress; the material is said to *fatigue*. Fig. 1.24 shows the forms of the *stress-endurance* curves for steel, aluminium light alloy and titanium alloy. The endurance to fatigue at a given stress level is the number of complete cycles of loading to that stress level required to bring about fracture of the material. Failure of a material after a large number of cycles of tensile stress occurs with little, or no, permanent set; fractures show the characteristics of brittle materials. Fatigue is primarily a problem of repeated tensile stresses; this is due probably to the fact that microscopic cracks in a material can propagate more easily when the material is stressed in tension. In the case of steels it is found that there is a critical stress—called the *endurance limit*—below which fluctuating stresses cannot cause a fatigue failure; titanium alloys show a similar

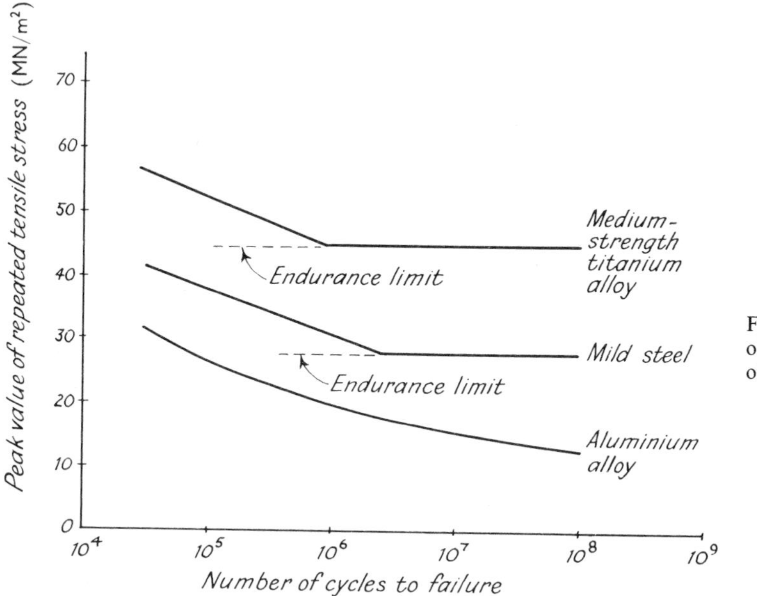

Fig. 1.24 Comparison of the fatigue strengths of metals under repeated tensile stresses.

phenomenon. No such endurance limit has been found for other non-ferrous metals and other materials.

FURTHER PROBLEMS
(answers on page 397)

1.17 The piston rod of a double-acting hydraulic cylinder is 20 cm diameter and 4 m long. The piston has a diameter of 40 cm, and is subjected to 10 MN/m^2 water pressure on one side and 3 MN/m^2 on the other. On the return stroke these pressures are interchanged. Estimate the maximum stress occurring in the piston-rod, and the change of length of the rod between two strokes, allowing for the area of piston-rod on one side of the piston. Take $E = 200$ GN/m^2. (*RNC*)

1.18 A uniform steel rope 250 m long hangs down a shaft. Find the elongation of the first 125 m at the top if the density of steel is 7840 kg/m^3 and Young's modulus is 200 GN/m^2. (*Cambridge*)

1.19 A steel wire, 150 m long, weighs 20 N per metre length. It is placed on a horizontal floor and pulled slowly along by a horizontal force applied to one end. If this force measures 600 N, estimate the increase in length of the wire due to its being towed, assuming a uniform coefficient of function. Take the density of steel as 7840 kg/m^3 and Young's modulus as 200 GN/m^2. (*RNEC*)

1.20 The hoisting rope for a mine shaft is to lift a cage of weight W. The rope is of variable section so that the stress on every section is equal to σ when the rope is fully extended. If ρ is the density of the material of the rope, show that the cross-sectional area A at a height z above the cage is

$$A = (W/\sigma)e^{\rho g z/\sigma}$$

1.21 To enable two walls, 10 m apart, to give mutual support they are stayed together by a 2·5 cm diameter steel tension rod with screwed ends, plates and nuts. The rod is heated to 100° C when the nuts are screwed up. If the walls yield, relatively, by 0·5 cm when the rod cools to 15° C, find the pull of the rod at that temperature. The coefficient of linear expansion of steel is $\alpha = 1{\cdot}2 \times 10^{-5}$ per ° C, and Young's modulus $E = 200$ GN/m^2. (*RNEC*)

1.22 A steel tube 3 cm diameter, 0·25 cm thick and 4 m long, is covered and lined throughout with copper tubes 0·2 cm thick. The three tubes are firmly joined together at their ends. The compound tube is then raised in temperature by 100° C. Find the stresses in the steel and copper, and the increase in length of the tube; also, what must be the magnitude of the forces which, applied to the ends of the tube, will prevent its expansion? Assume $E = 200\ \text{GN/m}^2$ for steel and $E = 110\ \text{GN/m}^2$ for copper; the coefficients of linear expansion of steel and copper are $1{\cdot}2 \times 10^{-5}$ per ° C and $1{\cdot}9 \times 10^{-5}$ per ° C, respectively.

2 Pin-jointed frames

2.1 Introduction

In problems of stress analysis we discriminate between two types: in the first the forces in a structure can be determined by considering only its statical equilibrium. Fig. 2.1 shows a rigid beam *BD* supported by two vertical wires *BF* and *DG*; the beam

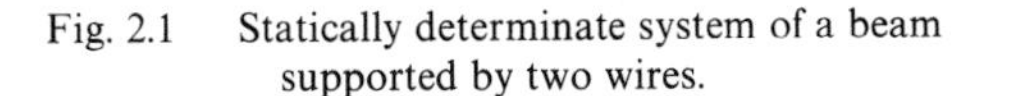

Fig. 2.1 Statically determinate system of a beam supported by two wires.

carries a force of $4W$ at *C*. We suppose the wires extend by negligibly small amounts, so that the geometrical configuration of the structure is practically unaffected; then for equilibrium the forces in the wires must be $3W$ in *BF* and *W* in *DG*. As the forces in the wires are known, it is a simple matter to calculate their extensions and hence to determine the displacement of any point of the beam. The calculation of the forces in the wires and their extensions is possible considering only the statical equilibrium of the system; the structure of Fig. 2.1 is said to be *statically determinate.* If, however, the rigid beam be supported by three wires, with an additional wire, say, between *H* and *J*, Fig. 2.1, then the forces in the three wires cannot be solved by considering statical equilibrium alone; this gives a second type of stress analysis problem, which is discussed more fully in a later stage; such a structure is *statically indeterminate.*

2.2 Statically determinate pin-jointed frames

By a *frame* we mean a structure which is composed of straight bars joined together at their ends. A *pin-jointed frame* is one in which no bending actions can be transmitted from one bar to another; ideally this could be achieved if the bars were joined together through pin-joints. If the frame has just sufficient bars to prevent collapse without the application of external forces it is *simply-stiff*; when there are more bars than this the frame is *redundant*. Definite relations exist which must be satisfied by the numbers of bars and joints if a frame is to be simply-stiff.

In the plane frame of Fig. 2.2, BC is one member. To locate the joint D relative to BC requires two members, BD and CD; to locate another joint F requires two further members, CF and DF. Obviously, for each new joint of the frame, two new members are required. If m be the total number of members, including BC, and j is the total number of joints, we must have

$$m = 2j - 3 \tag{2.1}$$

if the frame is to be simply-stiff.

When the frame is rigidly attached to a wall, say at B and C, BC is not part of the frame as such, and equation (2.1) becomes, omitting member BC, and joints B and C,

$$m = 2j \tag{2.2}$$

These conditions must be satisfied, but they may not necessarily ensure that the frame is simply-stiff. For example, the frames of Figs. 2.2 and 2.3 have the same numbers of

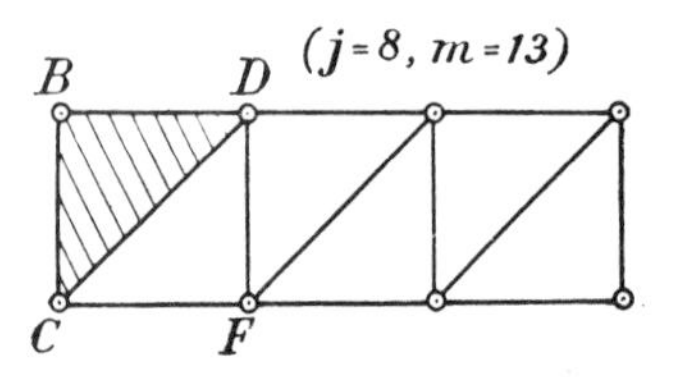

Fig. 2.2 Simply-stiff plane frame built up from a basic triangle BCD.

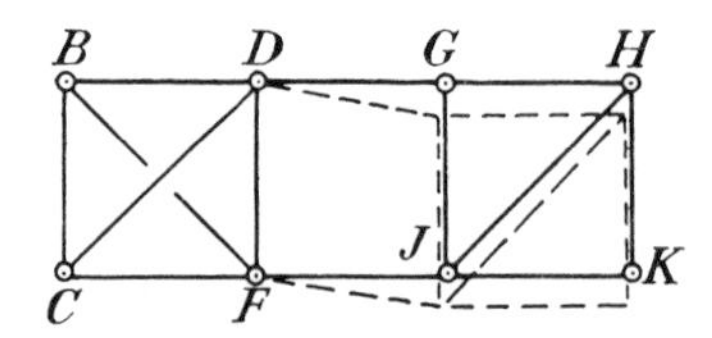

Fig. 2.3 Rearrangement of the members of Fig. 2.2 to give a mechanism.

members and joints; the frame of Fig. 2.2 is simply-stiff. The frame of Fig. 2.3 is *not* simply-stiff, since a mechanism can be formed with pivots at D, G, J, F. Thus, although a frame having j joints must have at least $(2j - 3)$ members, the mode of arrangement of these members is important.

For a pin-jointed space frame attached to three joints in a rigid wall, the condition for the frame to be simply-stiff is

$$m = 3j \tag{2.3}$$

where m is the total number of members, and j is the total number of joints, exclusive of the three joints in the rigid wall. When a space frame is not rigidly attached to a wall, the condition becomes

$$m = 3j - 6 \tag{2.4}$$

where m is the total number of members in the frame, and j the total number of joints.

2.3 Displacements of statically determinate frames

Estimating tension and compression forces in the members of a frame enables us to find the displacements of a pin-jointed framework under a given system of forces. We shall make use of a simple graphical method, which involves the construction of a *displacement diagram*; in Chapters 17 and 18 we shall discuss analytical methods giving more direct solutions of the displacements of a pin-jointed frame.

A simple triangular frame *BCD*, Fig. 2.4, consists of two members, *BD* and *CD*, which are pin-pointed to each other at *D*, and to a rigid foundation at *B* and *C*. The joint *D*

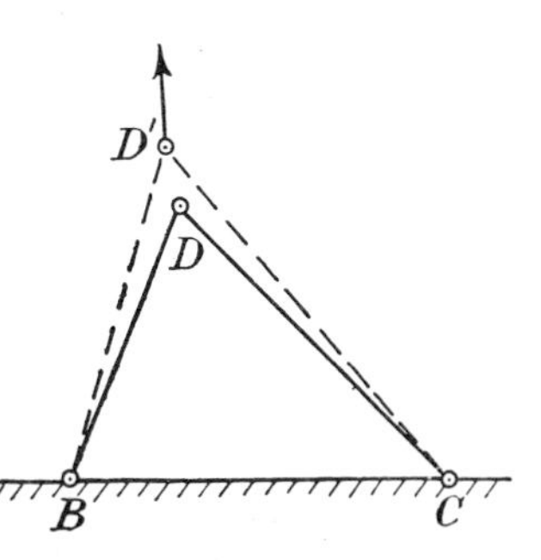

Fig. 2.4 Displacements in a simple triangular plane frame.

is acted upon by a force in the plane *BCD*. Axial forces are thereby set up in the members *BD* and *CD*; suppose the members change their lengths by small amounts, so that the joint *D* is displaced to *D'*. In practice most frames suffer very small distortions under working loads; the forces in the members can therefore be found with sufficient accuracy by assuming the original geometry of the frame to be unaffected. The frame suffers no gross distortions, in fact. If the bars of the frame of Fig. 2.4 remain elastic, the extensions of the bars may be calculated from the relation

$$e = \frac{Tl}{EA} \tag{2.5}$$

where T is the tensile load in a member, l is its unloaded length, E is Young's modulus, and A the cross-sectional area of the bar. Suppose e_1 and e_2 are, respectively, the extensions of the bars *BD* and *CD*.

Now, but for the constraint offered by *CD*, the extension of *BD* would displace *D* in the line *BD* produced by an amount e_1, Fig. 2.5. Similarly, but for *BD* the elongation of *CD* would displace *D* along *CD* produced by an amount e_2. On account of the restraint offered by the one bar to the other, the bars undergo rotations until the ends are coincident. If these rotations are very small, we can regard the ends as moving short distances at right angles to *BD* and *CD*, respectively. In constructing a *displacement diagram* we proceed as follows:

(i) Take a point *b*, Fig. 2.6, to represent *B*; since the displacement of *C* relative to *B* is zero, the point *b* also represents *C*, and is labelled *c* therefore in Fig. 2.6.

(ii) Draw bd_1 ($= e_1$) parallel to BD, and cd_2 ($= e_2$) parallel to CD. If either rod suffers a contraction, then the change of length is drawn in a sense opposite in direction to those of Figs. 2.5 and 2.6.

(iii) From the points d_1 and d_2 on the displacement diagram, draw lines at right angles to bd_1 and cd_2 to meet at d.

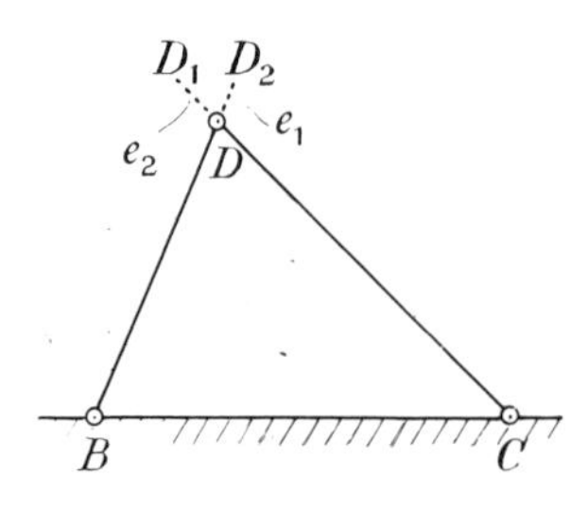

Fig. 2.5 Extensions of the bars.

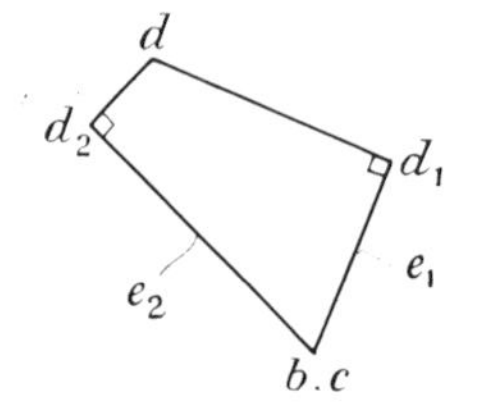

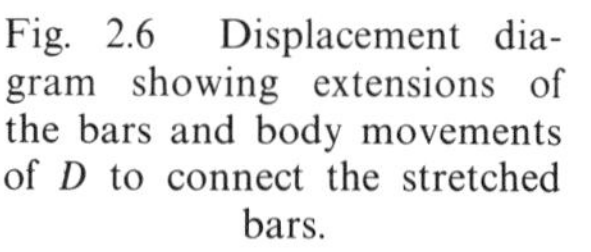

Fig. 2.6 Displacement diagram showing extensions of the bars and body movements of D to connect the stretched bars.

The displacement of D relative to B, or C, is then bd in the displacement diagram. The length of bd is the magnitude of the displacement, and the direction bd indicates the direction of the displacement.

In more complicated frameworks we have simply to repeat the construction for each triangle of the frame.

Problem 2.1: A pin-jointed, light-alloy frame carries a vertical load of 10 kN. If Young's modulus for the material is 70 GN/m^2, estimate the horizontal and vertical components of the displacement of the joint C. The cross-sectional area of each member is $0{\cdot}6 \times 10^{-3}$ m^2.

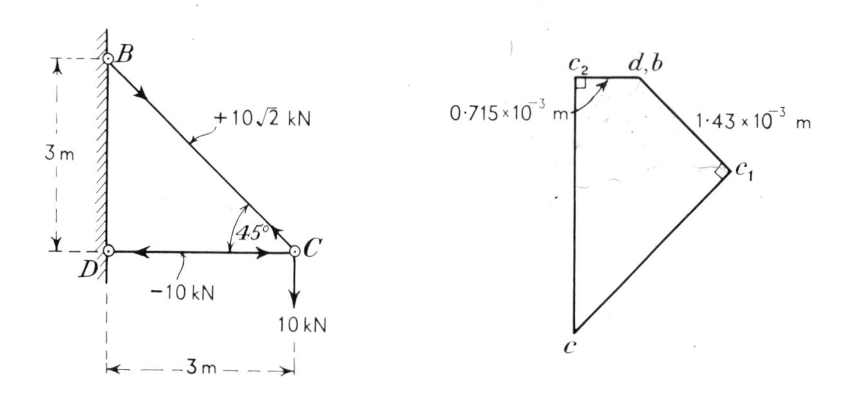

Solution

On considering equilibrium at joint C, the tensile force in BC is $10\sqrt{2}$ kN and the compressive force in DC is 10 kN. The elastic extension of BC is then

$$\frac{Tl}{EA} = \frac{(10\sqrt{2} \times 10^3)(3\sqrt{2})}{(70 \times 10^9)(0{\cdot}6 \times 10^{-3})} = 1{\cdot}43 \times 10^{-3}\ \text{m}$$

The contraction of DC is

$$\frac{Tl}{EA} = \frac{(10 \times 10^3)(3)}{(70 \times 10^9)(0{\cdot}6 \times 10^{-3})} = 0{\cdot}715 \times 10^{-3}\ \text{m}$$

The displacement diagram can now be constructed; in the diagram

$$bc_1 = 1{\cdot}43 \times 10^{-3}\ \text{m}, \qquad dc_2 = 0{\cdot}715 \times 10^{-3}\ \text{m}$$

The horizontal component of the displacement of C is

$$dc_2 = 0{\cdot}715 \times 10^{-3}\ \text{m}$$

The vertical component is

$$c_2c = (1{\cdot}43 \times 10^{-3})\sqrt{2} + 0{\cdot}715 \times 10^{-3} = 2{\cdot}74 \times 10^{-3}\ \text{m}$$

Problem 2.2: A pin-jointed frame is made up of members all of the same cross-sectional area A and Young's modulus E. Construct a displacement diagram for the loading conditions shown.

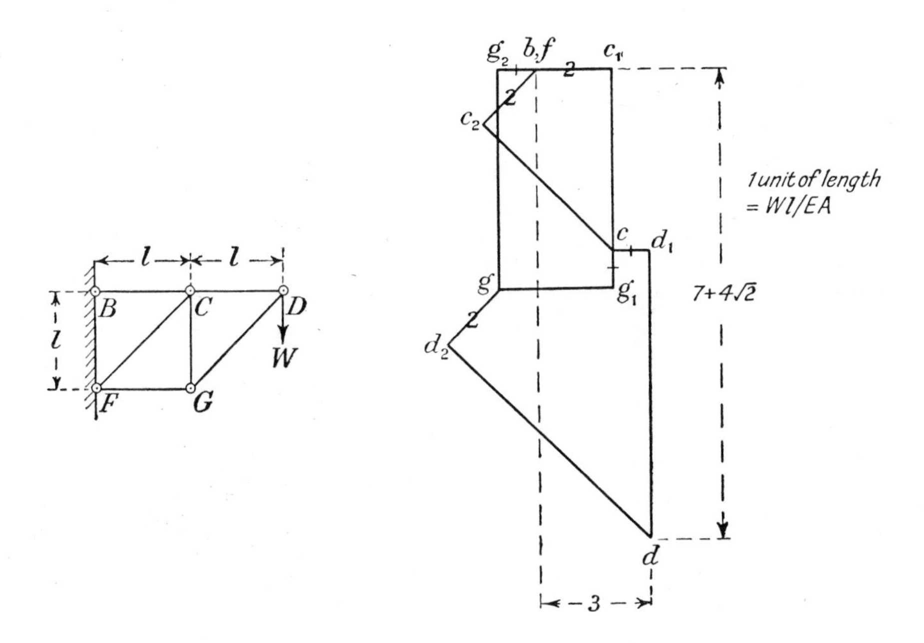

Solution

Firstly, find the axial forces in the members by considering equilibrium at each joint, or by some other suitable method; the elastic extension, or contraction, of each member of the frame can then be calculated; these calculations are summarised in the table. The displacement diagram can then be sketched taking Wl/EA as a unit of displacement.

Member	Axial load	Elastic extension
BC	$+2W$	$+2Wl/EA$
CD	$+\ W$	$+\ Wl/EA$
FG	$-\ W$	$-\ Wl/EA$
FC	$-\ W\sqrt{2}$	$-2Wl/EA$
GC	$+\ W$	$+\ Wl/EA$
GD	$-\ W\sqrt{2}$	$-2Wl/EA$

From the displacement diagram the vertical displacement of D is

$$(7 + 4\sqrt{2})\frac{Wl}{EA}$$

and the horizontal displacement of D is

$$3\frac{Wl}{EA}$$

The displacements of the other joints can be deduced.

2.4 Frames with non-linear members

The solution of the displacements of a framework by the method outlined in §2.3 is not restricted to frames in which the members obey Hooke's law. The method can also be applied to frames in which the members behave in a non-linear manner; the only essential requirement is that the distortions of the joints are small and not gross; this ensures that the internal loads in the members can be calculated by considering statics alone, i.e., these internal loads are statically determinate.

Problem 2.3: A pin-jointed, plane frame has members of equal lengths. The pin-joints are such that the members are eccentrically loaded by a small amount; if T is the tensile force in a member, its extension is

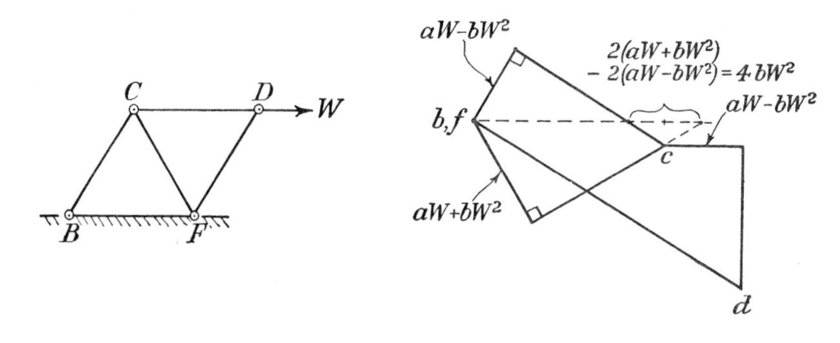

given approximately by

$$e = aT - bT^2$$

where a and b are positive constants. For a member in compression T is reckoned negative, and the resulting negative value of e is the contraction of the member; the values of a and b are the same for each member of the frame. Estimate the horizontal deflection of joint D.

Solution

The axial forces in the members of the frame are found first. The extensions of the members are then calculated from the relation

$$e = aT - bT^2$$

and summarised as follows:

Member	Tensile axial force	Extension
BC	$+W$	$aW - bW^2$
CD	$+W$	$aW - bW^2$
CF	$-W$	$-aW - bW^2$
DF	0	0

The displacement diagram can now be sketched; the horizontal displacement of C is

$$2(aW - bW^2) + 2bW^2 = 2aW$$

The horizontal displacement of D is therefore

$$2aW + (aW - bW^2) = 3aW - bW^2$$

The vertical displacement of D is then

$$(3aW - bW^2)/\sqrt{3}$$

2.5 Statically indeterminate problems

In §2.1 we mentioned a type of stress analysis problem in which internal stresses are not calculable on considering statical equilibrium alone; such problems are *statically indeterminate*. Consider the rigid beam BD of Fig. 2.7(i), which is supported on three wires; suppose the tensions in the wires are T_1, T_2 and T_3. Then for statical equilibrium of the beam we have

$$T_1 + T_2 + T_3 = 4W \tag{2.6}$$

and

$$T_1 - T_2 - 3T_3 = 0 \tag{2.7}$$

From these equations alone we cannot derive the values of the three tensile forces T_1, T_2, T_3; a third equation is found by discussing the extensions of the wires. If the wires extend amounts e_1, e_2, e_3, we must have from Fig. 2.7(ii) that

$$e_1 + e_3 = 2e_2 \tag{2.8}$$

since the beam BD is rigid. Suppose the wires are all of the same material and cross-sectional area, and that they remain elastic. Then we may write

$$e_1 = \lambda T_1, \qquad e_2 = \lambda T_2, \qquad e_3 = \lambda T_3 \tag{2.9}$$

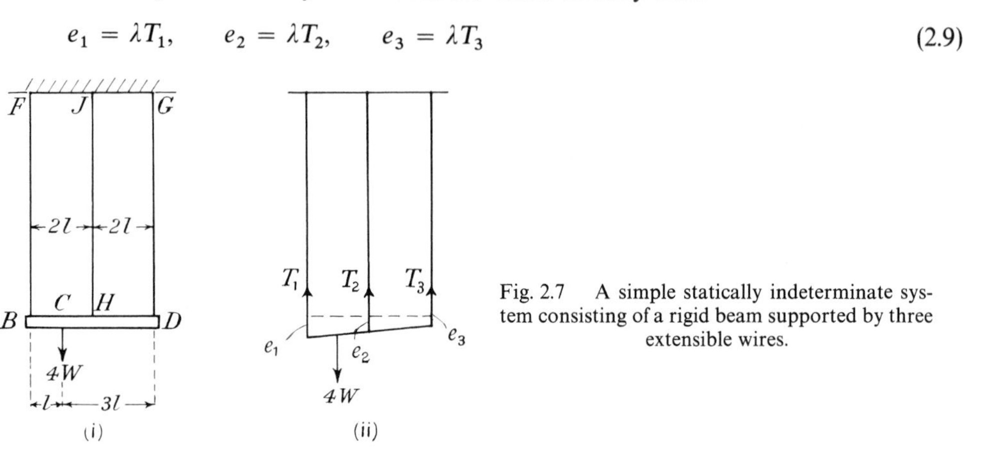

Fig. 2.7 A simple statically indeterminate system consisting of a rigid beam supported by three extensible wires.

where λ is a constant common to the three wires. Then equation (2.8) may be written

$$T_1 + T_3 = 2T_2 \tag{2.10}$$

The three equations (2.6), (2.7) and (2.10) then give

$$T_1 = \frac{7W}{12} \qquad T_2 = \frac{4W}{12} \qquad T_3 = \frac{W}{12} \tag{2.11}$$

Equation (2.8) is a condition which the extensions of the wires must satisfy; it is called a *strain compatibility* condition. Statically indeterminate problems are soluble if strain compatibilities are considered as well as statical equilibrium.

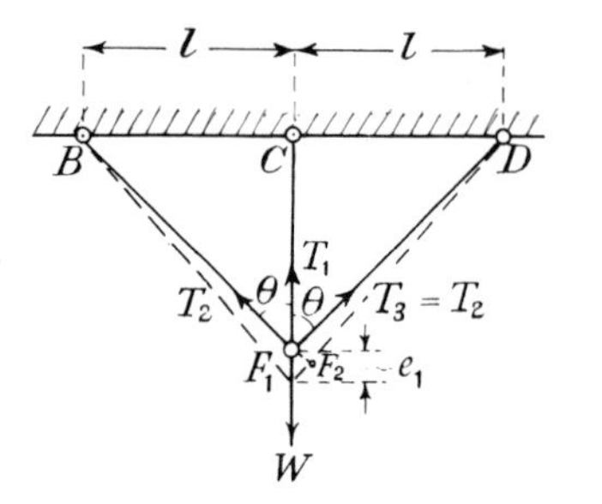

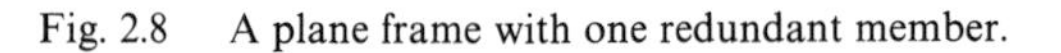

Fig. 2.8 A plane frame with one redundant member.

As another example of a statically indeterminate problem, consider the plane frame shown in Fig. 2.8; only two members are essential to locate the position of the joint F_1. If a third member be introduced it can be regarded as redundant; the forces in the bars meeting at F_1 are no longer calculable by statics alone. Consider F_1C as the redundant member, and let T_1 be the unknown tensile force in this redundant member. The tensile forces T_2, T_3 in the members F_1B, F_1D are related by the equilibrium equation

$$T_1 = W - (T_2 + T_3)\cos\theta \tag{2.12}$$

assuming no gross distortion of the whole frame.

Suppose members F_1B and F_1D are equal in all respects. Then $T_2 = T_3$, by symmetry. Let the subscript 1 refer to member F_1C, and subscript 2 to F_1B and F_1D. The joint F_1 is then displaced vertically downwards, and the extensions of the members are

$$e_1 = \frac{l_1 T_1}{E_1 A_1} \qquad e_2 = \frac{l_2 T_2}{E_2 A_2}$$

and from the diagram $e_2 = F_2F_1 = e_1 \cos\theta$, while $l_1 = l_2 \cos\theta$.

From these equations we have

$$\frac{T_2}{T_1} = \frac{E_2A_2}{E_1A_1}\cos^2\theta \tag{2.13}$$

From equations (2.12) and (2.13) we find

$$T_1 = \frac{W}{1 + 2\dfrac{E_2A_2}{E_1A_1}\cos^3\theta} \tag{2.14}$$

Then, from equations (2.12), (2.13) and (2.14)

$$T_2 = \frac{E_2A_2}{E_1A_1}\,\frac{W\cos^2\theta}{\left[1 + 2\dfrac{E_2A_2}{E_1A_1}\cos^3\theta\right]} \tag{2.15}$$

As an example, if $E_1A_1 = E_2A_2$, and $\theta = 45°$, the student should verify from first principles that $T_1 = 0{\cdot}586W$.

Problem 2.4: A pin-jointed frame is fabricated from three mild-steel members, each of cross-sectional area $0{\cdot}06 \times 10^{-3}$ m², and length 1·5 m. The material of each member is elastic in tension up to a yield stress of

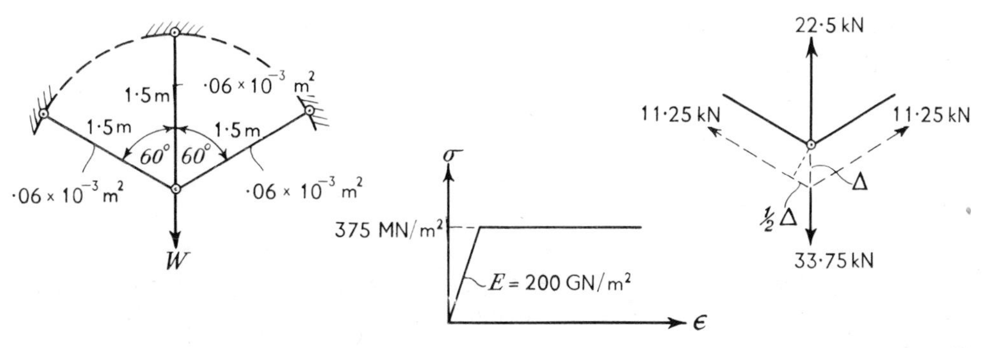

375 MN/m², Young's modulus being 200 GN/m². The material yields continuously at the yield stress. Estimate

(i) the value of W at which the yield stress is first attained in any of the three members, and
(ii) the value of W at which final collapse occurs.

Sketch a load-displacement curve for the frame.

Solution

Clearly the central vertical member is initially more heavily loaded than the inclined members. The central member first yields when its extension is

$$\frac{375 \times 10^6}{200 \times 10^9} \times 1{\cdot}5 = 0{\cdot}281 \text{ cm}$$

and when its tensile load is

$$0{\cdot}06 \times 10^{-3} \times 375 \times 10^6 = 22{\cdot}5 \text{ kN}$$

At the loaded joint the extension of the central member is always twice the extension of the inclined members. At the stage of initial yielding the tensile force in the inclined members is therefore

$$\tfrac{1}{2} \times 22{\cdot}5 \text{ kN} = 11{\cdot}25 \text{ kN}$$

The value of W at this stage is therefore 33·75 kN and its displacement is 0·281 cm.

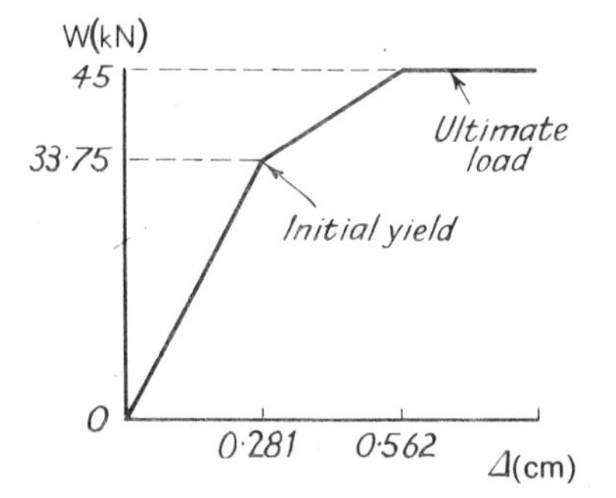

The inclined members yield when their extensions are equal to 0·281 cm. This implies a displacement of W of

$$2 \times 0{\cdot}281 \text{ cm} = 0{\cdot}562 \text{ cm}$$

The value of W at this stage is

$$W = 22{\cdot}5 + \tfrac{1}{2}(22{\cdot}5 + 22{\cdot}5) = 45{\cdot}0 \text{ kN}$$

FURTHER PROBLEMS
(answers on page 397)

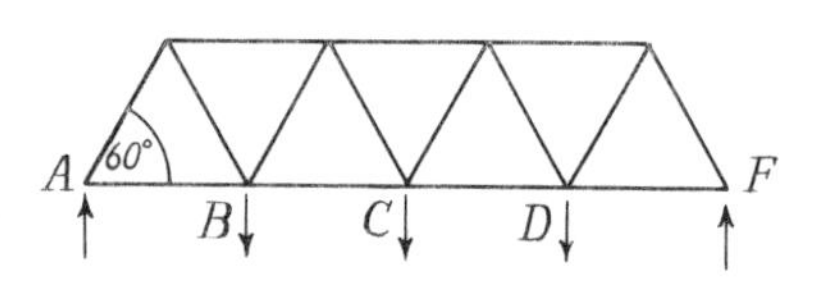

2.5 A Warren girder is supported on a hinge at A and on rollers at F. It carries three equal vertical loads at B, C and D, such that the tension members are stressed to 120 MN/m² and the compression members to 80 MN/m². Estimate the vertical deflection of C as a fraction of the span AF, taking Young's modulus as 200 GN/m². (*Bristol*)

2.6 Calculate the vertical deflection of the 30 kN load applied to a plane pin-jointed frame, each member of which has a cross-sectional area of 0·002 m², and a Young's modulus of 200 GN/m². (*London*)

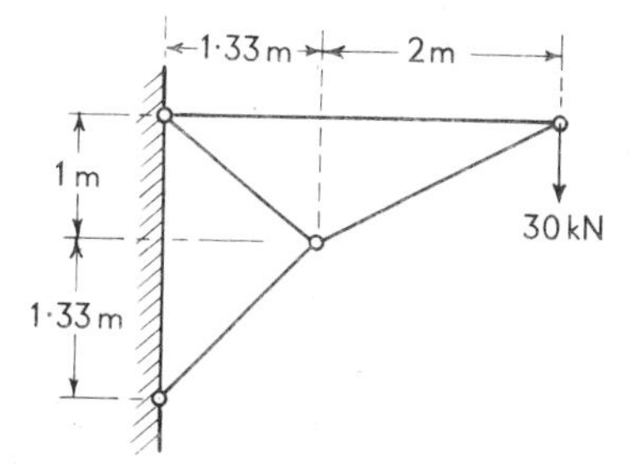

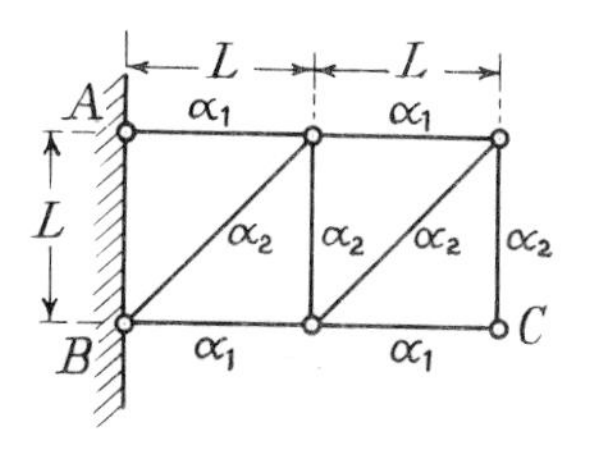

2.7 In a pin-jointed frame the four horizontal members are made of a material of coefficient of linear expansion α_1. The inclined and vertical members are made of a material of coefficient of linear expansion α_2. Estimate the vertical deflection of the joint C due to a temperature rise θ, assuming the distance between A and B to remain unchanged.

2.8 Three steel wires are joined at a point Q, and carry a vertical load of 100 kN at that point. Estimate the forces in the wires.

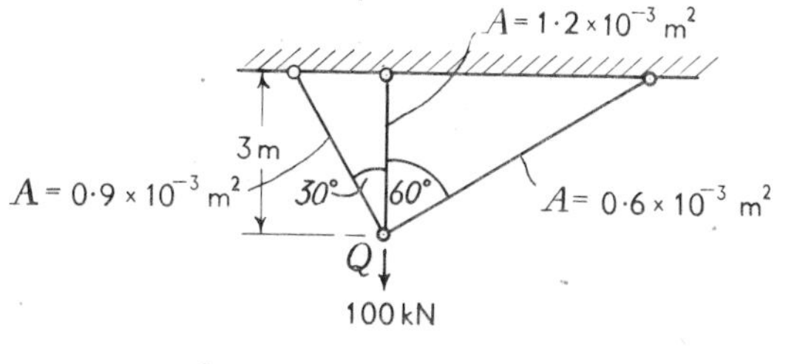

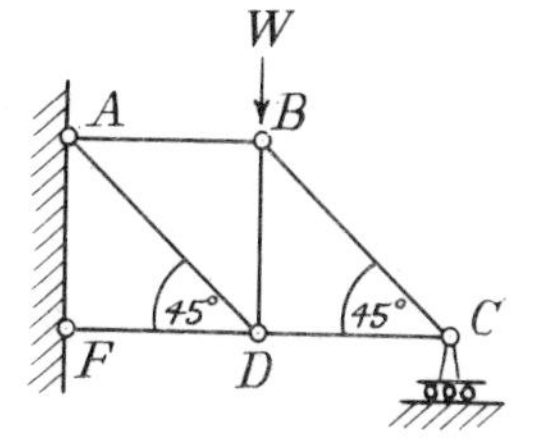

2.9 All the members of a pin-jointed truss are of the same material and cross-sectional area. The truss is subjected to a vertical load W at B. By drawing displacement diagrams, or otherwise, find the reaction at C. (*Cambridge*)

2.10 A frame consists of three hinged bars each of diameter 1·25 cm. If an initial compressive stress of 15 MN/m² exists in BD, what are the initial stresses in AD and CD? A vertical load is then applied at D. Find its maximum limit if the maximum stress in the bars is not to exceed 120 MN/m². (*Sheffield*)

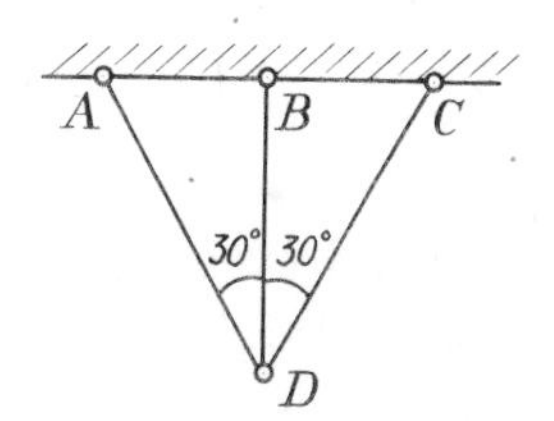

3 Shearing stresses

3.1 Introduction

In Chapter 1 we made a study of tensile and compressive stresses, which we called direct stresses. There is another type of stress which plays a vital rôle in the behaviour of materials, especially metals.

Consider a thin block of material, Fig. 3.1, which is glued to a table; suppose a thin plate is now glued to the upper surface of the block. If a horizontal force F is applied to the plate, the plate will tend to slide along the top of the block of material, and the block itself will tend to slide along the table. Provided the glued surfaces remain intact, the table resists the sliding of the block, and the block resists the sliding of the plate on its upper surface. If we consider the block to be divided by any imaginary horizontal plane, such as *ab*, the part of the block above this plane will be trying to slide over the part below the plane. The material on each side of this plane is said to be subjected to a *shearing* action; the stresses arising from these actions are called *shearing stresses.*

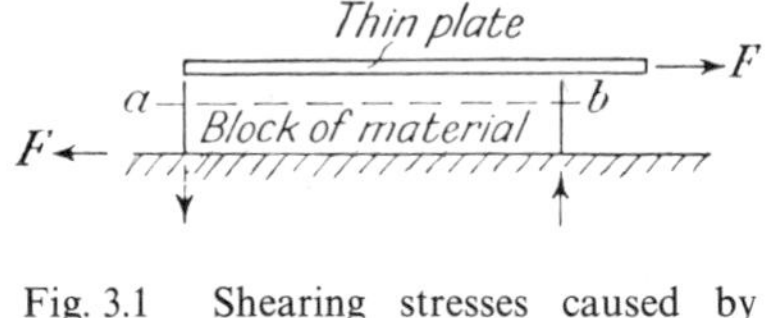

Fig. 3.1 Shearing stresses caused by shearing forces.

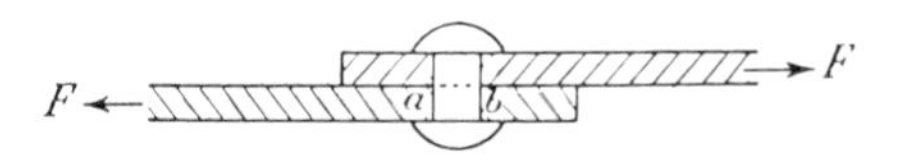

Fig. 3.2 Shearing stress in a rivet; shearing force F is transmitted over the face *ab* of the rivet.

Shearing stresses arise in many practical problems. Fig. 3.2 shows two flat plates held together by a single rivet, and carrying a tensile force F. We imagine the rivet divided into two portions by the plane *ab*; then the upper half of the rivet is tending to slide over the lower half, and a shearing stress is set up in the plane *ab*. Fig. 3.3 shows a circular shaft *a*, with a collar *c*, held in a bearing *b*, one end of the shaft being pushed with a force F; in this case there is, firstly, a tendency for the shaft to be pushed bodily through the collar, thereby inducing shearing stresses over the cylindrical surfaces *d* of the shaft and the collar; secondly, there is a tendency for the collar to push through the bearing, so that shearing stresses are set up on cylindrical surfaces such as *e* in the bearing. As a third example, consider the case of a steel bolt in the end of a bar of wood, Fig. 3.4, the bolt being pulled by forces F; suppose the grain of the wood runs parallel to the length of the bar; then if the forces F are large enough the block *abcd* will be pushed out, shearing taking place along the planes *ab* and *cd*.

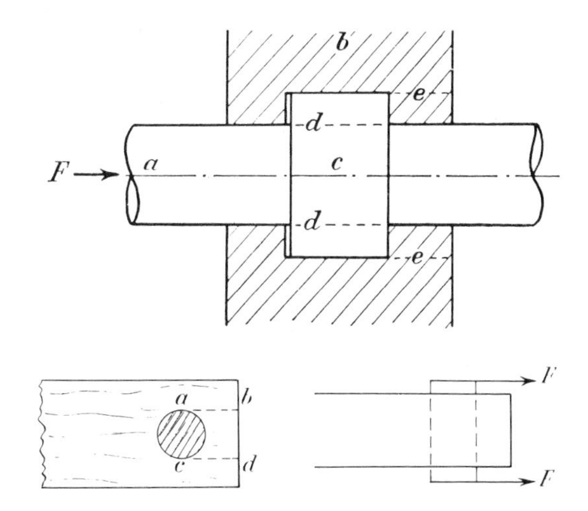

Fig. 3.3 Thrust on the collar of a shaft, generating shearing stress over the planes d.

Fig. 3.4 Tearing of the end of a timber member by a steel bolt, generating a shearing action on the planes ab, cd.

3.2 Measurement of shearing stress

Shearing stress on any surface is defined as the intensity of shearing force tangential to the surface. If the block of material of Fig. 3.1 has an area A over any section such as ab, the average shearing stress τ over the section ab is

$$\tau = \frac{F}{A} \tag{3.1}$$

In many cases the shearing force is not distributed uniformly over any section; if δF is the shearing force on any elemental area δA of a section, the shearing stress on that elemental area is

$$\tau = \underset{\delta A \to 0}{\text{Limit}} \frac{\delta F}{\delta A} = \frac{dF}{dA} \tag{3.2}$$

Problem 3.1: Three steel plates are held together by a 1·5 cm diameter rivet. If the load transmitted is 50 kN, estimate the shearing stress in the rivet.

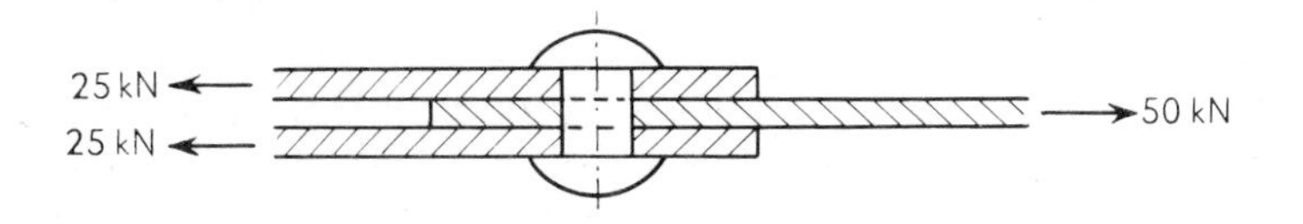

Solution

There is a tendency to shear across the planes in the rivet shown by broken lines. The area resisting shear is twice the cross-sectional area of the rivet; the area of the rivet is

$$A = \frac{\pi}{4}(0{\cdot}015)^2 = 0{\cdot}177 \times 10^{-3}\ \text{m}^2$$

The average shearing stress in the rivet is then

$$\tau = \frac{F}{A} = \frac{25 \times 10^3}{0{\cdot}177 \times 10^{-3}} = 141 \text{ MN/m}^2$$

Problem 3.2: Two steel rods are connected by a cotter joint. If the shearing strength of the steel used in the rods and the cotter is 150 MN/m², estimate which part of the joint is most prone to shearing failure.

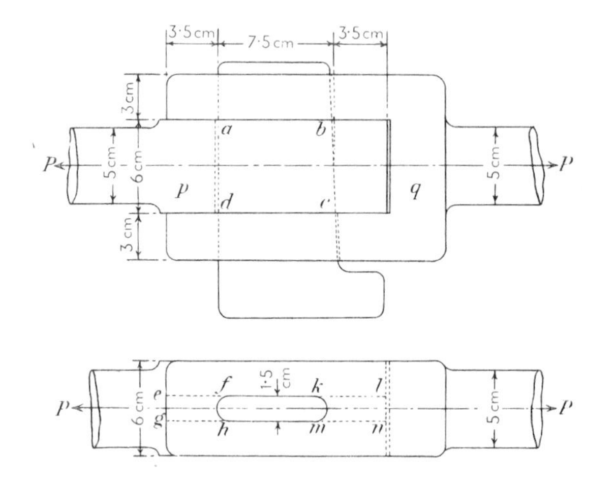

Solution

Shearing failure may occur in the following ways:

(i) Shearing of the cotter in the planes *ab* and *cd*. The area resisting shear is

$2(fkmh) = 2(0{\cdot}075)(0{\cdot}015) = 2{\cdot}25 \times 10^{-3}\ \text{m}^2$

For a shearing failure on these planes, the tensile force is

$P = \tau A = (150 \times 10^6)(2{\cdot}25 \times 10^{-3}) = 338\ \text{kN}$

(ii) By the cotter tearing through the ends of the socket *q*, i.e. by shearing in the planes *ef* and *gh*. The total area resisting shear is

$A = 4(0{\cdot}030)(0{\cdot}035) = 4{\cdot}20 \times 10^{-3}\ \text{m}^2$

For a shearing failure on these planes

$P = \tau A = (150 \times 10^6)(4{\cdot}20 \times 10^{-3}) = 630\ \text{kN}$

(iii) By the cotter tearing through the ends of the rod *p*, i.e., by shearing in the planes *kl* and *mn*. The total area resisting shear is

$A = 2(0{\cdot}035)(0{\cdot}060) = 4{\cdot}20 \times 10^{-3}\ \text{m}^2$

For a shearing failure on these planes

$P = \tau A = (150 \times 10^6)(4{\cdot}20 \times 10^{-3}) = 630\ \text{kN}$

Thus, the connection is most vulnerable to shearing failure in the cotter itself, as discussed in (i); the tensile load for shearing failure is 338 kN.

Problem 3.3: A lever is keyed to a shaft 4 cm diameter, the width of the key being 1·25 cm and its length 5 cm. What load P can be applied at an arm of $a = 1$ m if the average shearing stress in the key is not to exceed 60 MN/m²?

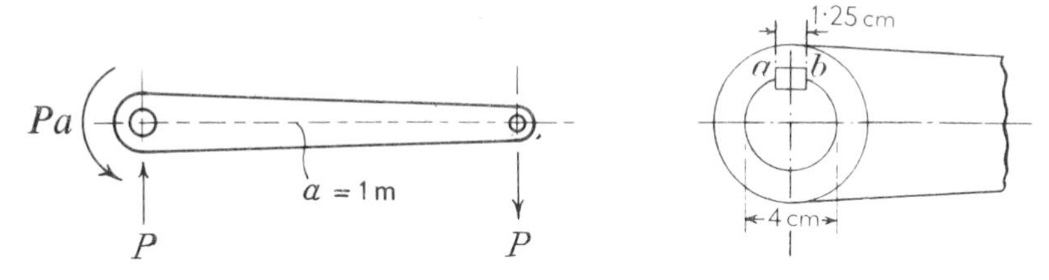

Solution

The torque applied to the shaft is Pa. If this is resisted by a shearing force F on the plane ab of the key, then

$$Fr = Pa$$

where r is the radius of the shaft. Then

$$F = \frac{Pa}{r} = \frac{P(1)}{(0{\cdot}02)} = 50P$$

The area resisting shear in the key is

$$A = 0{\cdot}0125 \times 0{\cdot}050 = 0{\cdot}625 \times 10^{-3}\ \text{m}^2$$

The permissible shearing force on the plane ab of the key is then

$$F = \tau A = (60 \times 10^6)(0{\cdot}625 \times 10^{-3}) = 37{\cdot}5\ \text{kN}$$

The permissible value of P is then

$$P = \frac{F}{50} = 750\ \text{N}$$

3.3 Complementary shearing stresses

Let us return now to the consideration of the block shown in Fig. 3.1. We have seen that horizontal planes, such as ab, are subjected to shearing stresses. In fact the state of stress is rather more complex than we have supposed, because for rotational equilibrium of the whole block an external couple is required to balance the couple due to the shearing forces F. Suppose the material of the block is divided into a number of rectangular elements, as shown by the full lines of Fig. 3.5. Under the actions of the shearing forces F, which together constitute a couple, the elements will tend to take up the positions shown by the broken lines in Fig. 3.5. It will be seen that there is a tendency for the vertical faces of the elements to slide over each other. Actually the ends of the elements do not slide over each other in this way, but the tendency to so do shows that the shearing stress in horizontal planes is accompanied by shearing stresses in vertical planes perpendicular to the applied shearing forces. This is true of all cases of shearing action: a given shearing stress acting on one plane is always accompanied by a *complementary shearing stress* on planes at right angles to the plane on which the given stress acts.

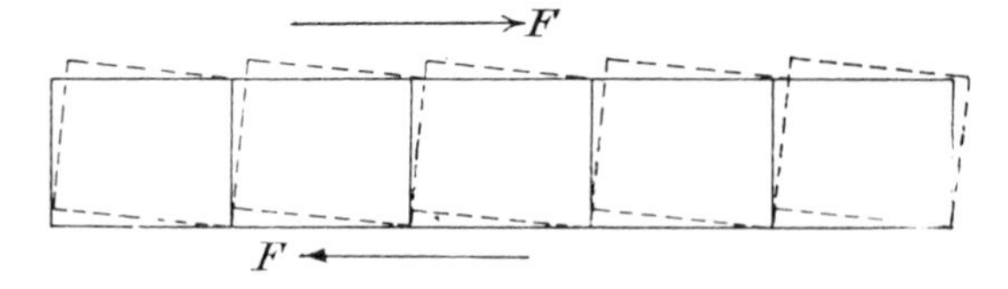

Fig. 3.5 Tendency for a set of disconnected blocks to rotate when shearing forces are applied.

Consider now the equilibrium of one of the elementary blocks of Fig. 3.5. Let τ_{xy} be the shearing stress on the horizontal faces of the element, and τ_{yx} the complementary shearing stress* on vertical faces of the element, Fig. 3.6. Suppose a is the length of the element, b its height, and that it has unit depth. The total shearing force on the upper and lower faces is then

$$\tau_{xy} \times a \times 1 = a\tau_{xy}$$

while the total shearing force on the end faces is

$$\tau_{yx} \times b \times 1 = b\tau_{yx}$$

For rotational equilibrium of the element we then have

$$(a\tau_{xy}) \times b = (b\tau_{yx}) \times a,$$

and thus

$$\tau_{xy} = \tau_{yx}$$

We see then that, whenever there is a shearing stress over a plane passing through a given line, there must be an *equal* complementary shearing stress on a plane perpendicular to the given plane, and passing through the given line. The directions of the two shearing stresses must be either both towards, or both away from, the line of intersection of the two planes in which they act.

It is extremely important to appreciate the existence of the complementary shearing stress, for its necessary presence has a direct effect on the maximum stress in the material, as we shall see later in Chapter 5.

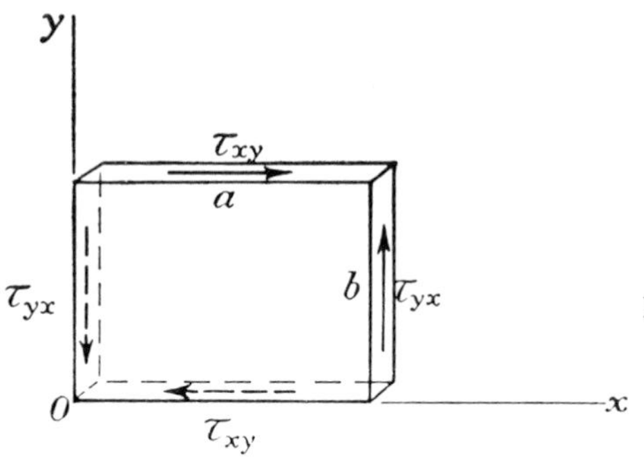

Fig. 3.6 Complementary shearing stresses over the faces of a block when they are connected.

* Notice that the first suffix x shows the direction, the second the plane on which the stress acts; thus τ_{xy} acts in direction of x axis on planes y = const.

3.4 Shearing strain

Shearing stresses in a material give rise to *shearing strains.* Consider a rectangular block of material, Fig. 3.7, subjected to shearing stresses τ in one plane. The shearing stresses distort the rectangular face of the block into a parallelogram. If the right-angles at the corners of the face change by amounts γ, then γ is the shearing strain. The angle γ is measured in radians, and is non-dimensional therefore.

Fig. 3.7 Shearing strain in a rectangular block; small values of γ lead to negligible change of volume in shear straining.

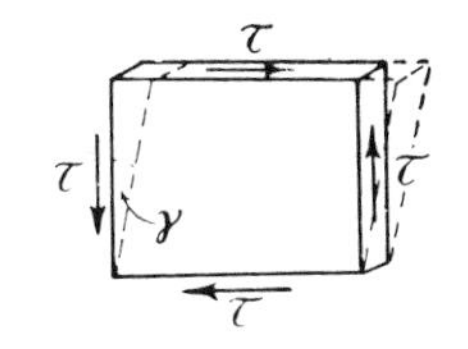

For many materials shearing strain is linearly proportional to shearing stress within certain limits. This linear dependence is similar to the case of direct tension and compression. Within the limits of proportionality

$$\tau = G\gamma \tag{3.3}$$

where G is the *shearing modulus*, and is similar to Young's modulus E, for direct tension and compression. For most materials E is about 2·5 times greater than G.

It should be noted that no volume changes occur as a result of shearing stresses acting alone. In Fig. 3.7 the volume of the strained block is approximately equal to the volume of the original rectangular prism if the angular strain γ is small.

3.5 Strain energy due to shearing actions

In shearing the rectangular prism of Fig. 3.7, the forces acting on the upper and lower faces undergo displacements. Work is done, therefore, during these displacements. If the strains are kept within the elastic limit the work done is recoverable, and is stored in the form of strain energy. Suppose all edges of the prism of Fig. 3.7 are of unit length; then the prism has unit volume, and the shearing forces on the sheared faces are τ. Now suppose τ is increased a small amount, causing a small increment of shearing strain $\delta\gamma$. The work done on the prism during this small change is $\tau\,\delta\gamma$, since the force τ moves through a distance $\delta\gamma$. The total work done in producing a shearing strain γ is then

$$\int_0^{\gamma} \tau\, d\gamma$$

While the material remains elastic, we have from equation (3.3) that $\tau = G\gamma$, and the work done is stored as strain energy; the strain energy is therefore

$$\int_0^{\gamma} \tau\, d\gamma = \int_0^{\gamma} G\gamma\, d\gamma = \tfrac{1}{2}G\gamma^2 \tag{3.4}$$

per unit volume. In terms of τ this becomes

$$\tfrac{1}{2}G\gamma^2 = \frac{\tau^2}{2G} \tag{3.5}$$

FURTHER PROBLEMS

(answers on page 397)

3.4 Rivet holes 2·5 cm diameter are punched in a steel plate 1 cm thick. The shearing strength of the plate is 300 MN/m². Find the average compressive stress in the punch at the time of punching.

3.5 The diameter of the bolt circle of a flanged coupling for a shaft 12·5 cm diameter is 37·5 cm. There are six bolts 2·5 cm diameter. What power can be transmitted at 150 rev/min if the shearing stress in the bolts is not to exceed 60 MN/m²?

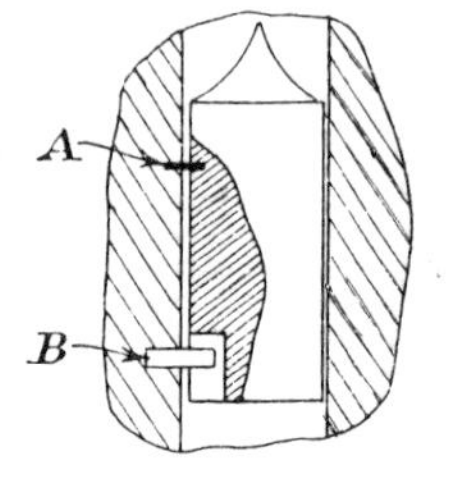

3.6 A pellet carrying the striking needle of a fuse has a mass of 0·1 kg; it is prevented from moving longitudinally relative to the body of the fuse by a copper pin A of diameter 0·05 cm. It is prevented from turning relative to the body of the fuse by a steel stud B. A fits loosely in the pellet so that no stress comes on A due to rotation. If the copper shears at 150 MN/m², find the retardation of the shell necessary to shear A. (*RNC*)

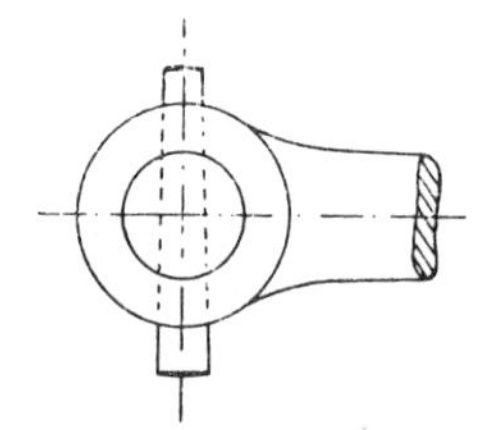

3.7 A lever is secured to a shaft by a taper pin through the boss of the lever. The shaft is 4 cm diameter and the mean diameter of the pin is 1 cm. What torque can be applied to the lever without causing the average shearing stress in the pin to exceed 60 MN/m².

3.8 A cotter joint connects two circular rods in tension. Taking the tensile strength of the rods as 350 MN/m², the shearing strength of the cotter 275 MN/m², the permissible bearing pressure between surfaces in contact 700 MN/m², the shearing strength of the rod ends 185 MN/m², calculate suitable dimensions for the joint so that it may be equally strong against the possible types of failure. Take the thickness of the cotter $= d/4$, and the taper of the cotter 1 in 48.

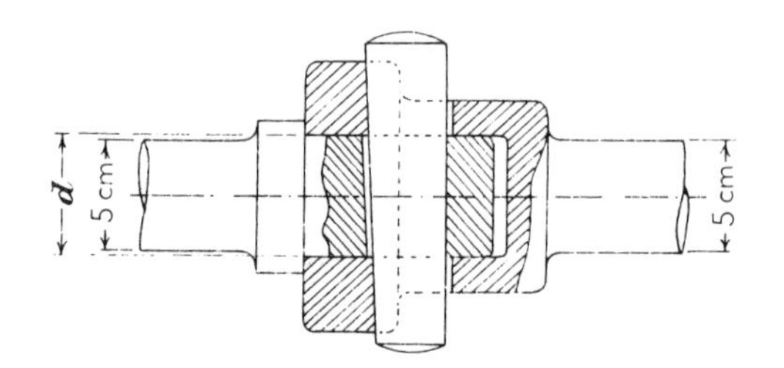

3.9 A horizontal arm, capable of rotation about a vertical shaft, carries a mass of 2·5 kg, bolted to it by a 1 cm bolt at a distance 50 cm from the axis of the shaft. The axis of the bolt is vertical. If the ultimate shearing strength of the bolt is 50 MN/m², at what speed will the bolt snap? (*RNEC*)

3.10 A copper disc 10 cm diameter and 0·0125 cm thick, is fitted in the casing of an air compressor, so as to blow and safeguard the cast-iron case in the event of a serious compressed air leak. If pressure inside the case is suddenly built up by a burst cooling coil, calculate at what pressure the disc will blow out, assuming that failure occurs by shear round the edges of the disc, and that copper will normally fail under a shearing stress of 120 MN/m². (*RNEC*)

4 Joints and connections

4.1 Importance of connections

Many engineering structures and machines consist of components suitably connected through carefully designed joints. In metallic materials, these joints may take a number of different forms, as for example welded joints, bolted joints and riveted joints. In general such joints are stressed in complex ways, and it is not usually possible to calculate stresses accurately because of the geometrical discontinuities in the region of a joint. For this reason, good design of connections is a mixture of stress analysis and experience of the behaviour of actual joints; this is particularly true of connections subjected to repeated loads.

Bolted joints are widely used in structural steel work, and recently the performance of such joints has been greatly improved by the introduction of high-tensile, friction-grip bolts. Welded joints are widely used in steel structures, as for example in ship construction. Riveted joints are still widely used in aircraft-skin construction in light-alloy materials. Epoxy resin glues have also been used in the aeronautical field to bond metals.

4.2 Modes of failure of simple bolted and riveted joints

One of the simplest types of joint between two plates of material is a bolted or riveted lap joint, Fig. 4.1. We shall discuss the forms of failure of the joint assuming it is bolted,

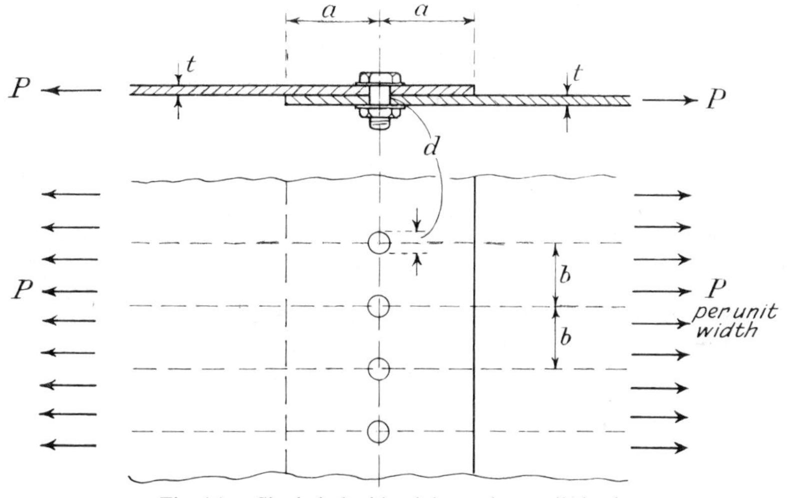

Fig. 4.1 Single-bolted lap joint under tensile load.

but the analysis can be extended in principle to the case of a riveted connection. Consider a joint between two wide plates, Fig. 4.1; suppose the plates are each of thickness t, and that they are connected together with a single line of bolts, giving a total overlap of breadth $2a$. Suppose also that the bolts are each of diameter d, and that their centres are a distance b apart along the line of bolts; the line of bolts is a distance a from the edge of each plate. It is assumed that a bolt fills a hole, so that the holes in the plates are also of diameter d.

We consider all possible simple modes of failure when each plate carries a tensile load of P per unit width of plate:

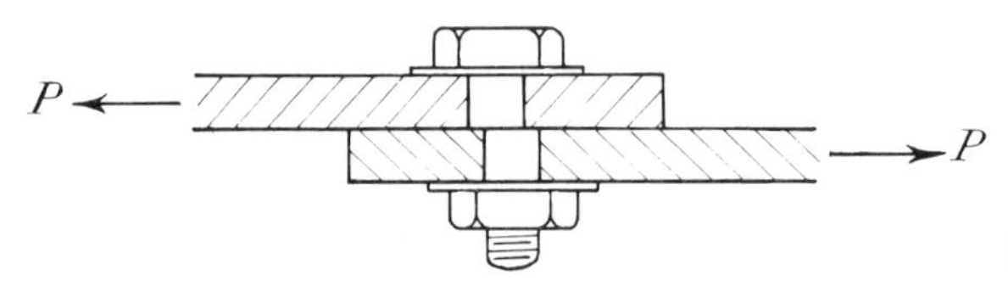

Fig. 4.2 Failure by shearing of the bolts.

(i) The bolts may fail by shearing, as shown in Fig. 4.2; if τ_1 is the maximum shearing stress the bolts will withstand, the total shearing force required to shear a bolt is

$$\tau_1 \times \left(\frac{\pi d^2}{4}\right)$$

Now, the load carried by a single bolt is Pb, so that a failure of this type occurs when

$$Pb = \tau_1\left(\frac{\pi d^2}{4}\right)$$

This gives

$$P = \frac{\pi d^2 \tau_1}{4b} \tag{4.1}$$

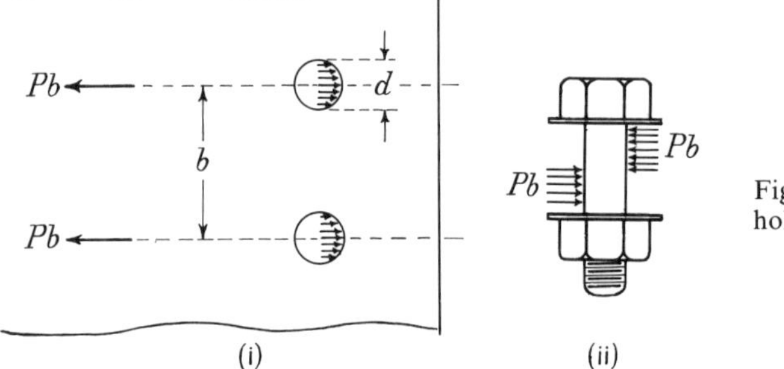

Fig. 4.3 (i) Bearing pressure on the holes of the upper plate. (ii) Bearing pressures on a bolt.

(ii) The bearing pressure between the bolts and the plates may become excessive; the total bearing load taken by a bolt is Pb, Fig. 4.3, so that the average bearing

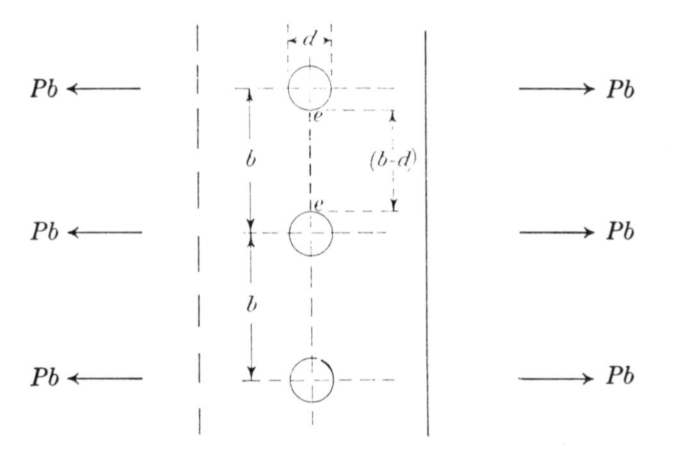

Fig. 4.4 Tensile failures in the plates.

pressure between a bolt and its surrounding hole is

$$\frac{Pb}{td}$$

If p_b is the pressure at which either the bolt or the hole fails in bearing, a failure of this type occurs when

$$P = \frac{p_b td}{b} \tag{4.2}$$

(iii) Tensile failures may occur in the plates; clearly the most heavily stressed regions of the plates are on sections such as *ee*, Fig. 4.4, through the line of bolts. The average tensile stress on the reduced area of plate through this section is

$$\frac{Pb}{(b - d)t}$$

If the material of the plate has an ultimate tensile stress of σ_{ult}, then a tensile failure occurs when

$$P = \frac{\sigma_{\text{ult}} t(b - d)}{b} \tag{4.3}$$

(iv) Shearing of the plates may occur on planes such as *cc*, Fig. 4.5, with the result that the whole block of material *cccc* is sheared out of the plate. If τ_2 is the maximum shearing stress of the material of the plates, this mode of failure occurs when

$$Pb = \tau_2 \times 2at$$

This gives

$$P = \frac{2at\tau_2}{b} \tag{4.4}$$

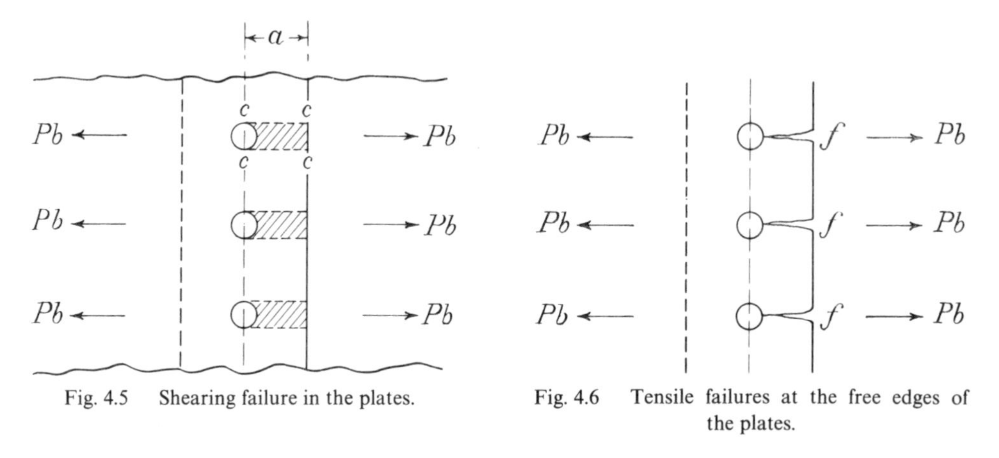

Fig. 4.5 Shearing failure in the plates.

Fig. 4.6 Tensile failures at the free edges of the plates.

(v) The plates may fail due to the development of large tensile stresses in the regions of points such as f, Fig. 4.6. The failing load in this condition is difficult to estimate, and we do not attempt the calculation at this stage.

In riveted joints it is found from tests on mild-steel plates and rivets that if the centre of a rivet hole is not less than $1\frac{1}{2}$ times the rivet hole diameter from the edge of the plate, then failure of the plate by shearing, as discussed in (iv), or by tensile failure of the free edge, as discussed in (v), or by a combination of (iv) and (v), does not occur. Thus if, for mild-steel plates and rivets,

$$a \geqslant 1{\cdot}5d \qquad (4.5)$$

we can disregard the modes of failure discussed in (iv) and (v). In the case of wrought aluminium alloys, the corresponding value of a is

$$a \geqslant 2d \qquad (4.6)$$

We have assumed, in discussing the modes of failure, that all load applied to the two plates of Fig. 4.1 is transmitted in shear through the bolts or rivets. This is so only if there is a negligible frictional force between the two plates. If hot-driven rivets are used, appreciable frictional forces are set up on cooling; these forces play a vital part in the behaviour of the connection. With cold-driven rivets the frictional force is usually small, and may be neglected.

Problem 4.1: Two steel plates, each 1 cm thick, are connected by riveting them between cover plates each 0·6 cm thick. The rivets are 1·6 cm diameter. The tensile stress in the plates must not exceed 140 MN/m^2, and the shearing stress in the rivets must not exceed 75 MN/m^2. Find the proportions of the joint so that it shall be equally strong in shear and tension, and estimate the bearing pressure between the rivets and the plates.

Solution

Suppose b is the rivet pitch, and that P is the tensile load per metre carried by the connection. Then the tensile load on one rivet is Pb. The cover plates, taken together, are thicker than the main plates, and may be disregarded therefore, in the strength calculations. We imagine there is no restriction on the distance from the rivets to the extreme edges of the main plates and cover plates; we may disregard then any possibility of shearing or tensile failure on the free edges of the plates.

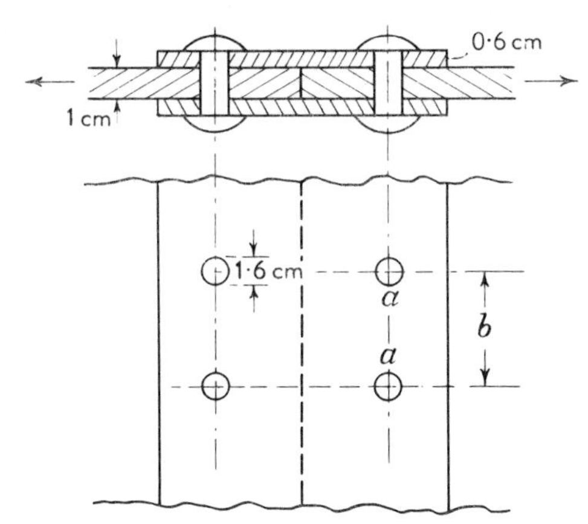

There are then two possible modes of failure:

(i) Tensile failure of the main plates may occur on sections such as *aa*. The area resisting tension is

$$0{\cdot}010(b - 0{\cdot}016)\ \text{m}^2$$

The permissible tensile load is, therefore,

$$Pb = (140 \times 10^6)[0{\cdot}010(b - 0{\cdot}016)]\ \text{N per rivet}$$

(ii) The rivets may fail by shearing. The area of each rivet is

$$\frac{\pi}{4}(0{\cdot}016)^2 = 0{\cdot}201 \times 10^{-3}\ \text{m}^2$$

The permissible load per rivet is then

$$Pb = 2(75 \times 10^6)(0{\cdot}201 \times 10^{-3})\ \text{N}$$

since each rivet is in double shear.

If the joint is equally strong in tension and shear, we have, from (i) and (ii),

$$(140 \times 10^6)[0{\cdot}010(b - 0{\cdot}016)] = 2(75 \times 10^6)(0{\cdot}201 \times 10^{-3})$$

This gives

$$b = 0{\cdot}038\ \text{m}$$

Now

$$Pb = 2(75 \times 10^6)(0{\cdot}201 \times 10^{-3}) = 30{\cdot}2\ \text{kN}$$

The average bearing pressure between the main plates and rivets is

$$\frac{30{\cdot}2 \times 10^3}{(0{\cdot}016)(0{\cdot}010)} = 189\ \text{MN/m}^2$$

4.3 Efficiency of a connection

After analysing the connection of Fig. 4.1, suppose we find that in the weakest mode of failure the carrying capacity of the joint is P_0. If the two plates were continuous through the connection, that is, if there were no overlap or bolts, the strength of the plates in tension would be

$$P_{ult} = \sigma_{ult} t$$

where σ_{ult} is the ultimate tensile stress of the material of the plates. The ratio

$$\eta = \frac{P_0}{P_{ult}} = \frac{P_0}{\sigma_{ult} t} \tag{4.7}$$

is known as the *efficiency* of the connection; clearly, η defines the extent to which the strength of the connection attains the full strength of the continuous plates.

Problem 4.2: What is the efficiency of the joint of Problem 4.1?

Solution

The permissible tensile load per rivet is 30·2 kN. For a continuous joint the tensile load which could be carried by a 3·8 cm width of main plate is

$$(0{\cdot}038)(0{\cdot}010)(140 \times 10^6) = 53{\cdot}2 \text{ kN}$$

Then

$$\eta = \frac{30{\cdot}2}{53{\cdot}2} = 0{\cdot}57, \text{ or } 57\%$$

4.4 Group-bolted and -riveted joints

When two members are connected by cover plates bolted or riveted in the manner shown in Fig. 4.7, the joint is said to be *group-bolted* or *-riveted*.

The greatest efficiency of the joint shown in Fig. 4.7 is obtained when the bolts or rivets are re-arranged in the form shown in Fig. 4.8, where it is supposed six bolts or rivets are required each side of the join. The loss of cross-section in the main members, on the line a, is that due to one bolt or rivet hole. If the load is assumed to be equally distributed among the bolts or rivets, the bolt or rivet on the line a will take one-sixth of the total load, so that the tension in the main plates, across b, will be $\frac{5}{6}$ths of the total. But this section is reduced by two bolt or rivet holes, so that, relatively, it is as strong as the section a, and so on: the reduction of the nett cross-section of the main plates increases as the load carried by these plates decreases. Thus a more efficient joint is obtained than when the bolts or rivets are arranged as in Fig. 4.7.

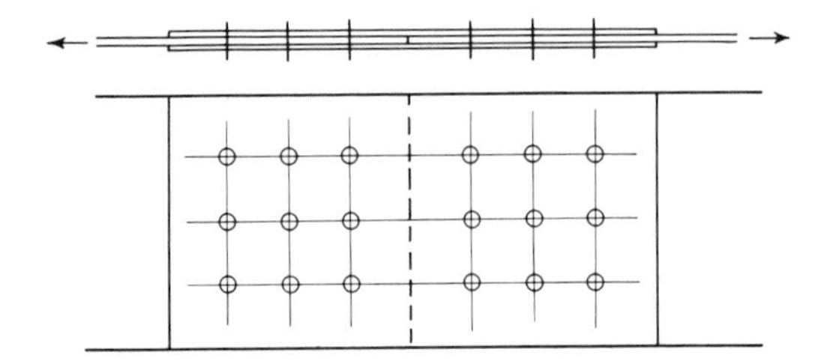

Fig. 4.7 A group-bolted or -riveted joint.

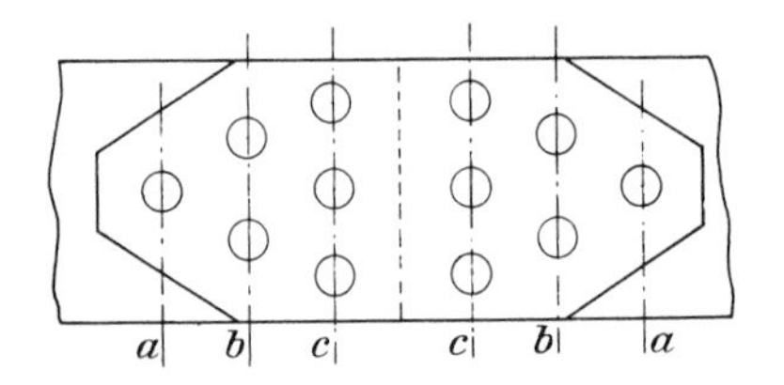

Fig. 4.8 Joint with tapered cover plates.

4.5 Eccentric loading of bolted and riveted connections

Structural connections are commonly required to transmit moments as well as axial forces. Fig. 4.9 shows the connection between a bracket and a stanchion; the bracket is attached to the stanchion through a system of six bolts or rivets, a vertical load P

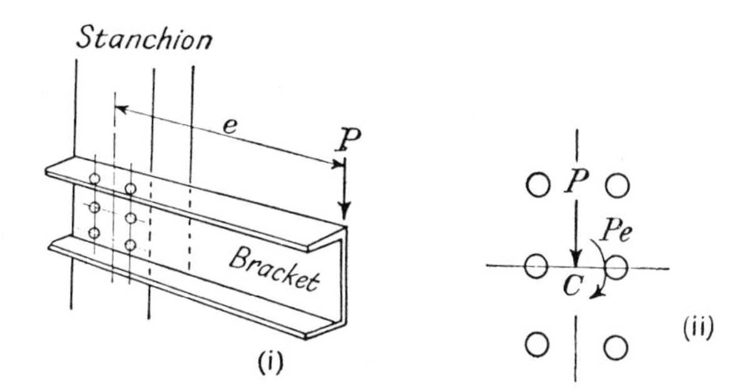

Fig. 4.9 Eccentrically loaded connection leading to a bending action on the group of bolts, as well as a shearing action.

is applied to the bracket. Suppose the bolts or rivets are all of the same diameter. The load P is then replaced by a parallel load P applied to the centroid C of the rivet system, together with a moment Pe about the centroid Fig. 4.9(ii); e is the perpendicular distance from C onto the line of action of P.

Consider separately the effects of the load P at C and the moment Pe. We assume that P is distributed equally amongst the bolts or rivets as a shearing force parallel to the line of action of P. The moment Pe is assumed to induce a shearing force F in any bolt or rivet perpendicular to the line joining C to the bolt or rivet; moreover the force

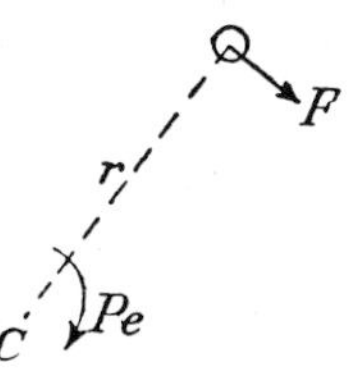

Fig. 4.10 Assumed forces on the bolts.

F is assumed to be proportional to the distance r from the bolt or rivet to C, (Fig. 4.10). For equilibrium we have

$$Pe = \sum Fr$$

If $F = kr$, where k is constant for all rivets, then

$$Pe = k \sum r^2$$

Thus we have

$$k = \frac{Pe}{\sum r^2}$$

The force on a rivet is

$$F = kr = \frac{Pe}{\sum r^2} \cdot r \tag{4.8}$$

The resultant force on a bolt or rivet is then the vector sum of the forces due to P and Pe.

Problem 4.3: A bracket is bolted to a vertical stanchion and carries a vertical load of 50 kN. Assuming that the total shearing stress in a bolt is proportional to the relative displacement of the bracket and the stanchion in the neighbourhood of the bolt, find the load carried by each of the bolts. (*Cambridge*)

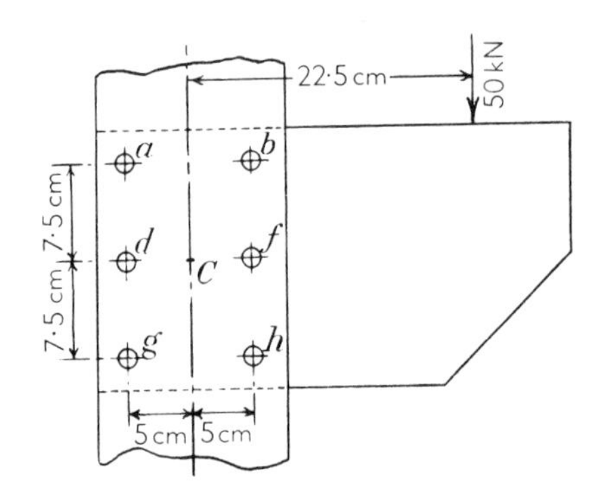

Solution

The centroid of the bolt system is at the point C. For bolt a

$$r = aC = [(0{\cdot}050)^2 + (0{\cdot}075)^2]^{\frac{1}{2}} = 0{\cdot}0902 \text{ m}$$

For bolt b,

$$r = bC = aC = 0{\cdot}0902 \text{ m}$$

For bolts d and f,

$$r = 0{\cdot}050 \text{ m}$$

For bolts g and h,

$$r = gC = aC = 0{\cdot}0902 \text{ m}$$

Then

$$\sum r^2 = 4(0{\cdot}0902)^2 + 2(0{\cdot}050)^2 = 0{\cdot}0376 \text{ m}^2$$

Now

$$e = 0{\cdot}225 \text{ m} \quad \text{and} \quad P = 50 \text{ kN}$$

Then

$$Pe = (0{\cdot}225)(50 \times 10^3) = 11{\cdot}25 \times 10^3 \text{ Nm}$$

The loads on the bolts a, b, g, h, due to the couple Pe alone, are then

$$\frac{Pe}{\sum r^2}r = \frac{11{\cdot}25 \times 10^3}{0{\cdot}0376}(0{\cdot}0902) = 28{\cdot}0 \text{ kN}$$

These loads are at right-angles to Ca, Cb, Cg, Ch, respectively. The corresponding loads on the bolts d and f are

$$\frac{Pe}{\sum r^2} r = \frac{11{\cdot}25 \times 10^3}{0{\cdot}0376}(0{\cdot}050) = 15{\cdot}0\,\text{kN}$$

perpendicular to Cd and Cf, respectively.

The load on each bolt due to the vertical shearing force of 50 kN alone is

$$\frac{50 \times 10^3}{6} = 8{\cdot}33 \times 10^3\ \text{N} = 8{\cdot}33\,\text{kN}$$

This force acts vertically downwards on each bolt. The resultant loads on all the rivets are found by drawing parallelograms of forces as follows:

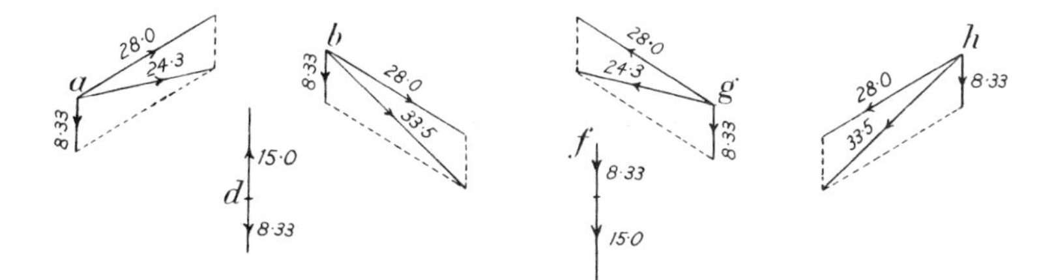

All force vectors are in KN. The resultant loads on the bolts are then as follows:

bolts	resultant load
a and g	24·3 kN
b and h	33·5 kN
d	6·7 kN
f	23·3 kN

4.6 Welded connections

Some metals used in engineering—such as steel and aluminium—can be deposited in a molten state between two components to form a joint, which is then called a welded connection. The metal deposited to form the joint is called the weld. Two types of weld are in common use—the *butt weld* and the *fillet weld*; Fig. 4.11 shows two plates

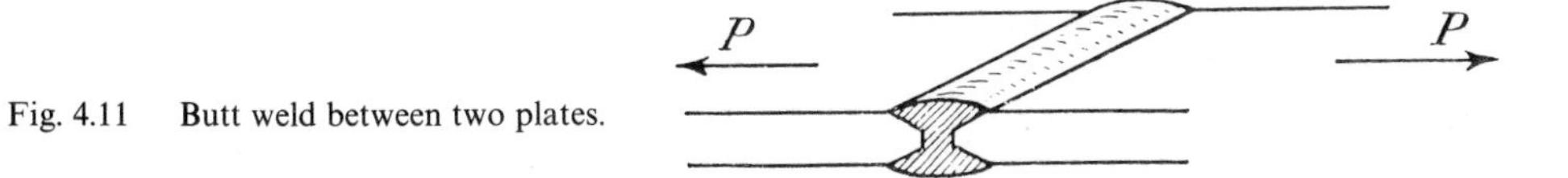

Fig. 4.11 Butt weld between two plates.

connected by a butt weld; the plates are tapered at the joint to give sufficient space for the weld material. If the plates carry a tensile load the weld material carries largely tensile stresses. Fig. 4.12 shows two plates connected by fillet welds; if the joint carries a tensile load the welds carry largely shearing stresses, although the state of stress in

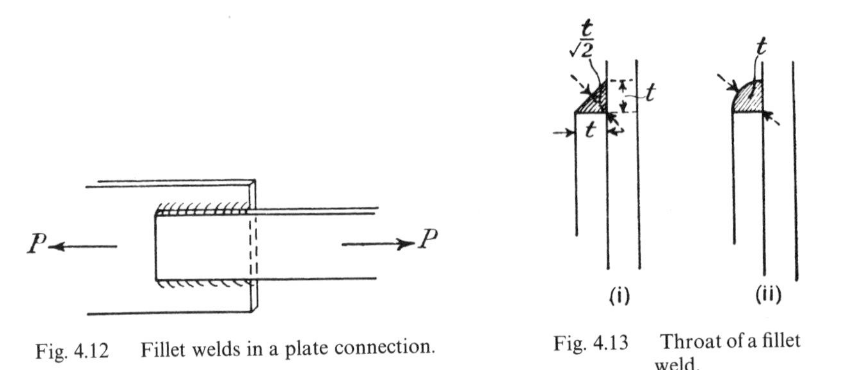

Fig. 4.12 Fillet welds in a plate connection.

Fig. 4.13 Throat of a fillet weld.

the welds is complex, and tensile stresses may also be present. Fillet welds of the type indicated in Fig. 4.12 transmit force between the two plates by shearing actions within the welds; if the weld has the triangular cross-section shown in Fig. 4.13(i), the shearing stress is greatest across the narrowest section of the weld, having a thickness $t/\sqrt{2}$. This section is called the *throat* of the weld. In Fig. 4.13(ii) the weld has the same thickness t at all sections. To estimate approximately the strength of the welds in Fig. 4.13 it is assumed that failure of the welds takes place by shearing across the throats of the welds.

Problem 4.4: A steel strip 5 cm wide is fillet-welded to a steel plate over a length of 7·5 cm and across the ends of the strip. The connection carries a tensile load of 100 kN. Find a suitable size of the fillet-weld if longitudinal welds can be stressed to 75 MN/m^2 and transverse welds to 100 MN/m^2.

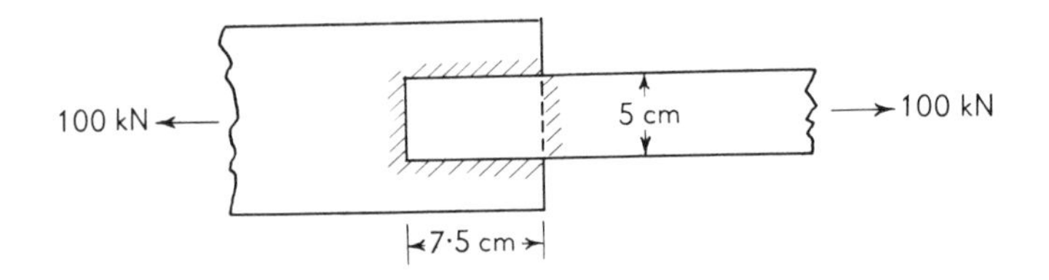

Solution

Suppose the throat thickness of the fillet-welds is t. Then the longitudinal welds carry a shearing force

$$\tau A = (75 \times 10^6)(0{\cdot}075 \times 2t) = (11{\cdot}25 \times 10^6)t \text{ N}$$

The transverse welds carry a shearing force

$$\tau A = (100 \times 10^6)(0{\cdot}050 \times 2t) = (10 \times 10^6)t \text{ N}$$

Then

$$(11{\cdot}25 \times 10^6)t + (10 \times 10^6)t = 100 \times 10^3$$

and, therefore,

$$t = \frac{100}{21{\cdot}25} \times 10^{-3} = 4{\cdot}71 \times 10^{-3} \text{ m} = 0{\cdot}471 \text{ cm}$$

The fillet size is then

$$t\sqrt{2} = 0{\cdot}67 \text{ cm}$$

Problem 4.5: Two metal plates of the same material and of equal breadth are fillet-welded at a lap joint. The one plate has a thickness t_1 and the other a thickness t_2. Compare the shearing forces transmitted through the welds, when the connection is under a tensile force P.

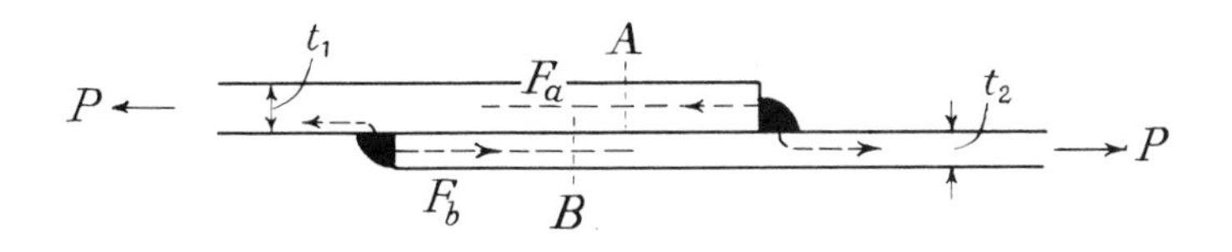

Solution

The sections of the plates between the welds will stretch by approximately the same amounts; thus, these sections will suffer the same strains and, since they are the same materials, they will also suffer the same stresses. If a shearing force F_a is transmitted by the one weld and a shearing force F_b by the other, then the tensile force over the section A in the one plate is F_a and over the section B in the other plate is F_b. If the plates have the same breadth and are to carry equal tensile stresses over the sections A and B, we have

$$\frac{F_a}{t_1} = \frac{F_b}{t_2}$$

and thus

$$\frac{F_a}{F_b} = \frac{t_1}{t_2}$$

We also have

$$F_a + F_b = P$$

and so

$$F_a = \frac{P}{1 + \dfrac{t_2}{t_1}} \quad \text{and} \quad F_b = \frac{P}{1 + \dfrac{t_1}{t_2}}$$

4.7 Welded connections under bending actions

Where a welded connection is required to transmit a bending moment we adopt a simple empirical method of analysis similar to that for bolted and riveted connections discussed in §4.5. We assume that the shearing stress in the weld is proportional to the distance of any part of the weld from the 'centroid' of the weld. Consider, for example, a plate which is welded to a stanchion and which carries a bending moment M in the plane of the welds, Fig. 4.14. We suppose the fillet-welds are of uniform thickness t round the circumference of a rectangle of sides a and b. At any point of the weld we take the shearing stress, τ, as acting normal to the line joining that point to the centroid C of the weld. If δA is an elemental area of weld at any point, then

$$M = \int \tau r \, dA$$

If

$$\tau = kr$$

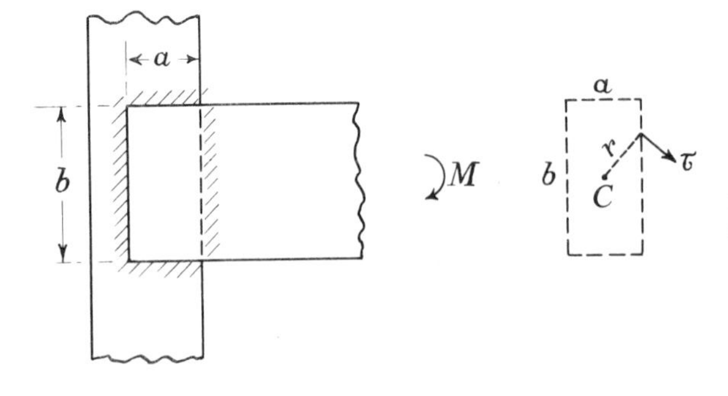

Fig. 4.14 A plate fillet-welded to a column, and transmitting a bending moment M.

then

$$M = \int kr^2 \, dA = kJ$$

where J is the polar second moment of area of the weld about the axis through C and normal to the plane of the weld. Thus

$$k = \frac{M}{J}$$

and

$$\tau = \frac{Mr}{J} \tag{4.9}$$

According to this simple empirical theory, the greatest stresses occur at points of the weld most remote from the centroid C.

Problem 4.6: Two steel plates are connected together by 0·5 cm fillet welds. Estimate the maximum shearing stress in the welds if the joint carries a bending moment of 2500 Nm.

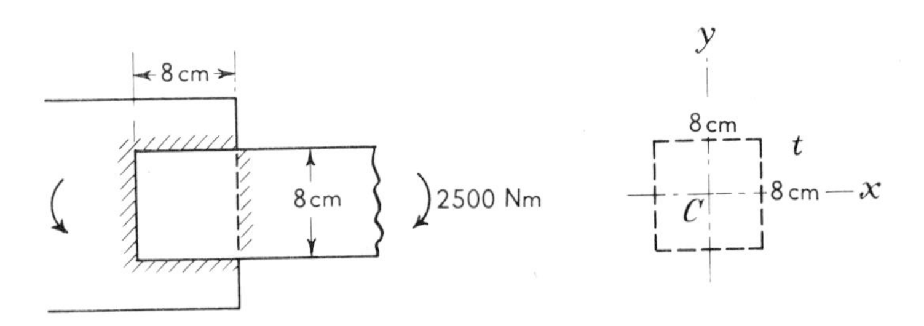

Solution

The centroid of the welds is at the centre of an 8 cm square. Suppose t is the throat thickness of the welds. The second moment of area of the weld about Cx or Cy is

$$I_x = I_y = 2[\tfrac{1}{12}(t)(0{\cdot}08)^3] + 2[(t)(0{\cdot}08)(0{\cdot}04)^2]$$
$$= (0{\cdot}341 \times 10^{-3})t \text{ m}^4$$

The polar second moment of area about an axis through C is then

$$J = I_x + I_y = 2(0{\cdot}341 \times 10^{-3})t = (0{\cdot}682 \times 10^{-3})t \text{ m}^4$$

Now $t = 0{\cdot}005/\sqrt{2}$ m, and so

$$J = 2{\cdot}41 \times 10^{-6} \text{ m}^4$$

The shearing stress in the weld at any radius r is

$$\tau = \frac{Mr}{J}$$

This is greatest at the corners of the square where it has the value

$$\tau_{max} = \frac{M}{J}\left(\frac{0{\cdot}08}{\sqrt{2}}\right) = \frac{2500}{2{\cdot}41 \times 10^{-6}}\left(\frac{0{\cdot}08}{\sqrt{2}}\right)$$
$$= 58{\cdot}6 \text{ MN/m}^2$$

FURTHER PROBLEMS
(answers on page 397)

4.7 Two plates, each 1 cm thick, are connected by riveting a single cover strap to the plates through two rows of rivets in each plate. The diameter of the rivets is 2 cm, and the distance between rivet centres along the breadth of the connection is 12·5 cm. Assuming the other unstated dimensions are adequate, calculate the strength of the joint per metre breadth, in tension, allowing 75 MN/m² shearing stress in the rivets and a tensile stress of 90 MN/m² in the plates. (*Cambridge*)

4.8 A flat steel bar is attached to a gusset plate by eight bolts. At the section AB the gusset plate exerts on the flat bar a vertical shearing force F and a counterclockwise couple M.

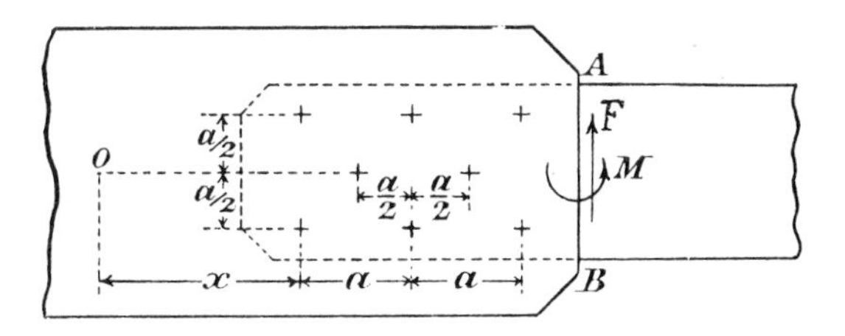

Assuming that the gusset plate, relative to the flat bar, undergoes a minute rotation about a point O on the line of the two middle rivets, also that the loads on the rivets are due to and proportional to the relative movement of the plates at the rivet holes, prove that

$$x = -a \, . \frac{4M + 3aF}{4M + 6aF}$$

Prove also that the horizontal and vertical components of the load on the top right-hand rivet are

$$\frac{2M + 3aF}{24a} \quad \text{and} \quad \frac{4M + 9aF}{24a}$$

respectively.

4.9 A steel strip of cross-section 5 cm by 1·25 cm is bolted to two copper strips, each of cross-section 5 cm by 0·9375 cm, there being two bolts on the line of pull. Show that, neglecting friction and the deformation of the bolts, a pull applied to the joint will be shared by the bolts in the ratio 3 to 4. Assume that E for steel is twice E for copper.

4.10 Two flat bars are riveted together using cover plates, x being the pitch of the rivets in a direction at right angles to the plane of the figure. Assuming that the rivets themselves do not deform, show that the load taken by the rivets (1) is $tPx/(t + 2t')$, and that the rivets (2) are free from load.

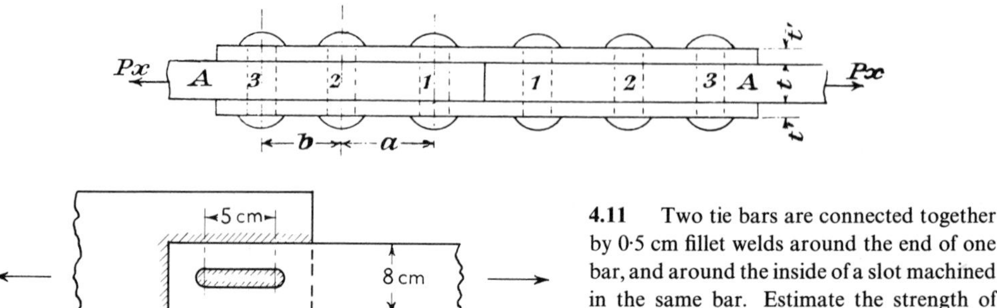

5 cm
8 cm
10 cm

4.11 Two tie bars are connected together by 0·5 cm fillet welds around the end of one bar, and around the inside of a slot machined in the same bar. Estimate the strength of the connection in tension if the shearing stresses in the welds are limited to 75 MN/m^2.

4.12 A bracket plate is welded to the face of a column and carries a vertical load P. Determine the value of P such that the maximum shearing stress in the 1 cm welds is 75 MN/m^2. (*Bristol*)

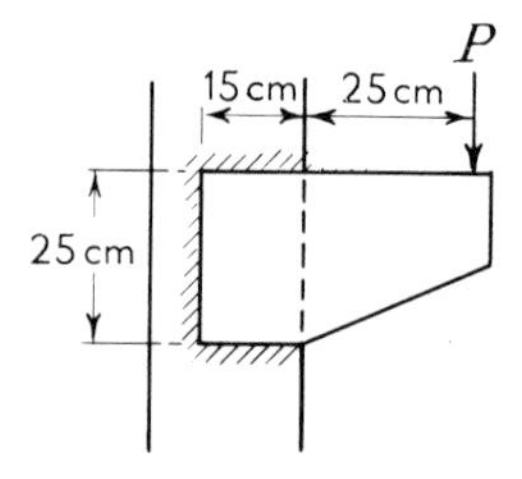

5 Analysis of stress and strain

5.1 Introduction

Up to the present we have confined our attention to considerations of simple direct and shearing stresses. But in most practical problems we have to deal with combinations of these stresses.

The strengths and elastic properties of materials are determined usually by simple tensile and compressive tests. How are we to make use of the results of such tests when we know that stress in a given practical problem is compounded of a tensile stress in one direction, a compressive stress in some other direction, and a shearing stress in a third direction? Clearly we cannot make tests of a material under all possible combinations of stress to determine its strength. It is essential, in fact, to study stresses and strains in more general terms; the analysis which follows should be regarded as having a direct and important bearing on practical strength problems, and is not merely a display of mathematical ingenuity.

5.2 Shearing stresses in a tensile test specimen

A long uniform bar, Fig. 5.1, has a rectangular cross-section of area A. The edges of the bar are parallel to perpendicular axes Ox, Oy, Oz. The bar is uniformly stressed

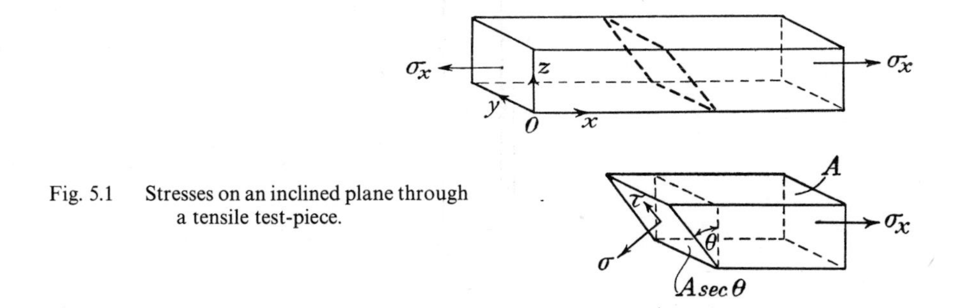

Fig. 5.1 Stresses on an inclined plane through a tensile test-piece.

in tension in the x-direction, the tensile stress on a cross-section of the bar normal to Ox being σ_x. Consider the stresses acting on an inclined cross-section of the bar; an inclined plane is taken at an angle θ to the yz-plane. The resultant force at the end cross-section of the bar is

$$P = A\sigma_x$$

acting parallel to Ox. For equilibrium the resultant force parallel to Ox on an inclined cross-section is also $P = A\sigma_x$. At the inclined cross-section in Fig. 5.1, resolve the force $A\sigma_x$ into two components—one perpendicular, and the other tangential, to the inclined cross-section, the latter component acting parallel to the xz-plane. These two components have the values

$$A\sigma_x \cos\theta \qquad A\sigma_x \sin\theta$$

respectively. The area of the inclined cross-section is

$$A \sec\theta$$

so that the normal and tangential stresses acting on the inclined cross-section are

$$\sigma = \frac{A\sigma_x \cos\theta}{A \sec\theta} = \sigma_x \cos^2\theta \tag{5.1}$$

$$\tau = \frac{A\sigma_x \sin\theta}{A \sec\theta} = \sigma_x \cos\theta \sin\theta \tag{5.2}$$

σ is the *direct stress* and τ the *shearing stress* on the inclined plane. It should be noted that the stresses on an inclined plane are not simply the resolutions of σ_x perpendicular and tangential to that plane; the important point in Fig. 5.1 is that the area of an inclined cross-section of the bar is different from that of a normal cross-section. The shearing stress τ may be written in the form

$$\tau = \sigma_x \cos\theta \sin\theta = \tfrac{1}{2}\sigma_x \sin 2\theta$$

At $\theta = 0°$ the cross-section is perpendicular to the axis of the bar, and $\tau = 0$; τ increases as θ increases until it attains a maximum value of $\frac{1}{2}\sigma_x$ at $\theta = 45°$; τ then diminishes as θ increases further until it is again zero at $\theta = 90°$. Thus on any inclined cross-section of a tensile test-piece, shearing stresses are always present; the shearing stresses are greatest on planes at 45° to the longitudinal axis of the bar.

Problem 5.1: A bar of cross-section 2·25 cm by 2·25 cm is subjected to an axial pull of 20 kN. Calculate the normal stress and shearing stress on a plane the normal to which makes an angle of 60° with the axis of the bar, the plane being perpendicular to one face of the bar.

Solution

We have $\theta = 60°$, $P = 20$ kN and $A = 0{\cdot}507 \times 10^{-3}$ m^2. Then

$$\sigma_x = \frac{20 \times 10^3}{0{\cdot}507 \times 10^{-3}} = 39{\cdot}4 \text{ MN/m}^2$$

The normal stress on the oblique plane is

$$\sigma = \sigma_x \cos^2 60° = (39{\cdot}4 \times 10^6)\tfrac{1}{4} = 9{\cdot}85 \text{ MN/m}^2$$

The shearing stress on the oblique plane is

$$\tfrac{1}{2}\sigma_x \sin 120° = \tfrac{1}{2}(39{\cdot}4 \times 10^6)\frac{\sqrt{3}}{2} = 17{\cdot}05 \text{ MN/m}^2$$

5.3 Strain figures in mild-steel; Lüder's lines

If a tensile specimen of mild steel is well polished and then stressed, it will be found that, when the specimen yields, a pattern of fine lines appears on the polished surface; these lines intersect roughly at right-angles to each other, and at 45° approximately to the longitudinal axis of the bar; these lines were first observed by Lüder in 1854. Lüder's lines on a tensile specimen of mild steel are shown in Fig. 5.2. These strain

Fig. 5.2 Lüder's lines in the yielding of a steel bar in tension.

figures suggest that yielding of the material consists of slip along the planes of greatest shearing stress; a single line represents a slip band, containing a large number of metal crystals.

5.4 Failure of materials in compression

Shearing stresses are also developed in a bar under uniform compression. The failure of some materials in compression is due to the development of critical shearing stresses on planes inclined to the direction of compression. Fig. 5.3 shows two failures of compressed timbers; failure is due primarily to breakdown in shear on planes inclined to the direction of compression.

Fig. 5.3 Failures of compressed specimens of timber, showing breakdown of the material in shear.

5.5 General two-dimensional stress system

A *two-dimensional* stress system is one in which the stresses at any point in a body act in the same plane. Consider a thin rectangular block of material, *abcd*, two faces of which are parallel to the *xy*-plane, Fig. 5.4. A two-dimensional state of stress exists if the stresses on the remaining four faces are parallel to the *xy*-plane. In general, suppose the *forces* acting on the faces are P, Q, R, S, parallel to the *xy*-plane, Fig. 5.4. Each of

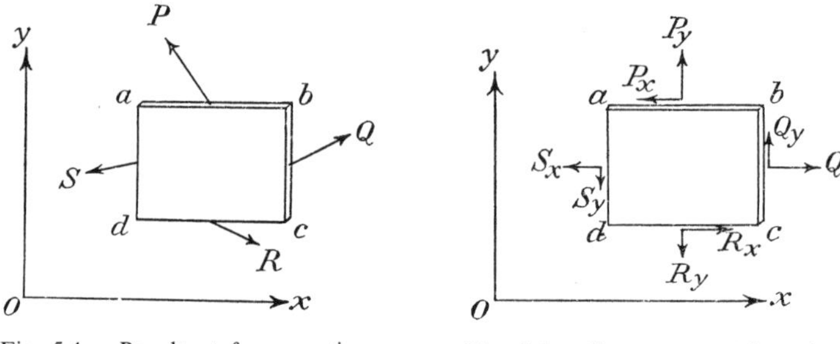

Fig. 5.4 Resultant forces acting on the faces of a 'two-dimensional' rectangular block.

Fig. 5.5 Components of resultant forces parallel to Ox and Oy.

these forces can be resolved into components P_x, P_y, etc., Fig. 5.5. The perpendicular components give rise to direct stresses, and the tangential components to shearing stresses.

The system of *forces* in Fig. 5.5 is now replaced by its equivalent system of *stresses*; the rectangular block of Fig. 5.6 is in a uniform state of two-dimensional stress; over the two faces parallel to Ox are direct and shearing stresses σ_y and τ_{xy}, respectively, over the two faces parallel to Oy are direct and shearing stresses σ_x, and τ_{yx}, respectively. The thickness is assumed to be 1 unit of length, for convenience, the other sides having lengths a and b. Equilibrium of the block in the x- and y-directions is already ensured; for rotational equilibrium of the block in the xy-plane we must have

$$[\tau_{xy}(a \times 1)] \times b = [\tau_{yx}(b \times 1)] \times a$$

Thus

$$(ab)\tau_{xy} = (ab)\tau_{yx}$$

or

$$\tau_{xy} = \tau_{yx} \tag{5.3}$$

Then the shearing stresses on perpendicular planes are equal and complementary as we found in the simpler case of pure shear in §3.3.

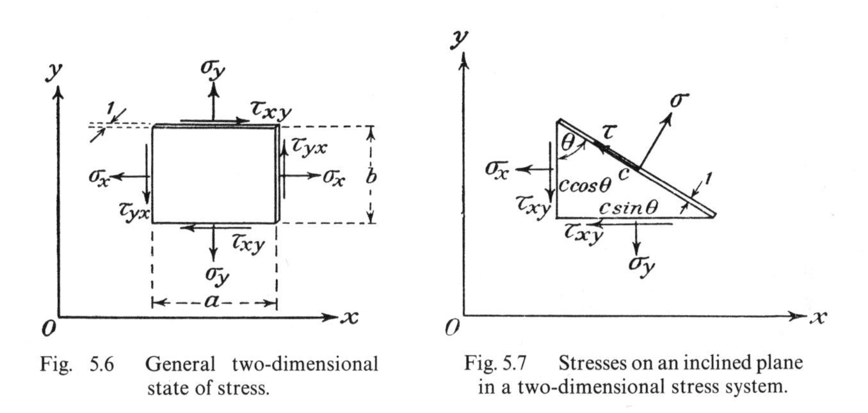

Fig. 5.6 General two-dimensional state of stress.

Fig. 5.7 Stresses on an inclined plane in a two-dimensional stress system.

5.6 Stresses on an inclined plane

Consider the stresses acting on an inclined plane of the uniformly stressed rectangular block of Fig. 5.6; the inclined plane makes an angle θ with Oy, and cuts off a 'triangular' block, Fig. 5.7. The length of the hypotenuse is c, and the thickness of the block is taken again as one unit of length, for convenience. The values of direct stress, σ, and shearing stress, τ, on the inclined plane are found by considering equilibrium of the triangular block. The direct stress acts along the normal to the inclined plane. Resolve the forces on the three sides of the block parallel to this normal: then

$$\sigma(c \,.\, 1) = \sigma_x(c \cos\theta \,.\, \cos\theta) + \sigma_y(c \sin\theta \,.\, \sin\theta) + \tau_{xy}(c \cos\theta \,.\, \sin\theta) + \tau_{xy}(c \sin\theta \,.\, \cos\theta)$$

This gives

$$\sigma = \sigma_x \cos^2\theta + \sigma_y \sin^2\theta + 2\tau_{xy} \sin\theta \cos\theta \tag{5.4}$$

Now resolve forces in a direction parallel to the inclined plane:

$$\tau \, . \, (c \, . \, 1) = -\sigma_x(c \cos\theta \, . \, \sin\theta) + \sigma_y(c \sin\theta \, . \, \cos\theta) + \tau_{xy}(c \cos\theta \, . \, \cos\theta) - \tau_{xy}(c \sin\theta \, . \, \sin\theta)$$

This gives

$$\tau = -\sigma_x \cos\theta \sin\theta + \sigma_y \sin\theta \cos\theta + \tau_{xy}(\cos^2\theta - \sin^2\theta) \tag{5.5}$$

The expressions for σ and τ are written more conveniently in the forms:

$$\sigma = \tfrac{1}{2}(\sigma_x + \sigma_y) + \tfrac{1}{2}(\sigma_x - \sigma_y)\cos 2\theta + \tau_{xy}\sin 2\theta \tag{5.6}$$

$$\tau = -\tfrac{1}{2}(\sigma_x - \sigma_y)\sin 2\theta + \tau_{xy}\cos 2\theta \tag{5.7}$$

The shearing stress τ vanishes when

$$\tfrac{1}{2}(\sigma_x - \sigma_y)\sin 2\theta = \tau_{xy}\cos 2\theta$$

that is, when

$$\tan 2\theta = \frac{2\tau_{xy}}{\sigma_x - \sigma_y} \tag{5.8}$$

or when

$$2\theta = \tan^{-1}\frac{2\tau_{xy}}{\sigma_x - \sigma_y} \quad \text{or} \quad \tan^{-1}\frac{2\tau_{xy}}{\sigma_x - \sigma_y} + 180^\circ$$

These may be written

$$\theta = \tfrac{1}{2}\tan^{-1}\frac{2\tau_{xy}}{\sigma_x - \sigma_y} \quad \text{or} \quad \tfrac{1}{2}\tan^{-1}\frac{2\tau_{xy}}{\sigma_x - \sigma_y} + 90^\circ \tag{5.9}$$

In a two-dimensional stress system there are thus two planes, separated by 90°, on which the shearing stress is zero. These planes are called the *principal planes*, and the corresponding values of σ are called the *principal stresses*. The direct stress σ is a maximum when

$$\frac{d\sigma}{d\theta} = -(\sigma_x - \sigma_y)\sin 2\theta + 2\tau_{xy}\cos 2\theta = 0$$

that is, when

$$\tan 2\theta = \frac{2\tau_{xy}}{\sigma_x - \sigma_y}$$

which is identical with equation (5.8), defining the directions of the principal stresses; thus the *principal stresses* are also the *maximum* and *minimum direct stresses* in the material.

5.7 Values of the principal stresses

The directions of the principal planes are given by equation (5.8). For any two-dimensional stress system, in which the values of σ_x, σ_y and τ_{xy} are known, $\tan 2\theta$ is calculable; two values of θ, separated by 90°, can then be found. The principal stresses are then calculated by substituting these values of θ into equation (5.6).

Alternatively, the principal stresses can be calculated more directly without finding the principal planes. Earlier we defined a principal plane as one on which there is no

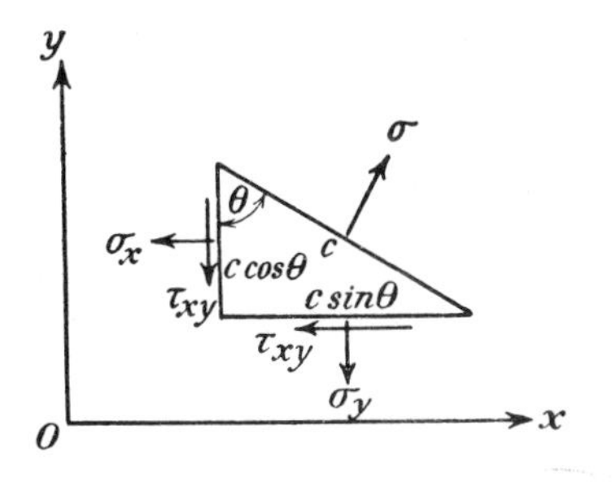

Fig. 5.8 A principal stress acting on an inclined plane; there is no shearing stress τ associated with a principal stress σ.

shearing stress; in Fig. 5.8 it is assumed that no shearing stress acts on a plane at an angle θ to Oy. For equilibrium of the triangular block in the x-direction,

$$\sigma(c \cos \theta) - \sigma_x(c \cos \theta) = \tau_{xy}(c \sin \theta)$$

and so

$$\sigma - \sigma_x = \tau_{xy} \tan \theta \tag{5.10}$$

For equilibrium of the block in the y-direction

$$\sigma(c \sin \theta) - \sigma_y(c \sin \theta) = \tau_{xy}(c \cos \theta)$$

and thus

$$\sigma - \sigma_y = \tau_{xy} \cot \theta \tag{5.11}$$

On eliminating θ between equations (5.10) and (5.11),

$$(\sigma - \sigma_x)(\sigma - \sigma_y) = \tau_{xy}^2$$

This equation is quadratic in σ; the solutions are

$$\begin{aligned} \sigma_1 &= \tfrac{1}{2}(\sigma_x + \sigma_y) + \tfrac{1}{2}\sqrt{(\sigma_x - \sigma_y)^2 + 4\tau_{xy}^2} \\ \sigma_2 &= \tfrac{1}{2}(\sigma_x + \sigma_y) - \tfrac{1}{2}\sqrt{(\sigma_x - \sigma_y)^2 + 4\tau_{xy}^2} \end{aligned} \tag{5.12}$$

which are the values of the principal stresses; these stresses occur on mutually perpendicular planes.

5.8 Maximum shearing stress

The principal planes define directions of zero shearing stress; on some intermediate plane the shearing stress attains a maximum value. The shearing stress is given by equation (5.7); τ attains a maximum value with respect to θ when

$$\frac{d\tau}{d\theta} = -(\sigma_x - \sigma_y)\cos 2\theta - 2\tau_{xy}\sin 2\theta = 0$$

i.e. when

$$\cot 2\theta = -\frac{2\tau_{xy}}{\sigma_x - \sigma_y}$$

The planes of maximum shearing stress are inclined then at 45° to the principal planes. On substituting this value of cot 2θ into equation (5.7), the maximum numerical value of τ is

$$\tau_{max} = \sqrt{[\tfrac{1}{2}(\sigma_x - \sigma_y)]^2 + [\tau_{xy}]^2} \qquad (5.13)$$

But from equations (5.12),

$$\sqrt{[\tfrac{1}{2}(\sigma_x - \sigma_y)]^2 + [\tau_{xy}]^2} = \sigma_1 - \tfrac{1}{2}(\sigma_x + \sigma_y) = \tfrac{1}{2}(\sigma_x + \sigma_y) - \sigma_2$$

where σ_1 and σ_2 are the principal stresses of the stress system. Then

$$2\sqrt{[\tfrac{1}{2}(\sigma_x - \sigma_y)]^2 + [\tau_{xy}]^2} = \sigma_1 - \sigma_2$$

and equation (5.13) becomes

$$\tau_{max} = \tfrac{1}{2}(\sigma_1 - \sigma_2) \qquad (5.14)$$

The maximum shearing stress is therefore half the difference between the principal stresses of the system.

Problem 5.2: At a point of a material the two-dimensional stress system is defined by

$\sigma_x = 60{\cdot}0$ MN/m², tensile

$\sigma_y = 45{\cdot}0$ MN/m², compressive

$\tau_{xy} = 37{\cdot}5$ MN/m², shearing

where σ_x, σ_y, τ_{xy} refer to Fig. 5.7. Evaluate the values and directions of the principal stresses. What is the greatest shearing stress?

Solution

Now, we have

$$\tfrac{1}{2}(\sigma_x + \sigma_y) = \tfrac{1}{2}(60{\cdot}0 - 45{\cdot}0) = 7{\cdot}5 \text{ MN/m}^2$$

$$\tfrac{1}{2}(\sigma_x - \sigma_y) = \tfrac{1}{2}(60{\cdot}0 + 45{\cdot}0) = 52{\cdot}5 \text{ MN/m}^2$$

Then, from equations (5.12),

$$\sigma_1 = 7{\cdot}5 + [(52{\cdot}5)^2 + (37{\cdot}5)^2]^{\frac{1}{2}} = 7{\cdot}5 + 64{\cdot}4 = 71{\cdot}9 \text{ MN/m}^2$$

$$\sigma_2 = 7{\cdot}5 - [(52{\cdot}5)^2 + (37{\cdot}5)^2]^{\frac{1}{2}} = 7{\cdot}5 - 64{\cdot}4 = -56{\cdot}9 \text{ MN/m}^2$$

From equation (5.8)

$$\tan 2\theta = \frac{2\tau_{xy}}{\sigma_x - \sigma_y} = \frac{37{\cdot}5}{52{\cdot}5} = 0{\cdot}714$$

Thus,

$$2\theta = \tan^{-1}(0{\cdot}714) = 35{\cdot}5^\circ \text{ or } 215{\cdot}5^\circ$$

Then

$$\theta = 17{\cdot}8^\circ \text{ or } 107{\cdot}8^\circ$$

From equation (5.14)

$$\tau_{max} = \tfrac{1}{2}(\sigma_1 - \sigma_2) = \tfrac{1}{2}(71{\cdot}9 + 56{\cdot}9) = 64{\cdot}4 \text{ MN/m}^2$$

This maximum shearing stress occurs on planes at 45° to those of the principal stresses.

5.9 Mohr's circle of stress

A geometrical interpretation of equations (5.6) and (5.7) leads to a simple method of stress analysis. Now, we have found already that

$$\sigma = \tfrac{1}{2}(\sigma_x + \sigma_y) + \tfrac{1}{2}(\sigma_x - \sigma_y)\cos 2\theta + \tau_{xy}\sin 2\theta \tag{5.6}$$

$$\tau = -\tfrac{1}{2}(\sigma_x - \sigma_y)\sin 2\theta + \tau_{xy}\cos 2\theta \tag{5.7}$$

Take two perpendicular axes $O\sigma$, $O\tau$, Fig. 5.9; on this coordinate system set off the points having coordinates (σ_x, τ_{xy}) and $(\sigma_y, -\tau_{xy})$, corresponding to the known stresses

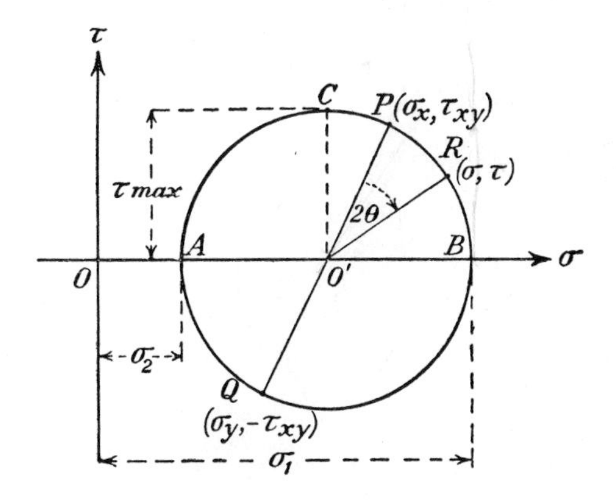

Fig. 5.9 Mohr's circle of stress. The points P, Q correspond to the stress states (σ_x, τ_{xy}), $(\sigma_y, -\tau_{xy})$ respectively, and are diametrally opposite; the state of stress (σ, τ) on a plane inclined at an angle θ to Oy is given by the point R.

in the x- and y-directions. The line PQ joining these two points is bisected by the $O\sigma$ axis at a point O'. With a centre at O', construct a circle passing through P and Q. The stresses σ and τ on a plane at an angle θ to Oy are found by setting off a radius of the circle at an angle 2θ to PQ, Fig. 5.9; 2θ is measured in a clockwise direction from $O'P$. The coordinates of the point $R(\sigma, \tau)$ give the direct and shearing stresses on the plane. We may write equations (5.6) and (5.7) in the forms

$$\sigma - \tfrac{1}{2}(\sigma_x + \sigma_y) = \tfrac{1}{2}(\sigma_x - \sigma_y)\cos 2\theta + \tau_{xy}\sin 2\theta$$

$$-\tau = \tfrac{1}{2}(\sigma_x - \sigma_y)\sin 2\theta - \tau_{xy}\cos 2\theta$$

Square each equation and add; then we have

$$[\sigma - \tfrac{1}{2}(\sigma_x + \sigma_y)]^2 + \tau^2 = [\tfrac{1}{2}(\sigma_x - \sigma_y)]^2 + [\tau_{xy}]^2 \tag{5.15}$$

Thus all corresponding values of σ and τ lie on a circle of radius

$$\sqrt{[\tfrac{1}{2}(\sigma_x - \sigma_y)]^2 + \tau_{xy}^2},$$

with its centre at the point $(\tfrac{1}{2}[\sigma_x + \sigma_y], 0)$, Fig. 5.9.

This circle defining all possible states of stress is known as *Mohr's circle of stress*; the principal stresses are defined by the points A and B, at which $\tau = 0$. The maximum shearing stress, which is given by the point C, is clearly the radius of the circle.

Problem 5.3: At a point of a material the stresses forming a two-dimensional system are

$$\sigma_x = 50 \text{ MN/m}^2 \qquad \sigma_y = 30 \text{ MN/m}^2 \qquad \tau_{xy} = 20 \text{ MN/m}^2$$

σ_x, σ_y, τ_{xy} having the directions shown in Fig. 5.7. Draw a Mohr's circle of stress, and deduce the values of the principal stresses and the maximum shearing stress in the plane of the stresses.

Solution

On the σ–τ diagram, construct a circle with the line joining (50, 20) and (30, −20) as diameter. The intercepts of the circle on the σ-axis are

$$\sigma_1 = 62{\cdot}4 \text{ MN/m}^2 \qquad \sigma_2 = 17{\cdot}6 \text{ MN/m}^2$$

which are the principal stresses. The maximum shearing stress is the radius of the circle:

$$\tau_{max} = \tfrac{1}{2}(44{\cdot}8 \times 10^6) = 22{\cdot}4 \text{ MN/m}^2$$

This is the maximum shearing stress in the plane of the applied stresses.

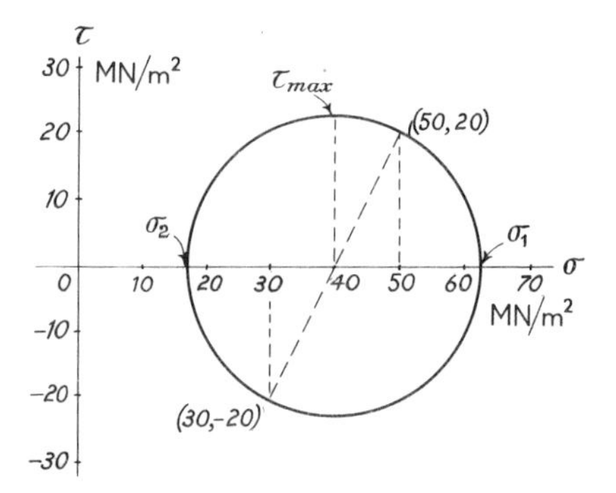

Problem 5.4: At a point of a material the two-dimensional state of stress is defined by

$$\sigma_x = 30 \text{ MN/m}^2, \qquad \sigma_y = -10 \text{ MN/m}^2, \qquad \tau_{xy} = 20 \text{ MN/m}^2$$

Find the principal stresses and the maximum shearing stress.

Solution

On the σ–τ diagram, construct a circle with the line joining (30, 20) and (−10, −20) as diameter. The intercepts of the circle on the σ-axis are

$$\sigma_1 = 38{\cdot}3 \text{ MN/m}^2 \qquad \sigma_2 = -18{\cdot}3 \text{ MN/m}^2$$

which are the principal stresses. The maximum shearing stress is the radius of the circle:

$$\tau_{max} = 28{\cdot}3\ \text{MN/m}^2$$

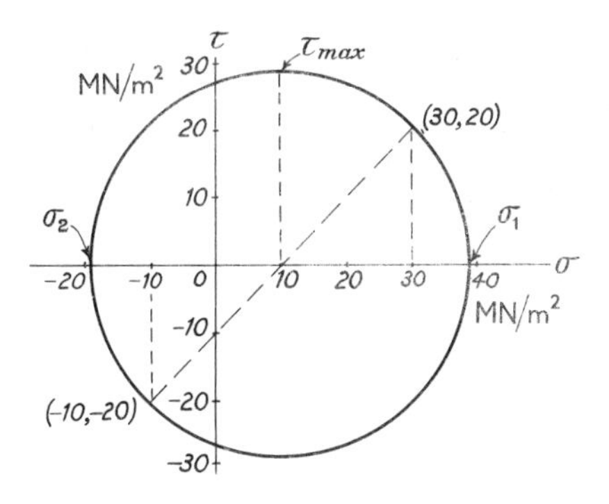

5.10 Strains in an inclined direction

For a two-dimensional system of strains the direct and shearing strains in any direction are known if the direct and shearing strains in two mutually perpendicular directions are given. Consider a rectangular element of material, $OABC$, in the xy-plane, Fig. 5.10; it is required to find the direct and shearing strains in the direction of the diagonal

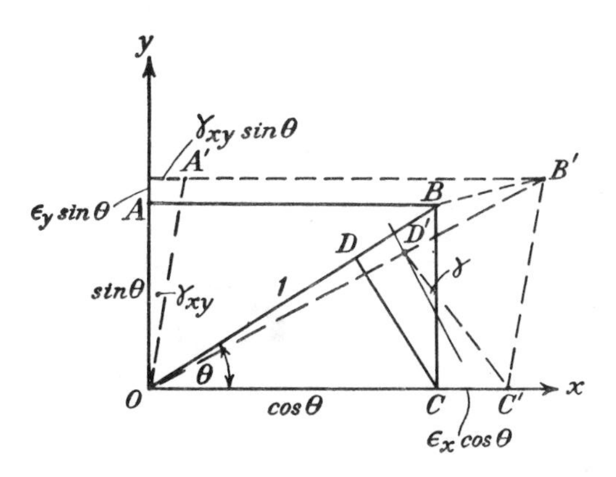

Fig. 5.10 Strains in an inclined direction; strains in the directions Ox, Oy and defined by ϵ_x, ϵ_y, γ_{xy}, lead to strains ϵ, γ along the inclined direction OB.

OB, when the direct and shearing strains in the directions Ox, Oy are given. Suppose ϵ_x is the strain in the direction Ox, ϵ_y the strain in the direction Oy, and γ_{xy} the shearing strain relative to Ox and Oy.

All the strains are considered to be small; in Fig. 5.10, if the diagonal OB of the rectangle is taken to be of unit length, the sides OA, OB are of lengths $\sin\theta$, $\cos\theta$, respectively, in which θ is the angle OB makes with Ox. In the strained condition OA extends a small amount $\epsilon_y \sin\theta$, OC extends a small amount $\epsilon_x \cos\theta$, and due to shearing strain OA rotates through a small angle γ_{xy}. If the point B moves to a point B', the movement of B parallel to Ox is

$$\epsilon_x \cos\theta + \gamma_{xy} \sin\theta$$

and the movement parallel to Oy is

$$\epsilon_y \sin\theta$$

Then the movement of B parallel to OB is

$$(\epsilon_x \cos\theta + \gamma_{xy} \sin\theta) \cos\theta + (\epsilon_y \sin\theta) \sin\theta$$

Since the strains are small this is equal to the extension of OB in the strained condition; but OB is of unit length, so that the extension is also the direct strain in the direction OB. If the direct strain in the direction OB is denoted by ϵ, then

$$\epsilon = (\epsilon_x \cos\theta + \gamma_{xy} \sin\theta) \cos\theta + (\epsilon_y \sin\theta) \sin\theta$$

This may be written in the form

$$\epsilon = \epsilon_x \cos^2\theta + \epsilon_y \sin^2\theta + \gamma_{xy} \sin\theta \cos\theta$$

and also in the form

$$\epsilon = \tfrac{1}{2}(\epsilon_x + \epsilon_y) + \tfrac{1}{2}(\epsilon_x - \epsilon_y) \cos 2\theta + \tfrac{1}{2}\gamma_{xy} \sin 2\theta \qquad (5.16)$$

This is similar in form to equation (5.6), defining the direct stress on an inclined plane; ϵ_x and ϵ_y replace σ_x and σ_y, respectively, and $\frac{1}{2}\gamma_{xy}$ replaces τ_{xy}.

To evaluate the shearing strain in the direction OB we consider the displacements of the point D, the foot of the perpendicular from C to OB, in the strained condition, Fig. 5.10. The point D is displaced to a point D'; we have seen that OB extends an amount ϵ, so that OD extends an amount

$$\epsilon \,.\, OD = \epsilon \cos^2\theta$$

During straining the line CD rotates anti-clockwise through a small angle

$$\frac{\epsilon_x \cos^2\theta - \epsilon \cos^2\theta}{\cos\theta \sin\theta} = (\epsilon_x - \epsilon) \cot\theta$$

At the same time OB rotates in a clockwise direction through a small angle

$$(\epsilon_x \cos\theta + \gamma_{xy} \sin\theta) \sin\theta - (\epsilon_y \sin\theta) \cos\theta$$

The amount by which the angle ODC diminishes during straining is the shearing strain γ in the direction OB. Thus

$$\gamma = -(\epsilon_x - \epsilon) \cot\theta - (\epsilon_x \cos\theta + \gamma_{xy} \sin\theta) \sin\theta + (\epsilon_y \sin\theta) \cos\theta$$

On substituting for ϵ from equation (5.16) we have

$$\gamma = -2(\epsilon_x - \epsilon_y) \cos\theta \sin\theta + \gamma_{xy}(\cos^2\theta - \sin^2\theta)$$

which may be written

$$\tfrac{1}{2}\gamma = -\tfrac{1}{2}(\epsilon_x - \epsilon_y)\sin 2\theta + \tfrac{1}{2}\gamma_{xy}\cos 2\theta \tag{5.17}$$

This is similar in form to equation (5.7) defining the shearing stress on an inclined plane; σ_x and σ_y in that equation are replaced by ϵ_x and ϵ_y, respectively, and τ_{xy} by $\frac{1}{2}\gamma_{xy}$.

5.11 Mohr's circle of strain

The direct and shearing strains in an inclined direction are given by relations which are similar to equations (5.6) and (5.7) for the direct and shearing stresses on an inclined plane. This suggests that the strains in any direction can be represented graphically in a similar way to the stress system. We may write equations (5.16) and (5.17) in the forms

$$\epsilon - \tfrac{1}{2}(\epsilon_x + \epsilon_y) = \tfrac{1}{2}(\epsilon_x - \epsilon_y)\cos 2\theta + \tfrac{1}{2}\gamma_{xy}\sin 2\theta$$

$$\tfrac{1}{2}\gamma = -\tfrac{1}{2}(\epsilon_x - \epsilon_y)\sin 2\theta + \tfrac{1}{2}\gamma_{xy}\cos 2\theta$$

Square each equation, and then add; we have

$$[\epsilon - \tfrac{1}{2}(\epsilon_x + \epsilon_y)]^2 + [\tfrac{1}{2}\gamma]^2 = [\tfrac{1}{2}(\epsilon_x - \epsilon_y)]^2 + [\tfrac{1}{2}\gamma_{xy}]^2$$

Thus all values of ϵ and $\frac{1}{2}\gamma$ lie on a circle of radius

$$\sqrt{[\tfrac{1}{2}(\epsilon_x - \epsilon_y)]^2 + [\tfrac{1}{2}\gamma_{xy}]^2}$$

with its centre at the point

$$[\tfrac{1}{2}(\epsilon_x + \epsilon_y), 0]$$

This circle defining all possible states of strain is usually called *Mohr's circle of strain*. For given values of ϵ_x, ϵ_y and γ_{xy} it is constructed in the following way: two mutually perpendicular axes, ϵ and $\frac{1}{2}\gamma$, are set up, Fig. 5.11; the points $(\epsilon_x, \frac{1}{2}\gamma_{xy})$ and $(\epsilon_y, -\frac{1}{2}\gamma_{xy})$

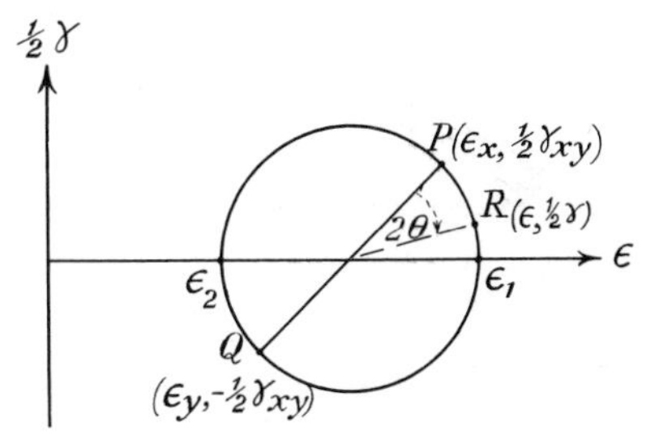

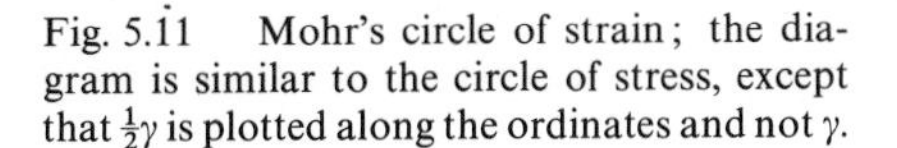
Fig. 5.11 Mohr's circle of strain; the diagram is similar to the circle of stress, except that $\frac{1}{2}\gamma$ is plotted along the ordinates and not γ.

are located; the line joining these points is a diameter of the circle of strain. The values of ϵ and $\frac{1}{2}\gamma$ in an inclined direction making an angle θ with Ox (Fig. 5.10) are given by the points on the circle at the ends of a diameter making an angle 2θ with PQ; the angle 2θ is measured clockwise.

We note that the maximum and minimum values of ϵ, given by ϵ_1 and ϵ_2 in Fig. 5.11, occur when $\frac{1}{2}\gamma$ is zero; ϵ_1, ϵ_2 are called *principal strains*, and occur for directions in which there is no shearing strain.

An important feature of this strain analysis is that we have *not* assumed that the strains are elastic; we have taken them to be small, however; with this limitation Mohr's circle of strain is applicable to both elastic and inelastic problems.

5.12 Elastic stress-strain relations

When a point of a body is acted upon by stresses σ_x and σ_y in mutually perpendicular directions, the strains are found by superposing the strains due to σ_x and σ_y acting

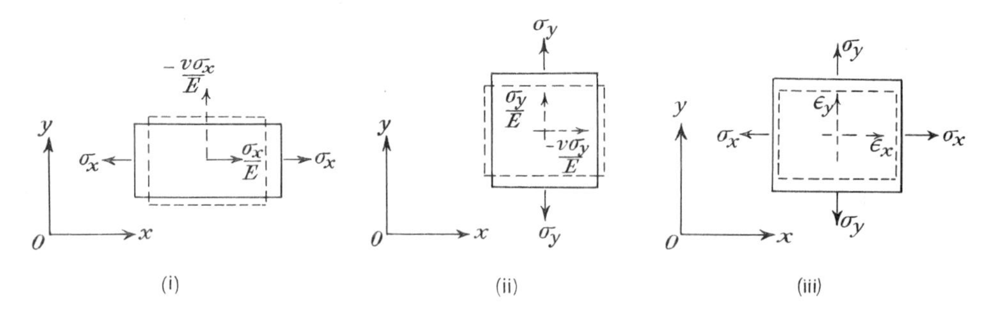

Fig. 5.12 Strains in a two-dimensional linear-elastic stress system; the strains can be regarded as compounded of two systems corresponding to uni-axial tension in the x and y directions.

separately. The rectangular element of material in Fig. 5.12(i) is subjected to a tensile stress σ_x in the x-direction; the tensile strain in the x-direction is

$$\frac{\sigma_x}{E}$$

and the compressive strain in the y-direction is

$$-\frac{v\sigma_x}{E}$$

in which E is Young's modulus, and v is Poisson's ratio (see §1.10). If the element is subjected to a tensile stress σ_y in the y-direction as in Fig. 5.12(ii), the compressive strain in the x-direction is

$$-\frac{v\sigma_y}{E}$$

and the tensile strain in the y-direction is

$$\frac{\sigma_y}{E}$$

These elastic strains are small, and the state of strain due to both stresses, σ_x and σ_y, acting simultaneously, as in Fig. 5.12(iii), is found by superposing the strains of Figs. 5.12(i) and (ii); taking tensile strain as positive and compressive strain as negative, the strains in the x- and y-directions are given, respectively, by

$$\epsilon_x = \frac{\sigma_x}{E} - \frac{\nu\sigma_y}{E}$$
$$\epsilon_y = \frac{\sigma_y}{E} - \frac{\nu\sigma_x}{E} \tag{5.18}$$

On multiplying each equation by E, we have

$$E\epsilon_x = \sigma_x - \nu\sigma_y$$
$$E\epsilon_y = \sigma_y - \nu\sigma_x \tag{5.19}$$

These are the elastic stress-strain relations for a two-dimensional system of direct stresses.

When a shearing stress τ_{xy} is present in addition to the direct stresses σ_x and σ_y, as in Fig. 5.13, the shearing stress τ_{xy} is assumed to have no effect on the direct strains ϵ_x

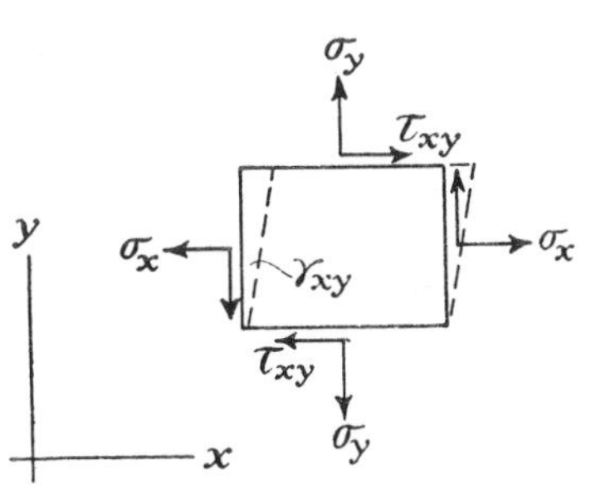

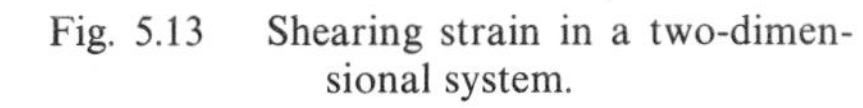

Fig. 5.13 Shearing strain in a two-dimensional system.

and ϵ_y caused by σ_x and σ_y. Similarly, the direct stresses σ_x and σ_y are assumed to have no effect on the shearing strain γ_{xy} due to τ_{xy}. When shearing stresses are present, as well as direct stresses, there is therefore an additional stress-strain relation having the form

$$\frac{\tau_{xy}}{\gamma_{xy}} = G$$

in which G is the shearing modulus. Then, in addition to equations (5.19) we have the relation

$$\tau_{xy} = G\gamma_{xy} \tag{5.20}$$

5.13 Principal stresses and strains

We have seen that in a two-dimensional system of stresses there are always two mutually perpendicular directions in which there are no shearing stresses; the direct stresses on these planes were referred to as principal stresses, σ_1 and σ_2. Since there are no shearing stresses in these two mutually perpendicular directions, there are also no shearing strains; for the principal directions the corresponding direct strains are given by

$$\begin{aligned} E\epsilon_1 &= \sigma_1 - \nu\sigma_2 \\ E\epsilon_2 &= \sigma_2 - \nu\sigma_1 \end{aligned} \tag{5.21}$$

The direct strains, ϵ_1, ϵ_2, are the principal strains already discussed in Mohr's circle of strain. It follows that the principal strains occur in directions parallel to the principal stresses.

5.14 Relation between E, G and ν

Consider an element of material subjected to a tensile stress σ_0 in one direction together with a compressive stress σ_0 in a mutually perpendicular direction, Fig. 5.14(i). The Mohr's circle for this state of stress has the form shown in Fig. 5.14(ii); the circle

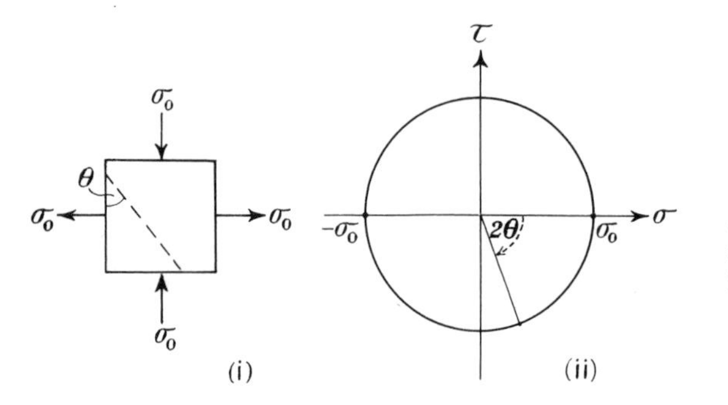

Fig. 5.14 (i) A stress system consisting of tensile and compressive stresses of equal magnitude, but acting in mutually perpendicular directions. (ii) Mohr's circle of stress for this system.

of stress has a centre at the origin and a radius of σ_0. The direct and shearing stresses on an inclined plane are given by the coordinates of a point on the circle; in particular we note that there is no direct stress when $2\theta = 90°$, that is, when $\theta = 45°$ in Fig. 5.14(i). Moreover when $\theta = 45°$, the shearing stress on this plane is of magnitude σ_0. We conclude then that a state of equal and opposite tension and compression, as indicated in Fig. 5.14(i), is equivalent, from the stress standpoint, to a condition of simple shearing in directions at 45°, the shearing stresses having the same magnitudes as the direct stresses σ_0 (Fig. 5.15).

If the material is elastic, the strains ϵ_x, ϵ_y caused by the direct stresses σ_0 are, from equations (5.18),

$$\epsilon_x = \frac{1}{E}(\sigma_0 + \nu\sigma_0) = \frac{\sigma_0}{E}(1 + \nu)$$

$$\epsilon_y = \frac{1}{E}(-\sigma_0 - \nu\sigma_0) = -\frac{\sigma_0}{E}(1 + \nu)$$

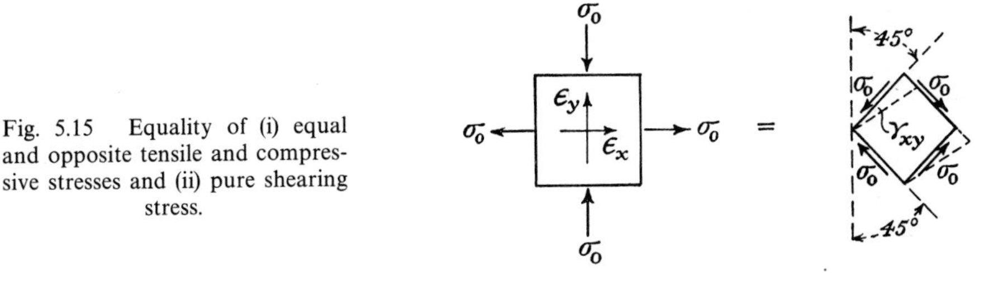

Fig. 5.15 Equality of (i) equal and opposite tensile and compressive stresses and (ii) pure shearing stress.

If the sides of the element are of unit length, the work done in distorting the element is

$$W = \tfrac{1}{2}\sigma_0\epsilon_x - \tfrac{1}{2}\sigma_0\epsilon_y = \frac{\sigma_0{}^2}{E}(1 + \nu) \tag{5.22}$$

per unit volume of the material.

In the state of pure shearing under stresses σ_0, the shearing strain is given by equation (5.20),

$$\gamma_{xy} = \frac{\sigma_0}{G}$$

The work done in distorting an element of sides unit length is

$$W = \tfrac{1}{2}\sigma_0\gamma_{xy} = \frac{\sigma_0{}^2}{2G} \tag{5.23}$$

per unit volume of the material. Since the one state of stress is equivalent to the other, the values of work done per unit volume of the material are equal. Then

$$\frac{\sigma_0{}^2}{E}(1 + \nu) = \frac{\sigma_0{}^2}{2G}$$

and hence

$$E = 2G(1 + \nu) \tag{5.24}$$

Thus ν can be calculated from measured values of E and G.

The shearing stress-strain relation is given by equation (5.20), which may now be written in the form

$$E\gamma_{xy} = 2(1 + \nu)\tau_{xy} \tag{5.25}$$

For most metals ν is approximately 0·3; then, approximately,

$$E = 2(1 + \nu)G = 2{\cdot}6G \tag{5.26}$$

Problem 5.5: From tests on a magnesium-alloy it is found that E is 45 GN/m² and G is 17 GN/m². Estimate the value of Poisson's ratio.

Solution

From equation (5.24),

$$\nu = \frac{E}{2G} - 1 = \tfrac{45}{34} - 1 = 1{\cdot}32 - 1$$

Then

$$\nu = 0{\cdot}32$$

Problem 5.6: A thin sheet of material is subjected to a tensile stress of 80 MN/m^2, in a certain direction. One surface of the sheet is polished, and on this surface fine lines are ruled to form a square of side 5 cm, one diagonal of the square being parallel to the direction of the tensile stresses. If $E = 200$ GN/m^2, and $\nu = 0{\cdot}3$, estimate the alteration in the lengths of the sides of the square, and the changes in the angles at the corners of the square.

Solution

The diagonal parallel to the tensile stresses increases in length by an amount

$$\frac{(80 \times 10^6)(0{\cdot}05\sqrt{2})}{200 \times 10^9} = 28{\cdot}3 \times 10^{-6}\ \text{m}$$

The diagonal perpendicular to the tensile stresses diminishes in length by an amount

$$0{\cdot}3(28{\cdot}3 \times 10^{-6}) = 8{\cdot}50 \times 10^{-6}\ \text{m}$$

The change in the corner angles is then

$$\frac{1}{0{\cdot}05}[(28{\cdot}3 + 8{\cdot}50)10^{-6}]\frac{1}{\sqrt{2}} = 52{\cdot}0 \times 10^{-3}\ \text{radians} = 0{\cdot}0405^\circ$$

The angles in the line of pull are diminished by this amount, and the others increased by the same amount. The increase in length of each side is

$$\frac{1}{2\sqrt{2}}[(28{\cdot}3 - 8{\cdot}50)10^{-6}] = 7{\cdot}00 \times 10^{-6}\ \text{m}$$

5.15 Strain 'rosettes'

To determine the stresses in a material under practical loading conditions, the strains are measured by means of small gauges; many types of gauges have been devised, but perhaps the most convenient is the electrical strain gauge, consisting of a short length of fine wire which is glued to the surface of the material. The resistance of the wire changes by small amounts as the wire is stretched, so that as the surface of the material is strained the gauge indicates a change of resistance which is measurable on a Wheatstone bridge. The lengths of wire resistance strain gauges can be as small as 0·5 cm, and they are therefore extremely useful in measuring local strains.

The state of strain at a point of a material is defined in the two-dimensional case if the direct strains, ϵ_x and ϵ_y, and the shearing strain, γ_{xy}, are known. Unfortunately, the shearing strain γ_{xy} is not readily measured; it is possible, however, to measure the direct strains in three different directions by means of strain gauges. Suppose ϵ_1, ϵ_2

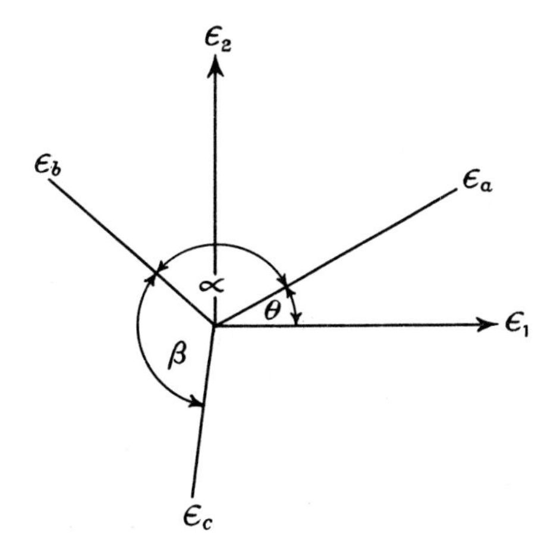

Fig. 5.16 Finding the principal strains in a two-dimensional system by recording three linear strains ϵ_a, ϵ_b, ϵ_c in the vicinity of a point.

are the principal strains in a two-dimensional system, Fig. 5.16. Then from equation (5.16) we have that the direct strains ϵ_a, ϵ_b, ϵ_c in directions inclined at θ, $(\theta + \alpha)$, $(\theta + \alpha + \beta)$ to ϵ_1 are

$$\begin{aligned} \epsilon_a &= \tfrac{1}{2}(\epsilon_1 + \epsilon_2) + \tfrac{1}{2}(\epsilon_1 - \epsilon_2)\cos 2\theta \\ \epsilon_b &= \tfrac{1}{2}(\epsilon_1 + \epsilon_2) + \tfrac{1}{2}(\epsilon_1 - \epsilon_2)\cos 2(\theta + \alpha) \\ \epsilon_c &= \tfrac{1}{2}(\epsilon_1 + \epsilon_2) + \tfrac{1}{2}(\epsilon_1 - \epsilon_2)\cos 2(\theta + \alpha + \beta) \end{aligned} \tag{5.27}$$

In practice the directions of the principal strains are not known usually; but if the three direct strains ϵ_a, ϵ_b, ϵ_c are measured in known directions, then the three unknowns in equations (5.27) are

$$\epsilon_1, \epsilon_2 \text{ and } \theta$$

Three strain gauges arranged so that $\alpha = \beta = 45°$ form a 45° rosette, Fig. 5.17. Equations (5.27) become

$$\begin{aligned} \epsilon_a &= \tfrac{1}{2}(\epsilon_1 + \epsilon_2) + \tfrac{1}{2}(\epsilon_1 - \epsilon_2)\cos 2\theta \\ \epsilon_b &= \tfrac{1}{2}(\epsilon_1 + \epsilon_2) - \tfrac{1}{2}(\epsilon_1 - \epsilon_2)\sin 2\theta \\ \epsilon_c &= \tfrac{1}{2}(\epsilon_1 + \epsilon_2) - \tfrac{1}{2}(\epsilon_1 - \epsilon_2)\cos 2\theta \end{aligned}$$

On eliminating θ from these equations we find that ϵ_1, ϵ_2 are the roots of the quadratic equation

?

$$\epsilon^2 - (\epsilon_a + \epsilon_c)\epsilon + [\epsilon_a\epsilon_c - \tfrac{1}{4}(2\epsilon_b - \epsilon_a - \epsilon_c)^2] = 0 \tag{5.28}$$

We have further that the directions of ϵ_1 and ϵ_2, relative to ϵ_a, ϵ_b and ϵ_c are given by the roots of the equation

$$\tan 2\theta = \frac{2\epsilon_b - \epsilon_a - \epsilon_c}{\epsilon_c - \epsilon_a} \tag{5.29}$$

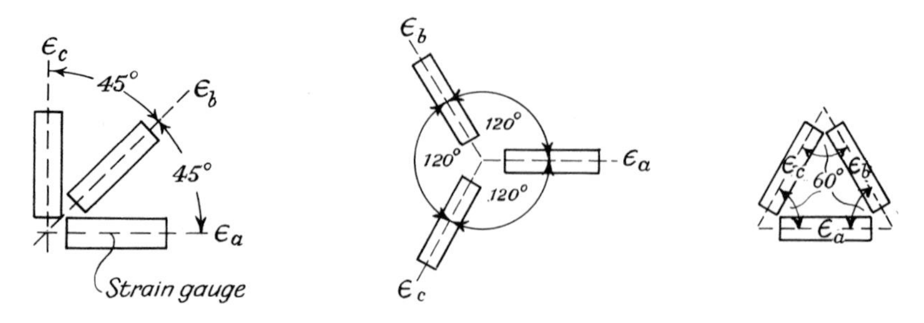

Fig. 5.17 A 45° strain rosette.

Fig. 5.18 Alternative arrangements of 120° rosettes.

θ is the angle between the directions of ϵ_1 and ϵ_a, and is measured clockwise from the direction of ϵ_1.

The alternative arrangements of gauges in Fig. 5.18 correspond to 120° rosettes. On putting $\alpha = \beta = 120°$ in equations (5.27), we have

$$\epsilon_a = \tfrac{1}{2}(\epsilon_1 + \epsilon_2) + \tfrac{1}{2}(\epsilon_1 - \epsilon_2)\cos 2\theta$$

$$\epsilon_b = \tfrac{1}{2}(\epsilon_1 + \epsilon_2) - \tfrac{1}{2}(\epsilon_1 - \epsilon_2)\left(\tfrac{1}{2}\cos 2\theta - \frac{\sqrt{3}}{2}\sin 2\theta\right) \tag{5.30}$$

$$\epsilon_c = \tfrac{1}{2}(\epsilon_1 + \epsilon_2) - \tfrac{1}{2}(\epsilon_1 - \epsilon_2)\left(\tfrac{1}{2}\cos 2\theta + \frac{\sqrt{3}}{2}\sin 2\theta\right)$$

The principal strains ϵ_1, ϵ_2 are the roots of the quadratic equation

$$3\epsilon^2 - 2(\epsilon_a + \epsilon_b + \epsilon_c)\epsilon - (\epsilon_a{}^2 + \epsilon_b{}^2 + \epsilon_c{}^2 - 2\epsilon_a\epsilon_b - 2\epsilon_b\epsilon_c - 2\epsilon_c\epsilon_a) = 0 \tag{5.31}$$

The directions of the principal strains are given by

$$\tan 2\theta = \frac{\sqrt{3}(\epsilon_b - \epsilon_c)}{2\epsilon_a - \epsilon_b - \epsilon_c} \tag{5.32}$$

When the principal strains ϵ_1 and ϵ_2 have been estimated, the corresponding principal stresses are deduced from the relations

$$E\epsilon_1 = \sigma_1 - v\sigma_2$$

$$E\epsilon_2 = \sigma_2 - v\sigma_1$$

These give

$$\sigma_1 = \frac{E}{1 - v^2}(\epsilon_1 + v\epsilon_2)$$

$$\sigma_2 = \frac{E}{1 - v^2}(\epsilon_2 + v\epsilon_1) \tag{5.33}$$

Obviously the values of E and v must be known before the stresses can be estimated.

5.16 Strain energy for a two-dimensional stress system

If σ_1, σ_2 are the principal stresses in a two-dimensional system, the corresponding principal strains for an elastic material are, from equations (5.21),

$$\epsilon_1 = \frac{1}{E}(\sigma_1 - \nu\sigma_2)$$

$$\epsilon_2 = \frac{1}{E}(\sigma_2 - \nu\sigma_1)$$

Consider a cube of material having sides of unit length, and therefore having also unit volume. The edges parallel to the direction of σ_1 extend amounts ϵ_1, and those parallel to the direction of σ_2 by amounts ϵ_2. The work done by the stresses σ_1 and σ_2 during straining is then

$$W = \tfrac{1}{2}\sigma_1\epsilon_1 + \tfrac{1}{2}\sigma_2\epsilon_2$$

per unit volume of material. On substituting for ϵ_1 and ϵ_2 we have

$$W = \tfrac{1}{2}\sigma_1\left[\frac{1}{E}(\sigma_1 - \nu\sigma_2)\right] + \tfrac{1}{2}\sigma_2\left[\frac{1}{E}(\sigma_2 - \nu\sigma_1)\right]$$

This is equal to the strain energy U *per unit volume*; thus

$$U = \frac{1}{2E}[\sigma_1{}^2 + \sigma_2{}^2 - 2\nu\sigma_1\sigma_2] \qquad (5.34)$$

5.17 Three-dimensional stress systems

In any two-dimensional stress system we found there were two mutually perpendicular directions in which only direct stresses, σ_1 and σ_2, acted; these were called the principal stresses. In any three-dimensional stress system we can always find three mutually perpendicular directions in which only direct stresses, σ_1, σ_2 and σ_3 in Fig. 5.19, are acting. No shearing stresses act on the faces of a rectangular block having its edges parallel to the axes 1, 2 and 3 in Fig. 5.19. These direct stresses are again called *principal stresses.*

If $\sigma_1 > \sigma_2 > \sigma_3$, then the three-dimensional stress system can be represented in the form of Mohr's circles, as shown in Fig. 5.20. Circle a passes through the points σ_1 and σ_2 on the σ-axis, and defines all states of stress on planes parallel to the axis 3, Fig. 5.19, but inclined to axes 1 and 2; similarly, circles b and c, Fig. 5.20, define stresses on planes parallel to axis 1 and axis 2, respectively. Circle c, having a diameter $\frac{1}{2}(\sigma_1 - \sigma_2)$, embraces the two smaller circles. For a plane inclined to all three axes the stresses are

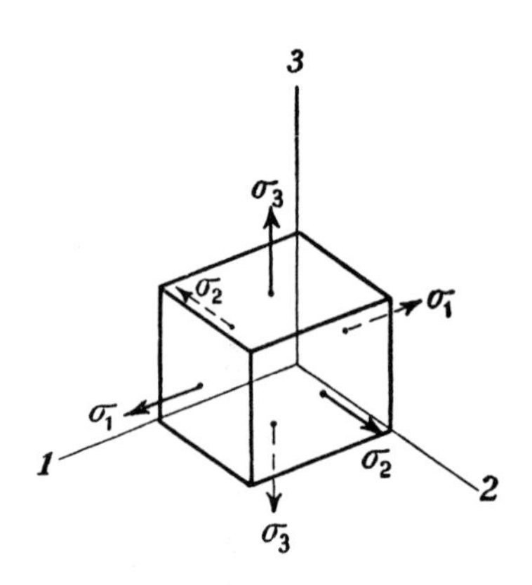

Fig. 5.19 Principal stresses in a three-dimensional system.

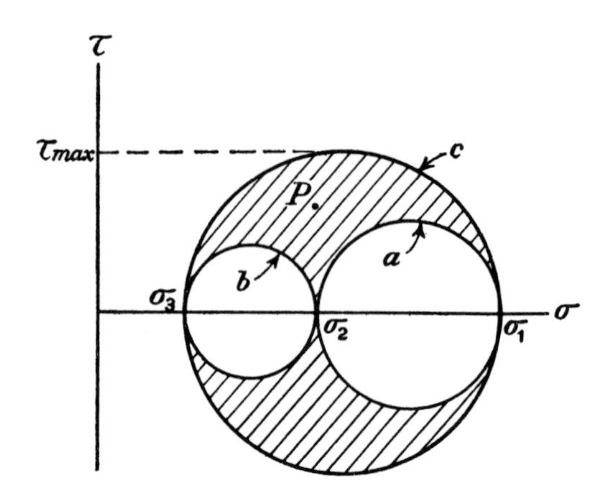

Fig. 5.20 Mohr's circle of stress for a three-dimensional system; circle a is the Mohr's circle of the two-dimensional system σ_1, σ_2; b corresponds to σ_2, σ_3 and c to σ_3, σ_1. The resultant direct and tangential stress on any plane through the point must correspond to a point P lying on or between the three circles.

defined by a point such as P within the shaded area in Fig. 5.20. The maximum shearing stress is

$$\tau_{max} = \tfrac{1}{2}(\sigma_1 - \sigma_3)$$

and occurs on a plane parallel to the axis 2.

From our discussion of three-dimensional stress systems we note that when one of the principal stresses, σ_3 say, is zero, Fig. 5.21, we have a two-dimensional system of

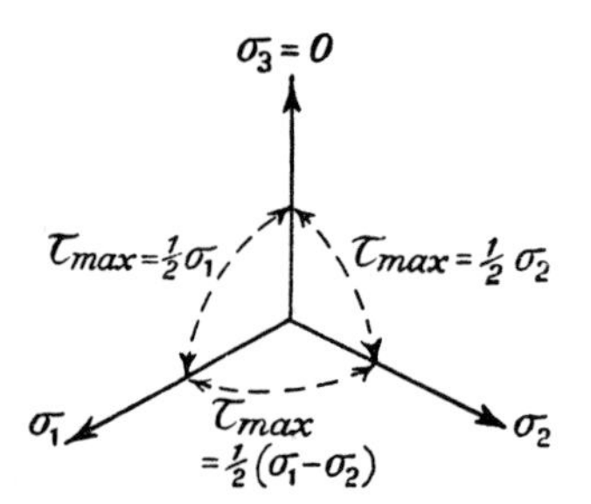

Fig. 5.21 Two-dimensional stress system as a particular case of a three-dimensional system with one of the three principal stresses equal to zero.

stresses σ_1, σ_2; the maximum shearing stresses in the planes 1–2, 2–3, 3–1 are, respectively,

$$\tfrac{1}{2}(\sigma_1 - \sigma_2), \quad \tfrac{1}{2}\sigma_1, \quad \tfrac{1}{2}\sigma_2$$

Suppose, initially, that σ_1 and σ_2 are both tensile and that $\sigma_1 > \sigma_2$; then the greatest of the three maximum shearing stresses is

$$\tfrac{1}{2}\sigma_1$$

which occurs in the 2–3 plane. If, on the other hand, σ_1 is tensile and σ_2 is compressive, the greatest of the maximum shearing stresses is

$$\tfrac{1}{2}(\sigma_1 - \sigma_2)$$

and occurs in the 1–2 plane. We conclude from this that the presence of a zero stress in a direction perpendicular to a two-dimensional stress system may have an important effect on the maximum shearing stresses in the material and cannot be disregarded therefore. The direct strains corresponding to σ_1, σ_2 and σ_3 for an elastic material are found by taking account of the Poisson ratio effects in the three directions; the principal strains in the directions 1, 2, 3 are, respectively,

$$\epsilon_1 = \frac{1}{E}(\sigma_1 - \nu\sigma_2 - \nu\sigma_3)$$

$$\epsilon_2 = \frac{1}{E}(\sigma_2 - \nu\sigma_3 - \nu\sigma_1)$$

$$\epsilon_3 = \frac{1}{E}(\sigma_3 - \nu\sigma_1 - \nu\sigma_2)$$

The strain energy stored per unit volume of the material is

$$U = \tfrac{1}{2}\sigma_1\epsilon_1 + \tfrac{1}{2}\sigma_2\epsilon_2 + \tfrac{1}{2}\sigma_3\epsilon_3$$

In terms of σ_1, σ_2, σ_3, this becomes

$$U = \frac{1}{2E}(\sigma_1{}^2 + \sigma_2{}^2 + \sigma_3{}^2 - 2\nu\sigma_1\sigma_2 - 2\nu\sigma_2\sigma_3 - 2\nu\sigma_3\sigma_1) \qquad (5.35)$$

5.18 Volumetric strain in a material under hydrostatic pressure

A material under the action of equal compressive stresses σ in three mutually perpendicular directions, Fig. 5.22, is subjected to a *hydrostatic pressure*, σ. The term hydrostatic is used because the material is subjected to the same stresses as would occur if it

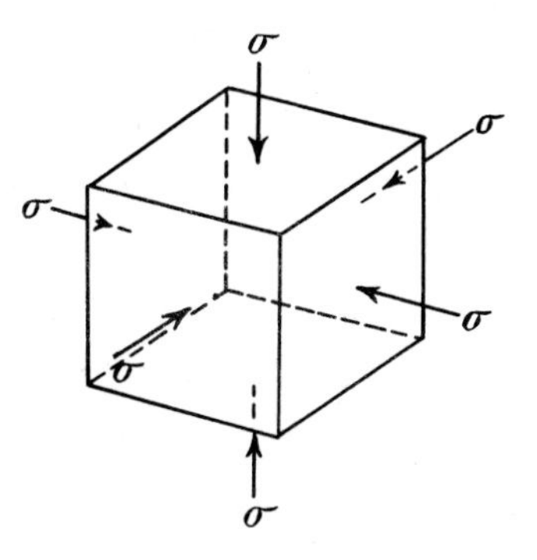

Fig. 5.22 Region of a material under a hydrostatic pressure.

were immersed in a fluid at a considerable depth. If the initial volume of the material is V_0, and if this diminishes an amount δV due to the hydrostatic pressure, the volumetric strain is

$$\frac{\delta V}{V_0}$$

The ratio of the hydrostatic pressure, σ, to the volumetric strain, $\delta V/V_0$, is called the *bulk modulus* of the material, and is denoted by K. Then

$$K = \frac{\sigma}{(\delta V/V_0)} \tag{5.36}$$

If the material remains elastic under hydrostatic pressure, the strain in each of the three mutually perpendicular directions is

$$\epsilon = -\frac{\sigma}{E} + \frac{\nu\sigma}{E} + \frac{\nu\sigma}{E}$$

$$= -\frac{\sigma}{E}(1 - 2\nu)$$

since there are two Poisson ratio effects on the strain in any of the three directions. If we consider a cube of material having sides of unit length in the unstrained condition, the volume of the strained cube is

$$(1 - \epsilon)^3$$

Now ϵ is small, so that this may be written approximately

$$1 - 3\epsilon$$

The change in volume of a unit volume is then

$$3\epsilon$$

which is therefore the volumetric strain. Then equation (5.36) gives

$$K = \frac{\sigma}{(\delta V/V_0)} = \frac{\sigma}{-3\epsilon} = \frac{E}{3(1 - 2\nu)}$$

We should expect the volume of a material to diminish under a hydrostatic pressure. In general, if K is always positive, we must have

$$1 - 2\nu > 0$$

or

$$\nu < \tfrac{1}{2}$$

Then Poisson's ratio is always less than $\frac{1}{2}$. For plastic strains of a metallic material there is a negligible change of volume, and Poisson's ratio is equal to $\frac{1}{2}$, approximately.

5.19 Strain energy of distortion

In the three-dimensional stress system of Fig. 5.19 we may consider the principal stress σ_1 to be the resultant of stresses

$$\tfrac{1}{3}(\sigma_1 + \sigma_2 + \sigma_3)$$

and stresses

$$\tfrac{1}{3}(2\sigma_1 - \sigma_2 - \sigma_3)$$

since

$$\tfrac{1}{3}(\sigma_1 + \sigma_2 + \sigma_3) + \tfrac{1}{3}(2\sigma_1 - \sigma_2 - \sigma_3) = \sigma_1$$

Similarly, we write

$$\sigma_2 = \tfrac{1}{3}(\sigma_1 + \sigma_2 + \sigma_3) + \tfrac{1}{3}(2\sigma_2 - \sigma_3 - \sigma_1)$$
$$\sigma_3 = \tfrac{1}{3}(\sigma_1 + \sigma_2 + \sigma_3) + \tfrac{1}{3}(2\sigma_3 - \sigma_1 - \sigma_2)$$

Now, the component $\tfrac{1}{3}(\sigma_1 + \sigma_2 + \sigma_3)$ which occurs in σ_1, σ_2 and σ_3, represents a *hydrostatic* tensile stress; the strains associated with this stress give rise to no distortion, i.e. a cube of material under stresses $\tfrac{1}{3}(\sigma_1 + \sigma_2 + \sigma_3)$ in three mutually perpendicular directions is strained into a cube. The remaining components of σ_1, σ_2, σ_3, are

$$\tfrac{1}{3}(2\sigma_1 - \sigma_2 - \sigma_3), \qquad \tfrac{1}{3}(2\sigma_2 - \sigma_3 - \sigma_1), \qquad \tfrac{1}{3}(2\sigma_3 - \sigma_1 - \sigma_2)$$

The strain energy associated with these stresses, which are the only stresses giving rise to distortion, is called the *strain energy of distortion.* The strains due to these distorting stresses are

$$\epsilon_1 = \frac{1}{3E}(1 + \nu)(2\sigma_1 - \sigma_2 - \sigma_3) = \frac{1}{6G}[(\sigma_1 - \sigma_2) + (\sigma_1 - \sigma_3)]$$

$$\epsilon_2 = \frac{1}{3E}(1 + \nu)(2\sigma_2 - \sigma_3 - \sigma_1) = \frac{1}{6G}[(\sigma_2 - \sigma_3) + (\sigma_2 - \sigma_1)]$$

$$\epsilon_3 = \frac{1}{3E}(1 + \nu)(2\sigma_3 - \sigma_1 - \sigma_2) = \frac{1}{6G}[(\sigma_3 - \sigma_1) + (\sigma_3 - \sigma_2)]$$

The strain energy of distortion is therefore

$$U_D = \frac{1}{36G}[(2\sigma_1 - \sigma_2 - \sigma_3)^2 + (2\sigma_2 - \sigma_3 - \sigma_1)^2 + (2\sigma_3 - \sigma_1 - \sigma_2)^2]$$

per unit volume. Then

$$U_D = \frac{1}{12G}[(\sigma_1 - \sigma_2)^2 + (\sigma_2 - \sigma_3)^2 + (\sigma_3 - \sigma_1)^2] \qquad (5.37)$$

For a two-dimensional stress system, σ_3 (say) $= 0$, and U_D reduces to

$$U_D = \frac{1}{12G}[(\sigma_1 - \sigma_2)^2 + \sigma_2{}^2 + \sigma_1{}^2]$$

or

$$U_D = \frac{1}{6G}[\sigma_1{}^2 - \sigma_1\sigma_2 + \sigma_2{}^2] \tag{5.38}$$

We shall see later that the strain energy of distortion plays an important part in the yielding of ductile materials under combined stresses.

5.20 Yielding of ductile materials under combined stresses

It was noted in §5.3 that when a polished bar of mild steel is loaded in tension, strain figures are observable on the surface of the bar after the yield point has been exceeded. The figures take the form of 'lines' inclined at about 45° to the axis of the bar; this direction corresponds to the planes of maximum shearing stress in the bar; the 'lines' are, in fact, bands of metal crystals shearing over similar bands. That yielding takes place in this way suggests that the crystal structure of the metal is relatively weak in shear; yielding takes the form of sliding of one crystal plane over another, and not the tearing apart of two crystal planes.

This form of behaviour—yielding by a shearing action—is typical of ductile materials. We note firstly that if a material is subjected to a hydrostatic pressure σ, the three principal stresses σ_1, σ_2, σ_3 in a three-dimensional system are each equal to σ. A state of stress of this sort exists in a solid sphere of material subjected to an external pressure σ, Fig. 5.23. Since the three principal stresses are equal in magnitude, there are no

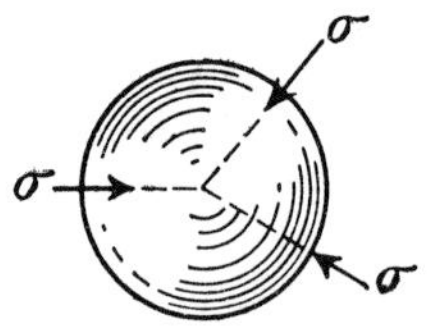

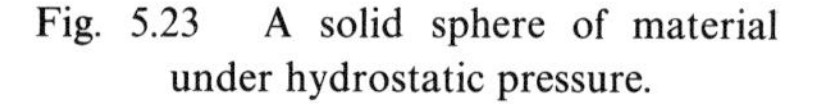
Fig. 5.23 A solid sphere of material under hydrostatic pressure.

shearing stresses in the material; if yielding is governed by the presence of shearing stresses on some planes in a material, then no yielding is theoretically possible when the material is under hydrostatic pressure.

For a two-dimensional stress system one of the three principal stresses of a three-dimensional system is zero. We consider now the yielding of a mild steel under different combinations of the principal stresses, σ_1 and σ_2, of a two-dimensional system; in discussing the problem we keep in mind the presence of a zero stress perpendicular to the plane of σ_1 and σ_2, Fig. 5.24.

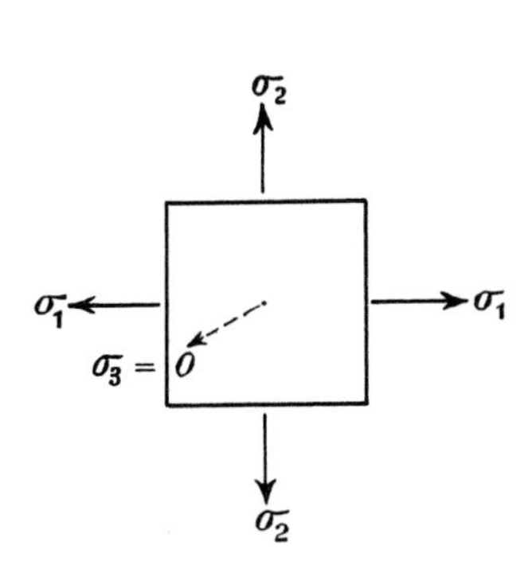

Fig. 5.24 In a two-dimensional stress system, one of the three principal stresses (σ_3 say) is zero.

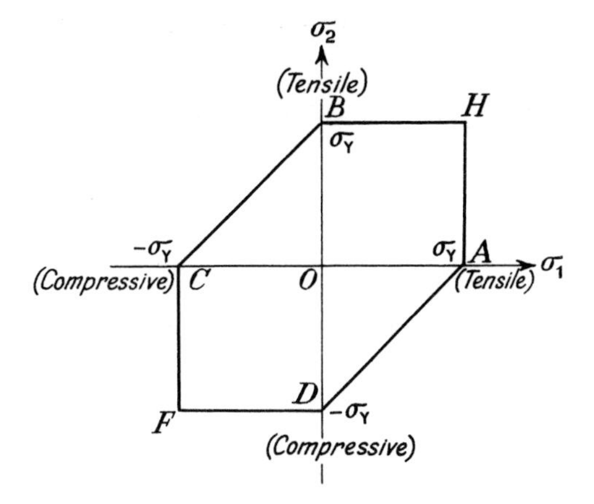

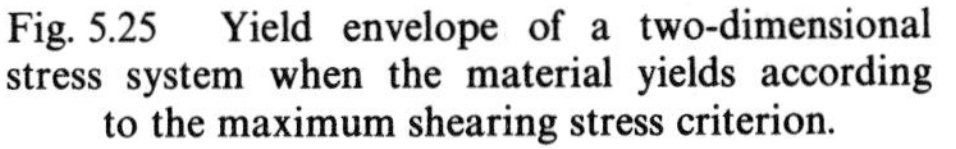

Fig. 5.25 Yield envelope of a two-dimensional stress system when the material yields according to the maximum shearing stress criterion.

Suppose firstly we conduct a simple tension test on the material; we may put $\sigma_2 = 0$, and yielding occurs when

$$\sigma_1 = \sigma_Y, \text{(say)}$$

This yielding condition corresponds to the point A in Fig. 5.25. If the material has similar properties in tension and compression, yielding under a compressive stress σ_1 occurs when $\sigma_1 = -\sigma_Y$; this condition corresponds to the point C in Fig. 5.25. We could, however, perform the tension and compression tests in the direction of σ_2, Fig. 5.24; if the material is isotropic—that is, it has the same properties in all directions—yielding occurs at the yield stress σ_Y; we can thus derive points B and D in the yield diagram, Fig. 5.25.

We consider now yielding of the material when both σ_1 and σ_2, Fig. 5.24, are present; we shall assume that yielding of the mild steel occurs when the maximum shearing stress attains a critical value; from the simple tensile test, the maximum shearing stress at yielding is

$$\tau_{max} = \tfrac{1}{2}\sigma_Y$$

which we shall take as the critical value. Suppose, firstly, that $\sigma_1 > \sigma_2$, and that both principal stresses are tensile; the maximum shearing stress is

$$\tau_{max} = \tfrac{1}{2}(\sigma_1 - 0) = \tfrac{1}{2}\sigma_1$$

and occurs in the 3–1 plane of Fig. 5.26; τ_{max} attains the critical value when

$$\tfrac{1}{2}\sigma_1 = \tfrac{1}{2}\sigma_Y, \quad \text{or} \quad \sigma_1 = \sigma_Y$$

Thus, yielding for these stress conditions is unaffected by σ_2. In Fig. 5.25, these stress conditions are given by the line AH. If we consider similarly the case when σ_1 and σ_2

are both tensile, but $\sigma_2 > \sigma_1$, yielding occurs when $\sigma_2 = \sigma_Y$, giving the line *BH* in Fig. 5.25. By making the stresses both compressive we can derive in a similar fashion the lines *CF* and *DF* of Fig. 5.25.

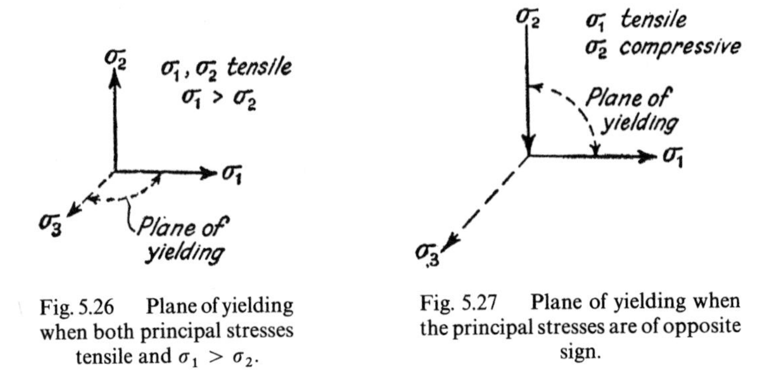

Fig. 5.26 Plane of yielding when both principal stresses tensile and $\sigma_1 > \sigma_2$.

Fig. 5.27 Plane of yielding when the principal stresses are of opposite sign.

But when σ_1 is tensile and σ_2 is compressive, Fig. 5.27, the maximum shearing stress occurs in the 1–2 plane, and has the value

$$\tau_{max} = \tfrac{1}{2}(\sigma_1 - \sigma_2)$$

Yielding occurs when

$$\tfrac{1}{2}(\sigma_1 - \sigma_2) = \tfrac{1}{2}\sigma_Y, \quad \text{or} \quad \sigma_1 - \sigma_2 = \sigma_Y$$

This corresponds to the line *AD* of Fig. 5.25. Similarly, when σ_1 is compressive and σ_2 is tensile, yielding occurs when

$$\sigma_2 - \sigma_1 = \sigma_Y$$

corresponding to the line *BC* of Fig. 5.25.

The hexagon *AHBCFD* of Fig. 5.25 is called a *yield locus*, since it defines all combinations of σ_1 and σ_2 giving yielding of the mild steel; for any state of stress within the hexagon the material remains elastic; for this reason the hexagon is also sometimes called a *yield envelope*. The *criterion of yielding* used in the derivation of the hexagon of Fig. 5.25 was that of maximum shearing stress; the use of this criterion was first suggested by Tresca in 1878.

Not all ductile metals obey the maximum shearing stress criterion; the yielding of some metals, including certain steels and alloys of aluminium, is governed by a critical value of the strain energy of distortion. For a two-dimensional stress system the strain energy of distortion per unit volume of the material is given by equation (5.38). In the simple tension test for which $\sigma_2 = 0$, say, yielding occurs when $\sigma_1 = \sigma_Y$. The critical value of U_D is therefore

$$U_D = \frac{1}{6G}[\sigma_1^2 - \sigma_1\sigma_2 + \sigma_2^2] = \frac{1}{6G}[\sigma_Y^2 - \sigma_Y(0) + 0^2] = \frac{\sigma_Y^2}{6G}$$

Then for other combinations of σ_1 and σ_2, yielding occurs when

$$\sigma_1{}^2 - \sigma_1\sigma_2 + \sigma_2{}^2 = \sigma_Y{}^2 \tag{5.39}$$

The yield locus given by this equation is an ellipse with major and minor axes inclined at 45° to the directions of σ_1 and σ_2, Fig. 5.28. This locus of yielding was first suggested by von Mises in 1913.

For a three-dimensional system the yield locus corresponding to the strain energy of distortion is of the form

$$(\sigma_1 - \sigma_2)^2 + (\sigma_2 - \sigma_3)^2 + (\sigma_3 - \sigma_1)^2 = \text{constant}$$

This relation defines the surface of a cylinder of circular cross-section, with its central axis on the line $\sigma_1 = \sigma_2 = \sigma_3$; the axis of the cylinder passes through the origin of the $\sigma_1\sigma_2\sigma_3$ coordinate system, and is inclined at equal angles to the axes σ_1, σ_2 and σ_3, Fig. 5.29. When σ_3 is zero, critical values of σ_1 and σ_2 lie on an ellipse in the σ_1-σ_2 plane, corresponding to the ellipse of Fig. 5.28.

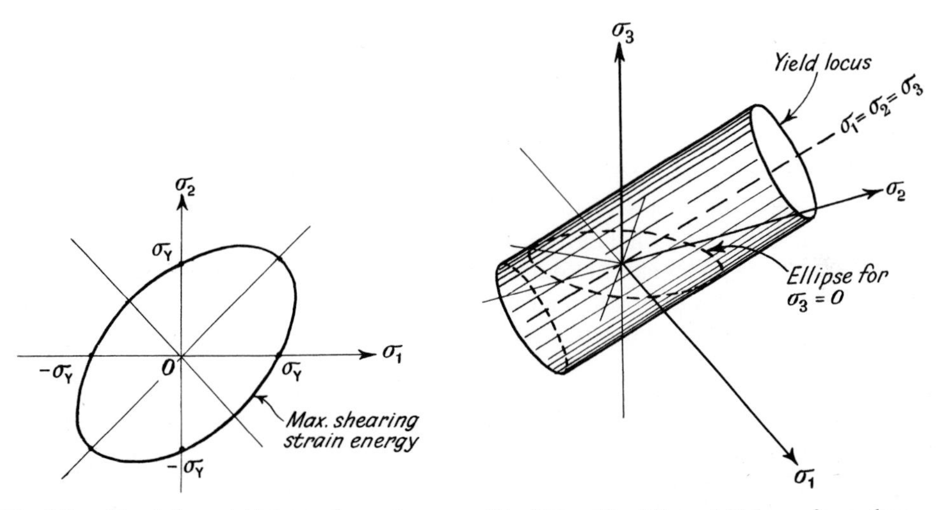

Fig. 5.28 The Mises yield locus for a two-dimensional system of stresses.

Fig. 5.29 The Mises yield locus for a three-dimensional stress system.

When a material obeys the maximum shearing stress criterion, the three-dimensional yield locus is a regular hexagonal cylinder with its central axis on the line $\sigma_1 = \sigma_2 = \sigma_3 = 0$, Fig. 5.30. When σ_3 is zero, the locus is an irregular hexagon, of the form already discussed in Fig. 5.25.

The surfaces of the yield loci in Figs. 5.29 and 5.30 extend indefinitely parallel to the line $\sigma_1 = \sigma_2 = \sigma_3$, which we call the hydrostatic stress line. Hydrostatic stress itself cannot cause yielding, and no yielding occurs at other stresses provided these fall within the cylinders of Figs. 5.29 and 5.30.

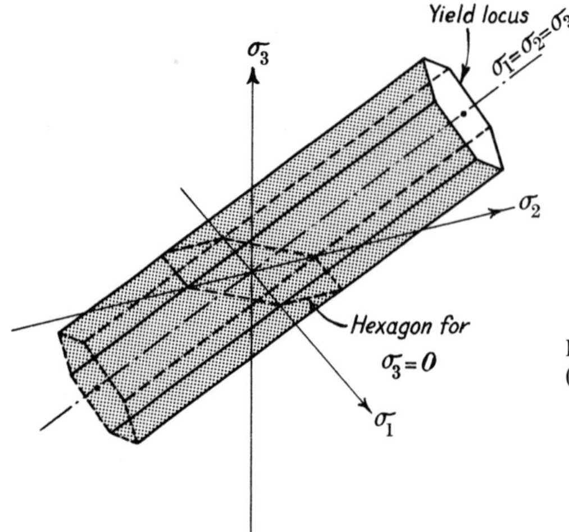

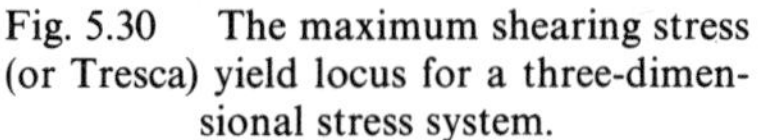

Fig. 5.30 The maximum shearing stress (or Tresca) yield locus for a three-dimensional stress system.

5.21 Elastic breakdown and failure of brittle materials

Unlike ductile materials the failure of brittle materials occurs at relatively low strains, and there is little, or no, permanent yielding on the planes of maximum shearing stress.

Some brittle materials, such as cast-iron and concrete, contain large numbers of microscopic cracks in their structures. These are believed to give rise to high stress concentrations, thereby causing local failure of the material. These stress concentrations are likely to have a greater effect in reducing tensile strength than compressive strength; a general characteristic of brittle materials is that they are relatively weak in tension. For this reason elastic breakdown and failure in a brittle material are governed largely

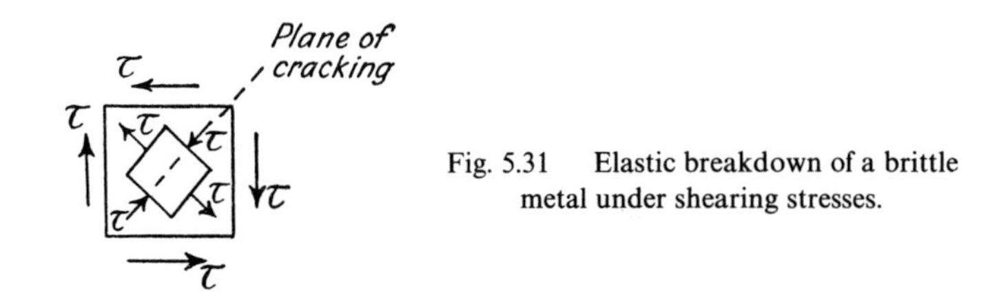

Fig. 5.31 Elastic breakdown of a brittle metal under shearing stresses.

by the maximum principal tensile stress; as an example of the application of this criterion consider a concrete: in simple tension the breaking stress is about 1·5 MN/m^2, whereas in compression it is found to be about 30 MN/m^2, or twenty times as great; in pure shear the breaking stress would be of the order of 1·5 MN/m^2, because the principal stresses are of the same magnitude, and one of these stresses is tensile, Fig. 5.31. Cracking in the concrete would occur on planes inclined at 45° to the directions of the applied shearing stresses.

FURTHER PROBLEMS
(answers on page 397)

5.7 A tie-bar of steel has a cross-section 15 cm by 2 cm, and carries a tensile load of 200 kN. Find the stress normal to a plane making an angle of 30° with the cross-section and the shearing stress on this plane. *(Cambridge)*

5.8 A rivet is under the action of a shearing stress of 60 MN/m^2 and a tensile stress, due to contraction, of 45 MN/m^2. Determine the magnitude and direction of the greatest tensile and shearing stresses in the rivet. *(RNEC)*

5.9 A propeller shaft is subjected to an end thrust producing a stress of 90 MN/m^2, and the maximum shearing stress arising from torsion is 60 MN/m^2. Calculate the magnitudes of the principal stresses. *(Cambridge)*

5.10 At a point in a vertical cross-section of a beam there is a resultant stress of 75 MN/m^2, which is inclined upwards at 35° to the horizontal. On the horizontal plane through the point there is only shearing stress. Find, in magnitude and direction, the resultant stress on the plane which is inclined at 40° to the vertical and 95° to the given resultant stress. *(Cambridge)*

5.11 A plate is subjected to two mutually perpendicular stresses, one compressive of 45 MN/m^2, the other tensile of 75 MN/m^2, and a shearing stress, parallel to these directions, of 45 MN/m^2. Find the principal stresses and strains, taking Poisson's ratio as 0·3 and $E = 200$ GN/m^2. *(Cambridge)*

5.12 At a point in a material the three principal stresses acting in directions Ox, Oy, Oz, have the values 75, 0, -45 MN/m^2, respectively. Determine the normal and shearing stresses for a plane perpendicular to the xz plane inclined at 30° to the xy plane. *(Cambridge)*

6 Thin shells under internal pressure

6.1 Thin cylindrical shell of circular cross-section

A problem in which combined stresses are present is that of a cylindrical shell under internal pressure. Suppose a long circular shell is subjected to an internal pressure p, which may be due to a fluid or gas enclosed within the cylinder, Fig. 6.1. The internal pressure acting on the long sides of the cylinder gives rise to a circumferential stress in the wall of the cylinder; if the ends of the cylinder are closed, the pressure acting on these ends is transmitted to the walls of the cylinder thus producing a longitudinal stress in the walls.

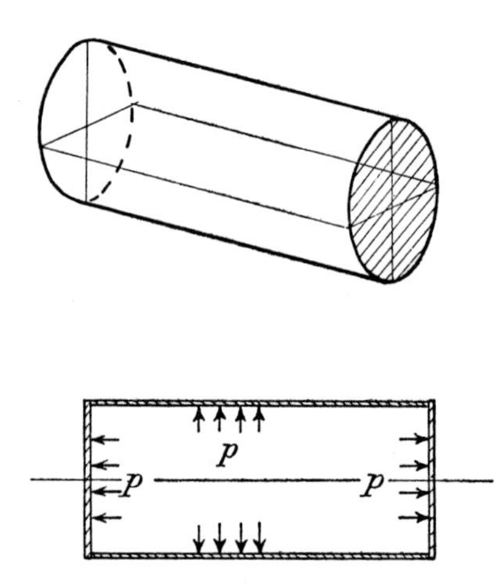

Fig. 6.1 Long thin cylindrical shell with closed ends under internal pressure.

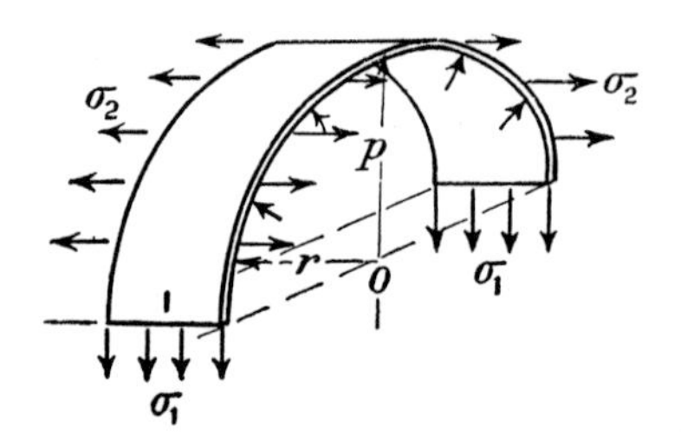

Fig. 6.2 Circumferential and longitudinal stresses in a thin cylinder with closed ends under internal pressure.

Suppose r is the mean radius of the cylinder, and that its thickness t is small compared with r. Consider a unit length of the cylinder remote from the closed ends, Fig. 6.2; suppose we cut this unit length with a diametral plane, as in Fig. 6.2. The tensile stresses acting on the cut sections are σ_1, acting circumferentially, and σ_2, acting longitudinally. There is an internal pressure p on the inside of the half-shell. Consider equilibrium of the half-shell in a plane perpendicular to the axis of the cylinder, as in Fig. 6.3; the total force due to the internal pressure p in the direction OA is

$$p \times (2r \times 1)$$

since we are dealing with a unit length of the cylinder. This force is opposed by the stresses σ_1; for equilibrium we must have

$$p \times (2r \times 1) = \sigma_1 \times 2(t \times 1)$$

Then

$$\sigma_1 = \frac{pr}{t} \tag{6.1}$$

We shall call this the *circumferential stress.*

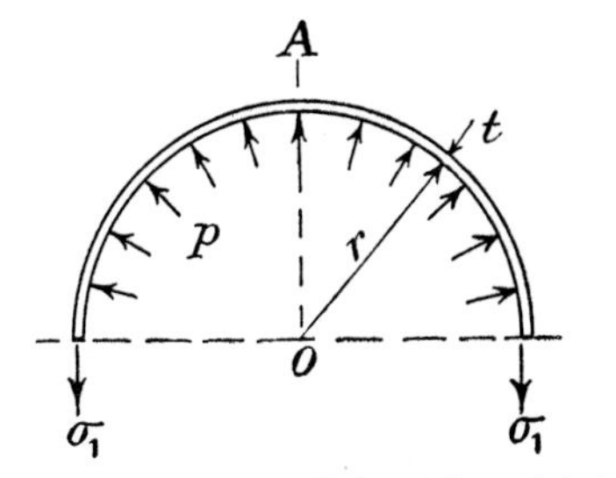

Fig. 6.3 Derivation of circumferential stress.

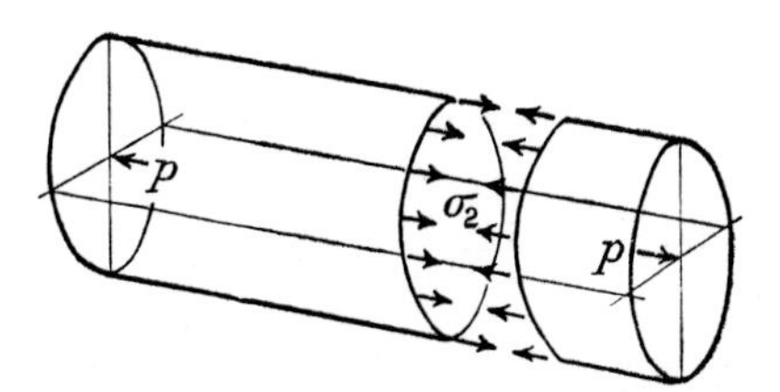

Fig. 6.4 Derivation of longitudinal stress.

Now consider any transverse cross-section of the cylinder remote from the ends, Fig. 6.4; the total longitudinal force on each closed end due to internal pressure is

$$p \times \pi r^2$$

At any section this is resisted by the internal stresses σ_2, Fig. 6.4. For equilibrium we must have

$$p \times \pi r^2 = \sigma_2 \times 2\pi rt$$

which gives

$$\sigma_2 = \frac{pr}{2t} \tag{6.2}$$

We shall call this the *longitudinal stress.* Thus the longitudinal stress, σ_2, is only half the circumferential stress, σ_1.

The stresses acting on an element of the wall of the cylinder consist of a circumferential stress σ_1, a longitudinal stress σ_2, and a radial stress p on the internal face of the element, Fig. 6.5. Since (r/t) is very much greater than unity, p is very small compared with σ_1 and σ_2. The state of stress in the wall of the cylinder approximates then to a simple

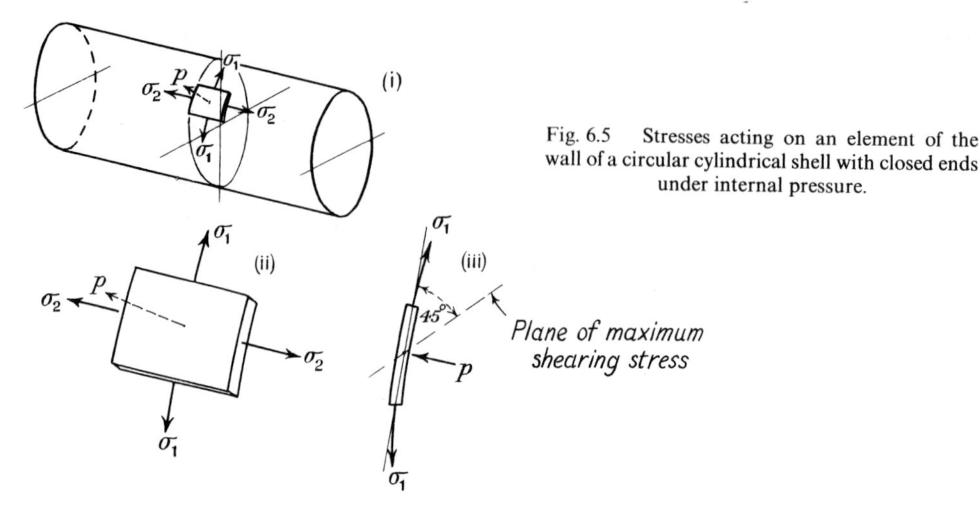

Fig. 6.5 Stresses acting on an element of the wall of a circular cylindrical shell with closed ends under internal pressure.

two-dimensional system with principal stresses σ_1 and σ_2. The maximum shearing stress in the plane of σ_1 and σ_2 is therefore

$$\tau_{max} = \tfrac{1}{2}(\sigma_1 - \sigma_2) = \frac{pr}{4t}$$

This is not, however, the maximum shearing stress in the wall of the cylinder, for, in the plane of σ_1 and p, the maximum shearing stress is

$$\tau_{max} = \tfrac{1}{2}(\sigma_1) = \frac{pr}{2t} \qquad (6.3)$$

since p is negligible compared with σ_1; again, in the plane of σ_2 and p, the maximum shearing stress is

$$\tau_{max} = \tfrac{1}{2}(\sigma_2) = \frac{pr}{4t}$$

The greatest of these maximum shearing stresses is given by equation (6.3); it occurs on a plane at 45° to the tangent and parallel to the longitudinal axis of the cylinder, Fig. 6.5(iii).

The circumferential and longitudinal stresses are accompanied by direct strains. If the material of the cylinder is elastic, the corresponding strains are given by

$$\left.\begin{aligned} \epsilon_1 &= \frac{1}{E}(\sigma_1 - \nu\sigma_2) = \frac{pr}{Et}(1 - \tfrac{1}{2}\nu) \\ \epsilon_2 &= \frac{1}{E}(\sigma_2 - \nu\sigma_1) = \frac{pr}{Et}(\tfrac{1}{2} - \nu) \end{aligned}\right\} \qquad (6.4)$$

The circumference of the cylinder increases therefore by a small amount $2\pi r\epsilon_1$; the increase in mean radius is therefore $r\epsilon_1$. The increase in length of a unit length of the cylinder is ϵ_2, so the change in internal volume of a unit length of the cylinder is

$$\delta V = \pi(r + r\epsilon_1)^2(1 + \epsilon_2) - \pi r^2$$

The volumetric strain is therefore

$$\frac{\delta V}{\pi r^2} = (1 + \epsilon_1)^2(1 + \epsilon_2) - 1$$

But ϵ_1 and ϵ_2 are small quantities, so the volumetric strain is

$$(1 + \epsilon_1)^2(1 + \epsilon_2) - 1 \fallingdotseq (1 + 2\epsilon_1)(1 + \epsilon_2) - 1$$
$$\fallingdotseq 2\epsilon_1 + \epsilon_2$$

In terms of σ_1 and σ_2 this becomes

$$2\epsilon_1 + \epsilon_2 = \frac{pr}{Et}[2(1 - \tfrac{1}{2}\nu) + (\tfrac{1}{2} - \nu)] = \frac{pr}{Et}(\tfrac{5}{2} - 2\nu) \qquad (6.5)$$

Problem 6.1: A thin cylindrical shell has an internal diameter of 20 cm, and is 0·5 cm thick. It is subjected to an internal pressure of 3·5 MN/m^2. Estimate the circumferential and longitudinal stresses if the ends of the cylinder are closed.

Solution

From equations (6.1) and (6.2),

$$\sigma_1 = \frac{pr}{t} = (3{\cdot}5 \times 10^6)(0{\cdot}1025)/(0{\cdot}005) = 71{\cdot}8 \text{ MN/m}^2$$

and

$$\sigma_2 = \frac{pr}{2t} = (3{\cdot}5 \times 10^6)(0{\cdot}1025)(0{\cdot}010) = 35{\cdot}9 \text{ MN/m}^2$$

Problem 6.2: If the ends of the cylinder in Problem 6.1 are closed by pistons sliding in the cylinder, estimate the circumferential and longitudinal stresses.

Solution

The effect of taking the end pressure on sliding pistons is to remove the force on the cylinder causing longitudinal stress. As in Problem 6.1, the circumferential stress is

$$\sigma_1 = 71{\cdot}8 \text{ MN/m}^2$$

but the longitudinal stress is zero.

Problem 6.3: A pipe of internal diameter 10 cm, and 0·3 cm thick is made of mild-steel having a tensile yield stress of 375 MN/m^2. What is the maximum permissible internal pressure if the stress factor on the maximum shearing stress is to be 4?

Solution

The greatest allowable maximum shearing stress is

$$\tfrac{1}{4}(\tfrac{1}{2} \times 375 \times 10^6) = 46{\cdot}9 \text{ MN/m}^2$$

The greatest shearing stress in the cylinder is

$$\tau_{max} = \frac{pr}{2t}$$

Then

$$p = \frac{2t}{r}(\tau_{max}) = \frac{2 \times 0{\cdot}003}{0{\cdot}0515} \times (46{\cdot}9 \times 10^6) = 5{\cdot}46 \text{ MN/m}^2$$

Problem 6.4: Two boiler plates, each 1 cm thick, are connected by a double-riveted butt joint with two cover plates, each 0·6 cm thick. The rivets are 2 cm diameter and their pitch is 0·90 cm. The internal diameter of the boiler is 1·25 m, and the pressure is 0·8 MN/m². Estimate the shearing stress in the rivets, and the tensile stresses in the boiler plates and cover plates.

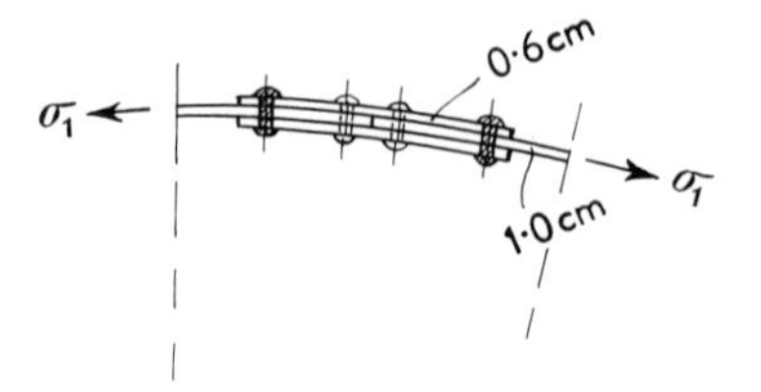

Solution

Suppose the rivets are staggered on each side of the joint. Then a single rivet takes the circumferential load associated with a $\frac{1}{2}(0{\cdot}090) = 0{\cdot}045$ m length of boiler. The load on a rivet is

$$[\tfrac{1}{2}(1{\cdot}25)](0{\cdot}045)(0{\cdot}8 \times 10^6) = 22{\cdot}5 \text{ kN}$$

Area of a rivet is

$$\frac{\pi}{4}(0{\cdot}02)^2 = 0{\cdot}314 \times 10^{-3} \text{ m}^2$$

The load of 22·5 kN is taken in double shear, and the shearing stress in the rivet is then

$$\tfrac{1}{2}(22{\cdot}5 \times 10^3)/(0{\cdot}314 \times 10^{-3}) = 35{\cdot}8 \text{ MN/m}^2$$

The rivet holes in the plates give rise to a loss in plate width of 2 cm in each 9 cm of rivet line. The effective area of boiler plate in a 9 cm length is then

$$(0{\cdot}010)(0{\cdot}090 - 0{\cdot}020) = (0{\cdot}010)(0{\cdot}070) = 0{\cdot}7 \times 10^{-3} \text{ m}^2$$

The tensile load taken by this area is

$$\tfrac{1}{2}(1{\cdot}25)(0{\cdot}090)(0{\cdot}8 \times 10^6) = 45{\cdot}0 \text{ kN}$$

The average circumferential stress in the boiler plates is therefore

$$\sigma_1 = \frac{45{\cdot}0 \times 10^3}{0{\cdot}7 \times 10^{-3}} = 64{\cdot}2 \text{ MN/m}^2$$

This occurs in the region of the riveted connection. Remote from the connection, the circumferential tensile stress is

$$\sigma_1 = \frac{pr}{t} = \frac{(0{\cdot}8 \times 10^6)(0{\cdot}625)}{(0{\cdot}010)} = 50{\cdot}0 \text{ MN/m}^2$$

In the cover plates, the circumferential tensile stress is

$$\frac{45{\cdot}0 \times 10^3}{2(0{\cdot}006)(0{\cdot}070)} = 53{\cdot}6 \text{ MN/m}^2$$

The longitudinal tensile stresses in the plates in the region of the connection are difficult to estimate; except very near to the rivet holes, the stress will be

$$\sigma_2 = \frac{pr}{2t} = 25{\cdot}0 \text{ MN/m}^2$$

Problem 6.5: A long steel tube, 7·5 cm internal diameter and 0·15 cm thick, has closed ends, and is subjected to an internal fluid pressure of 3 MN/m². If $E = 200$ GN/m², and $\nu = 0{\cdot}3$, estimate the percentage increase in internal volume of the tube.

Solution

The circumferential tensile stress is

$$\sigma_1 = \frac{pr}{t} = \frac{(3 \times 10^6)(0{\cdot}0383)}{(0{\cdot}0015)} = 76{\cdot}6 \text{ MN/m}^2$$

The longitudinal tensile stress is

$$\sigma_2 = \frac{pr}{2t} = 38{\cdot}3 \text{ MN/m}^2$$

The circumferential strain is

$$\epsilon_1 = \frac{1}{E}(\sigma_1 - \nu\sigma_2)$$

and the longitudinal strain is

$$\epsilon_2 = \frac{1}{E}(\sigma_2 - \nu\sigma_1)$$

The volumetric strain is then

$$2\epsilon_1 + \epsilon_2 = \frac{1}{E}[2\sigma_1 - 2\nu\sigma_2 + \sigma_2 - \nu\sigma_1]$$

$$= \frac{1}{E}[\sigma_1(2 - \nu) + \sigma_2(1 - 2\nu)]$$

Thus

$$2\epsilon_1 + \epsilon_2 = \frac{(76{\cdot}6 \times 10^6)[(2 - 0{\cdot}3) + \frac{1}{2}(1 - 0{\cdot}6)]}{200 \times 10^9}$$

$$= \frac{(76{\cdot}6 \times 10^6)(1{\cdot}9)}{(200 \times 10^9)} = 0{\cdot}727 \times 10^{-3}$$

The percentage increase in volume is therefore 0·0727%.

Problem 6.6: An air vessel, which is made of steel, is 2 m long; it has an external diameter of 45 cm and is 1 cm thick. Find the increase of external diameter and the increase of length when charged to an internal air pressure of 10 MN/m².

Solution

For steel, we take

$$E = 200 \text{ GN/m}^2, \qquad \nu = 0{\cdot}3$$

The mean radius of the vessel is $r = 0{\cdot}225$ m; the circumferential stress is then

$$\sigma_1 = \frac{pr}{t} = \frac{(10 \times 10^6)(0{\cdot}225)}{0{\cdot}010} = 22{\cdot}5 \text{ MN/m}^2$$

The longitudinal stress is

$$\sigma_2 = \frac{pr}{2t} = 11{\cdot}25 \text{ MN/m}^2$$

The circumferential strain is therefore

$$\epsilon_1 = \frac{1}{E}(\sigma_1 - \nu\sigma_2) = \frac{\sigma_1}{E}(1 - \tfrac{1}{2}\nu) = \frac{(22{\cdot}5 \times 10^6)(0{\cdot}85)}{200 \times 10^9}$$

$$= 0{\cdot}957 \times 10^{-3}$$

The longitudinal strain is

$$\epsilon_2 = \frac{1}{E}(\sigma_2 - \nu\sigma_1) = \frac{\sigma_1}{E}(\tfrac{1}{2} - \nu) = \frac{(22{\cdot}5 \times 10^6)(0{\cdot}2)}{200 \times 10^9}$$

$$= 0{\cdot}225 \times 10^{-3}$$

The increase in external diameter is then

$$0{\cdot}450(0{\cdot}957 \times 10^{-3}) = 0{\cdot}430 \times 10^{-3}\ \text{m}$$

$$= 0{\cdot}043\ \text{cm}$$

The increase in length is

$$2(0{\cdot}225 \times 10^{-3}) = 0{\cdot}450 \times 10^{-3}\ \text{m}$$

$$= 0{\cdot}045\ \text{cm}$$

Problem 6.7: A thin cylindrical shell is subjected to internal fluid pressure, the ends being closed by

(a) two watertight pistons attached to a common piston rod,
(b) flanged ends.

Find the increase in internal diameter in each case, given that the internal diameter is 20 cm, thickness is 0·5 cm, Poisson's ratio is 0·3, Young's modulus is 200 GN/m², and the internal pressure is 3·5 MN/m².

(RNC)

Solution

We have

$$p = 3{\cdot}5\ \text{MN/m}^2, \qquad r = 0{\cdot}1\ \text{m}, \qquad t = 0{\cdot}005\ \text{m}$$

In both cases the circumferential stress is

$$\sigma_1 = \frac{pr}{t} = \frac{(3{\cdot}5 \times 10^6)(0{\cdot}1)}{(0{\cdot}005)} = 70\ \text{MN/m}^2$$

(a) In this case there is no longitudinal stress. The circumferential strain is then

$$\epsilon_1 = \frac{\sigma_1}{E} = \frac{70 \times 10^6}{200 \times 10^9} = 0{\cdot}35 \times 10^{-3}$$

The increase of internal diameter is

$$0{\cdot}2(0{\cdot}35 \times 10^{-3}) = 0{\cdot}07 \times 10^{-3}\ \text{m} = 0{\cdot}007\ \text{cm}$$

(b) In this case the longitudinal stress is

$$\sigma_2 = \frac{pr}{2t} = 35\ \text{MN/m}^2$$

The circumferential strain is therefore

$$\epsilon_1 = \frac{1}{E}(\sigma_1 - \nu\sigma_2) = \frac{\sigma_1}{E}(1 - \tfrac{1}{2}\nu) = 0{\cdot}85\frac{\sigma_1}{E}$$

$$= 0{\cdot}85(0{\cdot}35 \times 10^{-3}) = 0{\cdot}298 \times 10^{-3}$$

The increase of internal diameter is therefore

$$0{\cdot}2(0{\cdot}298 \times 10^{-3}) = 0{\cdot}0596 \times 10^{-3}\ \text{m} = 0{\cdot}00596\ \text{cm}$$

6.2 Thin spherical shell

We consider next a thin spherical shell of mean radius r, and thickness t, which is subjected to an internal pressure p. Consider any diametral plane through the shell, Fig. 6.6; the total force normal to this plane due to p acting on a hemisphere is

$$p \times \pi r^2$$

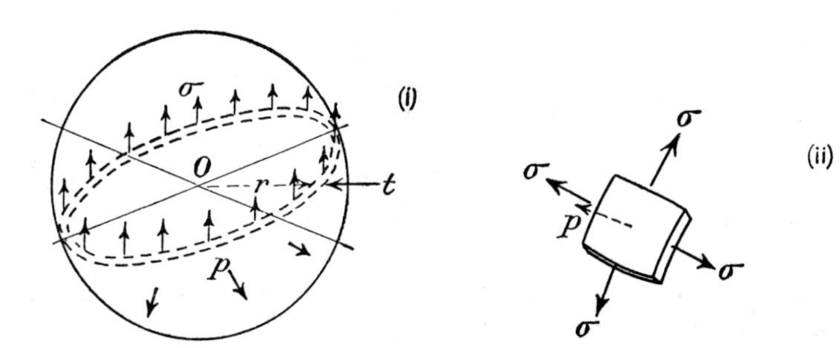

Fig. 6.6 Membrane stresses in a thin spherical shell under internal pressure.

This is opposed by a tensile stress σ in the walls of the shell. By symmetry σ is the same at all points of the shell; for equilibrium of the hemisphere we must have

$$p \times \pi r^2 = \sigma \times 2\pi r t$$

This gives

$$\sigma = \frac{pr}{2t} \qquad (6.6)$$

At any point of the shell the direct stress σ has the same magnitude in all directions in the plane of the surface of the shell; the state of stress is shown in Fig. 6.6(ii). Since p is small compared with σ, the maximum shearing stress occurs on planes at 45° to the tangent plane at any point.

If the shell remains elastic, the circumference of the sphere in any diametral plane is strained an amount

$$\epsilon = \frac{1}{E}(\sigma - \nu\sigma) = (1 - \nu)\frac{\sigma}{E} \qquad (6.7)$$

The volumetric strain of the enclosed volume of the sphere is therefore

$$3\epsilon = 3(1 - \nu)\frac{\sigma}{E} = 3(1 - \nu)\frac{pr}{2Et} \tag{6.8}$$

6.3 Cylindrical shell with hemispherical ends

Some pressure vessels are fabricated with hemispherical ends; this has the advantage of reducing the bending stresses in the cylinder when the ends are flat. Suppose the thicknesses t_1 and t_2 of the cylindrical section and the hemispherical end, respectively (Fig. 6.7), are proportioned so that the radial expansion is the same for both cylinder and hemisphere; in this way we eliminate bending stresses at the junction of the two parts.

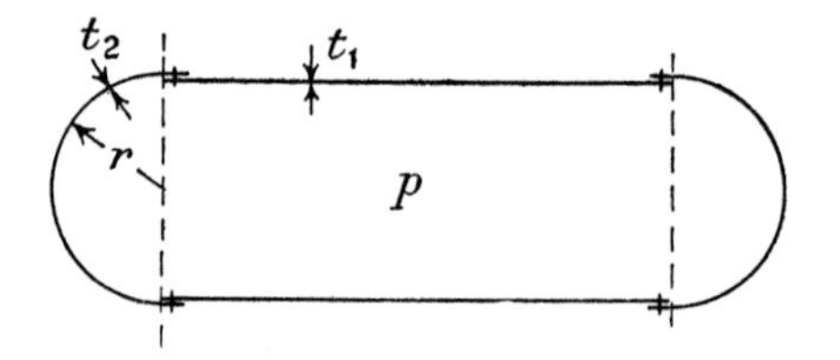

Fig. 6.7 Cylindrical shell with hemispherical ends so designed as to minimize the effects of bending stresses.

From equations (6.4), the circumferential strain in the cylinder is

$$\frac{pr}{Et_1}(1 - \tfrac{1}{2}\nu)$$

and from equation (6.7) the circumferential strain in the hemisphere is

$$(1 - \nu)\frac{pr}{2Et_2}$$

If these strains are equal, then

$$\frac{pr}{Et_1}(1 - \tfrac{1}{2}\nu) = \frac{pr}{2Et_2}(1 - \nu)$$

This gives

$$\frac{t_1}{t_2} = \frac{2 - \nu}{1 - \nu} \tag{6.9}$$

For most metals ν is approximately 0·3, so an average value of (t_1/t_2) is $1{\cdot}7/0{\cdot}7 \doteqdot 2{\cdot}4$. The hemispherical end is therefore thinner than the cylindrical section.

FURTHER PROBLEMS
(answers on page 397)

6.8 A pipe has an internal diameter of 10 cm and is 0·5 cm thick. What is the maximum allowable internal pressure if the maximum shearing stress does not exceed 55 MN/m^2? Assume a uniform distribution of stress over the cross-section. *(Cambridge)*

6.9 A long boiler tube has to withstand an internal test pressure of 4 MN/m^2, when the mean circumferential stress must not exceed 120 MN/m^2. The internal diameter of the tube is 5 cm and the density is 7840 kg/m^3. Find the mass of the tube per metre run. *(RNEC)*

6.10 A long, steel tube, 7·5 cm internal diameter and 0·15 cm thick, is plugged at the ends and subjected to internal fluid pressure such that the maximum direct stress in the tube is 120 MN/m^2. Assuming $\nu = 0{\cdot}3$ and $E = 200$ GN/m^2, find the percentage increase in the capacity of the tube. *(RNC)*

6.11 A copper pipe 15 cm internal diameter and 0·3 cm thick is closely wound with a single layer of steel wire of diameter 0·18 cm, the initial tension of the wire being 10 N. If the pipe is subjected to an internal pressure of 3 MN/m^2 find the stress in the copper and in the wire (*a*) when the temperature is the same as when the tube was wound, (*b*) when the temperature throughout is raised 200° C. E for steel = 200 GN/m^2, E for copper = 100 GN/m^2, coefficient of linear expansion for steel = 11×10^{-6}, for copper 18×10^{-6} per 1° C. *(Cambridge)*

6.12 A thin spherical copper shell of internal diameter 30 cm and thickness 0·16 cm is just full of water at atmospheric pressure. Find how much the internal pressure will be increased if 25 cc of water are pumped in. Take $\nu = 0{\cdot}3$ for copper and $K = 2$ GN/m^2 for water. *(Cambridge)*

6.13 A spherical shell of 60 cm diameter is made of steel 0·6 cm thick. It is closed when just full of water at 15° C, and the temperature is then raised to 35° C. For this range of temperature, water at atmospheric pressure increases 0·0059 per unit volume. Find the stress induced in the steel. The bulk modulus of water is 2 GN/m^2, E for steel is 200 GN/m^2, and the coefficient of linear expansion of steel is 12×10^{-6} per 1° C, and Poisson's ratio = 0·3. *(Cambridge)*

7 Bending moments and shearing forces

7.1 Introduction

In Chapter 1 we discussed the stresses set up in a bar due to axial forces of tension and compression. When a bar carries lateral forces, two important types of loading action are set up at any section: these are a bending moment and a shearing force.

Consider firstly the simple case of a beam which is fixed rigidly at one end B and is quite free at its remote end D, Fig. 7.1; such a beam is called a *cantilever*, a familiar

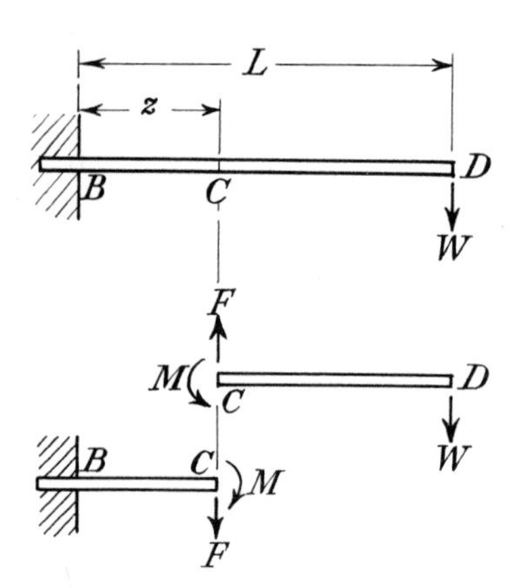

Fig. 7.1 Bending moment and shearing force in a simple cantilever beam.

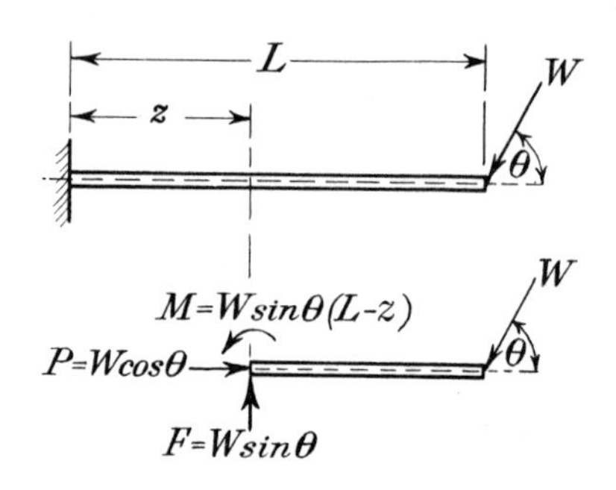

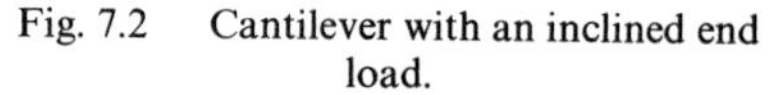

Fig. 7.2 Cantilever with an inclined end load.

example of which is a fishing rod held at one end. Imagine that the cantilever is horizontal, with one end B embedded in a wall, and that a lateral force W is applied at the remote end D. Suppose the cantilever is divided into two lengths by an imaginary section C; the lengths BC and CD must individually be in a state of statical equilibrium. If we neglect the mass of the cantilever itself, the loading actions over the section C of CD balance the actions of the force W at C. The length CD of the cantilever is in equilibrium if we apply an upwards vertical force F and an anti-clockwise couple M at C; F is equal in magnitude to W, and M is equal to $W(L - z)$, where z is measured from B. The force F at C is called a *shearing force*, and the couple M a *bending moment*.

But at the imaginary section C of the cantilever, the actions F and M on CD are provided by the length BC of the cantilever. In fact, equal and opposite actions F and M

are applied by CD to BC. For the length BC, the actions at C are a downwards shearing force F, and a clockwise couple M.

When the cantilever carries external loads which are not applied normally to the axis of the beam, Fig. 7.2, axial forces are set up in the beam. If W is inclined at an angle θ to the axis of the beam, Fig. 7.2, the axial thrust in the beam at any section is

$$P = W \cos \theta \tag{7.1}$$

The bending moment and shearing force at a section a distance z from the built-in end are

$$M = W(L - z) \sin \theta \qquad F = W \sin \theta \tag{7.2}$$

7.2 Concentrated and distributed loads

A concentrated load on a beam is one which can be regarded as acting wholly at one point of the beam. For the purposes of calculation such a load is localized at a point of the beam; in reality this would imply an infinitely large bearing pressure on the beam at the point of application of a concentrated load. All loads must be distributed in practice over perhaps only a small length of beam, thereby giving a finite bearing pressure. Concentrated loads arise frequently on a beam where the beam is connected to other transverse beams.

In practice there are many examples of distributed loads: they arise when a wall is built on a girder; they occur also in many problems of fluid pressure, such as wind pressure on a tall building, and aerodynamic forces on an aircraft wing.

7.3 Relation between the intensity of loading, the shearing force, and bending moment in a straight beam

Consider a straight beam under any system of lateral loads and external couples, Fig. 7.3; an elemental length δz of the beam at a distance z from one end is acted upon by an external lateral load, and internal bending moments and shearing forces. Suppose the external lateral loads are distributed so that the intensity of loading on the elemental

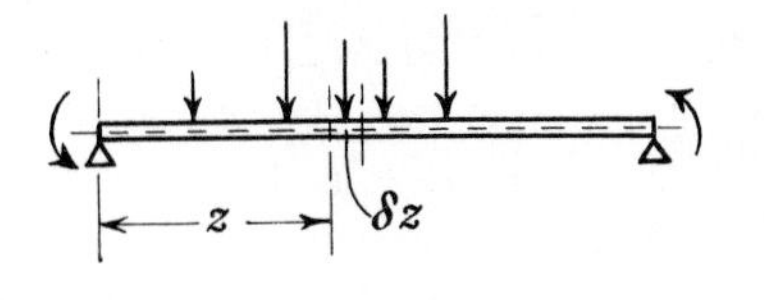

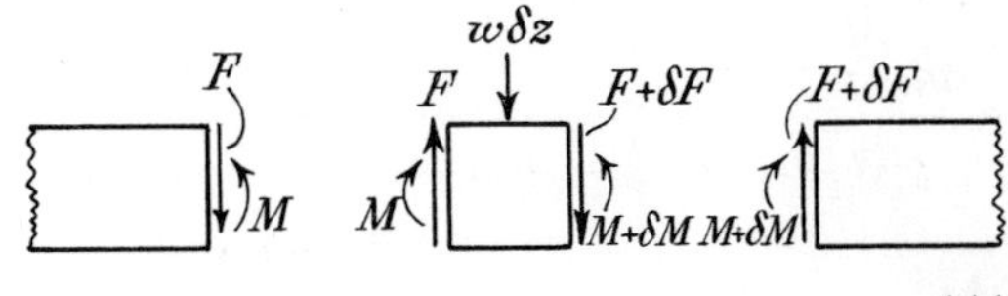

Fig. 7.3 Shearing and bending actions on an elemental length of a straight beam.

length δz is w. Then the external vertical force on the element is $w\,\delta z$, Fig. 7.3; this is reacted by an internal bending moment M and shearing force F on one face of the element, and $M + \delta M$ and $F + \delta F$ on the other face of the element. For vertical equilibrium of the element we have

$$(F + \delta F) - F + w\,\delta z = 0$$

If δz is infinitesimally small,

$$\frac{dF}{dz} = -w \tag{7.3}$$

Suppose this relation is integrated between the limits z_1 and z_2, then

$$\int_{z=z_1}^{z=z_2} dF = -\int_{z_1}^{z_2} w\,dz$$

If F_1 and F_2 are the shearing forces at $z = z_1$ and $z = z_2$ respectively, then

$$(F_2 - F_1) = -\int_{z_1}^{z_2} w\,dz$$

or

$$F_1 - F_2 = \int_{z_1}^{z_2} w\,dz \tag{7.4}$$

Then, the decrease of shearing force from z_1 to z_2 is equal to the area below the load distribution curve over this length of the beam, or the difference between F_1 and F_2 is the net lateral load over this length of the beam.

Furthermore, for rotational equilibrium of the elemental length δz,

$$(F + \delta F)\,\delta z - (M + \delta M) + M - w\,\delta z(\tfrac{1}{2}\delta z) = 0$$

Then, to the first order of small quantities,

$$F\,\delta z - \delta M = 0$$

Then, in the limit as δz approaches zero,

$$\frac{dM}{dz} = F \tag{7.5}$$

On integrating between the limits $z = z_1$ and $z = z_2$, we have

$$\int_{z=z_1}^{z=z_2} dM = \int_{z_1}^{z_2} F\,dz$$

Thus

$$M_2 - M_1 = \int_{z_1}^{z_2} F\,dz \tag{7.6}$$

where M_1 and M_2 are the values of M at $z = z_1$ and $z = z_2$, respectively. Then the increase of bending moment from z_1 to z_2 is the area below the shearing force curve for that length of the beam.

Equations (7.4) and (7.6) are extremely useful for finding the bending moments and shearing forces in beams with irregularly distributed loads. From equation (7.4) the shearing force F at a section distance z from one end of the beam is

$$F = F_1 - \int_{z_1}^{z} w \, dz \tag{7.7}$$

On substituting this value of F into equation (7.6),

$$M_2 - M_1 = \int_{z_1}^{z_2} \left\{ F_1 - \int_{z_1}^{z} w \, dz \right\} dz$$

Thus

$$M_2 = M_1 + F_1(z_2 - z_1) - \int_{z_1}^{z_2} \left\{ \int_{z_1}^{z} w \, dz \right\} dz \tag{7.8}$$

From equation (7.5) we have that the bending moment M has a stationary value when the shearing force F is zero. Equations (7.3) and (7.5) give

$$\frac{d^2M}{dz^2} = \frac{dF}{dz} = -w \tag{7.9}$$

For the directions of M, F and w considered in Fig. 7.3, M is *mathematically* a maximum, since d^2M/dz^2 is negative; the significance of the word mathematically will be made clearer in §7.8.

All the relations developed in this section are merely statements of statical equilibrium, and are therefore true independently of the state of the material of the beam.

7.4 Sign conventions for bending moments and shearing forces

The bending moments on the elemental length δz of Fig. 7.3 tend to make the beam concave on its upper surface and convex on its lower surface; such bending moments are sometimes called *sagging* bending moments, while moments in the opposite directions are called *hogging* bending moments. The shearing forces on the elemental length tend to rotate the element in a *clockwise* sense. In deriving the equations in this section it is assumed implicitly, therefore, that

(i) *downwards* vertical loads are positive,
(ii) *sagging* bending moments are positive, and
(iii) *clockwise* shearing forces are positive.

These sign conventions are shown in Fig. 7.4. Any other system of sign conventions can be used, provided the signs of the loads, bending moments and shearing forces are considered when equations (7.3) and (7.5) are applied to any particular problem.

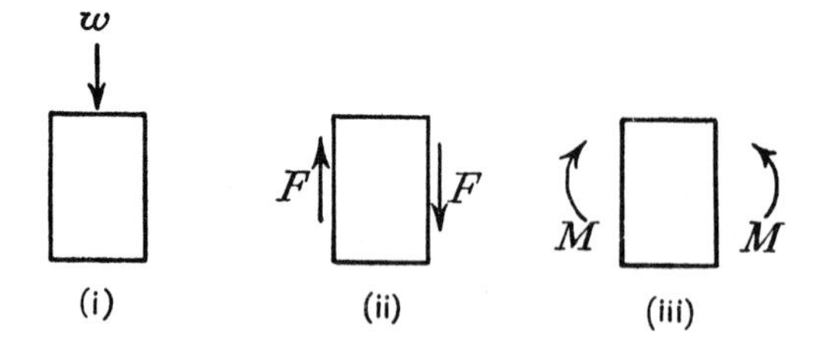

Fig. 7.4 Positive values of w, F and M. (i) Downwards vertical loading. (ii) Clockwise shearing forces. (iii) Sagging bending moment.

Figures which show graphically the variations of bending moment and shearing force along the length of a beam are called *bending moment diagrams* and *shearing force diagrams*. Sagging bending moments are considered positive, and clockwise shearing forces taken as positive. The two quantities are plotted above the centre line of the beam when positive, and below when negative. Before we can calculate the stresses and deformations of beams, we must be able to find the bending moment and shearing force at any section.

7.5 Cantilevers

A cantilever is a beam supported at one end only; for example, the beam already discussed in §7.1, and shown in Fig. 7.1, is held rigidly at B. Consider first the cantilever shown in Fig. 7.5, which carries a concentrated lateral load W at the free end. The bending moment at a section a distance z from B is

$$M = -W(L - z)$$

the negative sign occurring since the moment is hogging. The variation of bending moment is linear. The shearing force at any section is

$$F = +W$$

the shearing force being positive since it is clockwise. The shearing force is constant

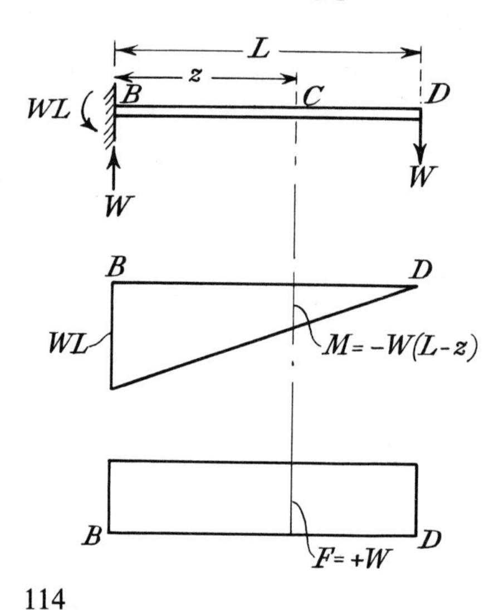

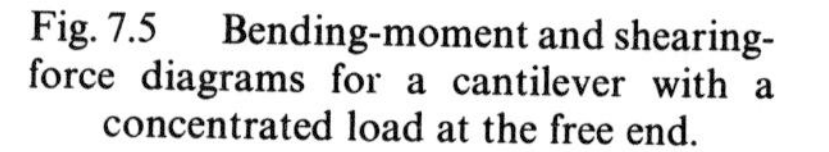
Fig. 7.5 Bending-moment and shearing-force diagrams for a cantilever with a concentrated load at the free end.

throughout the length of the cantilever. We note that

$$\frac{dM}{dz} = W = F$$

Further $dF/dz = 0$, since there are no lateral loads between B and D.

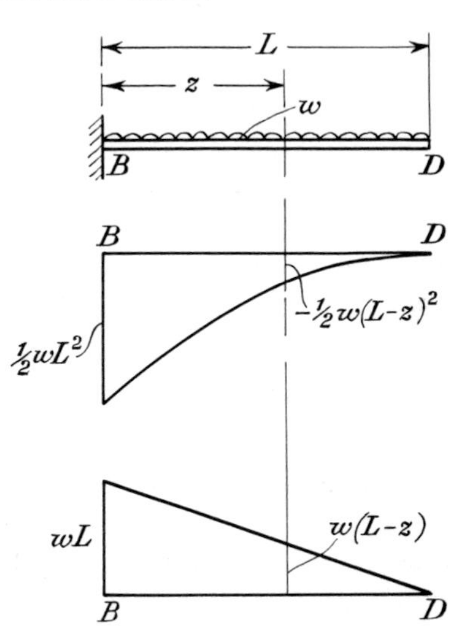

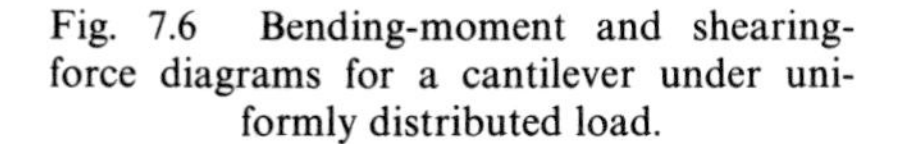
Fig. 7.6 Bending-moment and shearing-force diagrams for a cantilever under uniformly distributed load.

Now consider a cantilever carrying a uniformly-distributed downwards vertical load of intensity w, Fig. 7.6. The shearing force at a distance z from B is

$$F = +w(L - z)$$

The bending moment at a distance z from B is

$$M = -\tfrac{1}{2}w(L - z)^2$$

The shearing force varies linearly and the bending moment parabolically along the length of the beam. We see that

$$\frac{dM}{dz} = w(L - z) = +F$$

Problem 7.1: A cantilever 5 m long carries a uniformly distributed vertical load of 480 N per metre from C to H, and a concentrated vertical load of 1000 N at its mid-length, D. Construct the shearing force and bending moment diagrams.

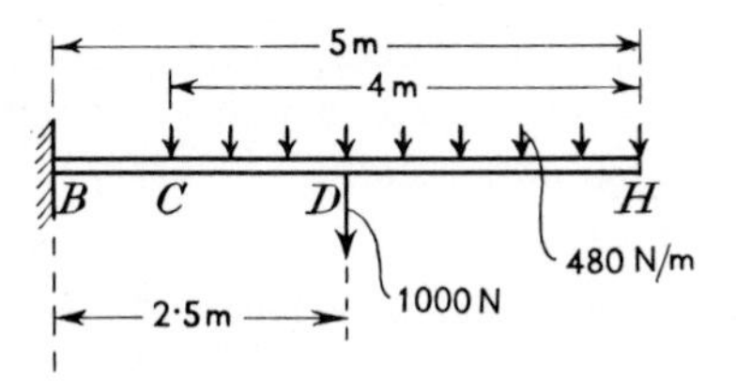

Solution

The shearing force due to the distributed load increases uniformly from zero at H to $+1920$ N at C, and remains constant at $+1920$ N from C to B; this is shown by the lines (i). Due to the concentrated load at D, the shearing force is zero from H to D, and equal to $+1000$ N from D to B, as shown by lines (ii). Adding the two together we get the total shearing force shown by lines (iii).

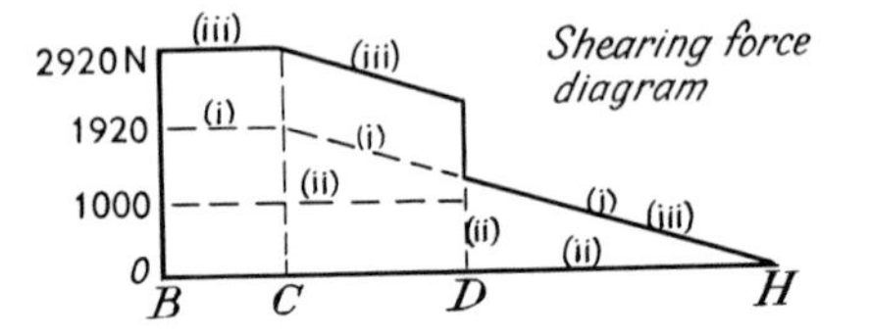

The bending moment due to the distributed load increases parabolically from zero at H to

$$-\tfrac{1}{2}(480)(4)^2 = -3840 \text{ Nm}$$

at C. The total load on CH is 1920 N with its centre of gravity 3 m from B; thus the bending moment at B due to this load is

$$-(1920)(3) = -5760 \text{ Nm}$$

From C to B the bending moment increases uniformly, giving lines (i). The bending moment due to the concentrated load increases uniformly from zero at D to

$$-(1000)(2{\cdot}5) = -2500 \text{ Nm}$$

at B, as shown by lines (ii). Combining (i) and (ii), the total bending moment is given by (iii).

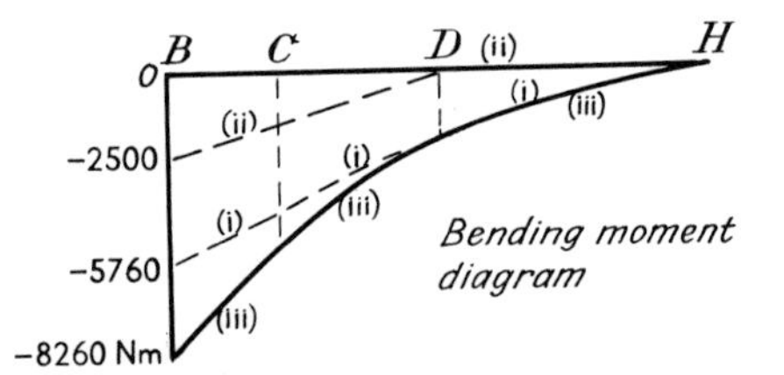

7.6 Cantilever with non-uniformly distributed load

Where a cantilever carries a distributed lateral load of variable intensity, we can find the bending moments and shearing forces from equations (7.4) and (7.6). When the loading intensity w cannot be expressed as a simple analytic function of z, equations (7.4) and (7.6) can be integrated numerically.

Problem 7.2: A cantilever of length 10 m carries a distributed lateral load of varying intensity w N per metre length. Construct curves of shearing force and bending moment in the cantilever.

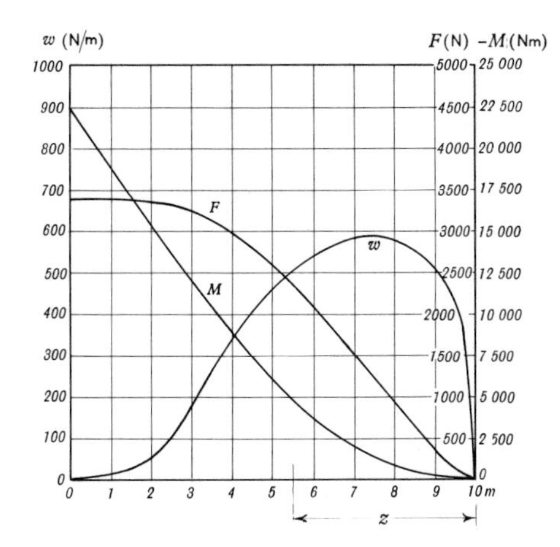

Solution

If z is the distance from the free end of the cantilever, the shearing force at a distance z from the free end is

$$F = \int_0^z w \, dz$$

We find first the shearing force F by numerical integration of the w-curve. The greatest shearing force occurs at the built-in end, and has the value

$$F_{max} \doteqdot 3400 \text{ N}$$

The bending moment at a section a distance z from the free end is

$$M = -\int_0^z F \, dz$$

and is found therefore by numerical integration of the F-curve. The greatest bending moment occurs at the built-in end, and has the value

$$M_{max} \doteqdot 22\,500 \text{ Nm}$$

7.7 Simply-supported beams

By *simply-supported* we mean that the supports are of such a nature that they do not apply any resistance to bending of a beam; for instance, knife-edges or frictionless pins perpendicular to the plane of bending cannot transmit couples to a beam. The remarks concerning bending moments and shearing forces, which were made in §7.5 in relation to cantilevers, apply equally to beams simply-supported at each end, or with any conditions of end support.

As an example, consider the beam shown in Fig. 7.7, which is simply-supported at B and C, and carries a vertical load W a distance a from B. If the ends are simply-supported no bending moments are applied to the beam at B and C. By taking moments about B and C we find that the reactions at these supports are

$$\frac{W}{L}(L - a) \quad \text{and} \quad \frac{Wa}{L}$$

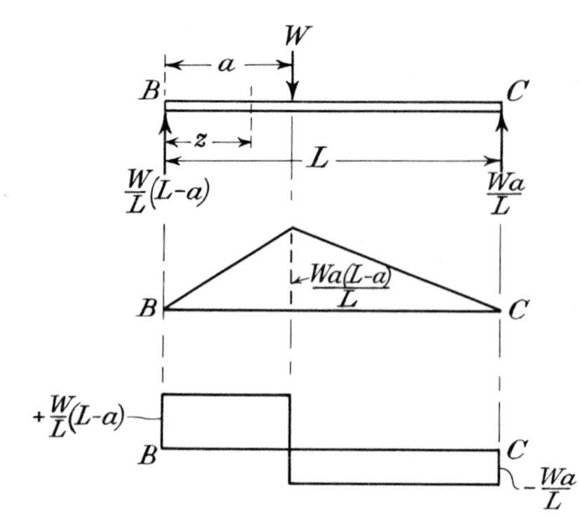

Fig. 7.7 Bending-moment and shearing-force diagrams for a simply-supported beam with a single concentrated lateral load.

respectively. Now consider a section of the beam a distance z from B; if $z < a$, the bending moment and shearing force are

$$M = +\frac{Wz}{L}(L - a), \qquad F = +\frac{W}{L}(L - a)$$

If $z > a$,

$$M = +\frac{Wz}{L}(L - a) - W(z - a) = +\frac{Wa}{L}(L - z)$$

$$F = -\frac{Wa}{L}$$

The bending moment and shearing force diagrams show discontinuities at $z = a$; the maximum bending moment occurs under the load W, and has the value

$$M_{\max} = \frac{Wa}{L}(L - a) \tag{7.10}$$

The simply-supported beam of Fig. 7.8 carries a uniformly-distributed load of intensity w. The vertical reactions at B and C are $\frac{1}{2}wL$. Consider a section at a distance z from B. The bending moment at this section is

$$\begin{aligned} M &= \tfrac{1}{2}wLz - \tfrac{1}{2}wz^2 \\ &= \tfrac{1}{2}wz(L - z) \end{aligned}$$

and the shearing force is

$$\begin{aligned} F &= +\tfrac{1}{2}wL - wz \\ &= w(\tfrac{1}{2}L - z) \end{aligned}$$

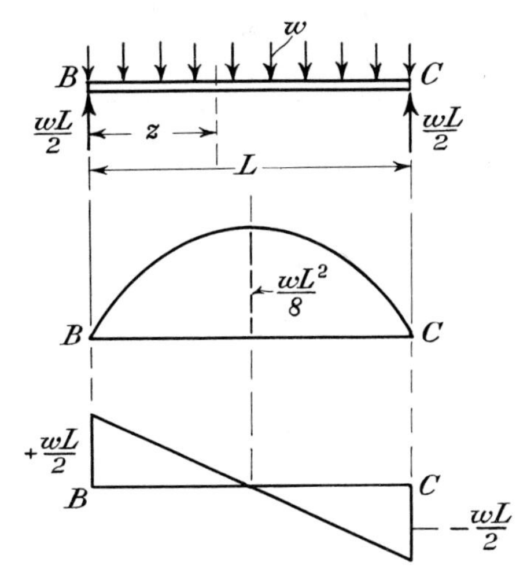

Fig. 7.8 Bending-moment and shearing-force diagrams for a simply-supported beam with a uniformly distributed lateral load.

The bending moment is a maximum at $z = \frac{1}{2}L$, where

$$M_{max} = \frac{wL^2}{8} \tag{7.11}$$

At $z = \frac{1}{2}L$, we note that

$$\frac{dM}{dz} = +F = 0$$

Problem 7.3: A simply-supported beam carries concentrated lateral loads at C and D, and a uniformly distributed lateral load over the length DF. Construct the bending moment and shearing force diagrams.

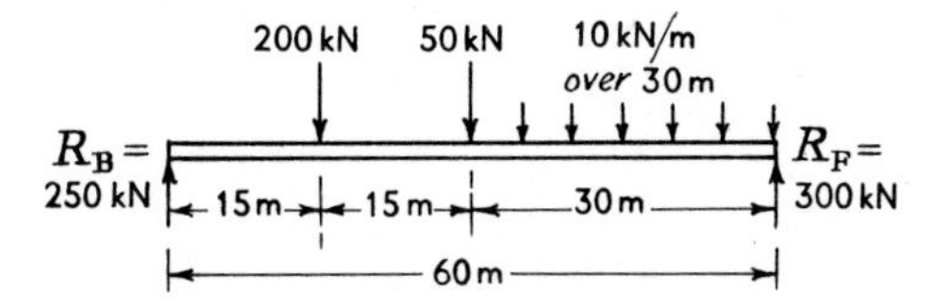

Solution

First we calculate the vertical reactions at B and F. On taking moments about F,

$$60R_B = (200 \times 10^3)(45) + (50 \times 10^3)(30) + (300 \times 10^3)(15) = 15\,000 \times 10^3$$

Then

$$R_B = 250 \text{ kN}$$

and

$$R_F = (200 \times 10^3) + (50 \times 10^3) + (300 \times 10^3) - R_B = 300 \text{ kN}$$

The bending moment varies linearly between B and C, and between C and D, and parabolically from D to F. The maximum bending moment is 4·5 MNm, and occurs at D. The maximum shearing force is 300 kN, and occurs at F.

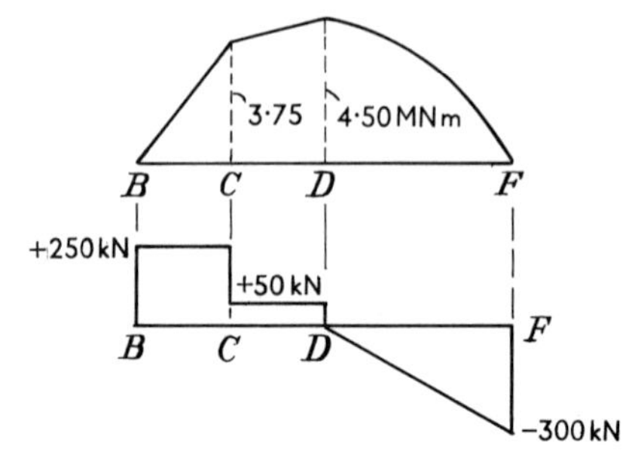

Problem 7.4: A beam rests on knife-edges at each end, and carries a clockwise moment M_B at B, and an anti-clockwise moment M_C at C. Construct bending moment and shearing force diagrams for the beam.

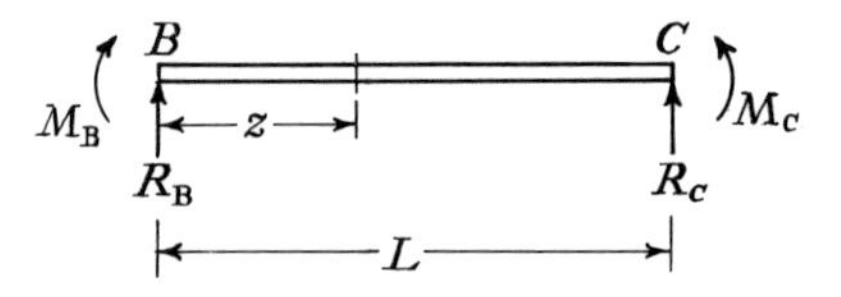

Solution

Suppose R_B and R_C are the vertical reactions at B and C; then for statical equilibrium of the beam

$$R_B = -R_C = \frac{1}{L}(M_C - M_B)$$

The shearing force at all sections is then

$$F = R_B = \frac{1}{L}(M_C - M_B)$$

The bending moment a distance z from B is

$$M = M_B + R_B z = \frac{M_B}{L}(L - z) + \frac{M_C z}{L}$$

so M varies linearly between B and C.

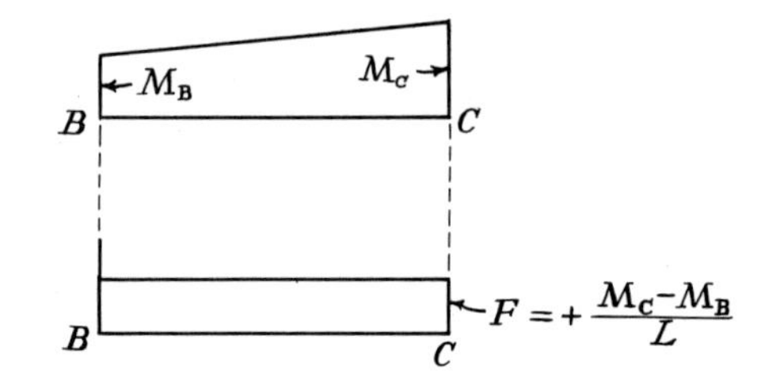

Problem 7.5: A simply-supported beam carries a couple M_0 applied at a point distant a from B. Construct bending moment and shearing force diagrams for the beam.

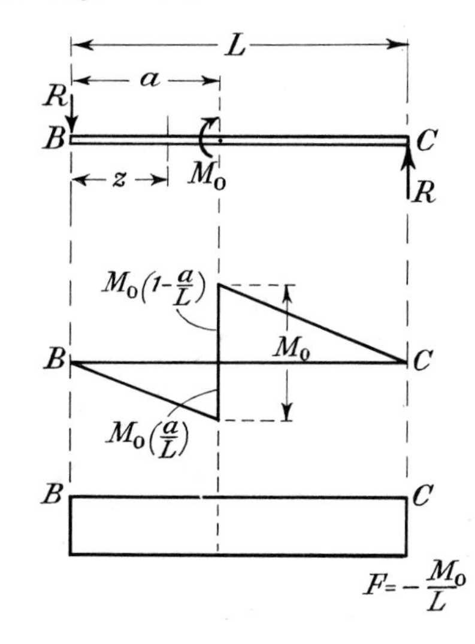

Solution

The vertical reactions R at B and C are equal and opposite. For statical equilibrium of BC,

$$M_0 = RL, \quad \text{or} \quad R = \frac{M_0}{L}$$

The shearing force at all sections is

$$F = -R = -\frac{M_0}{L}$$

The bending moment at $z < a$ is

$$M = -Rz = -\frac{M_0 z}{L}$$

and for $z > a$,

$$M = -Rz + M_0 = M_0\left(1 - \frac{z}{L}\right)$$

7.8 Simply-supported beam carrying a uniformly distributed load and end couples

Consider a simply-supported beam BC, carrying a uniformly distributed load w per unit length, and couples M_B and M_C applied to the ends, Fig. 7.9(i). The reactions R_B and R_C can be found directly by taking moments about B and C in turn; we have

$$R_B = \frac{wL}{2} - \frac{1}{L}(M_B - M_C)$$

$$R_C = \frac{wL}{2} + \frac{1}{L}(M_B - M_C) \qquad (7.12)$$

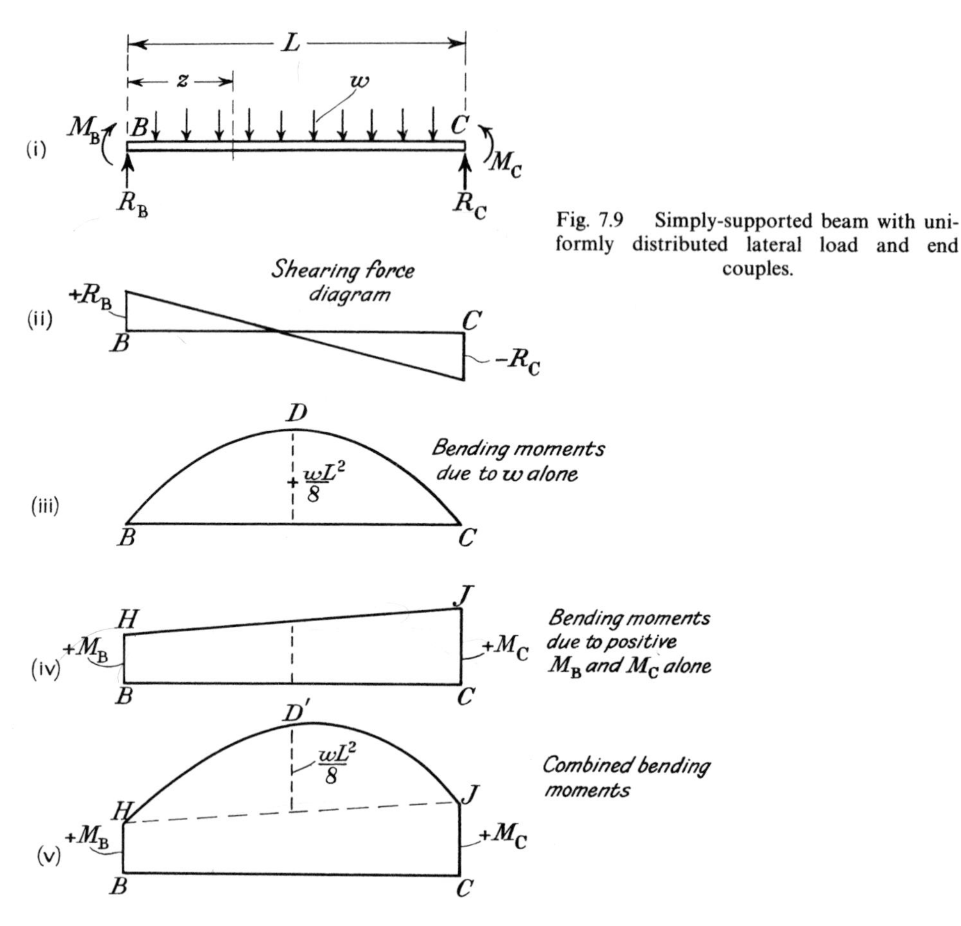

Fig. 7.9 Simply-supported beam with uniformly distributed lateral load and end couples.

These give the shearing forces at the ends of the beam, and the shearing force at any point of the beam can be deduced, Fig. 7.9(ii).

In discussing bending moments we consider the total loading actions on the beam as the superposition of a uniformly distributed load and end couples; the distributed load gives rise to a parabolic bending moment curve, BDC in Fig. 7.9(iii), while the end couples M_B and M_C give the straight line HJ, Fig. 7.9(iv). The combined effects of the lateral load and the end couples, give the curve $BHD'JC$, Fig. 7.9(v). The bending moment at a distance z from B is

$$M = \tfrac{1}{2}wz(L - z) + \frac{M_B}{L}(L - z) + \frac{M_C z}{L} \tag{7.13}$$

The 'maximum' bending moment occurs when

$$\frac{dM}{dz} = \tfrac{1}{2}w(L - 2z) - \frac{M_B}{L} + \frac{M_C}{L} = 0$$

that is, when

$$z = \tfrac{1}{2}L - \frac{1}{wL}(M_B - M_C)$$

The value of M for this value of z is

$$M_{\max} = \tfrac{1}{8}wL^2 + \tfrac{1}{2}(M_B + M_C) + \frac{1}{2wL^2}(M_B - M_C)^2 \tag{7.14}$$

This, however, is only a mathematical 'maximum'; if M_B or M_C is negative, the numerically greatest bending moment may occur at B or C. Care should therefore be taken to find the truly greatest bending moment in the beam.

7.9 Points of inflection

When either, or both, of the end couples in Fig. 7.9 is reversed in direction, there is at least one section of the beam where the bending moment is zero. In Fig. 7.10 the

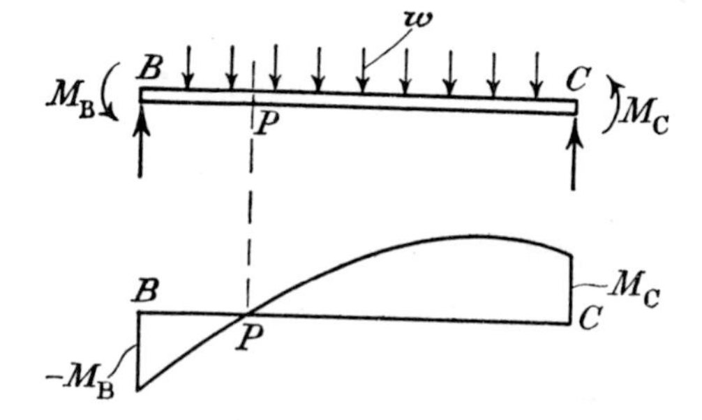

Fig. 7.10 Single point of inflection in a beam.

end couple M_B is applied in an anti-clockwise direction; the bending moment at a distance z from B is

$$M = \tfrac{1}{2}wz(L - z) - \frac{M_B}{L}(L - z) + \frac{M_C z}{L} \tag{7.15}$$

and this is zero when

$$z^2 - zL\left(1 + \frac{2}{wL^2}[M_B + M_C]\right) + \frac{2M_B}{w} = 0 \tag{7.16}$$

The distance PB is the relevant root of this quadratic equation.

When the end couple M_C is also reversed in direction, Fig. 7.11, there are two points, P and Q, in the beam at which the bending moment is zero. The distances of P and Q

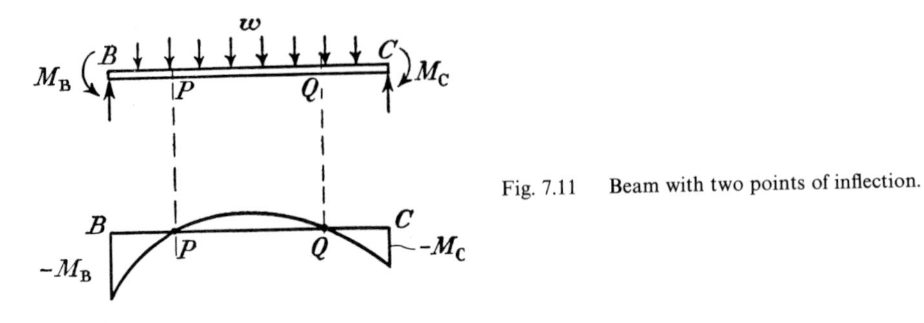

Fig. 7.11 Beam with two points of inflection.

from B are given by the roots of the equation

$$z^2 - zL\left[1 + \frac{2}{wL^2}(M_B - M_C)\right] + \frac{2M_B}{w} = 0 \tag{7.17}$$

The distance PQ is

$$2\sqrt{\frac{L^2}{4} - \left(\frac{M_B + M_C}{w}\right) + \left(\frac{M_B - M_C}{wL}\right)^2} \tag{7.18}$$

The points P and Q are called *points of inflexion*, or *points of contraflexure;* as we shall see later the curvature of the deformed beam changes sign at these points.

7.10 Simply-supported beam with a uniformly distributed load over part of a span

The beam $BCDF$, shown in Fig. 7.12, carries a uniformly distributed vertical load w per unit length over the portion CD. On taking moments about B and F,

$$V_B = \frac{bw}{2L}(b + 2c), \qquad V_F = \frac{bw}{2L}(b + 2a) \tag{7.19}$$

The bending moments at C and D are

$$\begin{aligned} M_C &= aV_B = \frac{baw}{2L}(b + 2c) \\ M_D &= cV_F = \frac{bcw}{2L}(b + 2a) \end{aligned} \tag{7.20}$$

The bending moments in BC and FD vary linearly. The bending moment in CD, at a distance z from C, is

$$M = \left(1 - \frac{z}{b}\right)M_C + \frac{z}{b}M_D + \tfrac{1}{2}wz(b - z) \tag{7.21}$$

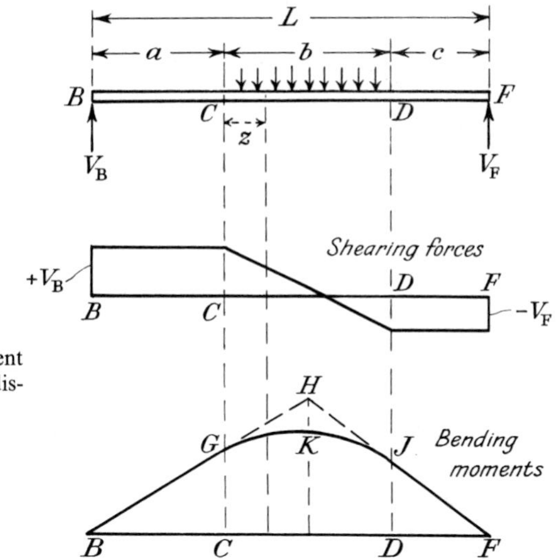

Fig. 7.12 Shearing-force and bending-moment diagrams for simply-supported beam with distributed load over part of the span.

Then

$$\frac{dM}{dz} = \frac{1}{b}(M_D - M_C) + \tfrac{1}{2}w(b - 2z)$$

On substituting for M_C and M_D from equations (7.20)

$$\frac{dM}{dz} = \frac{bw}{2L}(c - a) + \tfrac{1}{2}w(b - 2z)$$

At C, $z = 0$, and

$$\frac{dM}{dz} = \frac{bw}{2L}(b + 2c) = V_B$$

But V_B is the slope of the line BG in the bending moment diagram, so the curve of equation (7.21) is tangential to BG at G. Similarly, the curve of equation (7.21) is tangential to FJ at J. Between C and D the bending moment varies parabolically; the simplest method of constructing the bending moment diagram for CD is to produce BG and FJ to meet at H, and then to draw a parabola between G and J, having tangents BG and FJ.

7.11 Simply-supported beam with non-uniformly distributed load

Suppose a simply-supported beam of span L, Fig. 7.13, carries a lateral distributed load of variable intensity w. Then, from equation (7.4), if F is the shearing force a distance z from B,

$$F_0 - F = \int_0^z w\, dz$$

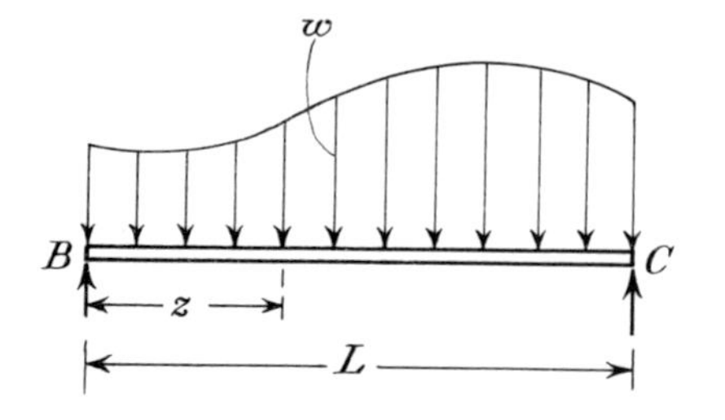

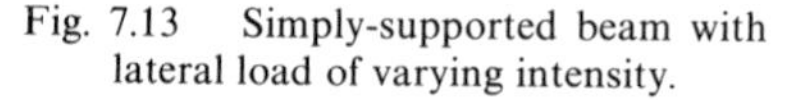
Fig. 7.13 Simply-supported beam with lateral load of varying intensity.

where F_0 is the shearing force at $z = 0$. Then

$$F = F_0 - \int_0^z w\,dz \tag{7.22}$$

Furthermore, from equation (7.6), the bending moment a distance z from B is

$$M = M_0 + F_0 z - \int_0^z \int_0^z w\,dz\,dz \tag{7.23}$$

where M_0 is the bending moment at $z = 0$. But since the beam is simply-supported at $z = 0$, we have $M_0 = 0$, and so

$$M = F_0 z - \int_0^z \int_0^z w\,dz\,dz$$

The end $z = L$ is also simply-supported, so for this end $M = 0$; then

$$F_0 L - \int_0^L \int_0^z w\,dz\,dz = 0$$

This gives

$$F_0 = \frac{1}{L} \int_0^L \int_0^z w\,dz\,dz \tag{7.24}$$

Equations (7.22), (7.23) and (7.24) may be used in the graphical solution of problems in which w is not an analytic function of z. The value of F_0 is found firstly from equation (7.24); numerical integrations then give the values of F and M, from equations (7.22) and (7.23), respectively.

7.12 A graphical method of drawing bending moment diagrams

The following graphical method is sometimes useful for dealing with a series of concentrated loads or a uniformly distributed load. In Fig. 7.14 five loads are shown acting at $C_1, C_2, \ldots$. First, draw a force diagram *abcdef*, beginning at the top with the force W_1. Then, taking a pole O on the right of the line *abcdef*, draw the funicular polygon $P_B P_1 P_2 P_3 P_4 P_5 P_D P_B$, and so determine the reactions V_D and V_B in the usual

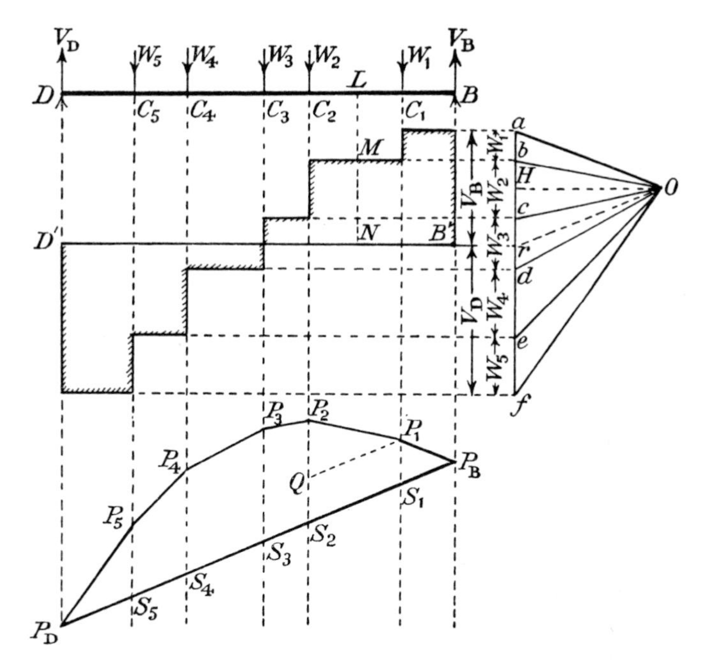

Fig. 7.14 Graphical method of determining the shape of the bending-moment diagram.

way: these are given by *fr* and *ra* in the force diagram. Then, with $P_D P_B$ as base the bending moment at any section is given by the height of the funicular polygon. The proof of this statement is as follows:

In the force diagram draw OH horizontal. The triangle $P_B P_1 S_1$ is similar to the triangle Oar since the sides of the two triangles are parallel. Therefore,

$$\frac{P_1 S_1}{P_B S_1} = \frac{ar}{rO} = \frac{V_B}{rO}$$

and thus

$$P_1 S_1 = \frac{P_B S_1 \times V_B}{rO} = \frac{BC_1 \times V_B}{OH} = \frac{M_1}{OH}$$

where M_1 is the bending moment at C_1.

Again, the triangle $P_1 P_2 Q$, where $P_1 Q$ is drawn parallel to $P_B P_D$, is similar to the triangle Obr. Therefore

$$\frac{P_2 Q}{QP_1} = \frac{br}{rO} = \frac{V_B - W_1}{rO}$$

and so

$$P_2 Q = (V_B - W_1)\frac{QP_1}{rO} = (V_B - W_1)\frac{C_1 C_2}{OH}$$

Then

$$P_2S_2 = P_2Q + P_1S_1$$
$$= \frac{(V_B - W_1)C_1C_2 + V_B(BC_1)}{OH}$$
$$= \frac{V_B(BC_2) - W_1(C_2C_1)}{OH} = \frac{M_2}{OH}$$

where M_2 is the bending moment at C_2. Thus

$$M_1 = OH \times P_1S_1$$
$$M_2 = OH \times P_2S_2$$

and so on for the bending moments at C_3, C_4, C_5. Also, we know that between these points the bending moment diagram consists of straight lines, so the funicular polygon gives the bending moment at any section.

If the beam is drawn to a scale of 1 m to s m, and the force diagram $abc \ldots f$ to a scale 1 m = w N, the scale of the funicular polygon or bending moments will be $1 = w \,.\, OH$ N $\times$ s m $= ws \,.\, OH$ Nm, where OH is measured in metres. Hence, if we wish the resulting bending moment diagram to be to a scale 1 m = m Nm, we must make

$$OH = \frac{m}{ws} \text{ metres}$$

If desired the bending moment diagram may be redrawn on a horizontal base. In the case of a beam built-in at D and free at B (i.e. a cantilever), O should be taken on the same horizontal level as a. The shearing force and bending moment diagrams will terminate on the vertical through C_1, and the horizontal line through P_1 will be the base of the bending moment diagram.

The shearing force diagram is drawn directly by projecting across horizontally from the line of forces, as shown in Fig. 7.14. Thus

$$MN = ra - ab - bc = V_B - W_1 - W_2$$
$$= \text{the shearing force at } L$$

To apply the method to distributed loads, the curve of loading must be divided into a number of vertical strips, and the distributed load is replaced by a series of concentrated loads given by the areas of the strips. When the funicular polygon for these loads has been drawn as above, a fair curve should be drawn touching this inside: this curve will be the bending moment diagram.

7.13 Plane curved beams

Consider a beam BCD, Fig. 7.15, which is curved in the plane of the figure. The beam is loaded so that no twisting occurs, and bending is confined to the plane of Fig. 7.15.

Suppose an imaginary cross-section of the beam is taken at C; statical equilibrium of the length CD of the beam is ensured if, in general, a force and a couple act at C; it is convenient to consider the resultant force at C as consisting of two components—an axial force P, acting along the centre line of the beam, and a lateral force F, acting along the normal to the centre line of the beam. The couple M at C acts about an axis perpendicular to the plane of bending and passing through the centre line of the beam. The actions at C on the length BC of the beam, Fig. 7.15, are equal and opposite to those at C on the length CD.

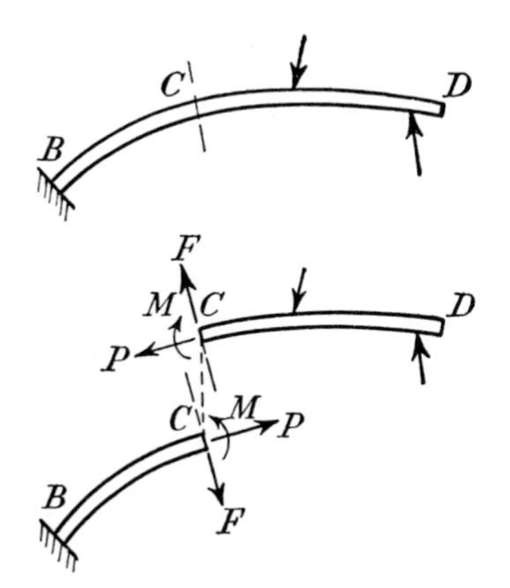

Fig. 7.15 Bending and shearing actions in a plane curved beam.

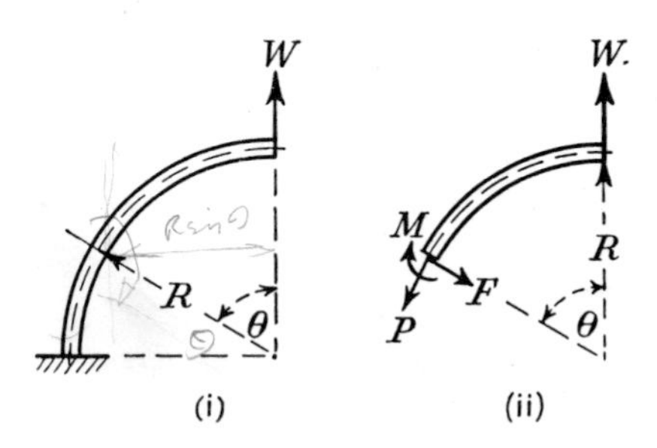

Fig. 7.16 Plane curved beam of circular form carrying an end load.

As before the couple M is the *bending moment* in the beam at C, and the lateral force F is the *shearing force.*

As an example, consider the beam of Fig. 7.16, which has a centre line of constant radius R. The beam carries a radial load W at its free end. Consider a section of the

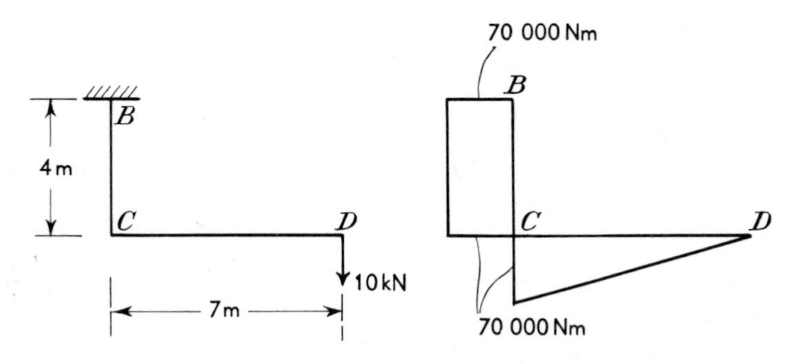

Fig. 7.17 Bending moments in a bracket.

beam at some angular position θ: for statical equilibrium of the length of the bar shown in Fig. 7.16(ii),

$$\begin{aligned} M &= WR \sin \theta \\ F &= W \cos \theta \\ P &= W \sin \theta \end{aligned} \qquad (7.25)$$

Consider again, the beam shown in Fig. 7.17, consisting of two straight limbs, BC and CD, connected at C. In CD the bending moment varies linearly, from zero at D to 70 000 Nm at C. In BC the bending moment is constant and equal to 70 000 Nm. In Fig. 7.17 the bending moments are plotted on the concave sides of the bent limbs; this is equivalent to following the sign convention of §7.4, that sagging bending moments are positive.

Problem 7.6: AB is a vertical post of a crane; the sockets at A and B offer no constraint against flexure. The horizontal arm CD is hinged to AB at C and supported by the strut FE which is freely hinged at its two extremities to AB and CD. Construct the bending moment diagrams for AB and CD. (*Cambridge*)

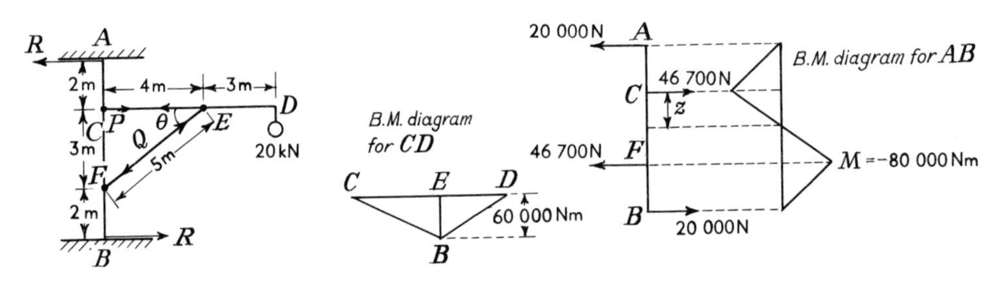

Solution

It is clear from considering the equilibrium of the whole crane that the horizontal reactions at A and B must be equal and opposite, and that the couple due to them must equal the moment of the 20 kN force. Let R be the magnitude of the horizontal reactions at A and B, then

$$7R = 7(20\,000)$$

and therefore

$$R = 20\,000 \text{ N}$$

Let P = the pull in CE, and Q = the thrust in FE. Then taking moments about C for the rod CD we have

$$4Q \sin \theta = 7(20\,000)$$

and therefore

$$Q = 58\,300 \text{ N}$$

Resolving horizontally for AB we have

$$P = Q \cos \theta = \tfrac{1}{4}(70\,000) \cot \theta = 46\,700 \text{ N}$$

The vertical reaction at $E = Q \sin \theta = 35\,000$ N.

We can now draw the bending moment diagrams for AB and CD, considering only the forces at right angles to each beam; let us take CD first. CD is a beam freely supported at C and E and loaded at D. The bending moment at $E = 3 \times 20\,000 = 60\,000$ Nm, to which value it rises uniformly from zero at D; from E to C the bending moment decreases uniformly to zero.

AB is supported at A and B and loaded with equal and opposite loads at C and F.
The bending moment at $C = (2)(20\,000) = 40\,000$ Nm.
The bending moment at $F = (2)(-20\,000) = -40\,000$ Nm.
At any point z between C and F, the bending moment is

$$M = 20\,000(z + 2) - 46\,700z = 40\,000 - 26\,700z$$

In the bending moment diagram positive bending moments are those which make the beam concave to the left, and are plotted to the left in the figure.

7.14 More general case of bending of a curved bar

In Fig. 7.18, OBC represents the centre line of a beam of any shape; the line OBC is curved in space in general. Suppose the beam carries any system of external loads; consider the actions over a section of the beam at B. For statical equilibrium of BC we

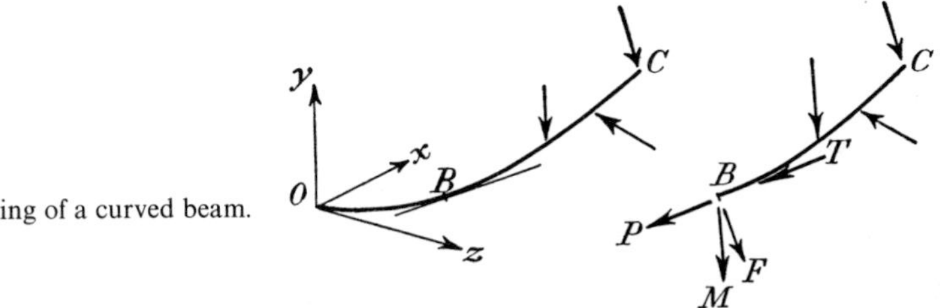

Fig. 7.18 Lateral loading of a curved beam.

require at B a force and a couple; the force is resolved into two components—an axial force P along the centre line of the beam, and a shearing force F normal to the centre line; the couple is resolved into two components—a torque T about the centre line of the beam, and a bending moment M about an axis perpendicular to the centre line. The axis of M is not necessarily coincident with the axis of F.

Problem 7.7: The centre line of a beam is curved in the plane xz with a radius a. Find the loading actions at any section of the beam when a concentrated load W is applied at C in a direction parallel to yO.

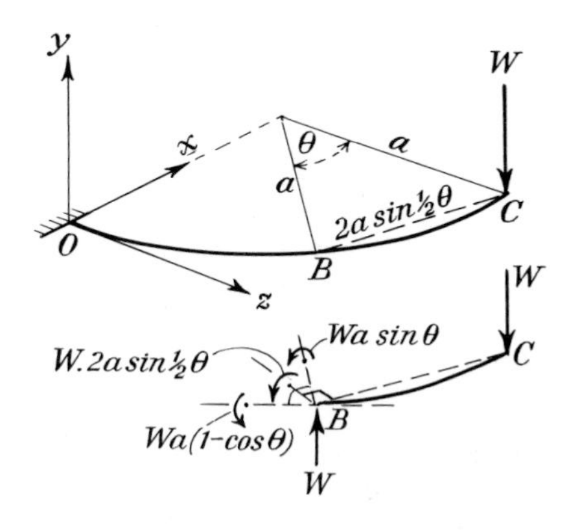

Solution

Consider any section at an angular position θ in the xz plane; there is no axial force on the centre line, and the shearing force at any section is W. The torque about the centre line is

$$W(a - a\cos\theta) = Wa(1 - \cos\theta)$$

The bending moment acts about the radius, and has the value

$$Wa \sin\theta$$

Problem 7.8: The axis of a beam consists of two lines BC and CD in a horizontal plane and at right angles to each other. Estimate the greatest bending moment and torque when the beam carries a vertical load of 10 kN at D.

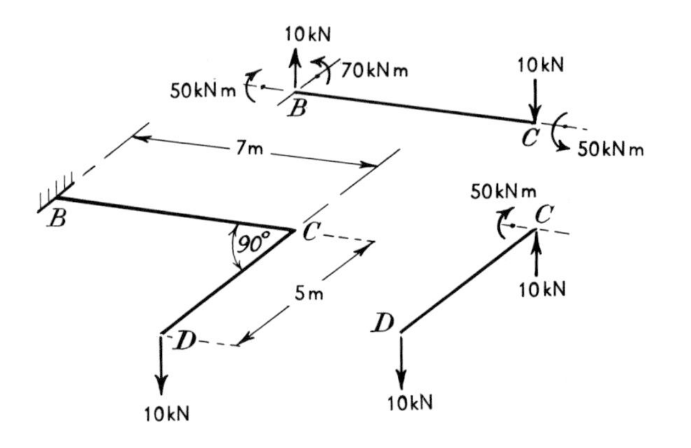

Solution

Consider the statical equilibrium of DC alone; there is no torque in DC, and the only internal actions at C in DC are a shearing force of 10 kN and a bending moment of 50 kNm. Now reverse the actions at C on DC and consider these reversed actions at C on BC. Equilibrium of BC is ensured if there is a shearing force of 10 kN at B, a bending moment of 70 kNm, and a torque of 50 kNm.

FURTHER PROBLEMS
(answers on page 397)

7.9 Draw the shearing-force and bending-moment diagrams for the following beams:

(i) A cantilever of length 20 m carrying a load of 10 kN at a distance of 15 m from the supported end.
(ii) A cantilever of length 20 m carrying a load of 10 kN uniformly distributed over the inner 15 m of its length.
(iii) A cantilever of length 12 m carrying a load of 8 kN, applied 5 m from the supported end, and a load of 2 kN/m over its whole length.
(iv) A beam, 20 m span, simply-supported at each end and carrying a vertical load of 20 kN at a distance 5 m from one support.
(v) A beam, 16 m span, simply-supported at each end and carrying a vertical load of 2·5 kN at a distance of 4 m from one support and the beam itself weighing 500 N per metre.

7.10 A pair of lock-gates are strengthened by two girders *AC* and *BC*. If the load on each girder amounts to 15 kN per metre run, find the bending moment at the middle of each girder. (*Cambridge*)

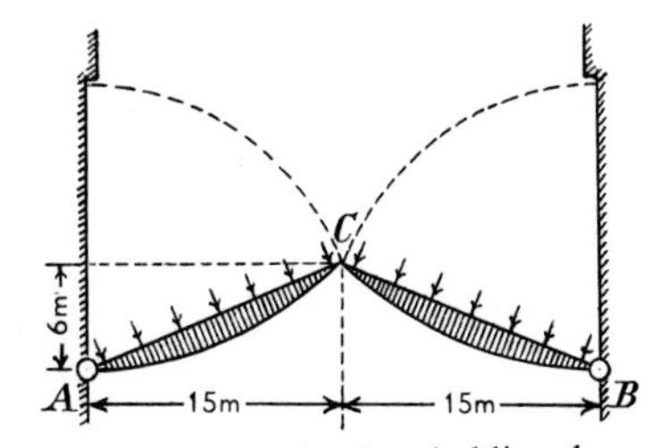

7.11 A girder *ABCDE* bears on a wall for a length *BC* and is prevented from overturning by a holding-down bolt at *A*. The packing under *BC* is so arranged that the pressure over the bearing is uniformly distributed and the 30 kN load may also be taken as a uniformly distributed load. Neglecting the mass of the beam, draw its bending moment and shearing force diagrams. (*Cambridge*)

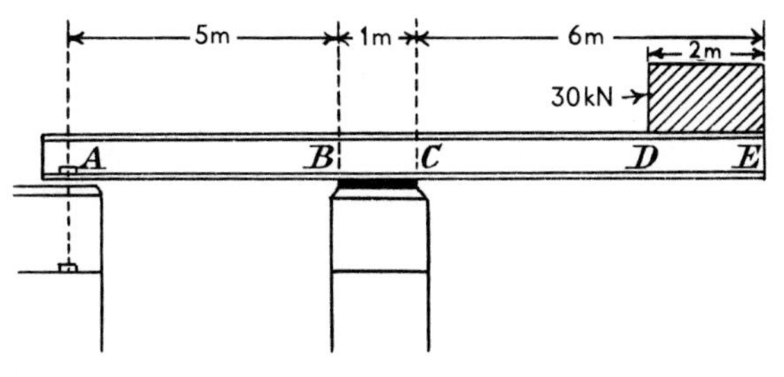

7.12 Draw the bending moment and shearing force diagrams for the beam shown. The beam is supported horizontally by the strut *DE*, hinged at one end to a wall, and at the other end to the projection *CD* which is firmly fixed at right angles to *AB*. The beam is freely hinged to the wall at *B*. The masses of the beam and strut can be neglected. (*Cambridge*)

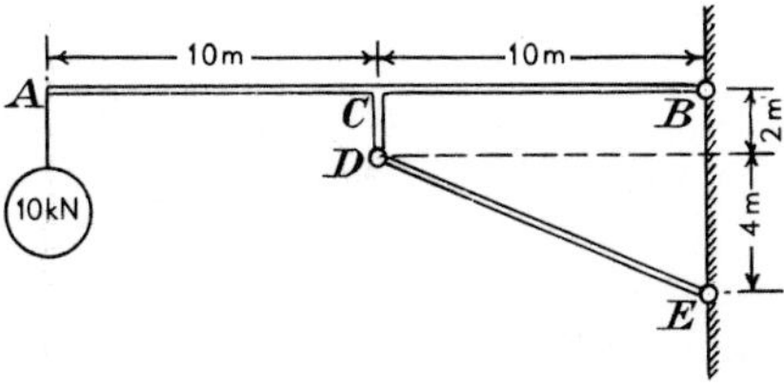

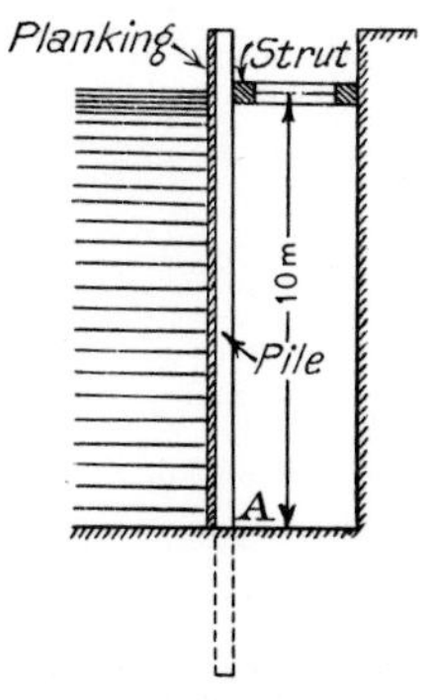

7.13 A timber dam is made of planking backed by vertical piles. The piles are built-in at the section *A* where they enter the ground and they are supported by horizontal struts whose centre lines are 10 m above *A*. The piles are spaced 1 m apart between centres and the depth of water against the dam is 10 m.

Assuming that the thrust in the strut is two-sevenths the total water pressure resisted by each pile, sketch the form of the bending moment and shearing force diagrams for a pile. Determine the magnitude of the bending moment at *A* and the position of the section which is free from bending moment. (*Cambridge*)

7.14 The load distribution (full lines) and upward water thrust (dotted lines) for a ship are given, the numbers indicating kN per metre run. Draw the bending moment diagram for the ship. (*Cambridge*)

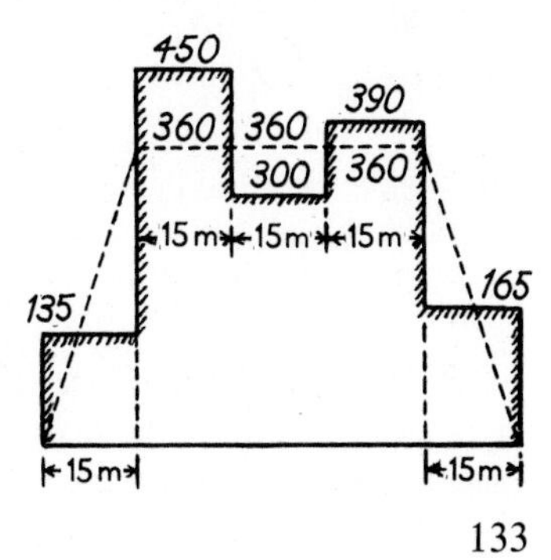

8 Bending moments and shearing forces due to slowly moving loads

8.1 Introduction

In the design of bridge girders it is frequently necessary to know the maximum bending moment and shearing force which each section will have to bear when a travelling load, such as a train, passes from one end of the bridge to the other. The diagrams which we have considered in Chapter 7 show the simultaneous values of the bending moment, or shearing force, for all sections of the beam with the loads in one fixed position; we shall now see how to construct a diagram which shows the greatest value of these quantities for all positions of the loads. These diagrams are called *maximum bending moment*, or *maximum shearing force*, diagrams.

We assume that the loads on a beam are moving slowly; then there are negligible inertia effects from the mass of the beam and any moving masses.

8.2 A single concentrated load traversing a beam

Suppose a single concentrated vertical load W travels slowly along a beam BC, which is simply-supported at each end, Fig. 8.1(i). If a is the distance of the load from B, the reactions at B and C are

$$R_B = \frac{W}{L}(L - a) \qquad R_C = \frac{Wa}{L}$$

The bending moment at a distance z from B, is

$$M = \frac{Wz}{L}(L - a) \quad \text{for} \quad z < a \tag{8.1}$$

$$M = \frac{Wa}{L}(L - z) \quad \text{for} \quad z > a \tag{8.2}$$

Consider the load rolling slowly from C to B: initially $z < a$, and the bending moment, given by equation (8.1), increases as a decreases: when $a = z$,

$$M = \frac{Wz}{L}(L - z) \tag{8.3}$$

As W proceeds further, we have $z > a$, and the bending moment, given by equation (8.2), decreases as a decreases further. Clearly, equation (8.3) is the greatest bending moment which can occur at the section; thus, for any section a distance z from B, the

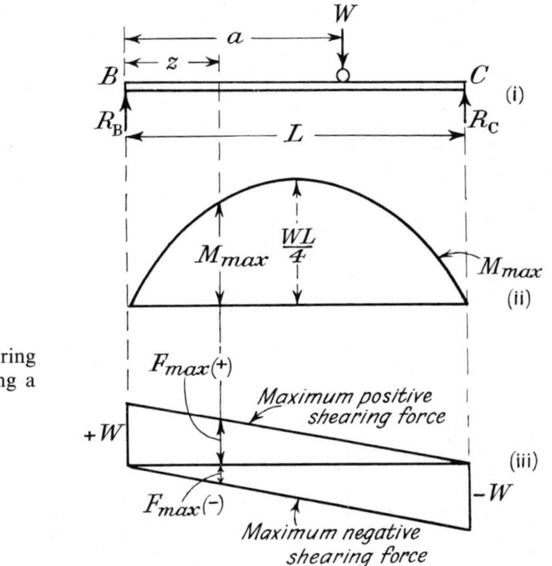

Fig. 8.1 Bending moments and shearing forces due to a rolling load traversing a simply-supported beam.

maximum bending moment which can be induced is

$$M_{\max} = \frac{Wz}{L}(L - z), \tag{8.4}$$

and this occurs when the load W is at that section of the beam. The variation of $M_{\max}$ for different values of z is shown in Fig. 8.1(ii); the curve of $M_{\max}$ is a parabola, attaining a peak value when $z = \frac{1}{2}L$, for which

$$M_{\max} = \frac{WL}{4}$$

The shearing force a distance z from B is

$$F = R_B = \frac{W}{L}(L - a) \quad \text{for} \quad z < a \tag{8.5}$$

$$F = -R_C = -\frac{Wa}{L} \quad \text{for} \quad z > a \tag{8.6}$$

Consider again a load rolling slowly from C to B; initially $z < a$, and the shearing force, given by equation (8.5) is positive and increases as a diminishes. The greatest positive shearing force occurs just before the load W passes the section under consideration; it has the value

$$F_{\max}(+) = \frac{W}{L}(L - z) \tag{8.7}$$

After the load has passed the section being considered, that is, when $z > a$, the shearing force, given by equation (8.6) is negative and decreases as a diminishes further. The greatest negative shearing force occurs when the load W has just passed the section at a distance z; it has the value

$$F_{max}(-) = -\frac{Wz}{L} \tag{8.8}$$

The variations of maximum positive and negative shearing forces are shown in Fig. 8.1(iii).

8.3 Uniformly-distributed load of sufficient length to cover the whole span

Suppose a long distributed load of uniform intensity w traverses a beam BCD, which is simply-supported at B and D, Fig. 8.2. Suppose the load is moving slowly from D

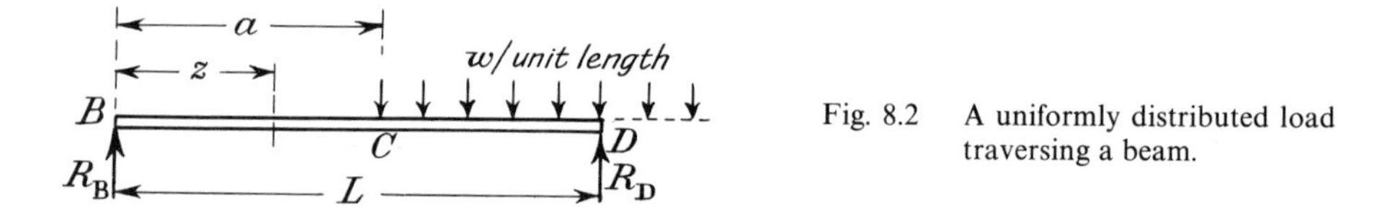

Fig. 8.2 A uniformly distributed load traversing a beam.

to B, the front end C being a distance a from B. Then, by taking moments about B and D, we have

$$R_B = \frac{w}{2L}(L - a)^2 \qquad R_D = \frac{w}{2L}(L^2 - a^2) \tag{8.9}$$

If $z < a$, the shearing force a distance z from B is

$$F = +R_B = \frac{w}{2L}(L - a)^2 \tag{8.10}$$

If $z > a$,

$$F = +R_B - w(z - a) = \frac{w}{2L}(L - a)^2 - w(z - a) \tag{8.11}$$

Consider the distributed load rolling across the span from D to B. Initially, when $z < a$, the shearing force is given by equation (8.10); as a decreases, F increases until it attains the value

$$F = \frac{w}{2L}(L - z)^2 \tag{8.12}$$

when the end C of the load reaches the section, that is, when $a = z$. As the load rolls further, $z > a$, and from equation (8.11), the shearing force decreases as a diminishes further; as a approaches zero, the shearing force becomes

$$F = \tfrac{1}{2}wL - wz \tag{8.13}$$

Clearly, F can become negative only if $z > \frac{1}{2}L$. We conclude then, that for the load rolling from D to B, the maximum positive shearing force is given by equation (8.12); negative shearing force is possible only if $z > \frac{1}{2}L$, and the maximum negative shearing force is then given by equation (8.13).

But consider now the possibility that the distributed load may roll onto the beam from the other end, B, Fig. 8.3; alternatively, we may consider Fig. 8.3 as showing the

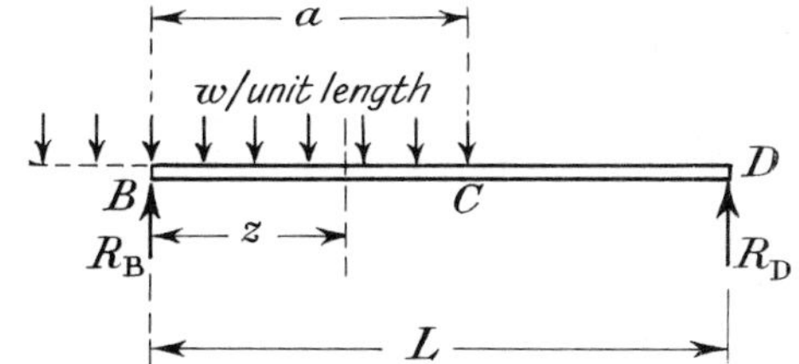

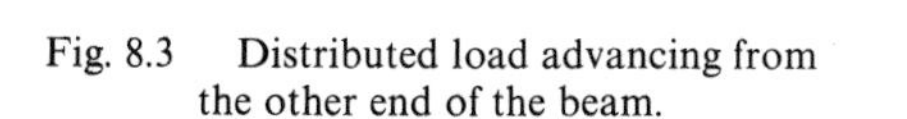
Fig. 8.3 Distributed load advancing from the other end of the beam.

end of the long distributed load of Fig. 8.2, rolling from D to B. The shearing force at a distance z from B is

$$F = \frac{wa}{2L}(2L - a) - wz \quad \text{for} \quad z < a \tag{8.14}$$

$$F = -\frac{wa^2}{2L} \quad \text{for} \quad z > a \tag{8.15}$$

Initially, when $a = L$, the shearing force given by equation (8.14) is

$$F = \tfrac{1}{2}wL - wz$$

This is clearly the same as that given by equation (8.13); it is positive if $z < \frac{1}{2}L$; as a decreases initially, F also decreases, until it reaches the value

$$F = -\frac{wz^2}{2L} \tag{8.16}$$

when $a = z$. Equation (8.16) defines the greatest negative shearing force at a section a distance z from B. As a decreases further, equation (8.16) shows that F diminishes from $-wz^2/2L$, when $a = z$, to zero when $a = 0$.

On plotting the relations given by equations (8.12), (8.13) and (8.16) we note that the greatest positive shearing forces are induced when the load is proceeding from the end D, as in Fig. 8.2; equation (8.12) gives the maximum positive shearing force, and is shown graphically in Fig. 8.4. The greatest negative shearing forces are induced when the load is proceeding from the end B, as in Fig. 8.3; equation (8.16) gives the maximum negative shearing forces in the beam. The greatest shearing forces implied by equation

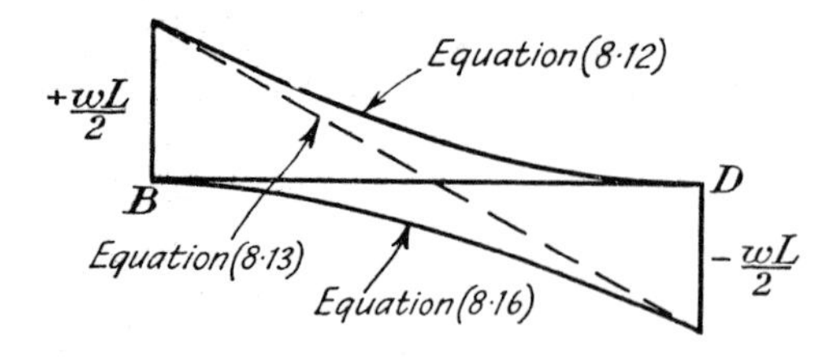

Fig. 8.4 Maximum positive and negative shearing forces caused by a distributed load.

(8.13) are not so great as those given by equations (8.12) and (8.16), and are not important therefore.

The bending moment on any section of the beam will have its greatest value when the whole span is loaded. Thus the diagram of maximum bending moment is identical with the bending moment diagram for a uniformly-distributed load of w per unit length over the whole span. The greatest attainable bending moment is $\frac{1}{8}wL^2$.

8.4 Two concentrated loads traversing a beam

When a four-wheeled vehicle crosses a bridge, the vertical loads on the bridge are effectively concentrated on the axles of the vehicle; the loads on the two axles form a simple system of two concentrated loads at a fixed distance apart.

Consider a beam simply-supported at each end, and carrying two vertical concentrated loads W_1 and W_2, which traverse the beam at a constant distance c apart, Fig. 8.5.

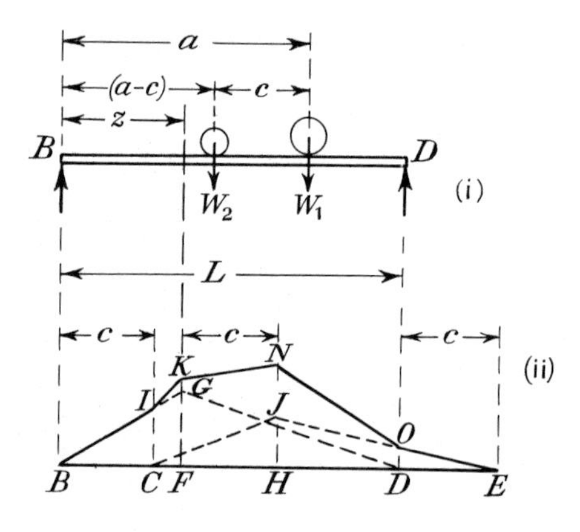

Fig. 8.5 (i) Simply-supported beam carrying two concentrated rolling loads. (ii) Bending moments at a distance z from B; distances along $BCDE$ define the position of the leading load W_1.

Take any section of the beam a distance z from B. Due to W_1 alone, the bending moment at a distance z from B varies according to the two straight lines BG and GD, Fig. 8.5(ii), where

$$FG = \frac{W_1 z}{L}(L - z)$$

Due to W_2 alone, the bending moment at a distance z from B is defined by CJ and JE; these lines are displaced horizontally in Fig. 8.5(ii) by an amount c, relative to the lines BG and GD. The ordinate HJ is given by

$$HJ = \frac{W_2 z}{L}(L - z)$$

The effect of W_1 and W_2 acting together is found by superposing the triangles BGD and CJE in Fig. 8.5(ii); this gives the set of straight lines $BIKNOE$. We note that the peak value at K corresponds to the load W_1 at the section being considered, while the peak value at N occurs when W_2 acts at the section. The greater of FK and HN is the maximum bending moment which can occur at the section. We have then that the greatest bending moment at any section occurs when one or other of the loads acts at that section.

When W_1 acts at the section, the bending moment at the section is

$$M_1 = \frac{W_1 z}{L}(L - z) + \frac{W_2}{L}(L - z)(z - c)$$

$$= \frac{1}{L}(L - z)[(W_1 + W_2)z - W_2 c] \quad (8.17)$$

when $z > c$. When W_2 acts at the section,

$$M_2 = W_2 \frac{z}{L}(L - z) + W_1 \frac{z}{L}(L - z - c)$$

$$= \frac{z}{L}[(W_1 + W_2)(L - z) - W_1 c] \quad (8.18)$$

when $c < L - z$. Equations (8.17) and (8.18) are plotted in Fig. 8.6; M_1 is zero at $z = L$, and at

$$z = \frac{W_2 c}{W_1 + W_2} \quad (8.19)$$

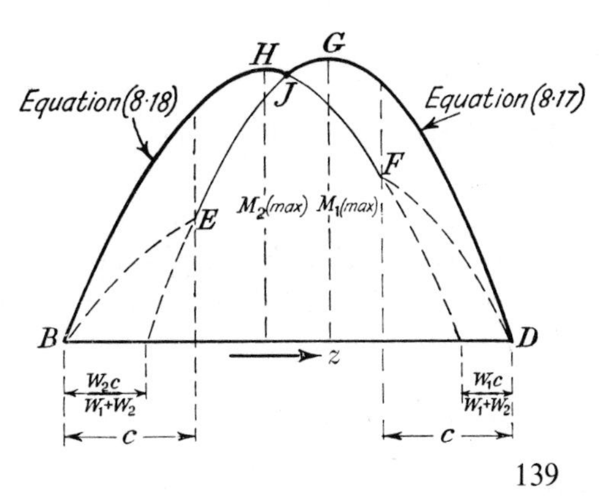

Fig. 8.6 Maximum bending moments due to two concentrated loads.

But at this latter value of z, we have $z < c$, and equation (8.17) is no longer valid; in fact, for the bending moments immediately below the section carrying W_1, we should have the curve $BEGD$, the portion BE corresponding to the load W_1 on the beam alone. M_1 has the maximum value

$$M_1(\max) = \frac{[L(W_1 + W_2) - cW_2]^2}{4L(W_1 + W_2)} \tag{8.20}$$

and occurs when

$$z = \tfrac{1}{2}L\left[1 + \frac{c}{L} \cdot \frac{W_2}{W_1 + W_2}\right] \tag{8.21}$$

Similarly, for the bending moment M_2, at the section carrying W_2, we have the curve $BHFD$ of Fig. 8.6; the portion FD is for the beam carrying the load W_2 alone. M_2 has the maximum value

$$M_2(\max) = \frac{[L(W_1 + W_2) - cW_1]^2}{4L(W_1 + W_2)} \tag{8.22}$$

and occurs when

$$z = \tfrac{1}{2}L\left[1 - \frac{c}{L} \cdot \frac{W_1}{W_1 + W_2}\right] \tag{8.23}$$

The curve defining the greatest bending moment at any section of the beam is given by $BHJGD$ in Fig. 8.6. $M_1(\max)$ is greater than $M_2(\max)$ if $W_1 > W_2$.

We have assumed above that the span of the beam is long compared with c, the distance apart of the loads. When c is of the order of the span L, the greatest bending moment may arise when the load W_1 alone is acting. Consider the case when $W_1 > W_2$; the maximum bending moment due to W_1 alone is $W_1L/4$. The maximum bending moment, M_1, due to W_1 and W_2, is greater than $W_1L/4$ if

$$\frac{[L(W_1 + W_2) - cW_2]^2}{4L(W_1 + W_2)} > \frac{W_1L}{4}$$

This condition gives

$$\frac{c}{L} < \frac{W_1}{W_2} + 1 - \sqrt{\left(\frac{W_1}{W_2}\right)^2 + \frac{W_1}{W_2}} \tag{8.24}$$

If this inequality is satisfied, the maximum bending moments due to the combined effects of W_1 and W_2 are the only relevant ones.

We must consider now the shearing forces. When both loads are to the right of a section distant z from B, Fig. 8.5(i), the shearing force at the section is

$$F = +\frac{W_1}{L}(L - a) + \frac{W_2}{L}(L - a + c) \tag{8.25}$$

Thus, as a diminishes, F increases, until it reaches the value

$$F = +\frac{W_1}{L}(L - z - c) + \frac{W_2}{L}(L - z) \tag{8.26}$$

when W_2 is just to the right of the section. When W_2 crosses the section, but W_1 is still to the right of the section,

$$F = +\frac{W_1}{L}(L - a) + \frac{W_2}{L}(L - a + c) - W_2 \tag{8.27}$$

When W_2 is just to the left of the section,

$$F = \frac{W_1}{L}(L - z - c) + \frac{W_2}{L}(L - z) - W_2 \tag{8.28}$$

Now, equation (8.26) represents a positive shearing force as W_2 approaches the section; the shearing force given by equation (8.28), as the load W_2 passes the section, is less than that given by equation (8.26), and could become negative if

$$\frac{W_1}{L}(L - z - c) + \frac{W_2}{L}(-z) < 0 \tag{8.29}$$

After W_2 has passed the section, and as W_1 approaches, the shearing force is

$$F = +\frac{W_1}{L}(L - z) + \frac{W_2}{L}(L - z + c) - W_2 \tag{8.30}$$

which may be negative if

$$\frac{W_1}{L}(L - z) - \frac{W_2}{L}(z - c) < 0 \tag{8.31}$$

When W_1 is to the left of the section, the shearing force is negative, and has the value

$$F = -W_1\frac{a}{L} - \frac{W_2}{L}(a - c) \tag{8.32}$$

This decreases numerically as a decreases; it is greatest numerically when W_1 is just to the left of the section. It then has the value

$$F = -W_1\frac{z}{L} - \frac{W_2}{L}(z - c) \tag{8.33}$$

Let us first consider the maximum positive shearing force; this is given either by equation (8.26) or equation (8.30). The maximum positive shearing force is given by equation (8.26) if

$$\frac{W_1}{L}(L - z - c) + \frac{W_2}{L}(L - z) > \frac{W_1}{L}(L - z) + \frac{W_2}{L}(L - z + c) - W_2$$

that is, if

$$\frac{c}{L} < \frac{W_2}{W_1 + W_2} \tag{8.34}$$

If W_1 and W_2 are of the same order, and if c is small compared with L, this inequality is usually satisfied; then the maximum positive shearing force is given by equation (8.26). Similarly, if

$$\frac{c}{L} < \frac{W_1}{W_1 + W_2} \tag{8.35}$$

the greatest negative shearing force is given by equation (8.33). The maximum positive and negative shearing forces which can occur at any section are shown in Fig. 8.7.

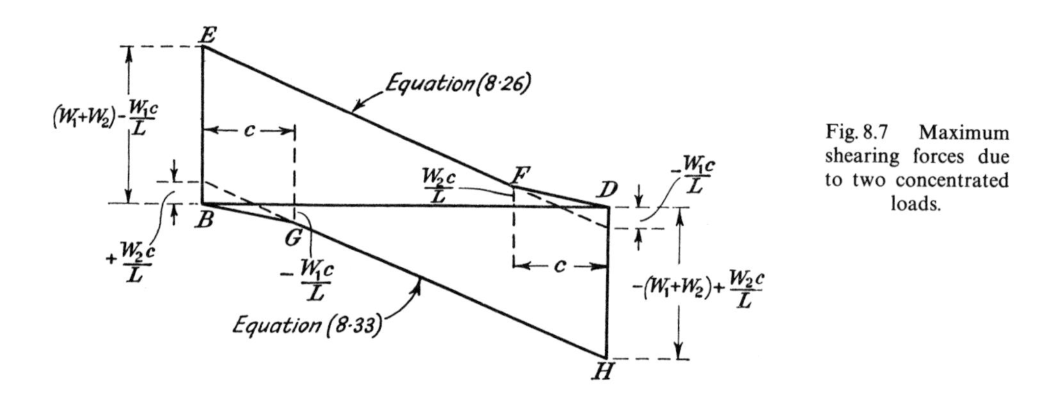

Fig. 8.7 Maximum shearing forces due to two concentrated loads.

Equations (8.26) and (8.33) are not valid over the whole span; near the ends of the beam one of the loads may roll off, so that the variation of positive maximum shearing force is given by EFD; similarly maximum negative shearing force is given by HGB.

8.5 Several concentrated loads

A simply-supported beam BC carries concentrated loads $W_1, W_2, W_3, \ldots, W_n$, which are constant distances apart, Fig. 8.8. For any position of the system of loads the maximum bending moment must occur at one of the loaded sections, since the variation of bending moment throughout the length of the beam is linear between loaded sections. Suppose M_r is the bending moment at the section of the beam carrying W_r, when that load is a distance z from B. Then

$$M_r = R_B z - (W_1 a_1 + W_2 a_2 + W_3 a_3 + \ldots + W_{r-1} a_{r-1}) \tag{8.36}$$

where $a_1, a_2, \ldots, a_{r-1}$ are the distances of the loads to the left of W_r from the line of

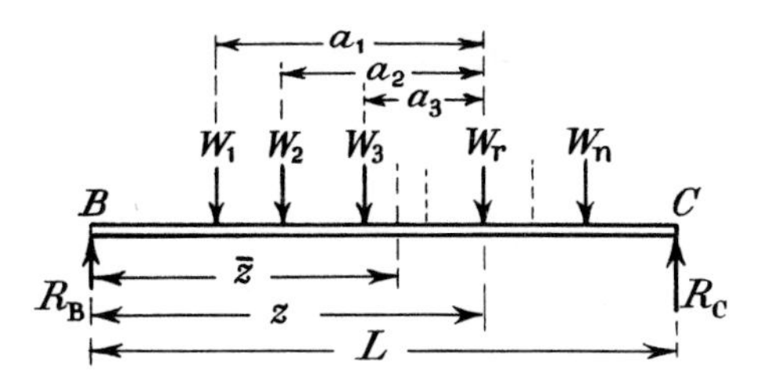

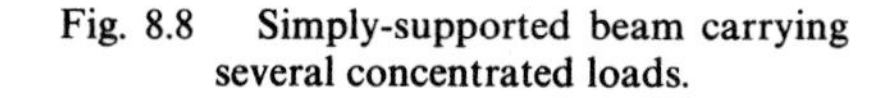
Fig. 8.8 Simply-supported beam carrying several concentrated loads.

action of W_r; M_r is a maximum when

$$\frac{dM_r}{dz} = 0$$

that is, when

$$z\frac{dR_B}{dz} + R_B = 0 \tag{8.37}$$

provided the value of z given by this equation does not imply that any of the loads is off the beam. Now,

$$R_B = \frac{1}{L}\sum_{r=1}^{n}(L - z + a_r)W_r \tag{8.38}$$

where $a_r = 0$, and $a_{r+1}, a_{r+2}, \ldots, a_n$ are reckoned negative. Then suppose all the loads advance a small distance δz to the right; the increase in the reaction at B is

$$\delta R_B = -\frac{1}{L}\sum_{r=1}^{n}\delta z \,.\, W_r$$

Then

$$\frac{dR_B}{dz} = -\frac{1}{L}\sum_{r=1}^{n}W_r \tag{8.39}$$

Suppose $\bar{z}$ is the distance to the centroid of the loads $W_1, W_2, W_3, \ldots, W_n$, defined as

$$\sum_{r=1}^{n}(a_r + \bar{z} - z)W_r = 0 \tag{8.40}$$

Then equation (8.38) may be written

$$R_B = \frac{1}{L}\sum_{r=1}^{n}(L - z + a_r)W_r = \frac{1}{L}\sum_{r=1}^{n}(L - \bar{z})W_r = \frac{L - \bar{z}}{L}\sum_{r=1}^{n}W_r \tag{8.41}$$

But from equations (8.37) and (8.39),

$$R_B = -z\frac{dR_B}{dz} = \frac{z}{L}\sum_{r=1}^{n}W_r \tag{8.42}$$

On comparing equations (8.41) and (8.42), we have that

$$L - \bar{z} = z$$

Then

$$\frac{L}{2} - \bar{z} = z - \frac{L}{2} \tag{8.43}$$

Hence M_r is a maximum when W_r and the centroid of all the loads are equidistant from the supports, or from the mid-length of the beam, provided all the loads remain in the span.

This result is of considerable help in examining the maximum bending moment, but for the rest we must be guided by general considerations, such as the fact that the greatest bending moment will usually occur in the middle portion of the beam.

Problem 8.1: A series of loads passes over a bridge of 15 m span from left to right; it is required to find the maximum bending moment which the bridge will have to bear.

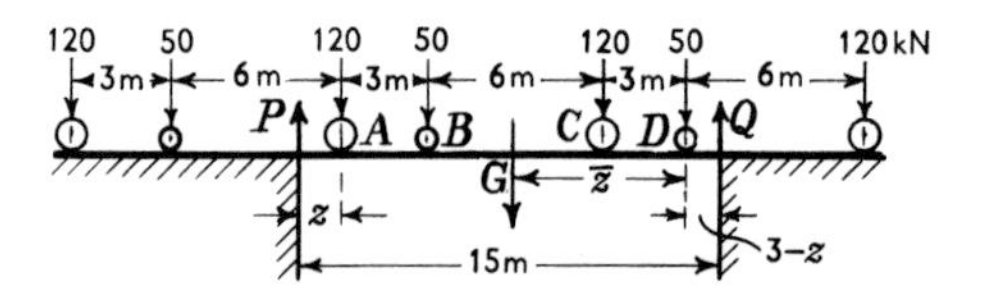

Solution

First find the horizontal position of the centroid of all the loads on the bridge: it will be seen that four is the greatest number of loads which can be on the bridge simultaneously.

Let $\bar{z}$ be the distance from D of the centroid of the loads A, B, C, D which are on the bridge. Then

$$\bar{z} = \frac{(120 \times 10^3)(12) + (50 \times 10^3)(9) + (120 \times 10^3)(3)}{(340 \times 10^3)} = 6{\cdot}63 \text{ m}$$

The maximum bending moment will occur under one of the loads, so we examine the bending moment under each load in turn. Maximum bending moment under load A: This will occur when A and G are equidistant from the supports, provided one of the loads does not go off the bridge. Referring to the diagram we see that this requires:

$$z = (3 - z) + \bar{z}$$

or

$$z = 4{\cdot}82 \text{ m}$$

But this will make D go off the bridge. Hence the above breaks down; with all the loads on the bridge the bending moment under A increases continuously as D gets nearer to the right-hand end. When D is only just on the bridge we have

$$15P = (6{\cdot}63)(340 \times 10^3) = 2{\cdot}25 \text{ MNm}$$

giving

$$P = 150 \text{ kN}$$

Then the bending moment under A is $3P = 450$ kNm.

Next consider the bending moment under the load B. It will have its greatest value when

$$z + 3 = (3 - z) + \bar{z}$$

or when

$$z = 3{\cdot}32 \text{ m}$$

This again makes D go off the bridge. When D is only just on the bridge we have seen that $P = 150$ kN, so then the bending moment under B is

$$(150 \times 10^3)(6) - (120 \times 10^3)(3) = 540 \text{ kNm}$$

Next consider the load C. The maximum bending moment under it will occur when

$$z + 9 = (3 - z) + \bar{z}$$

or when

$$z = 0{\cdot}32 \text{ m}$$

We then find

$$P = 211 \text{ kN} \qquad Q = 129 \text{ kN}$$

The bending moment under C is then

$$(6 - z)Q - 3(50 \times 10^3) = 583 \text{ kNm}$$

Proceeding in the same way we find that the bending moment under the load D is greatest when $z = -1{\cdot}18$ m, but its value is less than the greatest bending moment under C.

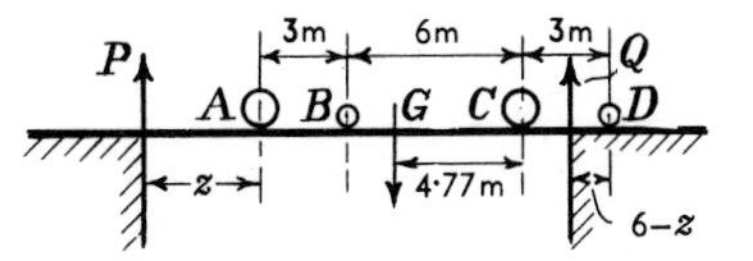

We must next consider the bending moments when D is off the bridge. The centroid of A, B and C is found to be 4·77 m from C. Proceeding exactly as above we find that

the greatest bending moment under A = 560 kNm

the greatest bending moment under B = 555 kNm

When we seek the value of z which will make the bending moment under C a maximum we find $z = 0{\cdot}88$ m, which brings D back on to the bridge again; this might be inferred from the above results.

Finally consider the bending moments before A has come on to the bridge. The only load which need be considered is C, and we find in the same way as before that the greatest bending moment under it is 600 kNm.

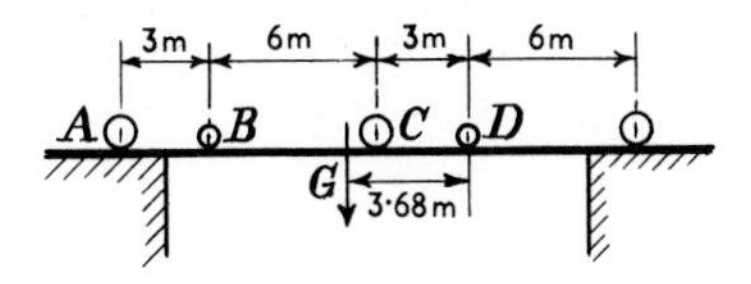

Comparing this with the figures above, it will be seen that this is the greatest of all the bending moments, and it should be noticed that it occurs when one of the loads, namely A, is off the bridge.

8.6 Influence lines of bending moment and shearing force

A curve which shows the value of the bending moment at a given section of a beam, for all positions of a travelling load, is called the bending-moment *influence line* for that section; similarly a curve which shows the shearing force at the section for all positions of the load is called the shearing force influence line for the section. The distinction between influence lines and maximum bending-moment (or shearing force) diagrams must be carefully noted: for a given load there will be only one maximum bending-moment diagram for the beam, but an infinite number of bending-moment influence lines, one for each section of the beam.

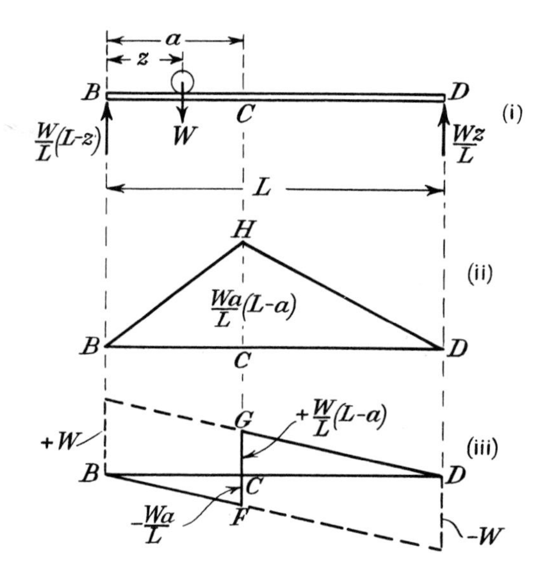

Fig. 8.9 (i) Single rolling load on a simply-supported beam. (ii) Bending-moment influence line for section C. (iii) Shearing-force influence line for section C.

Consider firstly a simply-supported beam, Fig. 8.9, carrying a single concentrated load, W. As the load rolls across the beam, the bending moments at a section C of the beam vary with the position of the load. Suppose W is a distance z from B; then the bending moment at a section C is given by

$$M = \frac{Wz}{L}(L - a) \quad \text{for} \quad z < a$$

and

$$M = \frac{Wa}{L}(L - z) \quad \text{for} \quad z > a$$

The first of these equations gives the straight line BH in Fig. 8.9(ii), and the second the line HD. The influence line for bending moments at C is then BHD; the bending moment is greatest when the load acts at the section.

Again, the shearing force at C is

$$F = -\frac{Wz}{L} \quad \text{for} \quad z < a$$

and

$$F = +\frac{W}{L}(L - z) \quad \text{for} \quad z > a$$

These relations give the lines $BFCGD$ for the shearing force influence line for C. There is an abrupt change of shearing force as the load W crosses the section C.

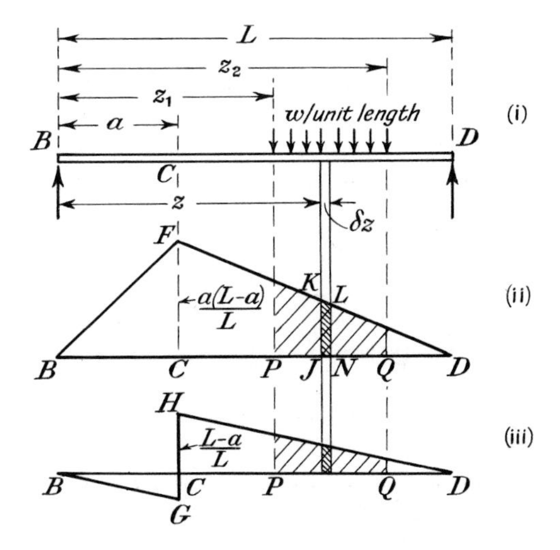

Fig. 8.10 (i) Simply-supported beam with distributed load covering part of the span. (ii) Deduction of bending moment from the influence line. (iii) Deduction of shearing force from the influence line.

Consider next a uniformly distributed load of w per unit length, which is advancing across the beam BD of Fig. 8.10(i); the influence line of bending moment at C for a *unit* rolling load is shown in Fig. 8.10(ii), and that for shearing force in Fig. 8.10(iii). Then

$$CF = \frac{a(L - a)}{L} \qquad CH = \frac{L - a}{L}$$

For a unit load at a distance z from B, the bending moment at C is given by the ordinate JK of the influence line. Then for a load $w\,\delta z$ acting over a small length δz of the beam, the bending moment at C is $(w\,\delta z \times JK)$, which is the area of the thin strip $JKLN$ multiplied by w. Clearly the total bending moment at C due to the total distributed load is the area of the influence line between sections P and Q, multiplied by w. Similar reasoning may be applied to the shearing force influence line to find the shearing force at C.

FURTHER PROBLEMS
(answers on page 398)

8.2 Two concentrated loads of 100 and 200 kN advance along a girder 20 m span, the distance between the loads being 8 m. Find the position of the section which has to support the greatest bending moment, and calculate the value of this bending moment. (*Cambridge*)

8.3 An engine advances across a bridge from left to right. The loads on the front and rear axles are 60 and 90 kN and the wheelbase is 3 m. The span of the bridge is 15 m. Construct the maximum bending-moment diagram for the bridge. (*Cambridge*)

8.4 Two concentrated loads of 120 kN and 80 kN, 3 m apart, advance across a horizontal girder of 15 m span. Draw to scale the maximum bending-moment diagram for this arrangement. (*Cambridge*)

8.5 A braced girder of 200 m span, divided into ten equal panels, supports a rolling load of 1 kN per metre run which may extend over the whole length of the girder. Show that the maximum positive and negative shearing forces, due to the rolling load, in the rth panel from the end, are $2(10 - r)^2$ and $2(r + 1)^2$. (*Cambridge*)

8.6 The loads on the front and rear axles of a motor lorry are 31·6 kN and 78·8 kN respectively, and the distance between them is 4·3 m. The lorry advances over a girder having a clear span of 25 m. Calculate the greatest bending moment set up and show that the equivalent uniformly distributed dead load is approximately 200 kN. (*Cambridge*)

8.7 A load of 100 kN, followed by another load of 50 kN at a distance of 10 m, advances across a girder of 100 m span. Obtain an expression for the maximum bending moment at a section of the girder distant z metres from an abutment. (*Cambridge*)

9 Longitudinal stresses in beams

9.1 Introduction

We have seen that when a straight beam carries lateral loads the actions over any cross-section of the beam comprise a bending moment and shearing force; we have also seen how to estimate the magnitudes of these actions. The next step in discussing the strength of beams is to consider the stresses caused by these actions.

As a simple instance consider a cantilever carrying a concentrated load W at its free end, Fig. 9.1. At sections of the beam remote from the free end the upper longitudinal

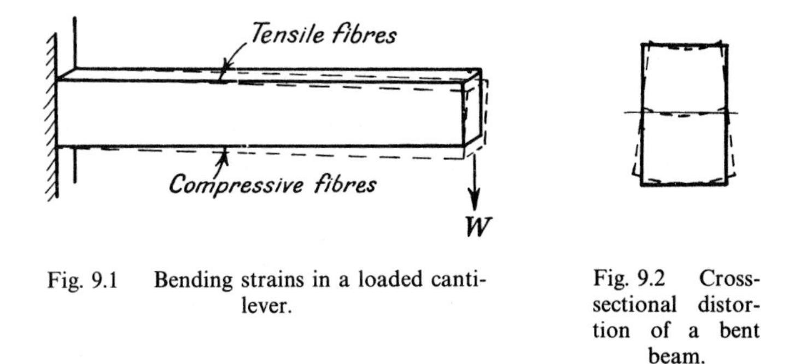

Fig. 9.1 Bending strains in a loaded cantilever.

Fig. 9.2 Cross-sectional distortion of a bent beam.

fibres of the beam are stretched, i.e. tensile stresses are induced; the lower fibres are compressed. There is thus a variation of direct stress throughout the depth of any section of the beam. In any cross-section of the beam, as in Fig. 9.2, the upper fibres which are stretched longitudinally contract laterally owing to the Poisson ratio effect, while the lower fibres extend laterally; thus the whole cross-section of the beam is distorted.

In addition to longitudinal direct stresses in the beam, there are also shearing stresses over any cross-section of the beam. In most engineering problems shearing *distortions* in beams are relatively unimportant; this is not true, however, of shearing *stresses*.

9.2 Pure bending of a rectangular beam

An elementary bending problem is that of a rectangular beam under end couples. Consider a straight uniform beam having a rectangular cross-section of breadth b and depth h, Fig. 9.3; the axes of symmetry of the cross-section are Cx, Cy. A long length

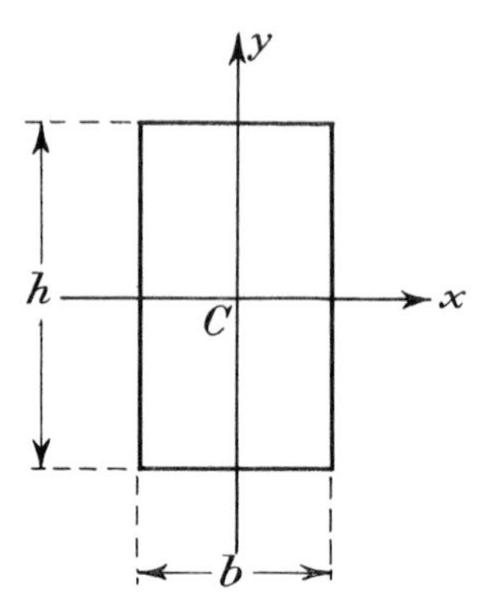

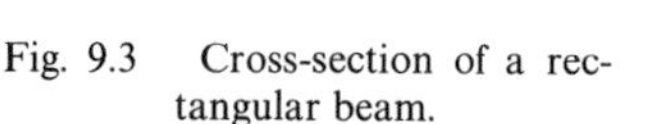

Fig. 9.3 Cross-section of a rectangular beam.

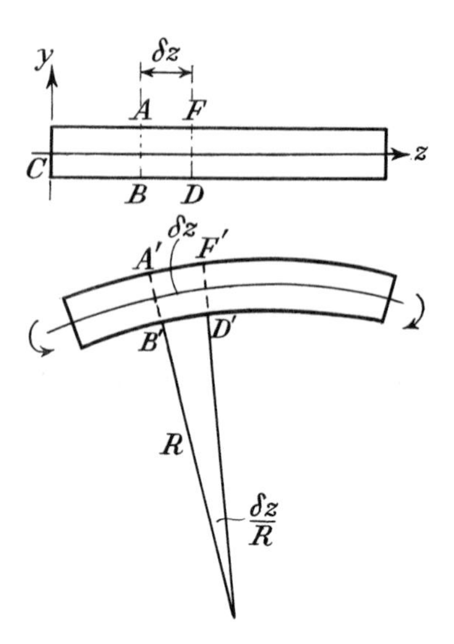

Fig. 9.4 Beam bent to a uniform radius of curvature R in the yz-plane.

of the beam is bent in the yz-plane, Fig. 9.4, in such a way that the longitudinal centroidal axis, Cz, remains unstretched and takes up a curve of uniform radius of curvature, R. We consider an elemental length δz of the beam, remote from the ends; in the unloaded condition, AB and FD are transverse sections at the ends of the elemental length, and these sections are initially parallel. In the bent form we assume that planes such as AB and FD remain flat planes; $A'B'$ and $F'D'$ in Fig. 9.4 are therefore cross-sections of the bent beam, but are no longer parallel to each other.

In the bent form, some of the longitudinal fibres, such as $A'F'$, are stretched, while others, such as $B'D'$, are compressed. The unstrained middle-surface of the beam is known as the *neutral surface*, and a line parallel to Cx in this surface is known as a *neutral axis*. Now consider an elemental fibre HJ of the beam, parallel to the longitudinal axis Cz, Fig. 9.5; this fibre is at a distance y from the neutral surface and on the tension side of the beam. The original length of the fibre HJ in the unstrained beam is δz; the strained length is

$$H'J' = (R + y)\frac{\delta z}{R}$$

since the angle between $A'B'$ and $F'D'$ in Figs. 9.4 and 9.5 is $(\delta z/R)$. Then during bending HJ stretches an amount

$$H'J' - HJ = (R + y)\frac{\delta z}{R} - \delta z = \frac{y}{R}\,\delta z$$

The longitudinal strain of the fibre HJ is therefore

$$\epsilon = \left(\frac{y}{R}\,\delta z\right)\bigg/\delta z = \frac{y}{R}$$

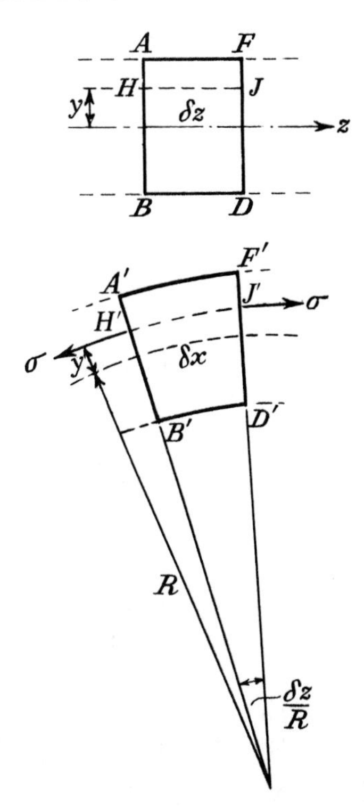

Fig. 9.5 Stresses on a bent element of the beam.

Then the longitudinal strain at any fibre is proportional to the distance of that fibre from the neutral surface; over the compressed fibres, on the lower side of the beam, the strains are of course negative.

If the material of the beam remains elastic during bending then the longitudinal stress on the fibre HJ is

$$\sigma = E\epsilon = \frac{Ey}{R} \tag{9.1}$$

The distribution of longitudinal stresses over the cross-section takes the form shown in Fig. 9.6; because of the symmetrical distribution of these stresses about Cx, there is no resultant longitudinal thrust on the cross-section of the beam. The resultant hogging moment is

$$M = \int_{-\frac{1}{2}h}^{+\frac{1}{2}h} \sigma b y \, dy \tag{9.2}$$

On substituting for σ from equation (9.2), we have

$$M = \frac{E}{R} \int_{-\frac{1}{2}h}^{+\frac{1}{2}h} b y^2 \, dy = \frac{EI_x}{R} \tag{9.3}$$

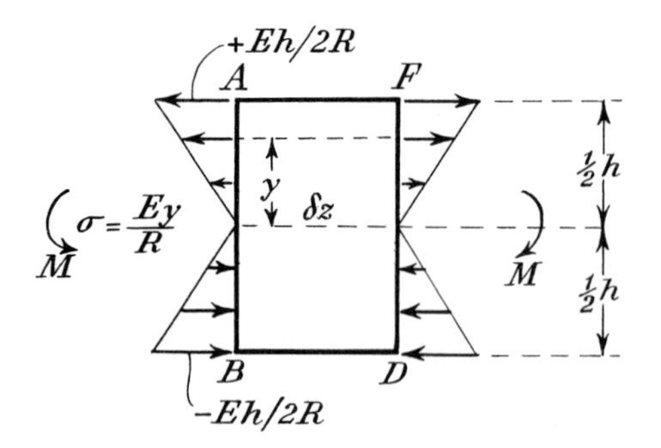

Fig. 9.6 Distribution of bending stresses giving zero resultant longitudinal force and a resultant couple M.

where I_x is the second moment of area of the cross-section about Cx. From equations (9.1) and (9.3) we have

$$\frac{\sigma}{y} = \frac{E}{R} = \frac{M}{I_x} \tag{9.4}$$

We deduce that a uniform radius of curvature, R, of the centroidal axis Cz can be sustained by end couples M, applied about the axes Cx at the ends of the beam.

Equation (9.3) implies a linear relation between M, the applied moment, and $(1/R)$, the curvature of the beam. The constant EI_x in this linear relationship is called the *bending stiffness* or sometimes the *flexural stiffness* of the beam; this stiffness is the product of Young's modulus, E, and the second moment of area, I_x, of the cross-section about the axis of bending.

Problem 9.1: A steel bar of rectangular cross-section, 10 cm deep and 5 cm wide, is bent in the planes of the longer sides. Estimate the greatest allowable bending moment if the bending stresses are not to exceed 150 MN/m² in tension and compression.

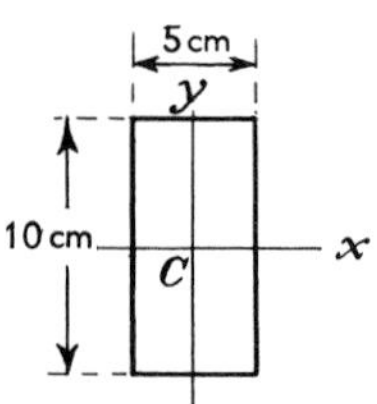

Solution

The bending moment is applied about Cx. The second moment of area about this axis is

$$I_x = \tfrac{1}{12}(0{\cdot}05)(0{\cdot}10)^3 = 4{\cdot}16 \times 10^{-6}\ \text{m}^2$$

The bending stress, σ, at a fibre a distance y from Cx is, by equation (9.4)

$$\sigma = \frac{My}{I_x}$$

where M is the applied moment. If the greatest stresses are not to exceed 150 MN/m², we must have

$$\frac{My}{I_x} \leqslant 150\ \text{MN/m}^2$$

The greatest bending stresses occur in the extreme fibres where $y = 5$ cm. Then

$$M \leqslant \frac{(150 \times 10^6)I_x}{(0{\cdot}05)} = \frac{(150 \times 10^6)(4{\cdot}16 \times 10^{-6})}{(0{\cdot}05)}$$

$$= 12\ 500\ \text{Nm}$$

The greatest allowable bending moment is therefore 12 500 Nm. (The second moment of area about Cy is

$$I_y = \tfrac{1}{12}(0{\cdot}10)(0{\cdot}05)^3 = 1{\cdot}04 \times 10^{-6}\ \text{m}^2$$

The greatest allowable bending moment about Cy is

$$M = \frac{(150 \times 10^6)I_y}{(0{\cdot}025)} = \frac{(150 \times 10^6)(1{\cdot}04 \times 10^{-6})}{(0{\cdot}025)}$$

$$= 6250\ \text{Nm}$$

which is only half that about Cx.)

9.3 Bending of a beam about a principal axis

In §9.2 we considered the bending of a straight beam of rectangular cross-section; this form of cross-section has two axes of symmetry. More generally we are concerned with sections having only one, or no, axis of symmetry.

Consider a long straight uniform beam having any cross-sectional form, Fig. 9.7; the axes Cx, Cy are *principal axes* of the cross-section. The principal axes of a cross-section are those centroidal axes for which the product second moments of area are zero. In Fig. 9.7, C is the centroid of the cross-section; Cz is the longitudinal centroidal axis.

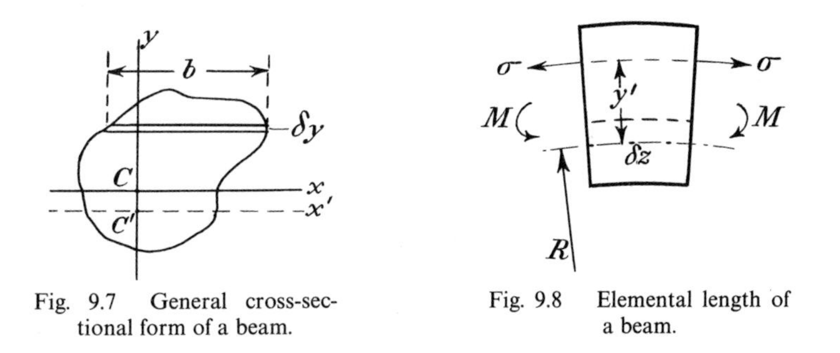

Fig. 9.7 General cross-sectional form of a beam.

Fig. 9.8 Elemental length of a beam.

When end couples M are applied to the beam, we assume as before that transverse sections of the beam remain plane during bending. Suppose further that, if the beam is bent in the yz-plane only, there is a neutral axis $C'x'$, Fig. 9.7, which is parallel to Cx and is unstrained; the radius of curvature of this neutral surface is R, Fig. 9.8. As before, the strain in a longitudinal fibre at a distance y' from $C'x'$ is

$$\epsilon = \frac{y'}{R}$$

If the material of the beam remains elastic during bending the longitudinal stress on this fibre is

$$\sigma = \frac{Ey'}{R}$$

If there is to be no resultant longitudinal thrust on the beam at any transverse section we must have

$$\int_A \sigma b \, dy' = 0$$

where b is the breadth of an elemental strip of the cross-section parallel to Cx, and the integration is performed over the whole cross-sectional area, A. But

$$\int_A \sigma b \, dy' = \frac{E}{R} \int_A y' b \, dy'$$

This can be zero only if $C'x$ is a centroidal axis; now, Cx is a principal axis, and is therefore a centroidal axis, so that $C'x'$ and Cx are coincident, and the neutral axis is Cx in any cross-section of the beam. The total moment about Cx of the internal stresses is

$$M = \int_A \sigma b y \, dy = \frac{E}{R} \int_A b y^2 \, dy$$

But $\int_A b y^2 \, dy$ is the second moment of area of the cross-section about Cx; if this is denoted by I_x, then

$$M = \frac{E I_x}{R} \qquad (9.5)$$

The stress in any fibre a distance y from Cx is

$$\sigma = \frac{Ey}{R} = \frac{My}{I_x} \qquad (9.6)$$

No moment about Cy is implied by this stress system, for

$$\int_A \sigma x \, dA = \frac{E}{R} \int_A xy \, dA = 0$$

since Cx, Cy are principal axes for which $\int_A xy \, dA$, or the product second moment of area, is zero; δA is an element of area of the cross-section.

9.4 Beams having two axes of symmetry in the cross-section

Many cross-sectional forms used in practice have two axes of symmetry; e.g. the I-section, and circular sections, Fig. 9.9, besides the rectangular beam already discussed.

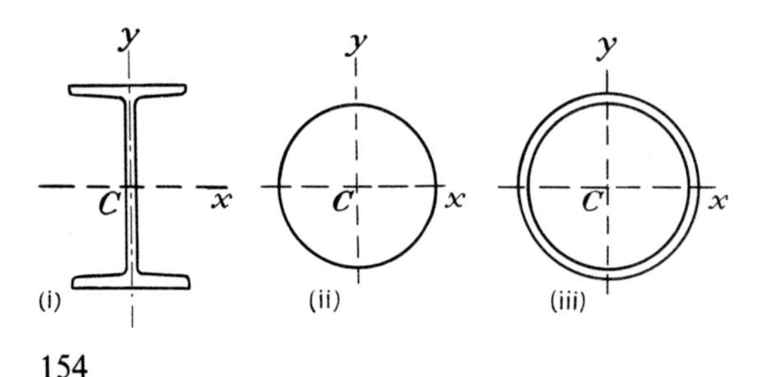

Fig. 9.9 (i) I-section beam. (ii) Solid circular cross-section. (iii) Hollow circular cross-section.

Now an axis of symmetry of a cross-section is also a principal axis; then for bending about the axis Cx we have, from equation (9.6),

$$\sigma = \frac{Ey}{R_x} = \frac{M_x y}{I_x} \tag{9.7}$$

where R_x is the radius of curvature in the yz-plane, M_x is the moment about Cx, and I_x is the second moment of area about Cx. Similarly for bending by a couple M_y about Cy,

$$\sigma = \frac{Ex}{R_y} = \frac{M_y x}{I_y} \tag{9.8}$$

where R_y is the radius of curvature in the xz-plane, and I_y is the second moment of area about Cy. The longitudinal centroidal axis is Cz. From equations (9.7) and (9.8) we see that the greatest bending stresses occur in the extreme longitudinal fibres of the beams.

Problem 9.2: A light-alloy I-beam of 10 cm overall depth has flanges of overall breadth 5 cm and thickness 0·625 cm, the thickness of the web is 0·475 cm. If the bending stresses are not to exceed 150 MN/m² in tension and compression estimate the greatest moments which may be applied about the principal axes of the cross-section.

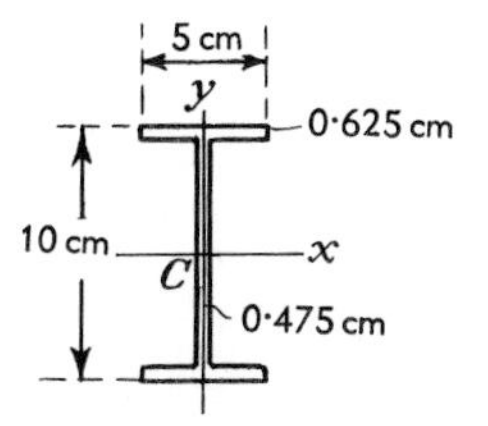

Solution

Consider, firstly, bending about Cx. The second moment of area about Cx is

$$\begin{aligned} I_x &= \tfrac{1}{12}(0{\cdot}00475)[0{\cdot}10 - 2(0{\cdot}00625)]^3 + 2\{\tfrac{1}{12}(0{\cdot}05)(0{\cdot}00625)^3 + (0{\cdot}05)(0{\cdot}00625)[0{\cdot}05 - \tfrac{1}{2}(0{\cdot}0625)]^2\} \\ &= [0{\cdot}266 + 1{\cdot}374 + 0{\cdot}002]10^{-6}\ \text{m}^4 = 1{\cdot}642 \times 10^{-6}\ \text{m}^4 \end{aligned}$$

The first term represents the contribution of the web; the second term is the intrinsic second moments of area of the flanges, that is, the second moments of area about their own centroidal axes; the last term is the product of the area of the flanges and the square of the distance of the centroid of the flange from Cx. The second term is usually negligible, and the value of I_x is dominated largely by the contributions of the flanges. The allowable moment is

$$M_x = \frac{\sigma I_x}{y} = \frac{(150 \times 10^6)(1{\cdot}642 \times 10^{-6})}{0{\cdot}05} = 4926\ \text{Nm}$$

Secondly, for bending about Cy,

$$I_y = \tfrac{1}{12}(0{\cdot}10 - 2[0{\cdot}00625]) + 2(\tfrac{1}{12})(0{\cdot}00625)(0{\cdot}05)^3\ \text{m}^4$$

The first term, which is the contribution of the web, is negligible compared with the second. With sufficient accuracy

$$I_y = 2(\tfrac{1}{12})(0{\cdot}00625)(0{\cdot}05)^3 = 0{\cdot}130 \times 10^{-6}\ \text{m}^4$$

The allowable moment about Cy is

$$M_y = \frac{\sigma I_y}{x} = \frac{(150 \times 10^6)(0{\cdot}130 \times 10^{-6})}{0{\cdot}025} = 780\ \text{Nm}$$

Problem 9.3: A steel scaffold tube has an external diameter of 5 cm, and a thickness of 0·5 cm. Estimate the allowable bending moment on the tube if the bending stresses are limited to 100 MN/m².

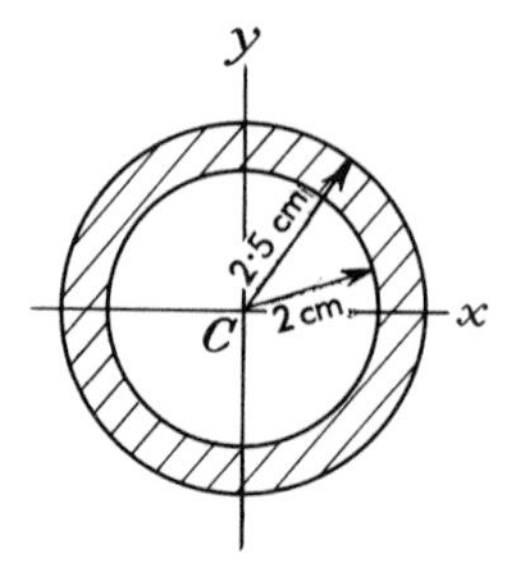

Solution

The second moment of area about a centroidal axis Cx is

$$I_x = \frac{\pi}{4}\,[(0{\cdot}025)^4 - (0{\cdot}020)^4] = 0{\cdot}181 \times 10^{-6}\ \text{m}^4$$

The allowable bending moment about Cx is

$$M_x = \frac{(100 \times 10^6)(0{\cdot}181 \times 10^{-6})}{0{\cdot}025} = 724\ \text{Nm}$$

9.5 Beams having only one axis of symmetry

Other common sections in use, as shown in Fig. 9.10, have only one axis of symmetry Cx. In each of these, Cx is the axis of symmetry, and Cx and Cy are both principal axes. When bending moments M_x and M_y are applied about Cx and Cy, respectively, the

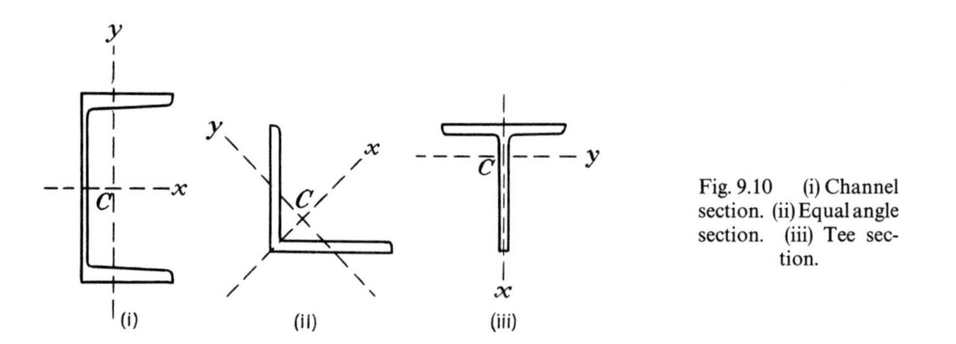

Fig. 9.10 (i) Channel section. (ii) Equal angle section. (iii) Tee section.

bending stresses are again given by equations (9.7) and (9.8). However, an important feature of beams of this type is that their behaviour in bending when shearing forces are also present is not as simple as that of beams having two axes of symmetry. This problem is discussed in Chapter 10.

Problem 9.4: A T-section of uniform thickness 1 cm has a flange breadth of 10 cm and an overall depth of 10 cm. Estimate the allowable bending moments about the principal axes if the bending stresses are limited to 150 MN/m².

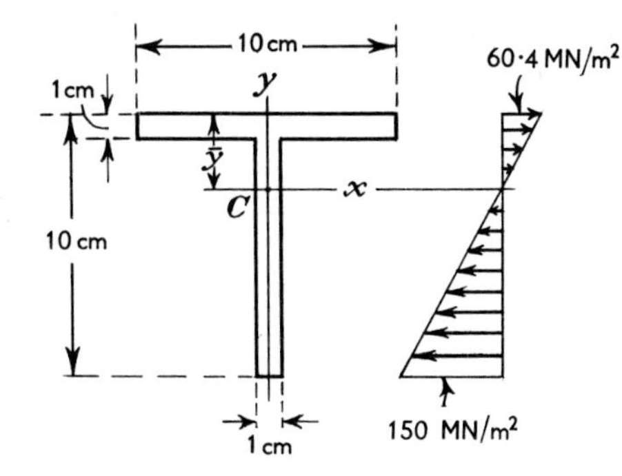

Solution

Suppose $\bar{y}$ is the distance of the principal axis Cx from the remote edge of the flange. The total area of the section is

$$A = (0{\cdot}10)(0{\cdot}01) + (0{\cdot}09)(0{\cdot}01) = 1{\cdot}90 \times 10^{-3}\ \text{m}^2$$

On taking first moments of areas about the upper edge of the flange,

$$A\bar{y} = (0{\cdot}10)(0{\cdot}01)(0{\cdot}005) + (0{\cdot}09)(0{\cdot}01)(0{\cdot}055) = 0{\cdot}0545 \times 10^{-3}\ \text{m}^3$$

Then

$$\bar{y} = \frac{0{\cdot}0545 \times 10^{-3}}{1{\cdot}9 \times 10^{-3}} = 0{\cdot}0287\ \text{m}$$

The second moment of area of the flange about Cx is

$$\tfrac{1}{12}(0{\cdot}10)(0{\cdot}01)^3 + (0{\cdot}10)(0{\cdot}01)(0{\cdot}0237)^2 = 0{\cdot}570 \times 10^{-6}\ \text{m}^4$$

The second moment of area of the web about Cx is

$$\tfrac{1}{12}(0{\cdot}01)(0{\cdot}09)^3 + (0{\cdot}09)(0{\cdot}01)(0{\cdot}0263)^2 = 1{\cdot}230 \times 10^{-6}\ \text{m}^4$$

Then

$$I_x = (0{\cdot}570 + 1{\cdot}230)10^{-6} = 1{\cdot}800 \times 10^{-6}\ \text{m}^4$$

For bending about Cx, the greatest bending stress occurs at the toe of the web, as shown in the figure. The maximum allowable moment is

$$M_x = \frac{(150 \times 10^6)(1{\cdot}800 \times 10^{-6})}{0{\cdot}0713} = 3790\ \text{Nm}$$

The bending stress in the extreme fibres of the flange is only 60·4 MN/m² at this bending moment. The second moment of area about Cy is

$$I_y = \tfrac{1}{12}(0{\cdot}01)(0{\cdot}10)^3 + \tfrac{1}{12}(0{\cdot}09)(0{\cdot}01)^3 = 0{\cdot}841 \times 10^{-6}\ \text{m}^4$$

The T-section is symmetrical about Cy, and for bending about this axis equal tensile and compressive stresses are induced in the extreme fibres of the flange; the greatest allowable moment is

$$M_y = \frac{(150 \times 10^6)(0{\cdot}841 \times 10^{-6})}{0{\cdot}05} = 2520\ \text{Nm}$$

9.6 More general case of pure bending

In the analysis of the preceding sections we have assumed either that the cross-section has two axes of symmetry, or that bending takes place about a principal axis. In the more general case we are interested in bending stresses in the beam when moments are applied about any axis of the cross-section. Consider a long uniform beam, Fig. 9.11,

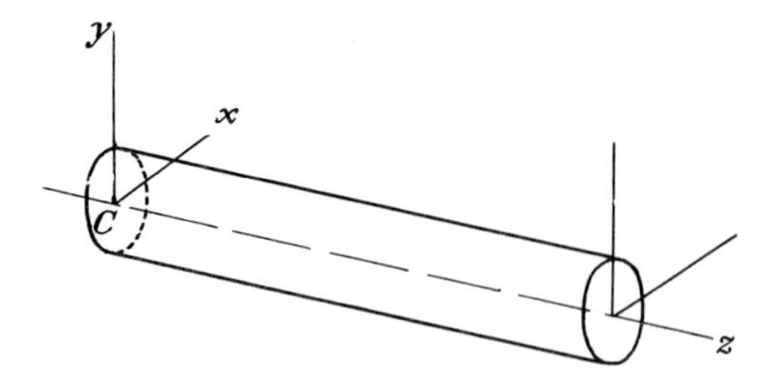

Fig. 9.11 Coordinate system for a beam of any cross-sectional form.

having any cross-section; the centroid of a cross-section is C, and Cz is the longitudinal axis of the beam; Cx, Cy are any two mutually perpendicular axes in the cross-section. The axes Cx, Cy, Cz are therefore centroidal axes of the beam.

We suppose first that the beam is bent in the yz-plane only, in such a way that the axis Cz takes up the form of a circular arc of radius R_x, Fig. 9.12. Suppose further there is no longitudinal strain of Cx; this axis is then a neutral axis. The strain at a distance y from the neutral axis is

$$\epsilon = \frac{y}{R_x}$$

If the material of the beam is elastic, the longitudinal stress in this fibre is

$$\sigma = \frac{Ey}{R_x}$$

Suppose δA is a small element of area of the cross-section of the beam acted upon by the direct stress σ, Figs. 9.12 and 9.13. Then the total thrust on any cross-section in the direction Cz is

$$\int_A \sigma \, dA = \frac{E}{R_x} \int_A y \, dA$$

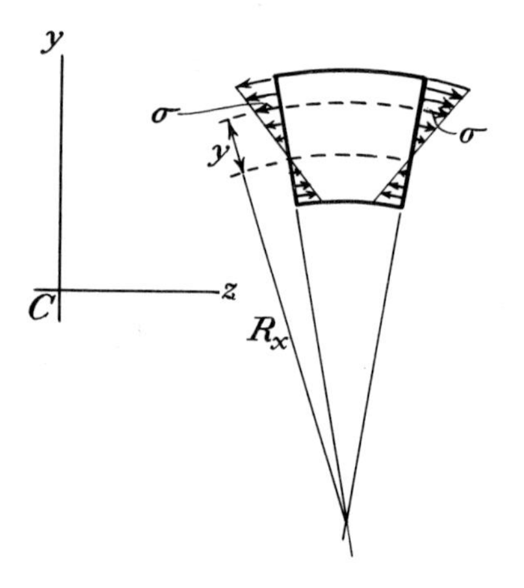

Fig. 9.12 Bending in the yz-plane.

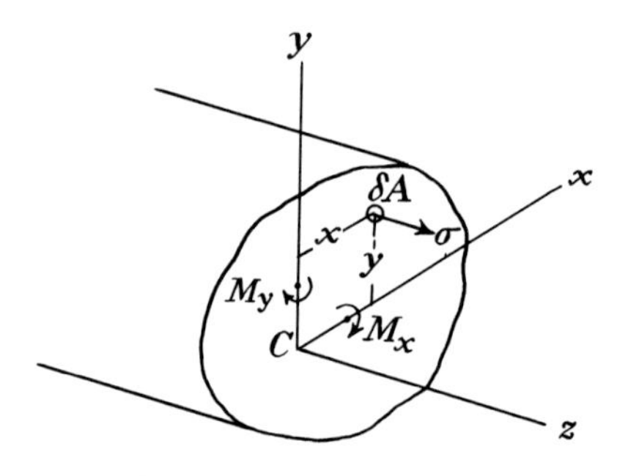

Fig. 9.13 Bending moments about the axes C_x, C_y.

where the integration is performed over the whole area A of the beam. But, since Cx is a centroidal axis, we have

$$\int_A y\,dA = 0$$

and no resultant longitudinal thrust is implied by the stresses σ. The moment about Cx due to the stresses σ is

$$M_x = \int_A \sigma y\,dA = \frac{E}{R_x}\int_A y^2\,dA = \frac{EI_x}{R_x} \tag{9.9}$$

where I_x is the second moment of area of the cross-section about Cx. For the resultant moment about Cy we have

$$M_y = \int_A \sigma x\,dA = \frac{E}{R_x}\int_A xy\,dA = \frac{EI_{xy}}{R_x} \tag{9.10}$$

where I_{xy} is the product second moment of area of the cross-section about Cx and Cy. Unless I_{xy} is zero, in which case Cx, Cy are the principal axes, bending in the yz-plane implies not only a couple M_x about the Cx axis, but also a couple M_y about Cy.

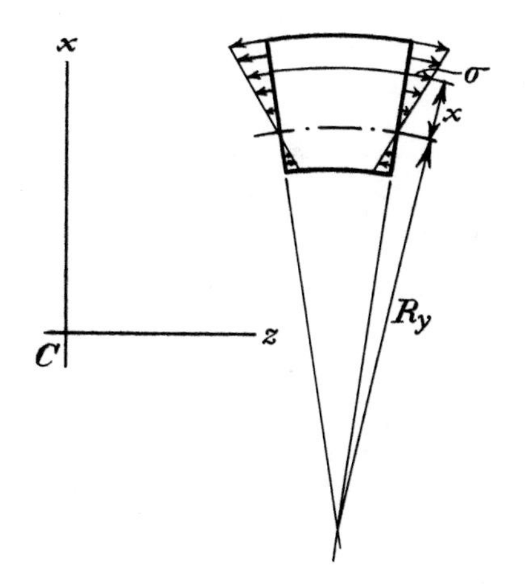

Fig. 9.14 Bending in the xz-plane.

When the beam is bent in the xz-plane only, Fig. 9.14, so that Cz again lies in the neutral surface, and takes up a curve of radius R_y, the longitudinal stress in a fibre a distance x from the neutral axis is

$$\sigma = \frac{Ex}{R_y}$$

The thrust implied by these stresses is again zero since

$$\int_A \sigma\,dA = \frac{E}{R_y}\int_A x\,dA = 0$$

because Cy is a centroidal axis of the cross-section. The bending moment about Cy due to stresses σ is

$$M_y = \int_A \sigma x \, dA = \frac{E}{R_y}\int_A x^2 \, dA = \frac{EI_y}{R_y} \tag{9.11}$$

where I_y is the second moment of area of the cross-section about Cy. Furthermore

$$M_x = \int_A \sigma y \, dA = \frac{E}{R_y}\int_A xy \, dA = \frac{EI_{xy}}{R_y} \tag{9.12}$$

where I_{xy} is again the product second moment of area.

If we now superimpose the two loading conditions, the total moments about the axes Cx and Cy, respectively, are

$$M_x = \frac{EI_x}{R_x} + \frac{EI_{xy}}{R_y} \tag{9.13}$$

$$M_y = \frac{EI_y}{R_y} + \frac{EI_{xy}}{R_x} \tag{9.14}$$

These equations may be rearranged in the forms

$$\frac{1}{R_x} = \frac{M_x I_y - M_y I_{xy}}{E(I_x I_y - I_{xy}^2)} \tag{9.15}$$

$$\frac{1}{R_y} = \frac{M_y I_x - M_x I_{xy}}{E(I_x I_y - I_{xy}^2)} \tag{9.16}$$

$(1/R_x)$, $(1/R_y)$ are the curvatures in the yz and xz planes caused by any set of moments M_x, M_y. If C_x, C_y are principal centroidal axes then $I_{xy} = 0$, and equations (9.15) and (9.16) reduce to

$$\frac{1}{R_x} = \frac{M_x}{EI_x}, \qquad \frac{1}{R_y} = \frac{M_y}{EI_y} \tag{9.17}$$

In general then we require a knowledge of three geometrical properties of the cross-section, namely, I_x, I_y and I_{xy}. The resultant longitudinal stress at any point (x, y) of the cross-section of the beam is

$$\sigma = \frac{Ex}{R_y} + \frac{Ey}{R_x} = \frac{x(M_y I_x - M_x I_{xy}) + y(M_x I_y - M_y I_{xy})}{(I_x I_y - I_{xy}^2)} \tag{9.18}$$

This stress is zero for points of the cross-section on the line

$$x(M_y I_x - M_x I_{xy}) + y(M_x I_y - M_y I_{xy}) = 0 \tag{9.19}$$

which is the equation of the unstressed fibre, or neutral axis, of the beam.

Problem 9.5: The I-section of Problem 9.2 is bent by couples of 2500 Nm about Cx and 500 Nm about Cy. Estimate the maximum bending stress in the cross-section, and find the equation of the neutral axis of the beam.

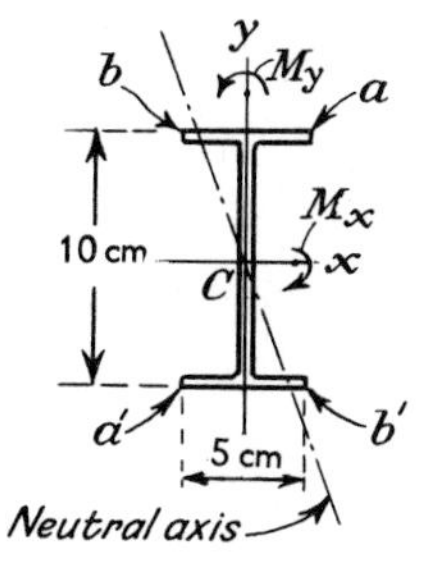

Solution

From Problem 9.2

$$I_x = 1{\cdot}642 \times 10^{-6}\ \mathrm{m}^2, \qquad I_y = 0{\cdot}130 \times 10^{-6}\ \mathrm{m}^2$$

For bending about Cx the bending stresses in the extreme fibres of the flanges are

$$\sigma = \frac{M_x y}{I_x} = \frac{(2500)(0{\cdot}05)}{1{\cdot}642 \times 10^{-6}} = 76{\cdot}1\ \mathrm{MN/m^2}$$

For bending about Cy the bending stresses at the extreme ends of the flanges are

$$\sigma = \frac{M_y x}{I_y} = \frac{(500)(0{\cdot}025)}{0{\cdot}130 \times 10^{-6}} = 96{\cdot}1\ \mathrm{MN/m^2}$$

On superposing the stresses due to the separate moments, the stress at the corner a is tensile, and of magnitude

$$\sigma_a = (76{\cdot}1 + 96{\cdot}1) = 172{\cdot}2\ \mathrm{MN/m^2}$$

The total stress at the corner a' is also $172{\cdot}2\ \mathrm{MN/m^2}$, but compressive. The total stress at the corner b is compressive, and of magnitude

$$\sigma_b = (96{\cdot}1 - 76{\cdot}1) = 20{\cdot}0\ \mathrm{MN/m^2}$$

The total stress at the corner b' is also $20{\cdot}0\ \mathrm{MN/m^2}$, but tensile. The equation of the neutral axis is given by

$$xM_y I_x + yM_x I_y = 0$$

Then

$$\frac{y}{x} = -\frac{M_y I_x}{M_x I_y} = -\frac{(500)(1{\cdot}642 \times 10^{-6})}{(2500)(0{\cdot}130 \times 10^{-6})} = -2{\cdot}53$$

The greatest bending stresses occur at points most remote from the neutral axis; these are the points a and a'. The greatest bending stresses are therefore $\pm 172{\cdot}2\ \mathrm{MN/m^2}$.

9.7 Elastic section modulus

For bending of a section about a principal axis Cx, the longitudinal bending stress at a fibre a distance y from Cx, due to a moment M_x, is from equation (9.18), (in which we put $I_{xy} = 0$ and $M_y = 0$),

$$\sigma = \frac{M_x y}{I_x}$$

where I_x is the second moment of area about Cx. The greatest bending stress occurs at the fibre most remote from Cx. If the distance to this extreme fibre is y_{max}, the maximum bending stress is

$$\sigma_{max} = \frac{M_x y_{max}}{I_x}$$

The allowable moment for a given value of σ_{max} is therefore

$$M_x = \frac{I_x \sigma_{max}}{y_{max}} \tag{9.20}$$

The geometrical quantity (I_x/y_{max}) is the *elastic section modulus*, and is denoted by Z_e. Then

$$M_x = Z_e \sigma_{max} \tag{9.21}$$

The allowable bending moment is therefore the product of a geometrical quantity, Z_e, and the maximum allowable stress, σ_{max}. The quantity $Z_e \sigma_{max}$ is frequently called the *elastic moment of resistance*.

Problem 9.6: A steel I-beam is to be designed to carry a bending moment of 10^5 Nm, and the maximum bending stress is not to exceed 150 MN/m^2. Estimate the required elastic section modulus, and find a suitable beam.

Solution

The required elastic section modulus is

$$Z_e = \frac{M}{\sigma} = \frac{10^5}{150 \times 10^6} = 0{\cdot}667 \times 10^{-3} \text{ m}^3$$

The elastic section modulus of a 22·8 cm by 17·8 cm standard steel I-beam about its axis of greatest bending stiffness is $0{\cdot}759 \times 10^{-3}$ m^3, which is a suitable beam.

9.8 Longitudinal stresses while shearing forces are present

The analysis of the preceding paragraphs deals with longitudinal stresses in beams under uniform bending moment. No shearing forces are present at cross-sections of the beam in this case.

When a beam carries lateral forces, bending moments may vary along the length of the beam. Under these conditions we may assume with sufficient accuracy in most engineering problems that the longitudinal stresses at any section are dependent only on the bending moment at that section, and are unaffected by the shearing force at that section.

Where a shearing force is present at the section of a beam, an elemental length of the beam undergoes a slight shearing distortion; these shearing distortions make a negligible contribution to the total deflection of the beam in most engineering problems.

Problem 9.7: A 4 m length of the I-beam of Problem 9.2 is simply-supported at each end. What maximum central lateral load may be applied if the bending stresses are not to exceed 150 MN/m^2?

Solution

Suppose W is the central load. If this is applied in the plane of the web, then bending takes place about Cx. The maximum bending moment is

$$M_x = \tfrac{1}{2}W(2) = W \text{ Nm}$$

From Problem 9.2,

$$I_x = 1{\cdot}642 \times 10^{-6} \text{ m}^4$$

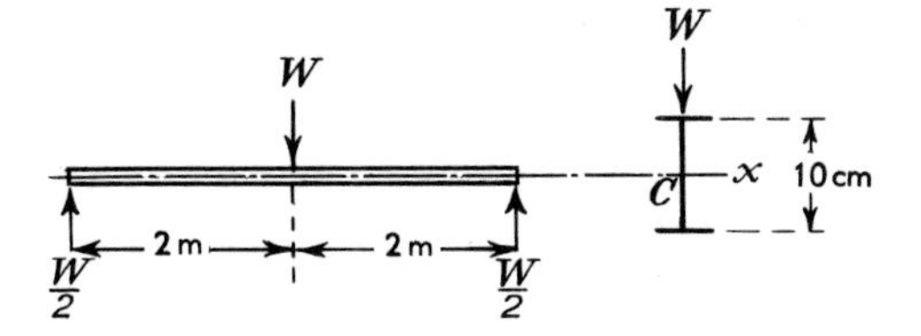

Then, the greatest bending stress is

$$\sigma = \frac{M_x y_{max}}{I_x} = \frac{(W)(0{\cdot}05)}{1{\cdot}642 \times 10^{-6}}$$

If this is equal to 150 MN/m^2, then

$$W = \frac{(150 \times 10^6)(1{\cdot}642 \times 10^{-6})}{0{\cdot}05} = 4920 \text{ N}$$

Problem 9.8: If the bending stresses are again limited to 150 MN/m^2, what total uniformly-distributed load may be applied to the beam of Problem 9.7?

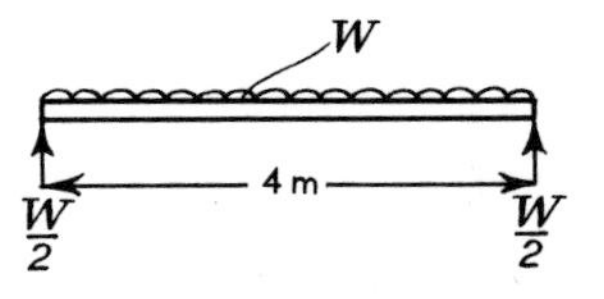

Solution

The maximum bending moment occurs at mid-span, and has the value

$$M_x = \frac{WL}{8} = \tfrac{1}{2}W \text{ Nm}$$

Then

$$\tfrac{1}{2}W = \frac{(150 \times 10^6)(1{\cdot}642 \times 10^{-6})}{0{\cdot}05} = 4920 \text{ N}$$

and

$$W = 9840 \text{ N}$$

9.9 Calculation of the principal second moments of area

In problems of bending involving beams of unsymmetrical cross-section we have frequently to find the principal axes of the cross-section.

Suppose Cx, Cy are any two centroidal axes of the cross-section of the beam, Fig. 9.15. If δA is an elemental area of the cross-section at the point (x, y), then the property of the axes Cx, Cy is that

$$\int_A x\, dA = \int_A y\, dA = 0$$

The second moments of area about the axes Cx, Cy, respectively, are

$$I_x = \int_A y^2\, dA, \qquad I_y = \int_A x^2\, dA \tag{9.22}$$

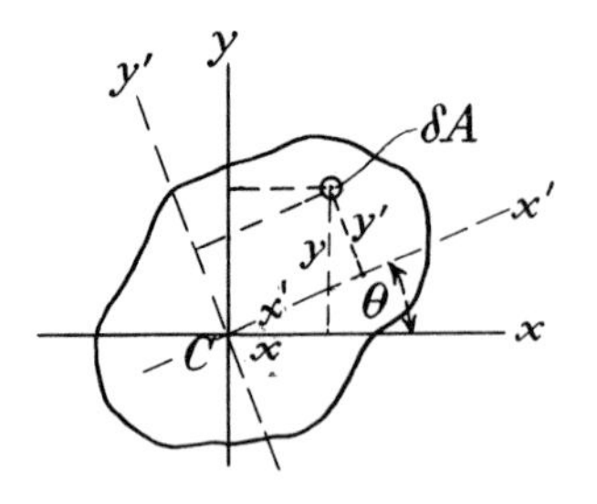

Fig. 9.15 Derivation of the principal axes of a section.

The product second moment of area is

$$I_{xy} = \int_A xy\, dA \tag{9.23}$$

Now consider two mutually perpendicular axes Cx', Cy', which are inclined at an angle θ to the axes Cx, Cy. A point having coordinates (x, y) in the xy-system, now has coordinates (x', y') in the $x'y'$-system. Further, we have

$$x' = x \cos\theta + y \sin\theta$$
$$y' = y \cos\theta - x \sin\theta$$

The second moment of area of the cross-section about Cx' is

$$I_{x'} = \int_A y'^2\, dA$$

which becomes

$$I_{x'} = \int_A (y \cos\theta - x \sin\theta)^2\, dA$$

This may be written

$$I_{x'} = \cos^2\theta \int_A y^2\, dA - 2 \cos\theta \sin\theta \int_A xy\, dA + \sin^2\theta \int_A x^2\, dA$$

But

$$\int_A y^2\, dA = I_x, \qquad \int_A x^2\, dA = I_y, \quad \text{and} \quad \int_A xy\, dA = I_{xy}$$

Then

$$I_{x'} = I_x \cos^2\theta - 2I_{xy} \cos\theta \sin\theta + I_y \sin^2\theta \tag{9.24}$$

Similarly, the second moment of area about Cy' is

$$I_{y'} = \int_A x'^2\, dA = \int_A (x \cos\theta + y \sin\theta)^2\, dA$$

Then

$$I_{y'} = I_y \cos^2\theta + 2I_{xy} \cos\theta \sin\theta + I_x \sin^2\theta \tag{9.25}$$

Finally, the product second moment of area about Cx' and Cy' is

$$I_{x'y'} = \int_A x'y'\, dA = \int_A (x \cos \theta + y \sin \theta)(y \cos \theta - x \sin \theta)\, dA$$

Then

$$I_{x'y'} = I_x \sin \theta \cos \theta + I_{xy}(\cos^2 \theta - \sin^2 \theta) - I_y \cos \theta \sin \theta \tag{9.26}$$

We note first that

$$I_{x'} + I_{y'} = I_x + I_y \tag{9.27}$$

that is, the sum of the second moments of area about any perpendicular axes is independent of θ. The sum is in fact the polar second moment of area, or the second moment of area about an axis through C, perpendicular to the xy-plane.

We may write equation (9.26) in the form

$$I_{x'y'} = \tfrac{1}{2}(I_x - I_y) \sin 2\theta + I_{xy} \cos 2\theta \tag{9.28}$$

The principal axes are defined as those for which $I_{x'y'} = 0$; then for the principal axes

$$\tfrac{1}{2}(I_x - I_y) \sin 2\theta + I_{xy} \cos 2\theta = 0$$

or

$$\tan 2\theta = \frac{2I_{xy}}{I_y - I_x} \tag{9.29}$$

This relation gives two values of θ differing by 90°. On making use of equation (9.27), we may write equations (9.24) and (9.25) in the forms

$$\begin{aligned} I_{x'} &= \tfrac{1}{2}(I_x + I_y) + \tfrac{1}{2}(I_x - I_y) \cos 2\theta - I_{xy} \sin 2\theta \\ I_{y'} &= \tfrac{1}{2}(I_x + I_y) - \tfrac{1}{2}(I_x - I_y) \cos 2\theta + I_{xy} \sin 2\theta \end{aligned} \tag{9.30}$$

Then

$$I_{x'}I_{y'} = \tfrac{1}{2}(I_x + I_y)^2 - \tfrac{1}{2}[(I_x - I_y) \cos 2\theta - 2I_{xy} \sin 2\theta]^2 \tag{9.31}$$

For the principal axes, $\tan 2\theta$ is defined by equation (9.29), and $I_{x'}I_{y'}$ reduces to

$$I_{x'}I_{y'} = I_xI_y - I_{xy}^2 \tag{9.32}$$

From equations (9.27) and (9.31) we see that $I_{x'}$, $I_{y'}$ are the roots of the quadratic equation

$$I^2 + (I_{x'} + I_{y'})I + I_{x'}I_{y'} = 0 \tag{9.33}$$

or

$$I^2 + (I_x + I_y)I + (I_xI_y - I_{xy}^2) = 0 \tag{9.34}$$

Then

$$I = \tfrac{1}{2}(I_x + I_y) \pm \sqrt{\tfrac{1}{4}(I_x + I_y)^2 - (I_xI_y - I_{xy}^2)} \tag{9.35}$$

which may be written

$$I = \tfrac{1}{2}(I_x + I_y) \pm \sqrt{\tfrac{1}{4}(I_x - I_y)^2 + I_{xy}^2} \tag{9.36}$$

Equations (9.30) may be written in the forms

$$\begin{aligned} I_{x'} - \tfrac{1}{2}(I_x + I_y) &= \tfrac{1}{2}(I_x - I_y) \cos 2\theta - I_{xy} \sin 2\theta \\ I_{x'y'} &= \tfrac{1}{2}(I_x - I_y) \sin 2\theta + I_{xy} \cos 2\theta \end{aligned} \tag{9.37}$$

Square each equation, and then add; we have

$$[I_{x'} - \tfrac{1}{2}(I_x + I_y)]^2 + [I_{x'y'}]^2 = [\tfrac{1}{2}(I_x - I_y)]^2 + [I_{xy}]^2 \tag{9.38}$$

Then $I_{x'}$, $I_{x'y'}$ lie on a circle of radius

$$\{[\tfrac{1}{2}(I_x - I_y)]^2 + [I_{xy}]^2\}^{\frac{1}{2}} \tag{9.39}$$

and centre

$$[\tfrac{1}{2}(I_x + I_y), 0] \tag{9.40}$$

in the $I_{x'}I_{x'y'}$ diagram. Suppose $OI_{x'}$, $OI_{x'y'}$ are mutually perpendicular axes; then equation (9.38) has the graphical representation shown in Fig. 9.16. To find the principal

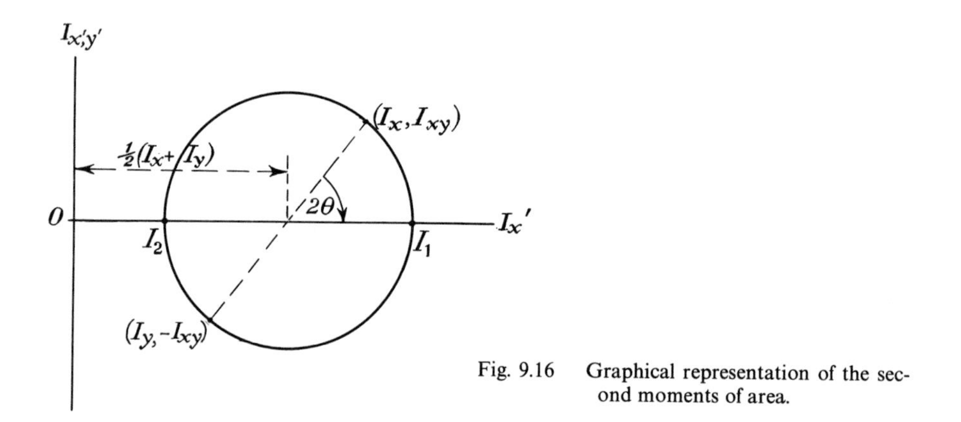

Fig. 9.16 Graphical representation of the second moments of area.

second moments of area, locate the points (I_x, I_{xy}) and $(I_y, -I_{xy})$ in the $(I_{x'}, I_{x'y'})$ plane. With the line joining these points as a diameter construct a circle. The principal second moments of area, I_1 and I_2, are given by the points where the circle cuts the axis $OI_{x'}$. Figure 9.16 might be referred to as the *circle of second moments of area.*

Problem 9.9: An unequal angle section of uniform thickness 0·5 cm has legs of lengths 6 cm and 4 cm. Estimate the positions of the principal axes, and the principal second moments of area.

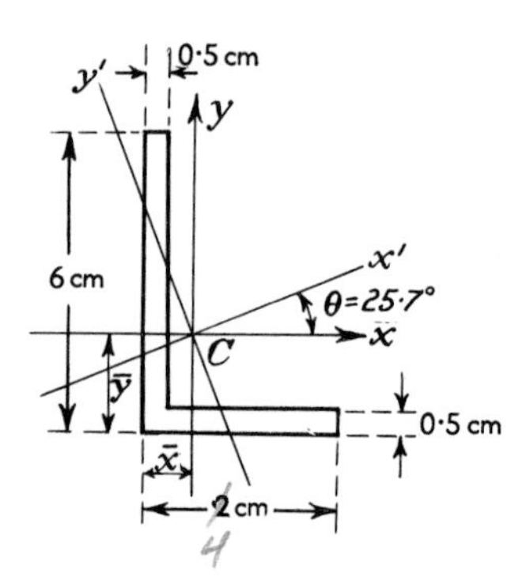

Solution

Firstly, find the position of the centroid of the cross-section:

Total area, $A = (0{\cdot}06)(0{\cdot}005) + (0{\cdot}035)(0{\cdot}005)$

$= 0{\cdot}475 \times 10^{-3}\ \text{m}^2$

Now,

$$A\bar{z} = (0{\cdot}055)(0{\cdot}005)(0{\cdot}0025) + (0{\cdot}04)(0{\cdot}005)(0{\cdot}02)$$
$$= 4{\cdot}69 \times 10^{-6}\ \text{m}^3$$

Then

$$\bar{z} = \frac{4{\cdot}69 \times 10^{-6}}{0{\cdot}475 \times 10^{-3}} = 9{\cdot}86 \times 10^{-3}\ \text{m}$$

Again,

$$A\bar{y} = (0{\cdot}035)(0{\cdot}005)(0{\cdot}0025) + (0{\cdot}06)(0{\cdot}005)(0{\cdot}03) = 9{\cdot}44 \times 10^{-6}\ \text{m}^3$$

Then

$$\bar{y} = \frac{9{\cdot}44 \times 10^{-6}}{0{\cdot}475 \times 10^{-3}} = 19{\cdot}85 \times 10^{-3}\ \text{m}$$

Now,

$$I_x = \tfrac{1}{3}(0{\cdot}005)(0{\cdot}06)^3 + \tfrac{1}{3}(0{\cdot}035)(0{\cdot}005)^3 - (0{\cdot}475 \times 10^{-3})(0{\cdot}01985)^2$$
$$= 0{\cdot}154 \times 10^{-6}\ \text{m}^4$$

and

$$I_y = \tfrac{1}{3}(0{\cdot}005)(0{\cdot}04)^3 + \tfrac{1}{3}(0{\cdot}055)(0{\cdot}005)^3 - (0{\cdot}475 \times 10^{-3})(0{\cdot}00986)^2$$
$$= 0{\cdot}063 \times 10^{-6}\ \text{m}^4$$

With the axes Cx, Cy having the positive directions shown,

$$I_{xy} = \iint xy\, dx\, dy$$
$$= \int_{-\bar{x}}^{0{\cdot}04-\bar{x}} x\, dx \int_{-\bar{y}}^{0{\cdot}005-\bar{y}} y\, dy + \int_{-\bar{x}}^{0{\cdot}005-\bar{x}} x\, dx \int_{-\bar{y}}^{0{\cdot}06-\bar{y}} y\, dy$$
$$= \tfrac{1}{4}\{[(0{\cdot}04 - \bar{x})^2 - (-\bar{x})^2][(0{\cdot}005 - \bar{y})^2 - (-\bar{y})^2] + [(0{\cdot}005 - \bar{x}) - (-\bar{x})^2][(0{\cdot}06 - \bar{y})^2 - (-\bar{y})^2]\}$$
$$= -0{\cdot}0568 \times 10^{-6}\ \text{m}^4$$

From equation (9.29),

$$\tan 2\theta = \frac{2 \times 0{\cdot}0568}{0{\cdot}063 - 0{\cdot}154} = 1{\cdot}25$$

Then

$$2\theta = 51{\cdot}3^\circ$$

and

$$\theta = 25{\cdot}7^\circ$$

From equations (9.36) the principal second moments of area are

$$\tfrac{1}{2}(I_x + I_y) \pm [\{\tfrac{1}{2}(I_x - I_y)\}^2 + I_{xy}^2]^{\frac{1}{2}} = (0{\cdot}1085 \pm 0{\cdot}0728)10^{-6}$$
$$= 0{\cdot}1813 \text{ or } 0{\cdot}0357 \times 10^{-6}\ \text{m}^4$$

9.10 Compound beams

In all the bending problems treated so far, the beams have been assumed to have solid cross-sections. Many structural sections are of this type, as for example I-sections and channel sections. In practice, however, many beams are built-up from simple component sections; Fig. 9.17(i) shows an I-section compounded of a standard I-beam and flat plates, the components being welded together; Fig. 9.17(ii) shows a compound box-section, consisting of channels to which flange plates are riveted. If the components of

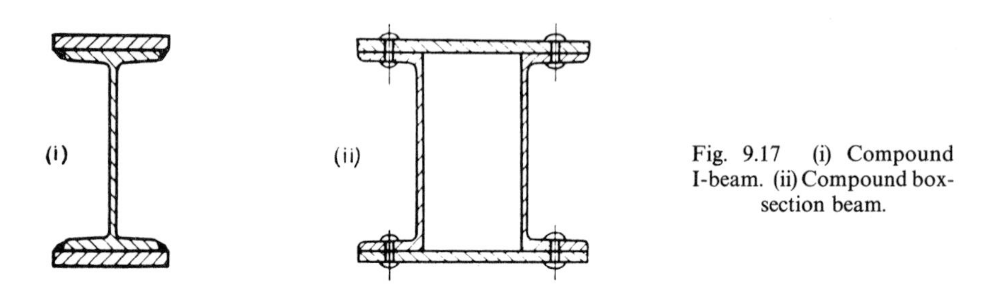

Fig. 9.17 (i) Compound I-beam. (ii) Compound box-section beam.

a compound section are of the same material, the bending stresses may be evaluated by the methods of the preceding sections, assuming that the components are rigidly connected to each other.

9.11 Elastic strain energy of bending

As couples are applied to a beam, strain energy is stored in the fibres. Consider an elemental length δz of a beam, which is bent about a principal axis Cx by a moment M_x, Fig. 9.18. During bending, the moments M_x at each end of the element are displaced

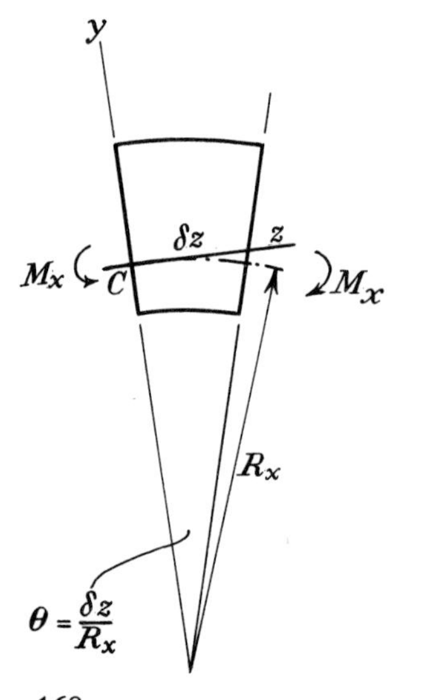

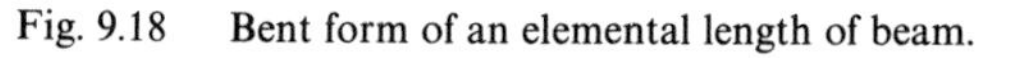

Fig. 9.18 Bent form of an elemental length of beam.

with respect to each other an angular amount

$$\theta = \frac{\delta z}{R_x} \tag{9.41}$$

where R_x is the radius of curvature in the yz-plane. But from equation (9.6)

$$M_x = \frac{EI_x}{R_x}$$

and thus

$$\theta = \frac{M_x \, \delta z}{EI_x} \tag{9.42}$$

Since there is a linear relation between θ and M_x, the total work done by the moments M_x during bending of the element is

$$\tfrac{1}{2}M_x\theta = \frac{M_x^2 \, \delta z}{2EI_x} \tag{9.43}$$

which is equal to the strain energy of bending of the element. For a uniform beam of length L under a moment M_x, constant throughout its length, the bending strain energy is then

$$U = \frac{M_x^2 L}{2EI_x} \tag{9.44}$$

When the bending moment varies along the length, the total bending strain energy is

$$U = \int_L \frac{M_x^2 \, dz}{2EI_x} \tag{9.45}$$

where the integration is carried out over the whole length L of the beam.

9.12 Change of cross-section in pure bending

In §9.1 we pointed out the change which takes place in the shape of the cross-section when a beam is bent. This change involves infinitesimal lateral strains in the beam. The upper and lower edges of a cross-section which was originally rectangular, are strained into concentric circular arcs with their centre on the opposite side of the beam

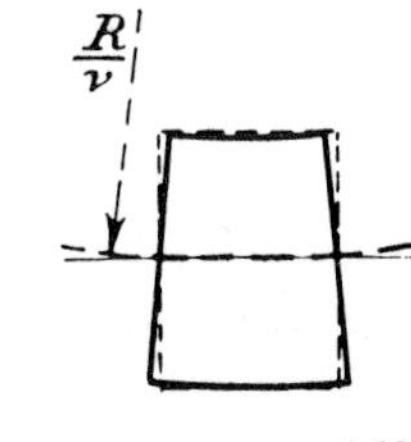

Fig. 9.19 Anti-elastic curvature in the cross-section of a bent rectangular beam.

to the axis of bending. The upper and lower surfaces of the beam then have *anticlastic curvature*, the general nature of the strain being as shown in Fig. 9.19. The anticlastic curvature effect can be readily observed by bending a flat piece of india-rubber. If the beam is bent to a mean radius R, we find that cross-sections are bent to a mean radius (R/v).

Problem 9.10: What load can a beam 4 m long carry at its centre, if the cross-section is a hollow square 30 cm by 30 cm outside and 4 cm thick, the permissible longitudinal stress being 75 MN/m²?

Solution

We must first find the second moment of area of the cross-section about its neutral axis. The inside is a square 22 cm by 22 cm. Then

$$\tfrac{1}{12}(0{\cdot}3^4 - 0{\cdot}22^4) = 0{\cdot}47 \times 10^{-3}\ \mathrm{m}^4$$

The length of the beam is 4 m; therefore if W N be a concentrated load at the middle, the maximum bending moment is

$$M_x = \frac{WL}{4} = W\ \mathrm{Nm}$$

Hence the maximum stress is

$$\sigma = \frac{M_x y}{I_x} = \frac{W(0{\cdot}15)}{0{\cdot}47 \times 10^{-3}}$$

If $\sigma = 75\ \mathrm{MN/m^2}$ we must therefore have

$$W = \frac{(75 \times 10^6)(0{\cdot}47 \times 10^{-3})}{0{\cdot}15} = 235\ \mathrm{kN}$$

Problem 9.11: Estimate the elastic section modulus and the maximum longitudinal stress in a built-up I-girder, with equal flanges carrying a load of 50 kN per metre run, with a clear span of 20 m. The web is of thickness 1·25 cm and the depth between flanges 2 m. Each flange consists of four 1 cm plates 65 cm wide, and is attached to the web by angle iron sections 10 cm by 10 cm by 1·25 cm thick. (*Cambridge*)

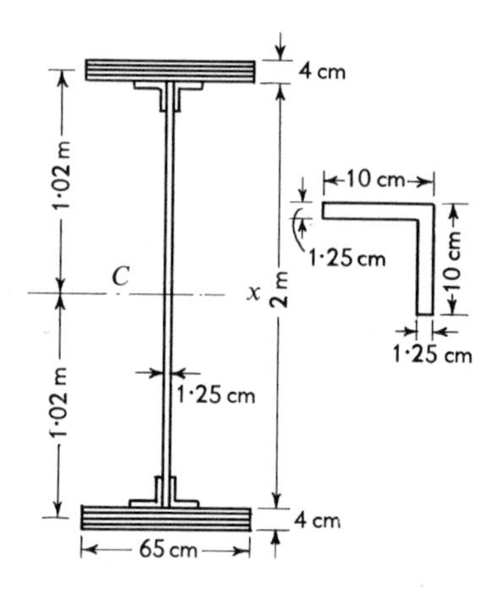

Solution

The second moment of area of each flange about Cx is

$$(0{\cdot}04)(0{\cdot}65)(1{\cdot}02)^2 = 0{\cdot}0270\ \mathrm{m}^4$$

The second moment of area of the web about Cx is

$$\tfrac{1}{12}(0{\cdot}0125)(2)^3 = 0{\cdot}0083 \text{ m}^4$$

The horizontal part of each angle section has an area $0{\cdot}00125 \text{ m}^2$, and its centroid is 0·994 m from the neutral axis. Hence the second moment of area of this about Cx is approximately

$$(0{\cdot}00125)(0{\cdot}994)^2 = 0{\cdot}0012 \text{ m}^4$$

The area of the vertical part of each angle section is $0{\cdot}001093 \text{ m}^2$, and its centroid is 0·944 m from the neutral axis. Therefore the corresponding second moment of area is approximately

$$(0{\cdot}001093)(0{\cdot}944)^2 = 0{\cdot}00097 \text{ m}^4$$

The second moment of area of the whole section of the angle section about Cx is then

$$0{\cdot}0012 + 0{\cdot}00097 = 0{\cdot}0022 \text{ m}^4$$

The second moment of area of the whole cross-section of the beam is then

$$I_x = 2(0{\cdot}0270) + (0{\cdot}0083) + 4(0{\cdot}0022)$$
$$= 0{\cdot}0711 \text{ m}^4$$

The elastic section modulus is therefore

$$Z_e = \frac{0{\cdot}0711}{1{\cdot}04} = 0{\cdot}0684 \text{ m}^3$$

The bending moment at the mid-span is

$$M_x = \frac{wL^2}{8} = \frac{(50)(20)^2}{8} = 2{\cdot}50 \text{ MNm}$$

The greatest longitudinal stress is then

$$\sigma = \frac{M_x}{Z_e} = \frac{2{\cdot}50 \times 10^6}{0{\cdot}0684} = 36{\cdot}6 \text{ MN/m}^2$$

FURTHER PROBLEMS
(answers on page 398)

9.12 A beam of I-section is 25 cm deep and has equal flanges 10 cm broad. The web is 0·75 cm thick and the flanges 1·25 cm thick. If the beam may be stressed in bending to 120 MN/m², what bending moment will it carry? (*Cambridge*)

9.13 The front-axle beam of a motor vehicle carries the loads shown. The axle is of I-section: flanges 7·5 cm by 2·5 cm, web 5 cm by 2·5 cm. Calculate the tensile stress at the bottom of the axle beam. (*Cambridge*)

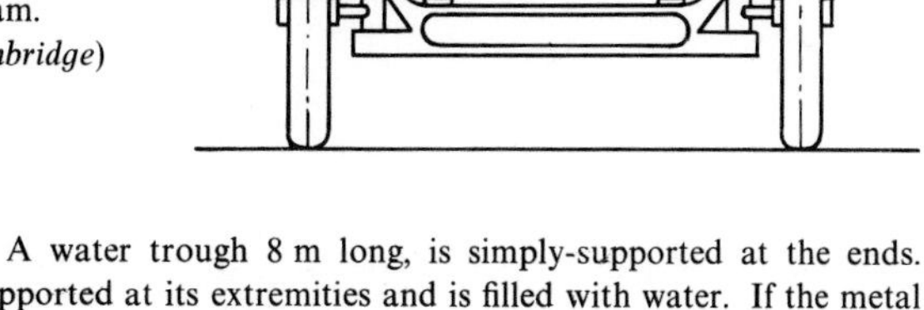

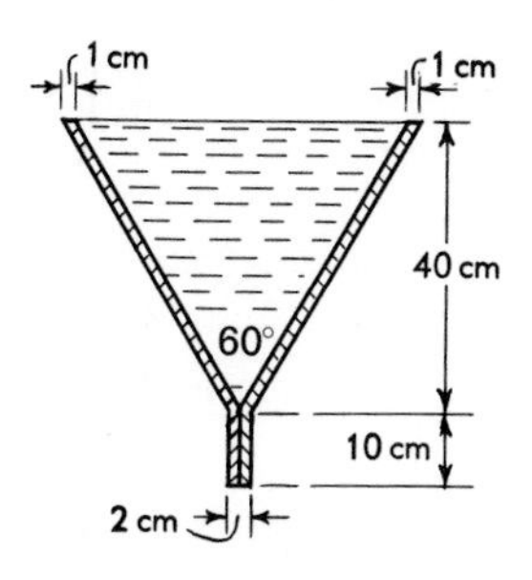

9.14 A water trough 8 m long, is simply-supported at the ends. It is supported at its extremities and is filled with water. If the metal has a density 7840 kg/m³, and the water a density 1000 kg/m³, calculate the greatest longitudinal stress for the middle cross-section of the trough. (*Cambridge*)

9.15 A built-up steel I-girder is 2 m deep over the flanges, each of which consists of four 1 cm plates, 1 m wide, riveted together. The web is 1 cm thick and is attached to the flanges by four 9 cm by 9 cm by 1 cm angle sections. The girder has a clear run of 30 m between the supports and carries a superimposed load of 60 kN per metre. Find the maximum longitudinal stress. (*Cambridge*)

9.16 A beam rests on supports 3 m apart and carries a load of 10 kN uniformly distributed. The beam is rectangular in section, 7·5 cm deep. How wide should it be if the skin-stress must not exceed 60 MN/m^2? (*RNEC*)

10 Shearing stresses in beams

10.1 Introduction

We referred earlier to the existence of longitudinal direct stresses in a cantilever with a lateral load at the free end; on a closer study we found that these stresses are distributed linearly over the cross-section of a beam carrying a uniform bending moment. In general we are dealing with bending problems in which there are shearing forces present at any cross-section, as well as bending moments. In practice we find that the longitudinal direct stresses in the beam are almost unaffected by the shearing force at any section, and are governed largely by the magnitude of the bending moment at that section. Consider again the bending of a cantilever with a concentrated lateral load F, at the free end, Fig. 10.1; suppose the beam is of rectangular cross-section. If we cut

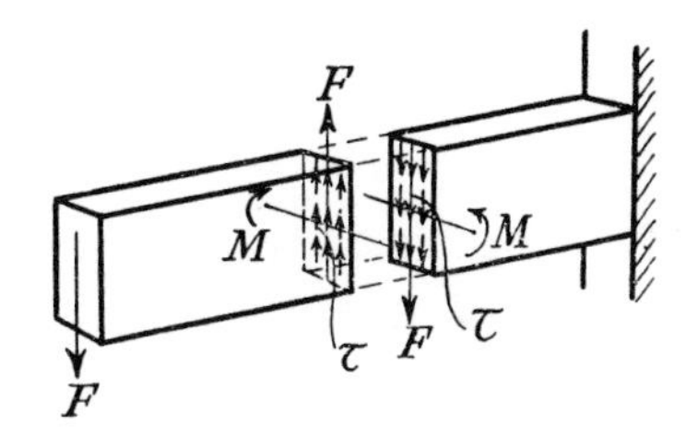

Fig. 10.1 Shearing actions in a cantilever carrying an end load.

the beam at any transverse cross-section, we must apply bending moments M and shearing forces F at the section to maintain equilibrium. The bending moment M is distributed over the cross-section in the form of longitudinal direct stresses, as already discussed. The shearing force F is distributed in the form of shearing stresses τ, acting tangentially to the cross-section of the beam; the form of the distribution of τ is dependent on the shape of the cross-section of the beam, and on the direction of application of the shearing force F. An interesting feature of these shearing stresses is that, since they give rise to complementary shearing stresses, we find that shearing stresses are also set up in longitudinal planes parallel to the axis of the beam.

10.2 Shearing stresses in a beam of narrow rectangular cross-section

We consider firstly the simple problem of a cantilever of *narrow* rectangular cross-section, carrying a concentrated lateral load F at the free end, Fig. 10.2; h is the depth of

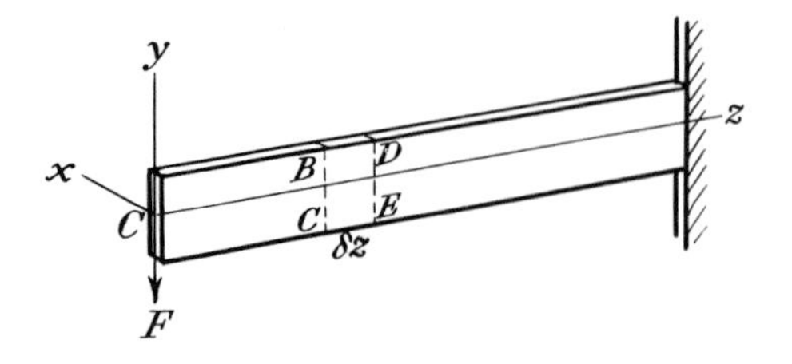

Fig. 10.2 Shearing stresses in a cantilever of narrow rectangular cross-section under end load.

the cross-section, and t is the thickness, Fig. 10.3; the depth is assumed to be large compared with the thickness. The load is applied in a direction parallel to the longer side h.

Consider an elemental length δz of the beam at a distance z from the loaded end. On the face BC of the element the hogging bending moment is

$$M = Fz$$

We suppose the longitudinal stress σ at a distance y from the centroidal axis Cx is the same as that for uniform bending of the element. Then

$$\sigma = \frac{My}{I_x} = \frac{Fyz}{I_x}$$

where I_x is the second moment of area about the centroidal axis of bending, Cx, which is also a neutral axis. On the face DE of the element the bending moment has increased to

$$M + \delta M = F(z + \delta z)$$

The longitudinal bending stress at a distance y from the neutral axis has increased correspondingly to

$$\sigma + \delta\sigma = \frac{F(z + \delta z)y}{I_x}$$

Now consider a depth of the beam contained between the upper extreme fibre BD, given by $y = \frac{1}{2}h$, and the fibre GH, given by $y = y_1$, Fig. 10.3(ii). The total longitudinal force on the face BG due to the bending stresses σ is

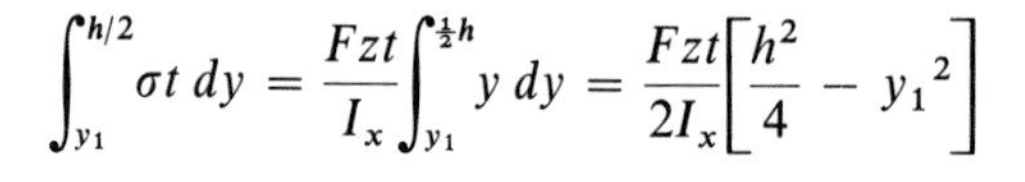

$$\int_{y_1}^{h/2} \sigma t\, dy = \frac{Fzt}{I_x}\int_{y_1}^{\frac{1}{2}h} y\, dy = \frac{Fzt}{2I_x}\left[\frac{h^2}{4} - {y_1}^2\right]$$

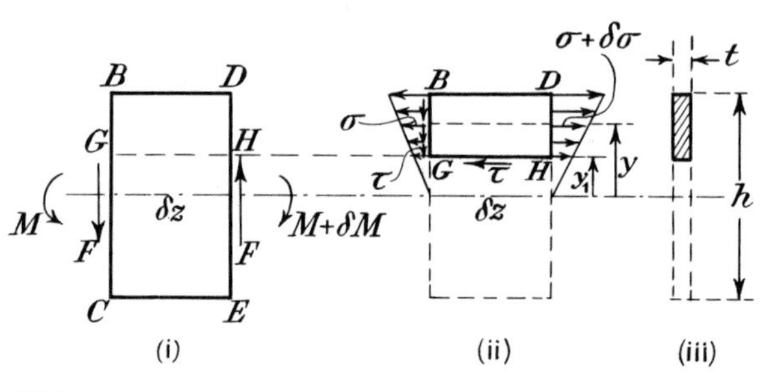

Fig. 10.3 Shearing actions on an elemental length of a beam of narrow rectangular cross-section.

By a similar argument we have that the total force on the face DH due to bending stresses $\sigma + \delta\sigma$ is

$$\frac{Ft}{2I_x}\left(\frac{h^2}{4} - y_1{}^2\right)(z + \delta z)$$

These longitudinal forces, which act in opposite directions, are not quite in balance; they differ by a small amount

$$\frac{Ft}{2I_x}\left(\frac{h^2}{4} - y_1{}^2\right)\delta z$$

Now the upper surface BD is completely free of shearing stress, and this out-of-balance force can only be equilibrated by a shearing force on the face GH. We suppose this shearing force is distributed uniformly over the face GH; the shearing stress on this face is then

$$\tau = \frac{Ft}{2I_x}\left(\frac{h^2}{4} - y_1{}^2\right)\delta z \div t\,\delta z$$

$$= \frac{F}{2I_x}\left(\frac{h^2}{4} - y_1{}^2\right) \tag{10.1}$$

This shearing stress acts on a plane parallel to the neutral surface of the beam; it gives rise therefore to a complementary shearing stress τ at a point of the cross-section a distance y_1 from the neutral axis, and acting tangentially to the cross-section. Our analysis gives then the variation of shearing stress over the depth of the cross-section. For this simple type of cross-section

$$I_x = \tfrac{1}{12}h^3 t,$$

and so

$$\tau = \frac{6F}{h^3 t}\left(\frac{h^2}{4} - y_1{}^2\right) = \frac{6F}{ht}\left[\frac{1}{4} - \left(\frac{y_1}{h}\right)^2\right] \tag{10.2}$$

We note firstly that τ is independent of z; this is so because the resultant shearing force is the same for all cross-sections, and is equal to F. The resultant shearing force implied by the variation of τ is

$$\int_{-h/2}^{+h/2} \tau t\, dy_1 = \frac{6F}{h}\int_{-h/2}^{+h/2}\left[\frac{1}{4} - \left(\frac{y_1}{h}\right)^2\right] dy_1 = F$$

The shearing stresses τ are sufficient then to balance the force F applied to every cross-section of the beam.

The variation of τ over the cross-section of the beam is parabolic, Fig. 10.4; τ attains a maximum value on the neutral axis of the beam, where $y_1 = 0$, and

$$\tau_{\max} = \frac{3F}{2ht} \tag{10.3}$$

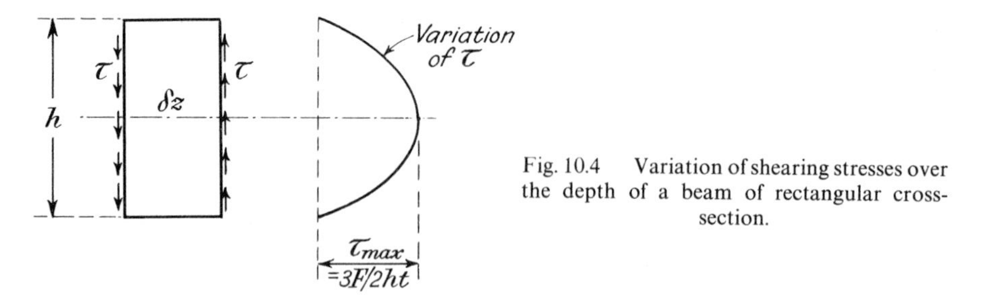

Fig. 10.4 Variation of shearing stresses over the depth of a beam of rectangular cross-section.

The shearing stresses must necessarily be zero at the extreme fibres since there can be no complementary shearing stresses in the longitudinal direction on the upper and lower surfaces of the beam.

In the case of a cantilever with a single concentrated load F at the free end the shearing force is the same for all cross-sections, and the distribution of shearing stresses is also the same for all cross-sections. In a more general case the shearing force is variable from one cross-section to another: in this case the value of F to be used is the shearing force at the section being considered.

10.3 Beam of any cross-section having one axis of symmetry

We are concerned generally with more complex cross-sectional forms than narrow rectangles. Consider a beam having a uniform cross-section which is symmetrical about Cy, Fig. 10.5. Suppose, as before, that the beam is a cantilever carrying an end load F

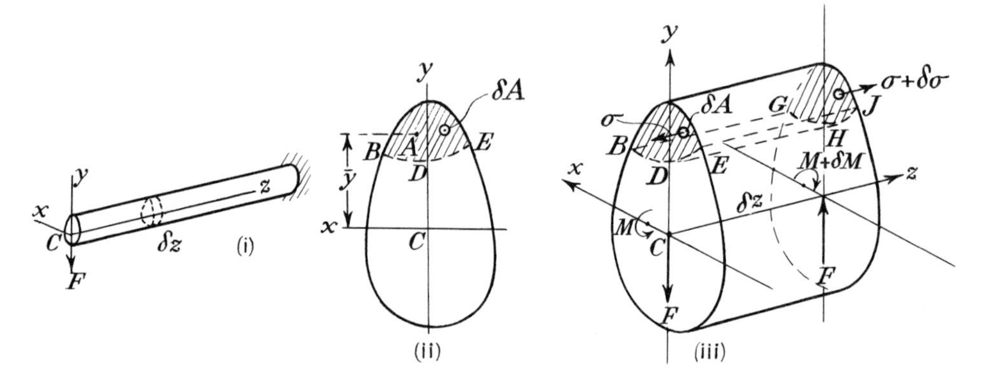

Fig. 10.5 Shearing stresses in a bent beam having one axis of symmetry.

acting parallel to Cy and passing through the centroid C of the cross-section. Then Cx is the axis of bending.

Consider an elemental length δz of the beam; on the near face of this element, which is at a distance z from the free end of the cantilever, the bending moment is

$$M = Fz$$

This gives rise to bending stresses in the cross-section; the longitudinal bending stress at a point of the cross-section a distance y from the neutral axis Cx is

$$\sigma = \frac{My}{I_x} = \frac{Fyz}{I_x}$$

Now consider a section of the element cut off by the cylindrical surface $BDEGHJ$, Fig. 10.5(iii), which is parallel to Cz. Suppose A is the area of each end of this cylindrical element; then the total longitudinal force on the end BDE due to bending stresses is

$$\int_A \sigma \, dA = \frac{Fz}{I_x} \int_A y \, dA$$

where δA is an element of the area A, and y is the distance of this element from the neutral axis Cx. The total longitudinal force on the remote end GHJ due to bending stresses is

$$\int_A (\sigma + \delta\sigma) \, dA = \frac{F}{I_x}(z + \delta z) \int_A y \, dA$$

since the bending moment at this section is

$$M + \delta M = F(z + \delta z)$$

The tension loads at the ends of the element $BDEGHJ$ differ by an amount

$$\frac{F \, \delta z}{I_x} \int_A y \, dA$$

If $\bar{y}$ is the distance of the centroid of the area A from Cx, then

$$\int_A y \, dA = A\bar{y}$$

The out-of-balance tension load is equilibrated by a shearing force over the cylindrical surface $BDEGHJ$. This shearing force is then

$$\frac{F \, \delta z}{I_x} \int_A y \, dA = \frac{F \, \delta z}{I_x} A\bar{y}$$

and acts along the surface $BDEGHJ$ and parallel to Cz. The total shearing force per unit length of the beam is

$$q = \frac{F \, \delta z}{I_x} A\bar{y} \div \delta z = \frac{FA\bar{y}}{I_x} \tag{10.4}$$

If b is the length of the curve BDE, or GHJ, then the average shearing stress over the surface $BDEGHJ$ is

$$\bar{\tau} = \frac{FA\bar{y}}{bI_x} \tag{10.5}$$

When b is small compared with the other linear dimensions of the cross-section we find that the shearing stress is nearly uniformly-distributed over the surfaces of the type $BDEGHJ$. This is the case in thin-walled beams, such as I-sections and channel sections.

10.4 Shearing stresses in an I-beam

As an application of the general method developed in the preceding paragraph, consider the shearing stresses induced in a thin-walled I-beam carrying a concentrated load F at the free end, acting parallel to Cy, Fig. 10.6. The cross-section has two axes of

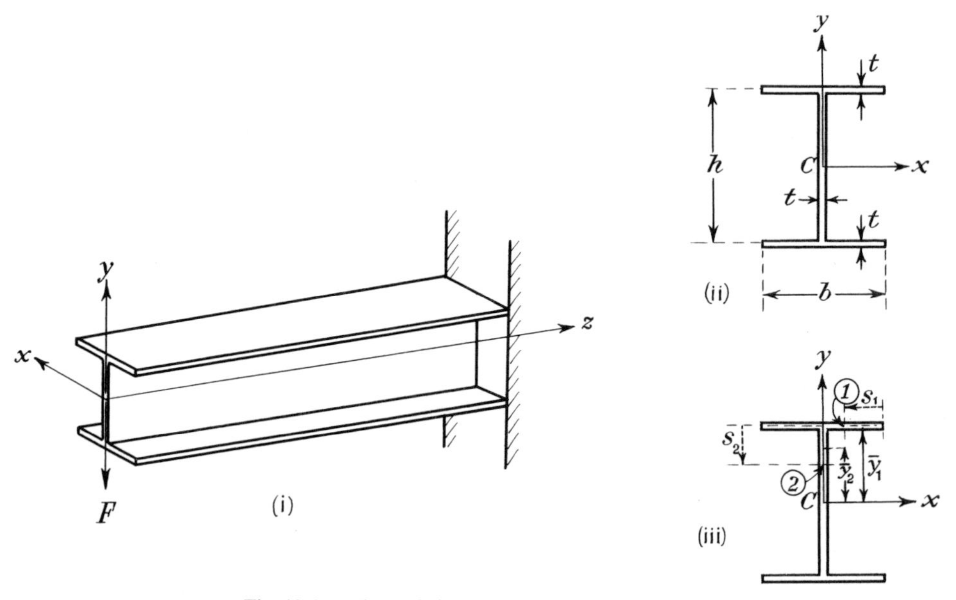

Fig. 10.6 Flexural shearing stresses in an I-beam.

symmetry—Cx and Cy; the flanges are of breadth b, and the distance between the centres of the flanges is h; the flanges and web are assumed to be of uniform thickness t.

Equation (10.4) gives the shearing force q per unit length of beam at any region of the cross-section. Consider firstly a point *1* of the flange at a distance s_1 from a free edge, Fig. 10.6(iii); the area of flange cut off by a section through the point *1* is

$$A = s_1 t$$

The distance of the centroid of this area from the neutral axis Cx is

$$\bar{y}_1 = \tfrac{1}{2}h$$

Then the shearing force at point *1* of the cross-section is

$$q = \frac{Fs_1 th}{2I_x} \tag{10.6}$$

If the wall thickness t is small compared with the other linear dimensions of the cross-section, we may assume that q is distributed uniformly over the wall thickness t; the shearing stress is then

$$\tau = \frac{q}{t} = \frac{Fs_1 h}{2I_x} \tag{10.7}$$

at point *1*. At the free edge, given by $s_1 = 0$, we have $\tau = 0$, since there can be no longitudinal shearing stress on a free edge of the cross-section. The shearing stress τ increases linearly in intensity as s_1 increases from zero to $\tfrac{1}{2}b$; at the junction of web and flanges $s_1 = \tfrac{1}{2}b$, and

$$\tau = \frac{Fbh}{4I_x} \tag{10.8}$$

Since the cross-section is symmetrical about Cy, the shearing stress in the adjacent flange also increases linearly from zero at the free edge.

Consider secondly a section through the web at the point *2* at a distance s_2 from the junctions of the flanges and web. In evaluating $A\bar{y}$ for this section we must consider the total area cut off by the section through the point *2*. However, we can evaluate $A\bar{y}$ for the component areas cut off by the section through the point *2*; we have

$$A\bar{y} = (bt)\tfrac{1}{2}h + (s_2 t)(\tfrac{1}{2}h - \tfrac{1}{2}s_2)$$
$$= \tfrac{1}{2}t[bh + s_2(h - s_2)]$$

Then

$$q = \frac{Ft}{2I_x}[bh + s_2(h - s_2)]$$

If this shearing force is assumed to be uniformly distributed as a shearing stress, then

$$\tau = \frac{q}{t} = \frac{F}{2I_x}[bh + s_2(h - s_2)] \tag{10.9}$$

At the junction of web and flanges $s_2 = 0$, and

$$\tau = \frac{Fbh}{2I_x} \tag{10.10}$$

At the neutral axis, $s_2 = \tfrac{1}{2}h$, and

$$\tau = \frac{Fbh}{2I_x}\left[1 + \frac{h}{4b}\right] \tag{10.11}$$

We notice that τ varies parabolically throughout the depth of the web, attaining a maximum value at $s_2 = \frac{1}{2}h$, the neutral axis, Fig. 10.7. In any cross-section of the beam the shearing stresses vary in the form shown; in the flanges the stresses are parallel to Cx, and contribute nothing to the total force on the section parallel to Cy.

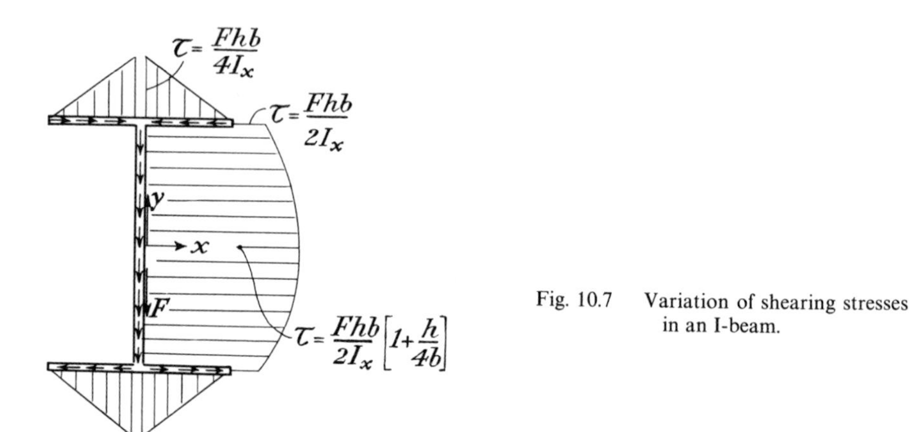

Fig. 10.7 Variation of shearing stresses in an I-beam.

At the junctions of the web and flanges the shearing stress in the web is twice the shearing stresses in the flanges. The reason for this is easily seen by considering the equilibrium conditions at this junction. Consider a unit length of the beam along the line of the junction, Fig. 10.8; the shearing stresses in the flanges are

$$\tau_f = \frac{Fbh}{4I_x} \tag{10.12}$$

while the shearing stress in the web we have estimated to be

$$\tau_w = \frac{Fbh}{2I_x} \tag{10.13}$$

For longitudinal equilibrium of a unit length of the junction of web and flanges, we have

$$2[\tau_f \times (t \times 1)] = \tau_w \times (t \times 1)$$

Fig. 10.8 Equilibrium of shearing forces at the junction of the web and flanges of an I-beam.

which gives

$$\tau_w = 2\tau_f \tag{10.14}$$

This is true, in fact, for the relations we have derived above; longitudinal equilibrium is ensured at any section of the cross-section in our treatment of the problem. If the flanges and web were of different thicknesses, t_f and t_w, respectively, the equilibrium condition at the junction would be

$$2\tau_f t_f = \tau_w t_w$$

Then

$$\frac{\tau_w}{\tau_f} = \frac{2t_f}{t_w} \tag{10.15}$$

The implication of this equilibrium condition is that at a junction, such as that of the flanges and web of an I-section, the sum of the shearing forces per unit length for the components meeting at that junction is zero when account is taken of the relevant directions of these shearing forces. For a junction

$$\sum \tau t = 0 \tag{10.16}$$

where τ is the shearing stress in an element at the junction, and t is the thickness of the element; the summation is carried out for all elements meeting at the junction.

For an I-section carrying a shearing force acting parallel to the web we see that the maximum shearing stress occurs at the middle of the web, and is given by equation (10.11). Now, I_x for the section is given approximately by

$$I_x = \tfrac{1}{12}h^3 t + \tfrac{1}{2}h^2 bt = \frac{h^3 t}{12}\left(1 + \frac{6b}{h}\right) \tag{10.17}$$

Then

$$\tau_{max} = \frac{6Fb}{h^2 t}\left[\frac{1 + h/4b}{1 + 6b/h}\right] \tag{10.18}$$

The total shearing force in the web of the beam parallel to Cy is F; if this were distributed uniformly over the depth of the web the average shearing stress would be

$$\tau_{av} = \frac{F}{ht} \tag{10.19}$$

Then for the particular case when $h = 3b$, we have

$$\tau_{max} = \frac{7}{6}\left(\frac{F}{ht}\right) \tag{10.20}$$

Then τ_{max} is only $\frac{1}{6}$th or about 17% greater than the mean shearing stress over the web.

Problem 10.1: The web of a girder of I-section is 45 cm deep and 1 cm thick; the flanges are each 22.5 cm wide by 1·25 cm thick. The girder at some particular section, has to withstand a total shearing force of 200 kN. Calculate the shearing stresses at the top and middle of the web. (Cambridge)

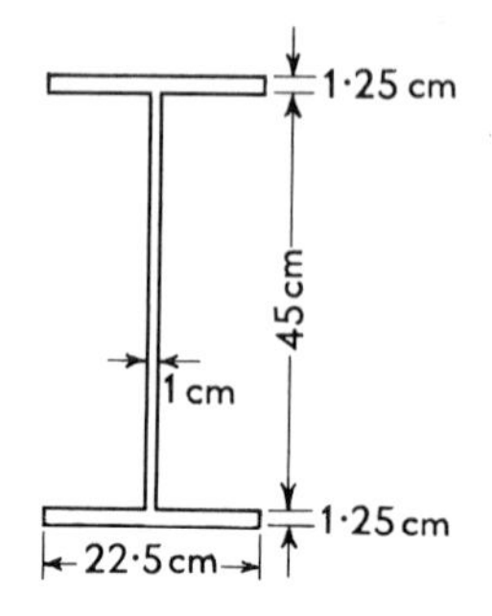

Solution

The second moment of area of the web about the centroidal axis is

$$\tfrac{1}{12}(0{\cdot}010)(0{\cdot}45)^3 = 0{\cdot}0760 \times 10^{-3}\ \text{m}^4$$

The second moment of area of each flange about the centroidal axis is

$$(0{\cdot}225)(0{\cdot}0125)(0{\cdot}231)^2 = 0{\cdot}150 \times 10^{-3}\ \text{m}^4$$

The total second moment of area is then

$$I_x = [0{\cdot}076 + 2(0{\cdot}150)]10^{-3} = 0{\cdot}376 \times 10^{-3}\ \text{m}^4$$

At a distance y above the neutral axis, the shearing stress from equation (10.9) is

$$\tau = \frac{F}{2I_x}[(bh + \tfrac{1}{4}h^2) - y^2]$$

$$= \frac{200 \times 10^3}{2 \times 0{\cdot}376 \times 10^{-3}}[(0{\cdot}225)(0{\cdot}4625) + \tfrac{1}{4}(0{\cdot}4625)^2 - y^2]$$

At the top of the web, we have $y = 0{\cdot}231$ m, and

$$\tau = 34{\cdot}6\ \text{MN/m}^2$$

While at the middle of the web, where $y = 0$, we have

$$\tau = 52{\cdot}2\ \text{MN/m}^2$$

10.5 Shearing stresses in compound beams

The results of the preceding analysis may be applied to the determination of the shearing stresses at the connection between the web and flanges of a compound beam. In the section of Fig. 10.9 the web is fillet-welded to the flanges to form a symmetrical I-beam. When a section *aa* is cut through the welds and the top of the web the shearing force, q, per unit length, over this section is carried entirely by the welds, since there is no direct connection between the web and flanges. An average shearing stress in the weld can therefore be evaluated.

When the component plates are riveted together, as shown in Fig. 10.10, the shearing force q is carried by the rivets at the junction of web and flanges. If the rivet pitch is known the load carried by each rivet can be estimated.

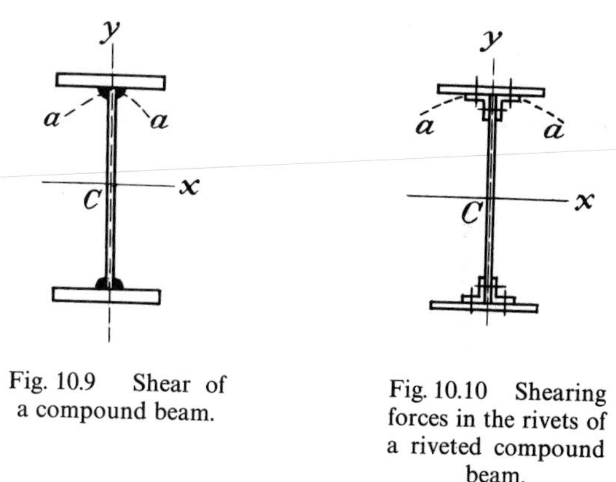

Fig. 10.9 Shear of a compound beam.

Fig. 10.10 Shearing forces in the rivets of a riveted compound beam.

10.6 Principal stresses in beams

We have shown how to find separately the longitudinal stress at any point in a beam due to bending moment, and the mean horizontal and vertical shearing stresses, but it does not follow that these are the greatest direct or shearing stresses. Within the limits of our present theory we can employ the formulae of §§ 5.7 and 5.8 to find the principal stresses and the maximum shearing stress.

We can draw, on a side elevation of the beam, lines showing the directions of the principal stresses. Such lines are called the *lines of principal stress*; they are such that the tangent at any point gives the direction of principal stress. As an example, the lines

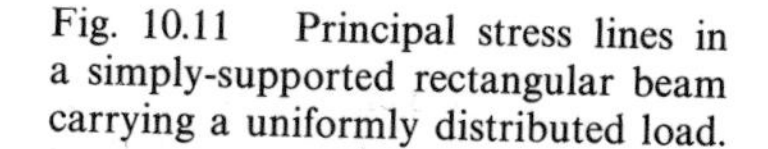

Fig. 10.11 Principal stress lines in a simply-supported rectangular beam carrying a uniformly distributed load.

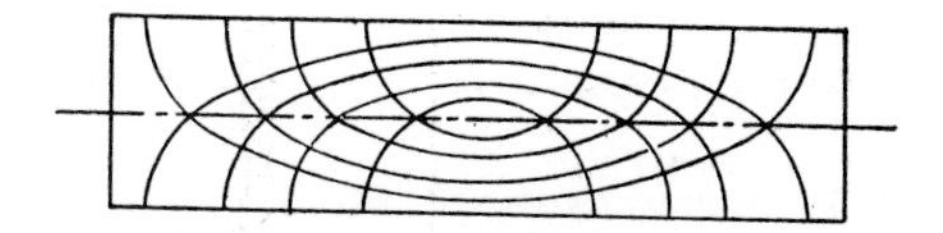

of principal stress have been drawn in Fig. 10.11 for a simply supported beam of uniform rectangular cross-section, carrying a uniformly distributed load. The stresses are a maximum where the tangents to the curves are parallel to the axis of the beam, and diminish to zero when the curves cut the faces of the beam at right angles. On the neutral axis, where the stress is one of shear, the principal stress curves cut the axis at 45°.

Problem 10.2: The flanges of an I-girder are 30 cm wide by 2·5 cm thick and the web is 60 cm deep by 1·25 cm thick. At a particular section the sagging bending moment is 500 kNm and the shearing force is 500 kN. Consider a point in the section at the top of the web and calculate for this point; (i) the longitudinal stress, (ii) the shearing stress, (iii) the principal stresses. (*Cambridge*)

Solution

First calculate the second moment of area about the neutral axis: the second moment of area of the web is

$$\tfrac{1}{12}(0{\cdot}0125)(0{\cdot}6)^3 = 0{\cdot}225 \times 10^{-3}\ \text{m}^4$$

The second moment of area of each flange is

$$(0{\cdot}3)(0{\cdot}025)(0{\cdot}3125)^2 = 0{\cdot}733 \times 10^{-3}\ \text{m}^4$$

The total second moment of area is then

$$I_x = [0{\cdot}225 + 2(0{\cdot}733)]10^{-3} = 1{\cdot}691 \times 10^{-3}\ \text{m}^4$$

Next, for a point at the top of the web,

$$A\bar{y} = (0{\cdot}3 \times 0{\cdot}025)(0{\cdot}3125) = 2{\cdot}34 \times 10^{-3}\ \text{m}^3$$

Then, for this point, with $M = 500$ kNm we have

$$\sigma = \frac{My}{I_x} = \frac{(500 \times 10^3)(0{\cdot}3)}{1{\cdot}691 \times 10^{-3}} = 88{\cdot}6\ \text{MN/m}^2\ \text{(compressive)}$$

$$\tau = \frac{FA\bar{y}}{I_x t} = \frac{(500 \times 10^3)(2{\cdot}34 \times 10^{-3})}{(1{\cdot}691 \times 10^{-3})(0{\cdot}0125)} = 55{\cdot}3\ \text{MN/m}^2$$

The principal stresses are then

$$\tfrac{1}{2}\sigma \pm [\tfrac{1}{4}\sigma^2 + \tau^2]^{\frac{1}{2}} = (-44{\cdot}3 \pm 70{\cdot}9)\ \text{MN/m}^2$$
$$= 26{\cdot}6 \text{ and } -115{\cdot}2\ \text{MN/m}^2$$

It should be noticed that the greater principal stress is about 30% greater than the longitudinal stress. At the top of the flange the longitudinal stress is -960 MN/m^2, so the greatest principal stress at the top of the web is 20% greater than the maximum longitudinal stress.

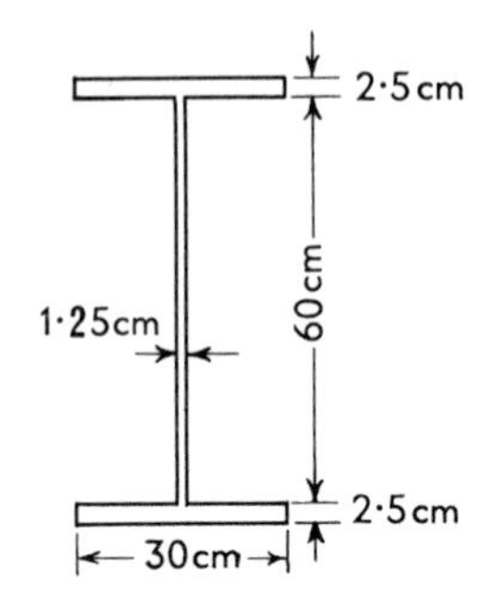

10.7 Superimposed beams

If we make a beam by placing one member on the top of another, Fig. 10.12, there will be a tendency, under the action of lateral loads, for the two members to slide over each other along the plane of contact *AB*, Fig. 10.12. Unless this sliding action is prevented

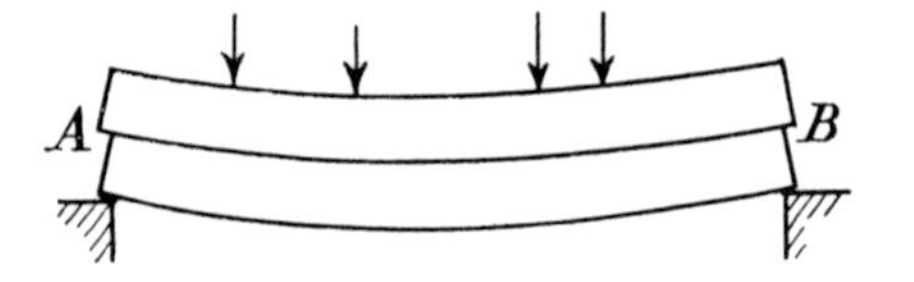

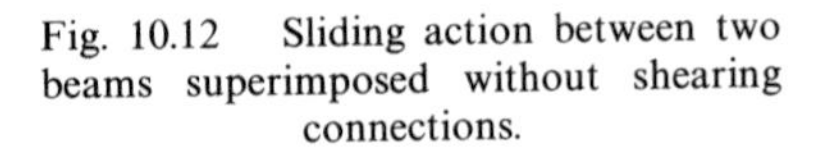

Fig. 10.12 Sliding action between two beams superimposed without shearing connections.

in some way, the one beam will act independently of the other; when there is no shearing connection between the beams along *AB*, the strength of the compound beam is the sum of the strengths of the separate beams.

However, if the sliding action is resisted, the compound beam behaves more nearly as a solid member; for elastic bending the permissible moment is proportional to the elastic section modulus. In the case of two equal beams of rectangular cross-section, the elastic section modulus of each beam is

$$\frac{bh^2}{6}$$

where b is the breadth and h is the depth of each beam. For two such beams, placed one on the other, without shearing connection, the elastic section modulus is

$$2 \times \tfrac{1}{6}bh^2 = \tfrac{1}{3}bh^2$$

If the two beams have a rigid shearing connection, the effective depth is $2h$, and the elastic section modulus is

$$\tfrac{1}{6}b(2h)^2 = \tfrac{2}{3}bh^2$$

The elastic section modulus, and therefore the permissible bending moment, is doubled by providing a shearing connection between the two beams. In the case of steel beams, the flanges along the plane of contact AB, may be riveted, bolted, or welded together.

10.8 Shearing stresses in a channel section; shear centre

We have discussed the general case of shearing stresses in the bending of a beam having an axis of symmetry in the cross-section; we assumed that the shearing forces were applied parallel to this axis of symmetry. This is a relatively simple problem to treat since there can be no twisting of the beam when a shearing force is applied parallel to the axis of symmetry. We consider now the case when the shearing force is applied at right angles to an axis of symmetry of the cross-section. Consider for example a channel section having an axis Cx of symmetry in the cross-section, Fig. 10.13; the section is of

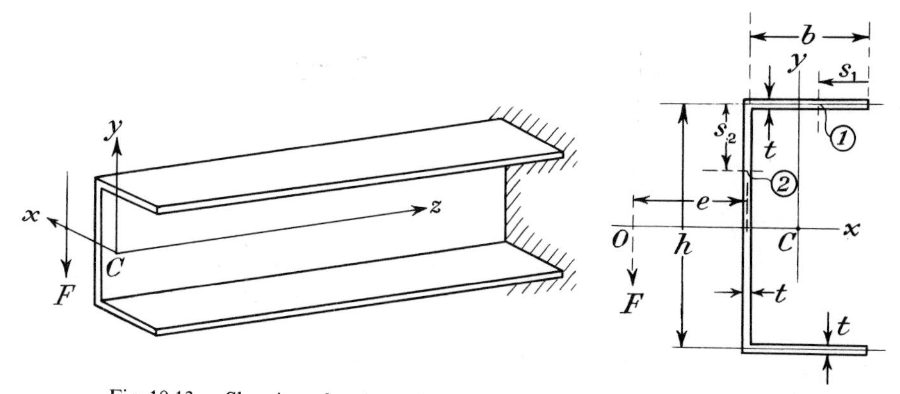

Fig. 10.13 Shearing of a channel cantilever.

Fig. 10.14 Shearing stress at any point of a channel beam.

uniform wall-thickness t, b is the total breadth of each flange, and h is the distance between the flanges; C is the centroid of the cross-section. Suppose the beam is supported at one end, and that a shearing force F is applied at the free end in a direction parallel to Cy. We apply this shearing force at a point O on Cx such that no torsion of the channel occurs, Fig. 10.14; if F is applied considerably to the left of C, twisting obviously will occur in a counter-clockwise direction; if F is applied considerably to the right then twisting occurs in a clockwise direction. There is some intermediate position of O for which no twisting occurs; as we shall see this position is not coincident with the centroid C.

The problem is greatly simplified if we assume that F is applied at a point O on Cx to give no torsion of the channel; suppose O is a distance e from the centre of the web, Fig. 10.14. Since at any section of the beam there are only bending actions present, we we can again use the relation

$$q = \frac{FA\bar{y}}{I_x} \tag{10.4}$$

At a distance s_1 from the free edge of a flange

$$q_1 = \frac{Fht}{2I_x} s_1$$

At a distance s_2 along the web from the junction of web and flange

$$q_2 = \frac{Ft}{2I_x} [bh + s_2(h - s_2)]$$

The shearing stress in the flange is

$$\tau_1 = \frac{q_1}{t} = \frac{Fh}{2I_x} s_1$$

and in the web is

$$\tau_2 = \frac{q_2}{t} = \frac{F}{2I_x} [bh + s_2(h - s_2)]$$

The shearing stress τ_1 in the flanges increases linearly from zero at the free edges to a maximum at the corners; the variation of shearing stress τ_2 in the web is parabolic in form, attaining a maximum value

$$\tau_{max} = \frac{Fbh}{2I_x}\left(1 + \frac{h}{4b}\right) \tag{10.21}$$

at the mid-depth of the web, Fig. 10.15. The shearing stresses τ_1 in the flanges imply total shearing forces of amounts

$$\frac{Fht}{2I_x}\int_0^b s_1\, ds_1 = \frac{Fb^2ht}{4I_x} \tag{10.22}$$

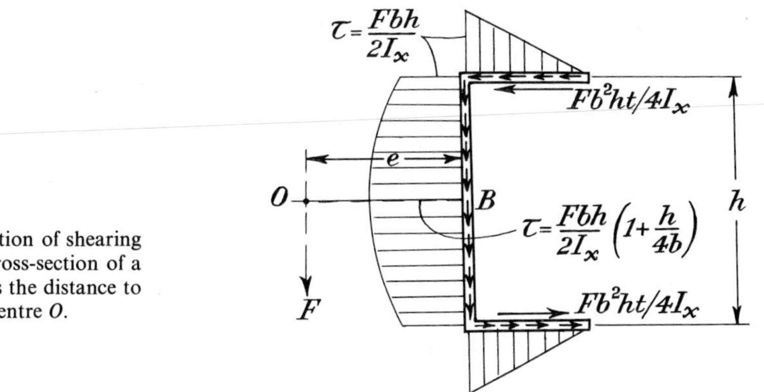

Fig. 10.15 Variation of shearing stresses over the cross-section of a channel beam; e is the distance to the stress centre O.

acting parallel to the centre lines of the flanges; the total shearing forces in the two flanges are in opposite directions. If the distribution of shearing stresses τ_1 and τ_2 is statically equivalent to the applied shearing force F, we have, on taking moments about B—the centre of the web—that

$$Fe = \frac{Fb^2ht}{4I_x} \, . \, h = \frac{Fb^2h^2t}{4I_x}$$

Then

$$e = \frac{b^2h^2t}{4I_x} \tag{10.23}$$

which, as we should expect, is independent of F. We note that O is remote from the centroid C of the cross-section; the point O is usually called the *shear centre*; it is the point of the cross-section through which the resultant shearing force must pass if bending is to occur without torsion of the beam.

FURTHER PROBLEMS
(answers on page 398)

10.3 A plate web girder consists of 4 plates, in each flange, of 30 cm width. The web is 60 cm deep, 2 cm thick and is connected to the flanges by 10 cm by 10 cm by 1·25 cm angles, riveted with 2 cm diameter rivets. Assuming the maximum bending moment to be 1000 kNm, and the shearing force to be 380 kN, obtain suitable dimensions for (i) the thickness of the flange plates, (ii) the pitch of the rivets. Take the tensile stress as 100 MN/m^2, and the shearing stress in the rivets as 75 MN/m^2. (*RNEC*)

10.4 In a small gantry for unloading goods from a railway waggon, it is proposed to carry the lifting tackle on a steel joist, 24 cm by 10 cm, of weight 320 N/m, supported at the ends, and of effective length 5 m. The equivalent dead load on the joist due to the load to be raised is 30 kN, and this may act at any point of the middle 4 m. By considering the fibre stress and the shear, examine whether the joist is suitable. The flanges are 10 cm by 1·2 cm, and the web is 0·75 cm thick. The allowable fibre stress is 115 MN/m^2, and the allowable shearing stress 75 MN/m^2. (*Cambridge*)

10.5 A girder of I-section has a web 60 cm by 1·25 cm and flanges 30 cm by 2·5 cm. The girder is subjected to a bending moment of 300 kNm and a shearing force of 1000 kN at a particular section. Calculate how much of the shearing force is carried by the web, and how much of the bending moment by the flanges. (*Cambridge*)

10.6 The shearing force at a given section of a built-up I-girder is 1000 kN and the depth of the web is 2 m. The web is joined to the flanges by fillet welds. Determine the thickness of the web plate and the thickness of the welds, allowing a shearing stress of 75 MN/m^2 in both the web and welds.

10.7 A thin metal pipe of mean radius R, thickness t, and length L, has its ends closed and is full of water. If the ends are simply-supported, estimate the form of the distribution of shearing stresses over a section near one support, ignoring the intrinsic weight of the pipe.

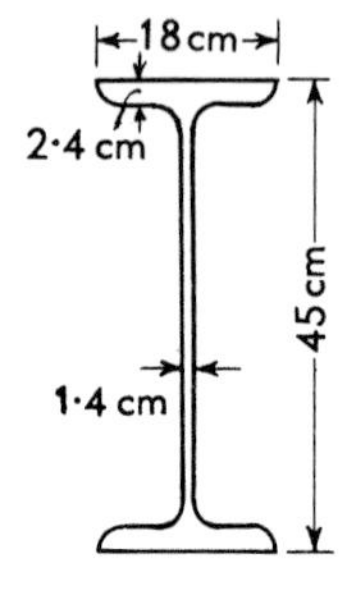

10.8 A compound girder consists of a 45 cm by 18 cm steel joist, of weight 1000 N/m, with a steel plate 25 cm by 3 cm welded to each flange. If the ends are simply-supported and the effective span is 10 m, what is the maximum uniformly distributed load which can be supported by the girder? What weld thicknesses are required to support this load?

Allowable longitudinal stress in plates = 110 MN/m^2

Allowable shearing stress in welds = 60 MN/m^2

Allowable shearing stress in web of girder = 75 MN/m^2

11 Beams of two materials

11.1 Introduction

Some beams used in engineering structures are composed of two materials. A timber joist, for example, may be reinforced by bolting steel plates to the flanges. Plain concrete has little or no tensile strength, and beams of this material are reinforced therefore with steel rods or wires in the tension fibres. In beams of these types there is a composite action between the two materials.

11.2 Transformed sections

The composite beam shown in Fig. 11.1 consists of a rectangular timber joist of breadth b and depth h, reinforced with two steel plates of depth h and thickness t. Con-

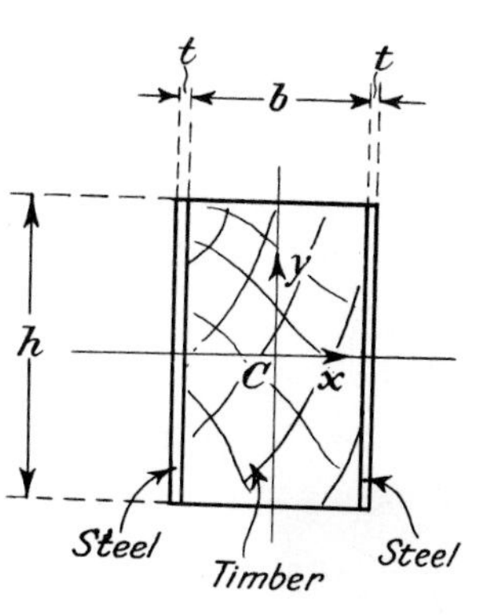

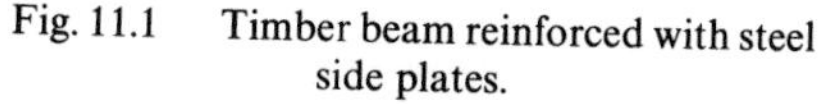

Fig. 11.1 Timber beam reinforced with steel side plates.

sider the behaviour of the composite beam under the action of a bending moment M applied about Cx; if the timber beam is bent into a curve of radius R, then, from equation (9.5), the bending moment carried by the timber beam is

$$M_t = \frac{(EI)_t}{R} \tag{11.1}$$

where $(EI)_t$ is the bending stiffness of the timber beam. If the steel plates are attached to the timber beam by bolting, or glueing, or some other means, the steel plates are bent to the same radius of curvature R as the timber beam. The bending moment carried by the two steel plates is then

$$M_s = \frac{(EI)_s}{R}$$

where $(EI)_s$ is the bending stiffness of the two steel plates. The total bending moment is then

$$M = M_t + M_s = \frac{1}{R}[(EI)_t + (EI)_s]$$

This gives

$$\frac{1}{R} = \frac{M}{(EI)_t + (EI)_s} \tag{11.2}$$

Clearly, the beam behaves as though the total bending stiffness EI were

$$EI = (EI)_t + (EI)_s \tag{11.3}$$

If E_t and E_s are the values of Young's modulus for timber and steel, respectively, and if I_t and I_s are the second moments of area about Cx of the timber and steel beams, respectively, we have

$$EI = (EI)_t + (EI)_s = E_t I_t + E_s I_s \tag{11.4}$$

Then

$$EI = E_t\left[I_t + \left(\frac{E_s}{E_t}\right)I_s\right] \tag{11.5}$$

If I_s is multiplied by (E_s/E_t), which is the ratio of Young's moduli for the two materials, then from equation (11.5) we see that the composite beam may be treated as wholly timber, having an equivalent second moment of area

$$I_t + \left(\frac{E_s}{E_t}\right)I_s \tag{11.6}$$

This is equivalent to treating the beam of Fig. 11.2(i) with reinforcing plates made of timber, but having thicknesses

$$\left(\frac{E_s}{E_t}\right) \times t$$

as shown in Fig. 11.2(ii); the equivalent timber beam of Fig. 11.2(ii) is the *transformed section* of the beam. In this case the beam has been transformed wholly to timber. Equally the beam may be transformed wholly to steel, as shown in Fig. 11.2(iii). For bending about Cx, the *breadths* of the component beams are factored to find the transformed section; the depth h of the beam is unaffected.

The bending stress σ_t in a fibre of the timber core of the beam a distance y from the neutral axis is

$$\sigma_t = M_t \frac{y}{I_t}$$

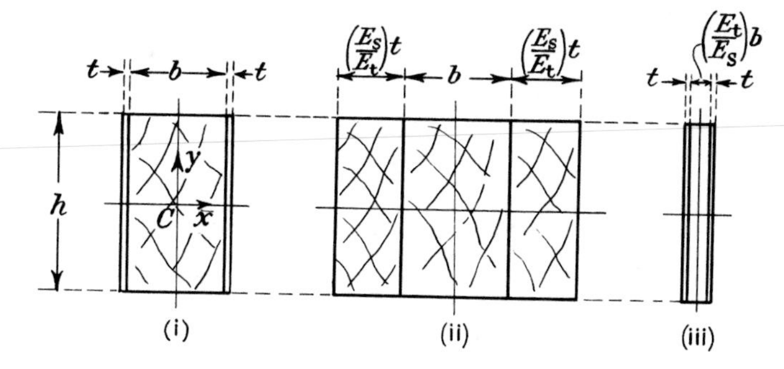

Fig. 11.2 (i) Composite beam of timber and steel bent about Cx. (ii) Equivalent timber beam. (iii) Equivalent steel beam.

Now, from equations (11.1) and (11.2)

$$M_t = \frac{(EI)_t}{R}, \; M = \frac{1}{R}[(EI)_t + (EI)_s]$$

and on eliminating R,

$$M_t = \frac{M}{1 + \dfrac{E_s I_s}{E_t I_t}} \tag{11.7}$$

Then

$$\sigma_t = \frac{My}{I_t\left(1 + \dfrac{E_s I_s}{E_t I_t}\right)} = \frac{My}{I_t + \left(\dfrac{E_s}{E_t}\right)I_s} \tag{11.8}$$

the bending stresses in the timber core are found therefore by considering the *total* bending moment M to be carried by the transformed timber beam of Fig. 11.2(ii). The longitudinal strain at the distance y from the neutral axis Cx is

$$\epsilon = \frac{\sigma_t}{E_t} = \frac{My}{E_t I_t + E_s I_s}$$

Then at the distance y from the neutral axis the stress in the steel reinforcing plates is

$$\sigma_s = E_s \epsilon = \frac{My}{I_s + \left(\dfrac{E_t}{E_s}\right)I_t} \tag{11.9}$$

since the strains in the steel and timber are the same at the same distance y from the neutral axis. This condition of equal strains is implied in the assumption made earlier that the steel and the timber components of the beam are bent to the same radius of curvature R.

Problem 11.1: A composite beam consists of a timber joist, 15 cm by 10 cm, to which reinforcing steel plates, $\frac{1}{2}$ cm thick, are attached. Estimate the maximum bending moment which may be applied about Cx, if the bending stress in the timber is not to exceed 5 MN/m^2, and that in the steel 120 MN/m^2. Take $E_s/E_t = 20$.

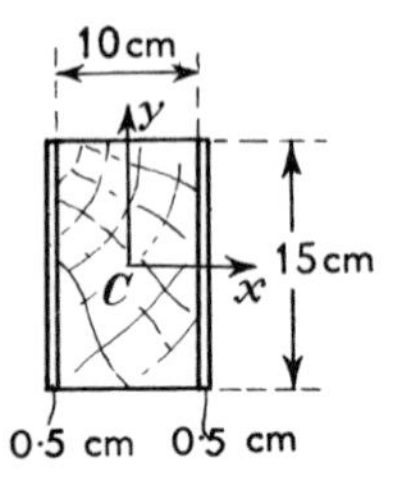

Solution

The maximum bending stresses occur in the extreme fibres. If the stress in the timber is 5 MN/m^2, the stress in the steel at the same distance from Cx is

$$5 \times 10^6 \times \frac{E_s}{E_t} = 100 \times 10^6 \text{ N/m}^2 = 100 \text{ MN/m}^2$$

Thus when the maximum timber stress is attained, the maximum steel stress is only 100 MN/m^2. If the maximum permissible stress of 120 MN/m^2 were attained in the steel, the stress in the timber would exceed 5 MN/m^2, which is not permissible. The maximum bending moment gives therefore a stress in the timber of 5 MN/m^2. The second moment of area about Cx of the equivalent timber beam is

$$I_x = \tfrac{1}{12}(0{\cdot}10)(0{\cdot}15)^3 + \tfrac{1}{12}(0{\cdot}010)(0{\cdot}15)^3$$
$$= 0{\cdot}0842 \times 10^{-3} \text{ m}^4$$

For a maximum stress in the timber of 5 MN/m^2, the moment is

$$M = \frac{(5 \times 10^6)(0{\cdot}0842 \times 10^{-3})}{0{\cdot}075} = 5610 \text{ Nm}$$

11.3 Timber beam with reinforcing steel flange plates

In §11.2 we discussed the composite bending action of a timber beam reinforced with steel plates over the depth of the beam. A similar bending problem arises when the timber joist is reinforced on its upper and lower faces with steel plates, as shown in Fig. 11.3(i); the timber web of the composite beam may be transformed into steel to give the equivalent steel section of Fig. 11.3(ii); alternatively, the steel flanges may be replaced by equivalent timber flanges to give the equivalent timber beam of Fig. 11.3(iii). The problem is then treated in the same way as the beam in §11.2; the stresses in the timber and steel are calculated from the second moment of area of the transformed timber and steel sections.

An important difference, however, between the composite actions of the beams of Figs. 11.2 and 11.3 lies in their behaviour under shearing forces. The two beams, used as cantilevers carrying end loads F, are shown in Fig. 11.4; for the timber joist reinforced over the depth, Fig. 11.4(i), there are no shearing actions between the timber and the steel plates, except near the loaded ends of the cantilever. However, for the joist of Fig. 11.4(ii), a shearing force is transmitted between the timber and the steel flanges at

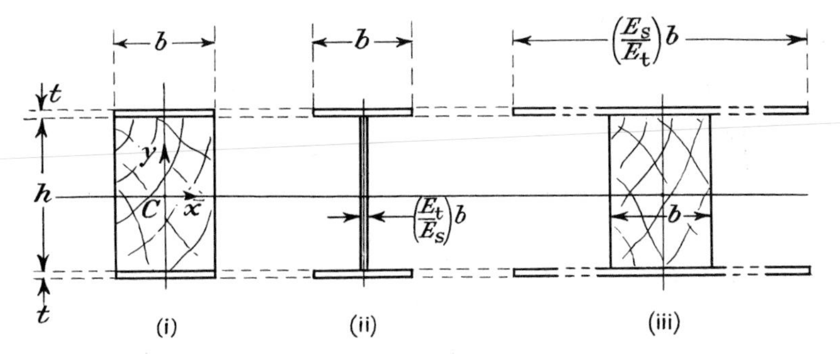

Fig. 11.3 (i) Timber beam with reinforcing steel flange plates. (ii) Equivalent steel I-beam. (iii) Equivalent timber I-beam.

all sections of the beam. In the particular case of thin reinforcing flanges, it is sufficiently accurate to assume that the shearing actions in the cantilever of Fig. 11.4(ii) are resisted largely by the timber joist; on considering the equilibrium of a unit length of the composite beam, equilibrium is ensured if a shearing force (F/h) per unit length of beam is transmitted between the timber joist and the reinforcing flanges, Fig. 11.5. This shearing

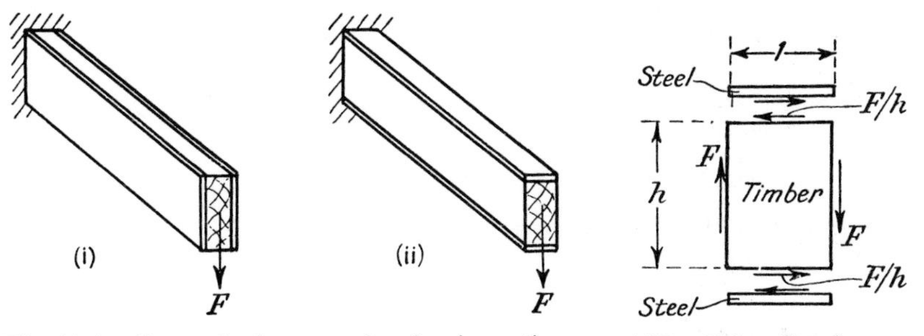

Fig. 11.4 Composite beams under shearing action, showing (i) steel and timber both resisting shear and (ii) timber alone resisting shear.

Fig. 11.5 Shearing actions in a timber joist with reinforcing steel flanges.

force must be carried by bolts, glue, or some other suitable means. The end deflections of the cantilevers shown in Fig. 11.4 may be difficult to estimate; this is due to the fact that account may have to be taken of the shearing distortions of the timber beams.

Problem 11.2: A timber joist 15 cm by 7·5 cm has reinforcing steel flange plates 1·25 cm thick. The composite beam is 3 m long, simply-supported at each end, and carries a uniformly distributed lateral load

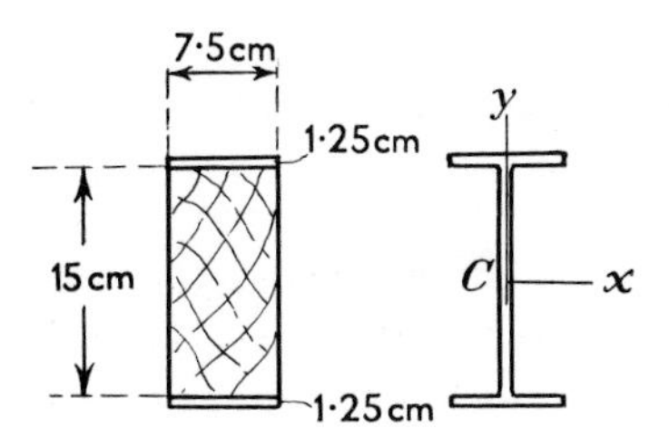

of 10 kN. Estimate the maximum bending stresses in the steel and timber, and the intensity of shearing force transmitted between the steel plates and the timber. Take $E_s/E_t = 20$.

Solution

The second moment of area of the equivalent steel section is

$$I_x = \frac{1}{20}[\frac{1}{12}(0{\cdot}075)(0{\cdot}15)^3] + 2[(0{\cdot}0125)(0{\cdot}075)^3] = 12{\cdot}1 \times 10^{-6}\ \text{m}^4$$

The maximum bending moment is

$$\frac{(10 \times 10^3)(3)}{8} = 3750\ \text{Nm}$$

The maximum bending stress in the steel is then

$$\sigma_s = \frac{(3750)(0{\cdot}0875)}{(12{\cdot}1 \times 10^{-6})} = 27{\cdot}1\ \text{MN/m}^2$$

The bending stress in the steel at the junction of web and flange is

$$\sigma_s = \frac{(3750)(0{\cdot}0750)}{(12{\cdot}1 \times 10^{-6})} = 23{\cdot}2\ \text{MN/m}^2$$

The stress in the timber at this junction is then

$$\sigma_t = \frac{E_t}{E_s} \times \sigma_s = \frac{1}{20}(23{\cdot}2) = 11{\cdot}6\ \text{MN/m}^2$$

On the assumption that the shearing forces at any section of the beam are taken largely by the timber, the shearing force between the timber and steel plates is

$$(5 \times 10^3)/(0{\cdot}15) = 33{\cdot}3\ \text{kN/m}$$

since the maximum shearing force in the beam is 5 kN.

11.4 Ordinary reinforced concrete

It was noted in Chapter 1 that concrete is a brittle material which is weak in tension. Consequently a beam composed only of concrete has little or no bending strength since cracking occurs in the extreme tension fibres in the early stages of loading. To

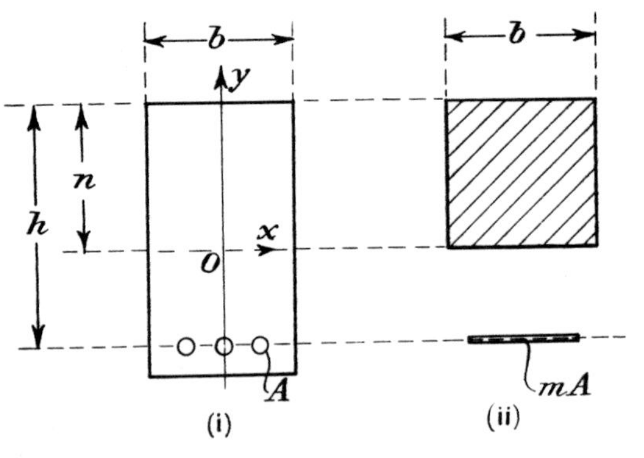

Fig. 11.6 Simple rectangular concrete beam with reinforcing steel in the tension flange.

overcome this weakness steel rods are embedded in the tension fibres of a concrete beam; if concrete is cast around a steel rod, on setting the concrete shrinks and grips the steel rod. It happens that the coefficients of linear expansion of concrete and steel are very nearly equal; consequently, negligible stresses are set up by temperature changes.

The bending of an ordinary reinforced concrete beam may be treated on the basis of transformed sections. Consider firstly the beam of rectangular cross-section shown in Fig. 11.6. The breadth of the concrete is b, and h is the depth of the steel reinforcement below the upper extreme fibres. The beam is bent so that tensile stresses occur in the lower fibres. The total area of cross-section of the steel reinforcing rods is A; the rods are placed longitudinally in the beam. The beam is now bent so that Ox becomes a neutral axis, compressive stresses being induced in the concrete above Ox. We assume that concrete below the neutral axis cracks in tension, and is therefore ineffectual; we neglect the contribution of the concrete below Ox to the bending strength of the beam. Suppose m is the ratio of Young's modulus of steel, E_s, to Young's modulus of concrete, E_c; then

$$m = \frac{E_s}{E_c} \tag{11.10}$$

If the area A of steel is transformed to concrete, its equivalent area is mA; the equivalent concrete beam then has the form shown in Fig. 11.6(ii). The depth of the neutral axis Ox below the extreme upper fibres is n. The equivalent concrete area mA on the tension side of the beam is concentrated approximately at a depth h.

We have that the neutral axis of the beam occurs at the centroid of the equivalent concrete beam; then

$$bn \times \tfrac{1}{2}n = mA(h - n)$$

Thus n is the root of the quadratic equation

$$\tfrac{1}{2}bn^2 + mAn - mAh = 0 \tag{11.11}$$

The relevant root is

$$n = \frac{mA}{b}\left(\sqrt{1 + \frac{2bh}{mA}} - 1\right) \tag{11.12}$$

The second moment of area of the equivalent concrete beam about its centroidal axis is

$$I_c = \tfrac{1}{3}bn^3 + mA(h - n)^2 \tag{11.13}$$

The maximum compressive stress induced in the upper extreme fibres of the concrete is

$$\sigma_c = \frac{Mn}{I_c} \tag{11.14}$$

$$\sigma_s = \frac{M(h - n)}{I_c} \times \frac{E_s}{E_c} = \frac{mM(h - n)}{I_c} \tag{11.15}$$

Problem 11.3: A rectangular concrete beam is 30 cm wide and 45 cm deep to the steel reinforcement. The direct stresses are limited to 115 MN/m² in the steel and 6·5 MN/m² in the concrete, and the modular ratio is 15. What is the area of steel reinforcement if both steel and concrete are fully stressed? Estimate the permissible bending moment for this condition.

Solution

From equations (11.14) and (11.15)

$$\sigma_s = \frac{M(h-n)}{\dfrac{bn^3}{3m} + A(h-n)^2} = 115\ \mathrm{MN/m^2}$$

and

$$\sigma_c = \frac{Mn}{\frac{1}{3}bn^3 + mA(h-n)^2} = 6{\cdot}5\ \mathrm{MN/m^2}$$

Then

$$\frac{M(h-n)}{115} = \frac{Mn}{6{\cdot}5\ m}$$

Hence

$$h - n = 1{\cdot}18n \quad \text{and} \quad \frac{n}{h} = \frac{1}{2{\cdot}18} = 0{\cdot}458$$

Then

$$n = 0{\cdot}458 \times 0{\cdot}45 = 0{\cdot}206\ \mathrm{m}$$

From equation (11.11)

$$\frac{2mA}{bh} = \frac{(n/h)^2}{1-(n/h)} = \frac{(0{\cdot}458)^2}{0{\cdot}542} = 0{\cdot}387$$

Then

$$A = 0{\cdot}387\frac{bh}{2m} = \frac{0{\cdot}387 \times 0{\cdot}30 \times 0{\cdot}45}{30} = 1{\cdot}74 \times 10^{-3}\ \mathrm{m^2}$$

Since the maximum allowable stresses of both the steel and the concrete are attained, the allowable bending moment may be evaluated on the basis of either the steel or the concrete stress. The second moment of area of the equivalent concrete beam is

$$I_c = \tfrac{1}{3}bn^3 + mA(h-n)^2 = \tfrac{1}{3}(0{\cdot}30)(0{\cdot}206)^3 + 15(0{\cdot}00174)(0{\cdot}244)^2 = 2{\cdot}42 \times 10^{-3}\ \mathrm{m^4}$$

The permissible bending moment is

$$M = \frac{\sigma_c I_c}{y_c} = \frac{(6{\cdot}5 \times 10^6)(2{\cdot}42 \times 10^{-3})}{(0{\cdot}206)} = 76{\cdot}4\ \mathrm{kNm}$$

Problem 11.4: A rectangular concrete beam has a breadth of 30 cm and is 45 cm deep to the steel reinforcement, which consists of two 2·5 cm diameter bars. Estimate the permissible bending moment if the stresses are limited to 115 MN/m² and 6·5 MN/m² in the steel and concrete, respectively, and if the modular ratio is 15.

Solution

The area of steel reinforcement is $A = 2(\pi/4)(0{\cdot}025)^2 = 0{\cdot}982 \times 10^{-3}\ \mathrm{m^2}$. From equation (11.12)

$$\frac{n}{h} = \frac{mA}{bh}\left[\sqrt{1 + \frac{2bh}{mA}} - 1\right]$$

Now

$$\frac{mA}{bh} = \frac{(15)(0{\cdot}982 \times 10^{-3})}{(30)(45)} = 0{\cdot}1091$$

Then

$$\frac{n}{h} = 0{\cdot}1091\left[\left(1 + \frac{2}{0{\cdot}1091}\right)^{\frac{1}{2}} - 1\right] = 0{\cdot}370$$

Thus

$$n = 0{\cdot}370h = 0{\cdot}167 \text{ m}$$

The second moment of area of the equivalent concrete beam is

$$\begin{aligned} I_c &= \tfrac{1}{3}bn^3 + mA(h - n)^2 \\ &= \tfrac{1}{3}(0{\cdot}30)(0{\cdot}167)^3 + 15(0{\cdot}982 \times 10^{-3})(0{\cdot}283)^2 \\ &= (0{\cdot}466 + 1{\cdot}180)10^{-3} \text{ m}^4 \\ &= 1{\cdot}646 \times 10^{-3} \text{ m}^4 \end{aligned}$$

If the maximum allowable concrete stress is attained, the permissible moment is

$$M = \frac{\sigma_c I_c}{n} = \frac{(6{\cdot}5 \times 10^6)(1{\cdot}646 \times 10^{-3})}{0{\cdot}167} = 64 \text{ kNm}$$

If the maximum allowable steel stress is attained, the permissible moment is

$$M = \frac{\sigma_s I_c}{m(h - n)} = \frac{(115 \times 10^6)(1{\cdot}646 \times 10^{-3})}{15(0{\cdot}283)} = 44{\cdot}6 \text{ kNm}$$

Steel is therefore the limiting material, and the permissible bending moment is

$$M = 44{\cdot}6 \text{ kNm}$$

Problem 11.5: A rectangular concrete beam, 30 cm wide, is reinforced on the tension side with four 2·5 cm diameter steel rods at a depth of 45 cm, and on the compression side with two 2·5 cm diameter rods at a depth of 5 cm. Estimate the permissible bending moment if the stresses in the concrete are not to exceed 6·5 MN/m² and in the steel 115 MN/m². The modular ratio is 15.

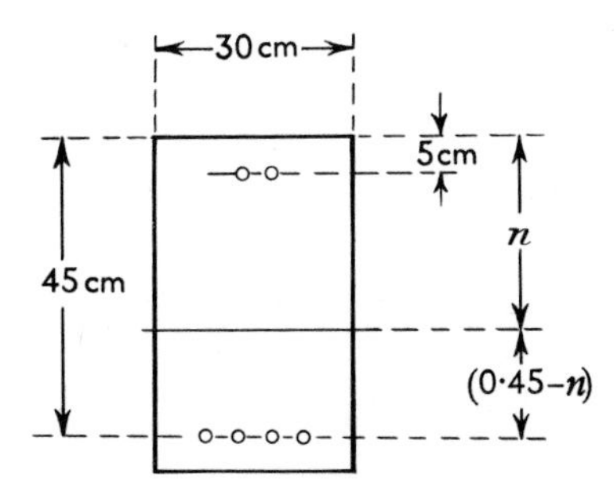

Solution

The area of steel reinforcement is $1{\cdot}964 \times 10^{-3}$ m² on the tension side, and $0{\cdot}982 \times 10^{-3}$ m² on the compression side. The cross-sectional area of the equivalent concrete beam is

$$(0{\cdot}30)n + (m - 1)(0{\cdot}000982) + m(0{\cdot}001964) = (0{\cdot}30n + 0{\cdot}0433) \text{ m}^2$$

The position of the neutral axis is given by

$$(0{\cdot}30n)(\tfrac{1}{2}n) + (m - 1)(0{\cdot}000982)(0{\cdot}05) + m(0{\cdot}001964)(0{\cdot}45) = (0{\cdot}30n + 0{\cdot}0433)n$$

This reduces to

$$n^2 - 0{\cdot}288n - 0{\cdot}093 = 0$$

giving

$$n = -0{\cdot}144 \pm 0{\cdot}337$$

The relevant root is $n = 0{\cdot}193$ m

The second moment of area of the equivalent concrete beam is

$$I_c = \tfrac{1}{3}(0{\cdot}30)n^3 + (m - 1)(0{\cdot}000982)(n - 0{\cdot}05)^2 + m(0{\cdot}001964)(0{\cdot}45 - n)^2$$

$$= (0{\cdot}720 + 0{\cdot}281 + 1{\cdot}950)10^{-3}$$

$$= 2{\cdot}95 \times 10^{-3}\ \text{m}^4$$

If the maximum allowable concrete stress is attained, the permissible moment is

$$M = \frac{\sigma_c I_c}{n} = \frac{(6{\cdot}5 \times 10^6)(2{\cdot}95 \times 10^{-3})}{0{\cdot}193} = 99{\cdot}3\ \text{kNm}$$

If the maximum allowable steel stress is attained, the permissible moment is

$$M = \frac{\sigma_s I_c}{m(0{\cdot}45 - n)} = \frac{(115 \times 10^6)(2{\cdot}95 \times 10^{-3})}{15(0{\cdot}257)} = 88{\cdot}0\ \text{kNm}$$

Thus, steel is the limiting material, and the allowable moment is 88·0 kNm

Problem 11.6: A steel I-section, 12·5 cm by 7·5 cm, is encased in a rectangular concrete beam of breadth 20 cm and depth 30 cm to the lower flange of the I-section. Estimate the position of the neutral axis of the composite beam, and find the permissible bending moment if the steel stress is not to exceed 115 MN/m² and the concrete stress 6·5 MN/m². The modular ratio is 15. The area of the steel beam is 0·00211 m² and its second moment of area about its minor axis is $5{\cdot}70 \times 10^{-6}$ m⁴.

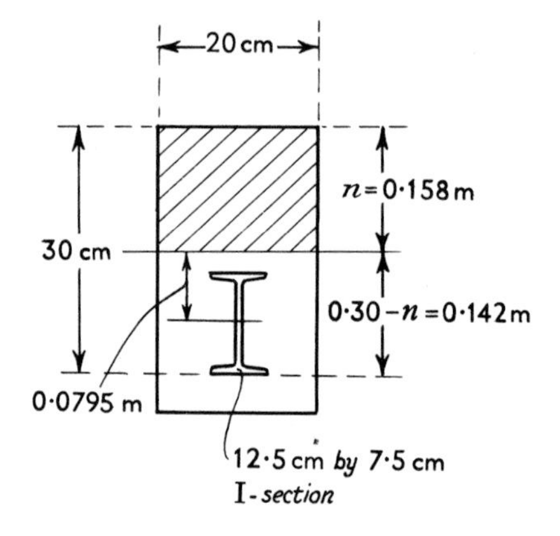

Solution

The area of the equivalent steel beam is

$$\frac{(0{\cdot}20)n}{15} + 0{\cdot}00211\ \text{m}^2$$

The position of the neutral axis is therefore

$$\left(\frac{0{\cdot}20n}{15} + 0{\cdot}00211\right)n = \left(\frac{0{\cdot}20}{15}\right)(\tfrac{1}{2}n) + (0{\cdot}00211)(0{\cdot}2375)$$

This reduces to

$$n^2 + 0{\cdot}316n - 0{\cdot}075 = 0$$

The relevant root of which is

$$n = 0{\cdot}158 \text{ m}$$

The second moment of area of the equivalent steel beam is

$$I_s = \frac{1}{3}\left(\frac{0{\cdot}20}{15}\right)(0{\cdot}158)^3 + (0{\cdot}00211)(0{\cdot}0795)^2 = 0{\cdot}0366 \times 10^{-3} \text{ m}^4$$

The allowable bending moment on the basis of the steel stress is

$$M = \frac{\sigma_s I_s}{(0{\cdot}30 - n)} = \frac{(115 \times 10^6)(0{\cdot}0366 \times 10^{-3})}{0{\cdot}142} = 29{\cdot}7 \text{ kNm}$$

If the maximum allowable concrete stress is 6·5 MN/m², the maximum allowable compressive stress in the equivalent steel beam is

$$m(6{\cdot}5 \times 10^6) = 97{\cdot}5 \text{ MN/m}^2$$

On this basis, the maximum allowable moment is

$$M = \frac{(97{\cdot}5 \times 10^6)(0{\cdot}0366 \times 10^{-3})}{0{\cdot}158} = 22{\cdot}6 \text{ kNm}$$

Concrete is therefore the limiting material, and the maximum allowable moment is

$$M = 22{\cdot}6 \text{ kNm}$$

Problem 11.7: A reinforced concrete T-beam contains $1{\cdot}25 \times 10^{-3}$ m² of steel reinforcement on the tension side. If the steel stress is limited to 115 MN/m² and the concrete stress to 6·5 MN/m², estimate the permissible bending moment. The modular ratio is 15.

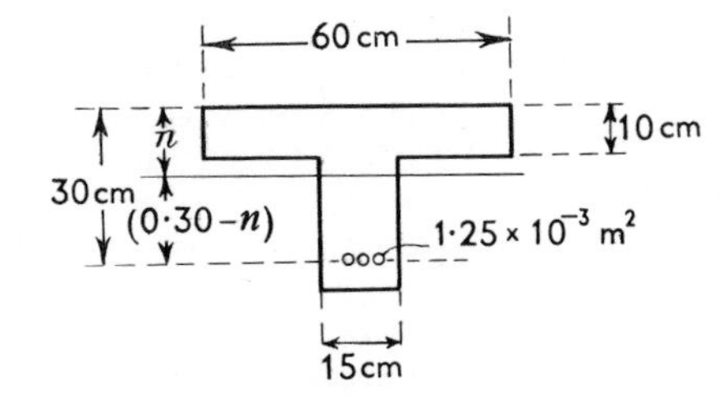

Solution

Suppose the neutral axis falls below the underside of the flange. The area of the equivalent concrete beam is

$$(0{\cdot}60)n - 0{\cdot}45(n - 0{\cdot}10) + (0{\cdot}00125)15 = 0{\cdot}15n + 0{\cdot}0638 \text{ m}^2$$

The position of the neutral axis is given by

$$(0{\cdot}60n)(\tfrac{1}{2}n) + (0{\cdot}00125)(15)(0{\cdot}30) - 0{\cdot}45(n - 0{\cdot}10)(\tfrac{1}{2})(n + 0{\cdot}10) = (0{\cdot}15n + {\cdot}0638)n$$

This reduces to

$$n^2 + 0{\cdot}850n - 0{\cdot}1044 = 0$$

the relevant root of which is $n = 0{\cdot}109$ m which agrees with our assumption earlier that the neutral axis lies below the flange.

The second moment of area of the equivalent concrete beam is

$$\begin{aligned} I_c &= \tfrac{1}{3}(0{\cdot}60)(n^3) - \tfrac{1}{3}(0{\cdot}45)(n - 0{\cdot}10)^3 + 0{\cdot}00125(15)(0{\cdot}30 - n)^2 \\ &= (0{\cdot}259 + 0{\cdot}000 + 0{\cdot}685)10^{-3}\ \text{m}^4 \\ &= 0{\cdot}944 \times 10^{-3}\ \text{m}^4 \end{aligned}$$

If the maximum allowable concrete stress is attained, the permissible moment is

$$M = \frac{\sigma_c I_c}{n} = \frac{(6{\cdot}5 \times 10^6)(0{\cdot}944 \times 10^{-3})}{0{\cdot}109} = 56{\cdot}3\ \text{kNm}$$

If the maximum allowable steel stress is attained, the permissible moment is

$$M = \frac{\sigma_s I_c}{m(0{\cdot}30 - n)} = \frac{(115 \times 10^6)(0{\cdot}944 \times 10^{-3})}{15(0{\cdot}191)} = 37{\cdot}9\ \text{kNm}$$

Steel is therefore the limiting material, and the permissible bending moment is 37·9 kNm.

FURTHER PROBLEMS
(answers on page 398)

11.8 A concrete beam of rectangular section is 10 cm wide and is reinforced with steel bars whose axes are 30 cm below the top of the beam. Estimate the required total area of the cross-section of the steel if the maximum compressive stress in the concrete is to be 7·5 MN/m^2 and the tensile stress in the steel is 135 MN/m^2 when the beam is subjected to pure bending. What bending moment would the beam withstand when in this condition? Assume that Young's modulus for steel is 15 times that for concrete and that concrete can sustain no tensile stresses. (*Cambridge*)

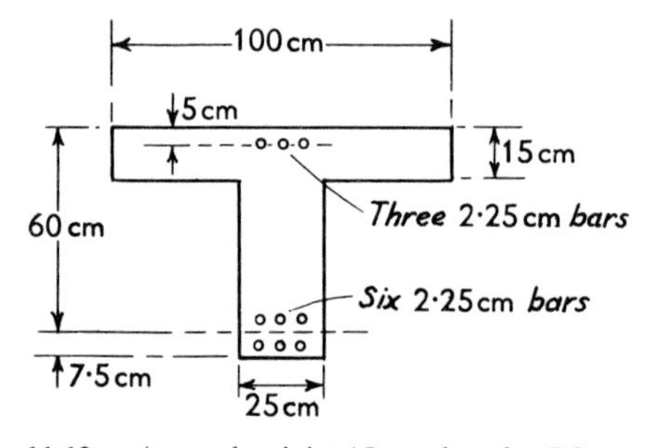

11.9 A reinforced concrete T-beam carries a uniformly distributed super-load on a simply-supported span of 8 m. The stresses in the steel and concrete are not to exceed 125 MN/m^2 and 7 MN/m^2, respectively. The modular ratio is 15, and the density of concrete is 2400 kg/m^3. Determine the permissible super-load. (*Nottingham*)

11.10 A wooden joist 15 cm deep by 7·5 cm wide is reinforced by glueing to its lower face a steel strip 7·5 cm wide by 0·3 cm thick. The joist is simply-supported over a span of 3 m, and carries a uniformly distributed load of 5000 N. Find the maximum direct stresses in the wood and steel and the maximum shearing stress in the glue. Take $E_s/E_t = 20$. (*Cambridge*)

11.11 A timber beam is 15 cm deep by 10 cm wide, and carries a central load of 30 kN at the centre of a 3 m span; the beam is simply-supported at each end. The timber is reinforced with flat steel plates 10 cm wide by 1·25 cm thick bolted to the upper and lower surfaces of the beam. Taking E for steel as 200 GN/m^2 and G for timber as 1 GN/m^2, estimate

(i) the maximum direct stress in the steel strips,
(ii) the average shearing stress in the timber,
(iii) the shearing load transmitted by the bolts,
(iv) the bending and shearing deflections at the centre of the beam.

12 Bending stresses and direct stresses combined

12.1 Introduction

Many instances arise in practice where a member undergoes bending combined with a thrust or pull. If a member carries a thrust, direct longitudinal stresses are set up; if a bending moment is now superimposed on the member at some section, additional longitudinal stresses are induced.

In this chapter we shall be concerned with the combined bending and thrust of short stocky members; in such cases the presence of a thrust does not lead to overall instability of the member. Buckling of beams under end thrust is discussed later in Chapter 20.

12.2 Combined bending and thrust of a stocky strut

Consider firstly a short column of rectangular cross-section, Fig. 12.1(i). The column carries an axial compressive load P, together with a bending moment M, at some section,

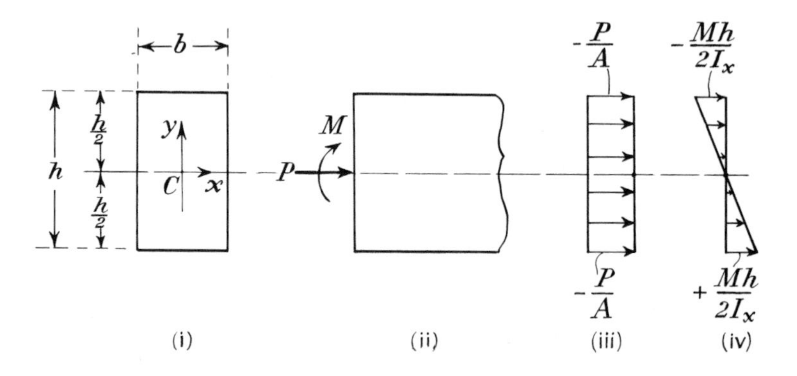

Fig. 12.1 Combined bending and thrust of a rectangular cross-section beam.

applied about the centroidal axis Cx. The area of the column is A, and I_x is the second moment of area about Cx. If P acts alone, the average longitudinal stress over the section is

$$-\frac{P}{A}$$

the stress being compressive. If the couple M acts alone, and if the material remains elastic, the longitudinal stress in any fibre a distance y from Cx is

$$-\frac{My}{I_x}$$

for positive values of y. We assume now that the combined effect of the thrust and the bending moment is the sum of the separate effects of P and M. The stresses due to P and M acting separately are shown in Fig. 12.1(iii) and (iv). On combining the two stress systems, the resultant stress in any fibre is

$$\sigma = -\frac{P}{A} - \frac{My}{I_x} \tag{12.1}$$

Clearly the greatest compressive stress occurs in the upper extreme fibres, and has the value

$$\sigma_{max} = -\frac{P}{A} - \frac{Mh}{2I_x} \tag{12.2}$$

In the lower fibres of the beam y is negative; in the extreme lower fibres

$$\sigma = -\frac{P}{A} + \frac{Mh}{2I_x} \tag{12.3}$$

which is compressive or tensile depending upon whether $(Mh/2I_x)$ is less than or greater than (P/A). The two possible types of stress distribution are shown in Fig. 12.2(i) and (ii). When $(Mh/2I_x) < (P/A)$, the stresses are compressive for all parts of the cross-section, Fig. 12.2(i). When $(Mh/2I_x) > (P/A)$, the stress is zero at a distance (PI_x/AM) below the centre line of the beam, Fig. 12.2(ii); this defines the position of the neutral axis of the

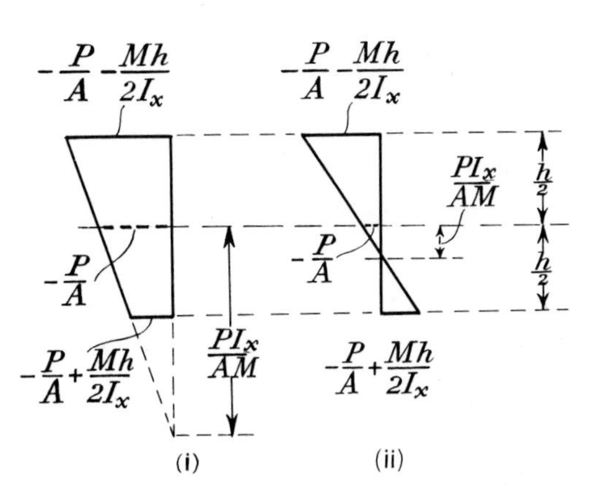

Fig. 12.2 Position of the neutral axis for combined bending and thrust.

column, or the axis of zero strain. In Fig. 12.2(i) the imaginary neutral axis is also a distance (PI_x/AM) from the centre line, but it lies outside the cross-section.

12.3 Eccentric thrust

We can use the analysis of §12.2 to find the stresses due to an eccentric thrust. The column of rectangular cross-section shown in Fig. 12.3(i) carries a thrust P, which can

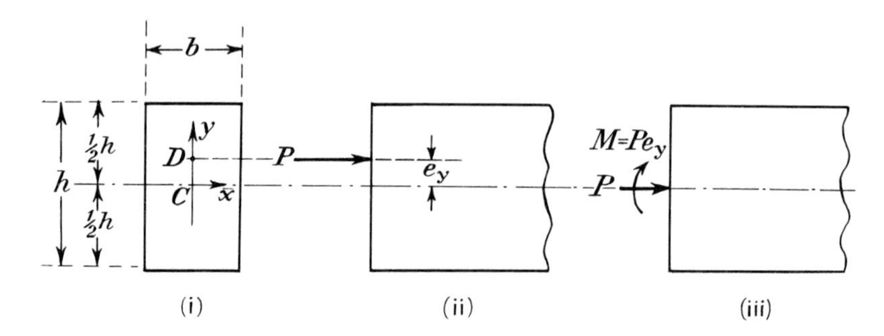

Fig. 12.3 Column of rectangular cross-section carrying an eccentric thrust.

be regarded as concentrated at the point D, which lies on the centroidal axis Cy, at a distance e_y from C, Fig. 12.3(ii). The eccentric load P is statically equivalent to an axial thrust P and a bending moment Pe_y applied about Cx, Fig. 12.3(iii). Then, from equation (12.1), the longitudinal stress at any fibre is

$$\sigma = -\frac{P}{A} - \frac{Pe_y}{I_x} = -\frac{P}{A}\left(1 + \frac{Ae_y y}{I_x}\right) \tag{12.4}$$

We are interested frequently in the condition that no tensile stresses occur in the column; clearly, tensile stresses are most likely to occur in the lowest extreme fibres, where

$$\sigma = -\frac{P}{A}\left(1 - \frac{Ae_y h}{2I_x}\right) \tag{12.5}$$

This stress is tensile if

$$\frac{Ae_y h}{2I_x} > 1 \tag{12.6}$$

that is, if

$$\frac{6e_y}{h} > 1$$

or

$$e_y > \tfrac{1}{6}h \tag{12.7}$$

Now suppose the thrust P is applied eccentrically about both centroidal axes, at a distance e_x from the axis Cy and a distance e_y from the axis Cx, Fig. 12.4. We replace the eccentric thrust P by an axial thrust P at C, together with couples Pe_y and Pe_x

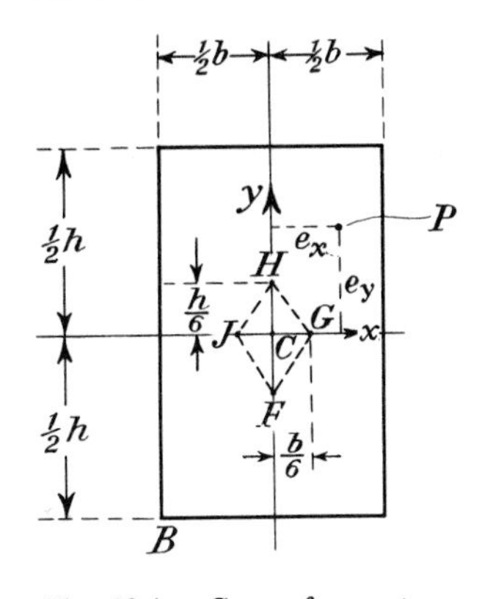

Fig. 12.4 Core of a rectangular cross-section.

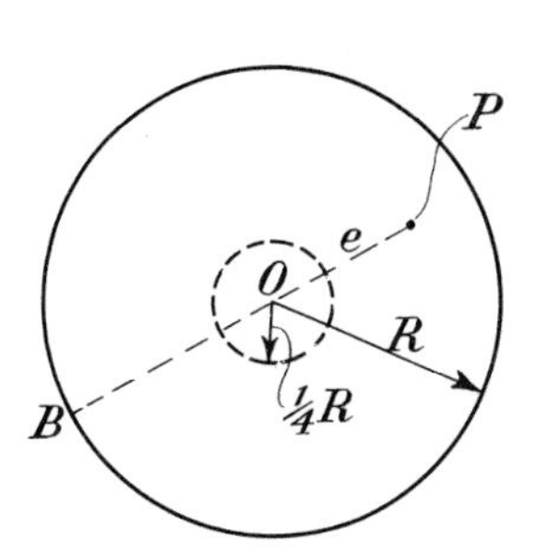

Fig. 12.5 Core of a circular cross-section.

about Cx and Cy, respectively. The resultant compressive stress at any fibre defined by coordinates (x, y) is

$$\sigma = -\frac{P}{A} - \frac{Pe_x x}{I_y} - \frac{Pe_y y}{I_x}$$

$$= -\frac{P}{A}\left[1 + \frac{Ae_x x}{I_y} + \frac{Ae_y y}{I_x}\right] \tag{12.8}$$

Suppose e_x and e_y are both positive; then a tensile stress is most likely to occur at the corner B of the rectangle. The stress at B is tensile when

$$1 - \frac{Ae_x b}{2I_y} - \frac{Ae_y h}{2I_x} < 0 \tag{12.9}$$

On substituting for A, I_x, and I_y, this becomes

$$1 - \frac{6e_x}{b} - \frac{6e_y}{h} < 0 \tag{12.10}$$

If P is applied at a point on the side of the line HG remote from C, this inequality is satisfied, and the stress at B becomes tensile, regardless of the value of P. Similarly, the lines HJ, JF and FG define limits of the point of application of P for the development of tensile stresses at the other three corners of the column. Clearly, if no tensile stresses are to be induced at all, the load P must not be applied outside the parallelogram $FGHJ$ in Fig. 12.4; the region $FGHJ$ is known as the *core of the section.* For the rectangular section of Fig. 12.4 the core is a parallelogram with diagonals of lengths $\frac{1}{3}h$ and $\frac{1}{3}b$.

For a column with a circular cross-section of radius R, Fig. 12.5, the tensile stress is most likely to develop at a point B on the perimeter diametrally opposed to the point of application of P. The stress at B is

$$\sigma = -\frac{P}{A} + \frac{PeR}{I} = -\frac{P}{A}\left(1 - \frac{AeR}{I}\right) \tag{12.11}$$

where I is the second moment of area about a diameter. Tensile stresses are developed if

$$\frac{AeR}{I} > 1 \tag{12.12}$$

On substituting for A and I, this becomes

$$\frac{4e}{R} > 1$$

or

$$e > \frac{R}{4} \tag{12.13}$$

The core of the section is then a circle of radius $\frac{1}{4}R$.

Problem 12.1: Find the maximum stress on the section AB of the cramp when a pressure of 7500 N is exerted by the screw. The section is rectangular 2·5 cm by 1 cm. (*Cambridge*)

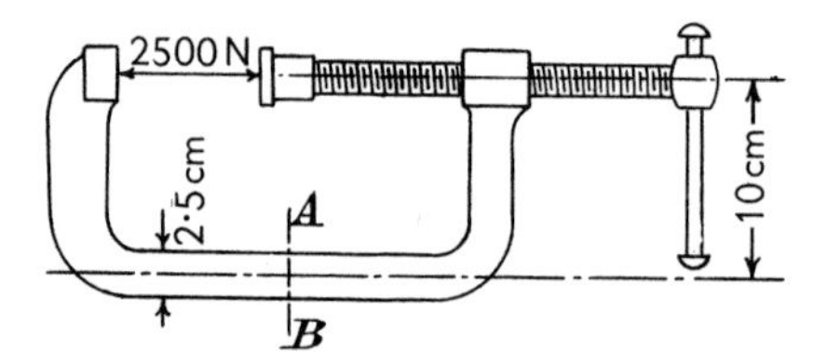

Solution

The section AB is subjected to a tension of 2500 N, and a bending moment (2500)(0·10) = 250 Nm. The area of the section = $0{\cdot}25 \times 10^{-3}$ m^2. The direct tensile stress = $(2500)/(0{\cdot}25 \times 10^{-3}) = 10$ MN/m^2. The second moment of area = $\frac{1}{12}(0{\cdot}01)(0{\cdot}025)^3 = 13{\cdot}02 \times 10^{-9}$ m^4. Therefore, the maximum bending stresses due to the couple of 250 Nm are equal to

$$\frac{(250)(0{\cdot}0125)}{(13{\cdot}02 \times 10^{-9})} = 240 \text{ MN/m}^2$$

Hence the maximum tensile stress on the section is

$$(240 + 10) = 250 \text{ MN/m}^2$$

The maximum compressive stress is

$$(240 - 10) = 230 \text{ MN/m}^2$$

Problem 12.2: A masonry pier has a cross-section 3 m by 2 m, and is subjected to a load of 1000 kN, the line of the resultant being 1·80 m from one of the shorter sides, and 0·85 m from one of the longer sides. Find the maximum tensile and compressive stresses produced. (*Cambridge*)

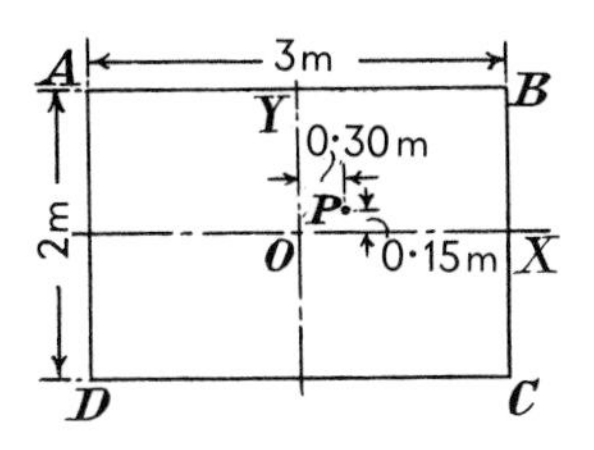

Solution

P represents the line of action of the thrust. The bending moments are

$$(0{\cdot}15)(1000 \times 10^3) = 150 \text{ kNm about } OX$$

$$(0{\cdot}30)(1000 \times 10^3) = 300 \text{ kNm about } OY$$

Now,

$$I_x = \tfrac{1}{12}(3)(2)^3 = 2 \text{ m}^4$$

$$I_y = \tfrac{1}{12}(2)(3) = 4{\cdot}5 \text{ m}^4$$

The cross-sectional area is

$$A = (3)(2) = 6 \text{ m}^2$$

For a point whose coordinates are (x, y) the compressive stress is

$$\sigma = -\frac{P}{A}\left(1 + \frac{Ae_x x}{I_y} + \frac{Ae_y y}{I_x}\right)$$

which gives

$$\sigma = -\frac{1000 \times 10^3}{6}\left(1 + \frac{x}{2{\cdot}5} + \frac{9y}{20}\right)$$

The compressive stress is a maximum at B, where $x = 1{\cdot}5$ m and $y = 1$ m. Then

$$\sigma_B = -\frac{10^6}{6}(1 + \tfrac{3}{5} + \tfrac{9}{20}) = -0{\cdot}342 \text{ MN/m}^2$$

The stress at D, where $x = -1{\cdot}5$ m and $y = -1$ m, is

$$\sigma_D = -\frac{10^6}{6}(1 - \tfrac{3}{5} - \tfrac{9}{20}) = +\,0{\cdot}008 \text{ MN/m}^2$$

which is the maximum tensile stress.

12.4 Pre-stressed concrete beams

The simple analysis of §12.2 is useful for problems of pre-stressed concrete beams. A concrete beam, unreinforced with steel, can withstand negligible bending loads since concrete is so weak in tension. But if the beam be pre-compressed in some way, the

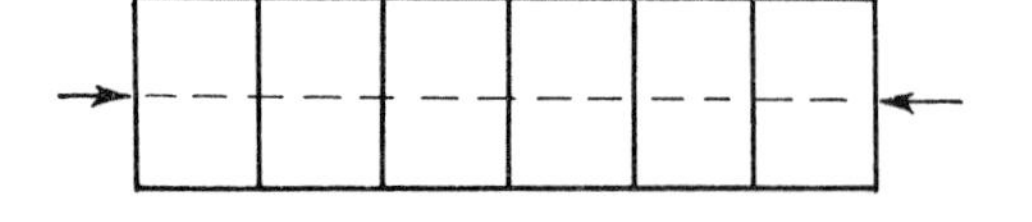

Fig. 12.6 Bending strength of a pre-compressed line of blocks.

tensile stresses induced by bending actions are countered by the compressive stresses already present. In Fig. 12.6, for example, a line of blocks carries an axial thrust; if this is sufficiently large, the line of blocks can be used in the same way as a solid beam.

Suppose a concrete beam of rectangular cross-section, Fig. 12.7, carries some system of lateral loads and is supported at its ends. An axial pre-compression P is applied at the ends. If M is the sagging moment at any cross-section, the greatest compressive stress occurs in the extreme top fibres, and has the value

$$\sigma = -\left(\frac{P}{A} + \frac{Mh}{2I_x}\right) \tag{12.14}$$

The stress in the extreme bottom fibres is

$$\sigma = -\left(\frac{P}{A} - \frac{Mh}{2I_x}\right) \tag{12.15}$$

Now suppose the maximum compressive stress in the concrete is limited to σ_1, and the maximum tensile stress to σ_2. Then we must have

$$\frac{P}{A} + \frac{Mh}{2I_x} \leqslant \sigma_1 \tag{12.16}$$

and

$$-\frac{P}{A} + \frac{Mh}{2I_x} \leqslant \sigma_2 \tag{12.17}$$

Then the design conditions are

$$\frac{Mh}{2I_x} \leqslant \sigma_1 - \frac{P}{A} \tag{12.18}$$

$$\frac{Mh}{2I_x} \leqslant \sigma_2 + \frac{P}{A} \tag{12.19}$$

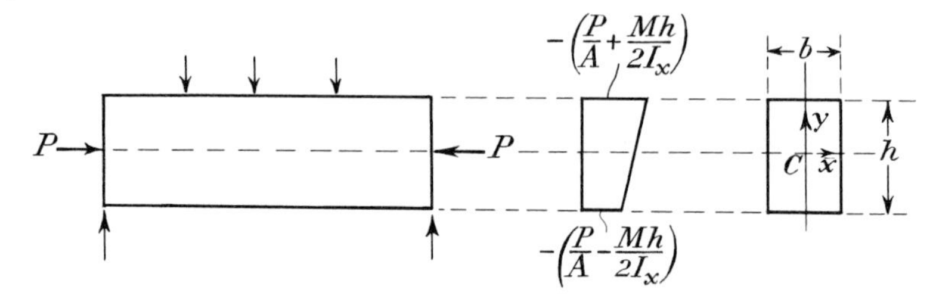

Fig. 12.7 Concrete beam with axial pre-compression.

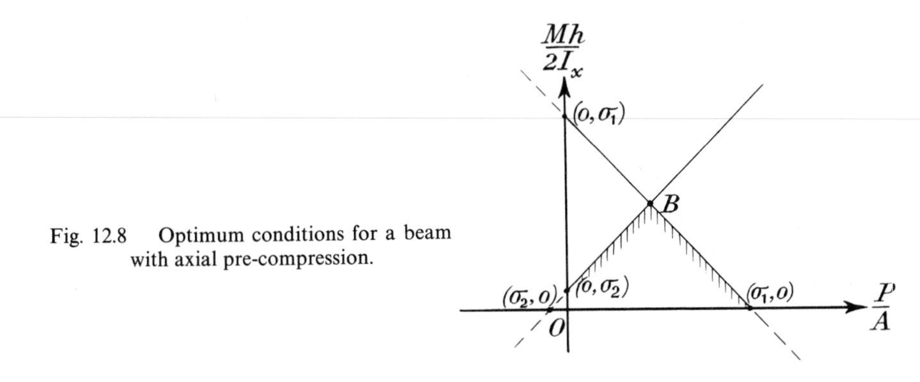

Fig. 12.8 Optimum conditions for a beam with axial pre-compression.

These two inequalities are shown graphically in Fig. 12.8, in which (P/A) is plotted against $(Mh/2I_x)$. Usually σ_2 is of the order of $\frac{1}{10}$th of σ_1. The optimum conditions satisfying both inequalities occur at the point B; the maximum bending moment which can be carried is given by

$$\frac{Mh}{I_x} = (\sigma_1 + \sigma_2) \tag{12.20}$$

that is,

$$M_{\max} = \frac{I_x}{h}(\sigma_1 + \sigma_2) \tag{12.21}$$

The required axial thrust for this load is

$$P = \tfrac{1}{2}A(\sigma_1 - \sigma_2) \tag{12.22}$$

Some advantage is gained by pre-compressing the beam eccentrically; in Fig. 12.9(i) a beam of rectangular cross-section carries a thrust P at a depth $\frac{1}{6}h$ below the centre line.

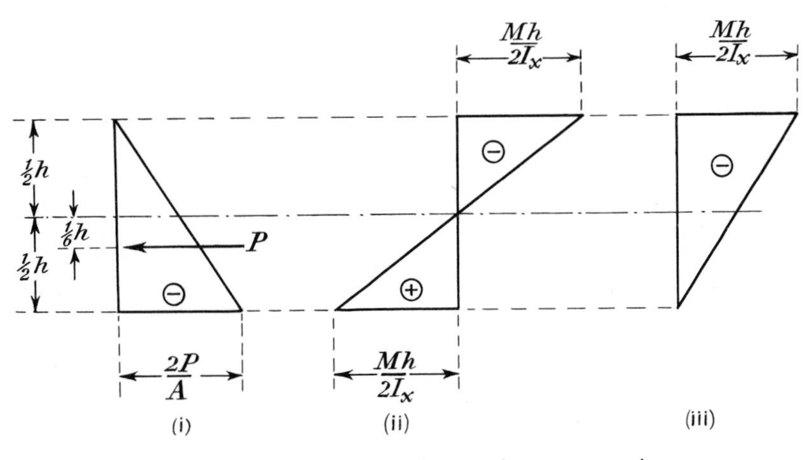

Fig. 12.9 Concrete beam with eccentric pre-compression.

As we saw in §12.3, this lies on the edge of the core of the cross-section, and no tensile stresses are induced. In the upper extreme fibres the longitudinal stress is zero, and in the lower extreme fibres the compressive stress is $2P/A$, Fig. 12.9(i). Now suppose a sagging bending moment M is superimposed on the beam; the extreme fibre stresses due to M are $(Mh/2I_x)$, tensile on the lower and compressive on the upper fibres, Fig. 12.9(ii). If

$$\frac{2P}{A} = \frac{Mh}{2I_x} \tag{12.23}$$

then the resultant stresses, Fig. 12.9(iii), are zero in the extreme lower fibres and a compressive stress of $(Mh/2I_x)$ in the extreme upper fibres. If this latter compressive stress does not exceed σ_1, the allowable stress in concrete, the design is safe. The maximum allowable value of M is

$$M = \frac{2I_x}{h}\sigma_1 \tag{12.24}$$

Since σ_2 in equation (12.21) is considerably less than σ_1,the bending moment given by equation (12.21) is approximately half that given by equation (12.24). Thus pre-compression by an eccentric load gives a considerably higher bending strength.

In practice the thrust is applied to the beam either externally through rigid supports, or by means of a stretched high-tensile steel wire passing through the beam and anchored at each end.

FURTHER PROBLEMS
(answers on page 398)

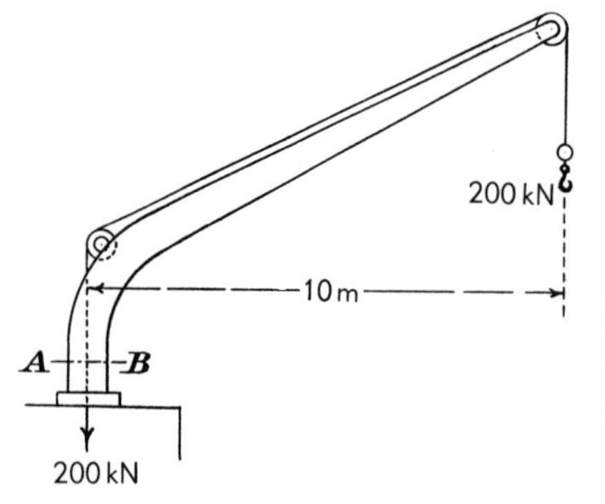

12.3 The single rope of a cantilever crane supports a load of 200 kN and passes over two pulleys and then vertically down the axis of the crane to the hoisting apparatus. The section AB of the crane is a hollow rectangle. The outside dimensions are 37·5 cm and 75 cm and the material is 2·5 cm thick all round, and the longer dimension is in the direction AB. Calculate the maximum tensile and compressive stresses set up in the section, and locate the position of the neutral axis. (*Cambridge*)

12.4 The horizontal cross-section of the cast-iron standard of a vertical drilling machine has the form shown. The line of thrust of the drill passes through P. Find the greatest value the thrust may have without the tensile stress exceeding 15 MN/m^2. What will be the stress along the face AB? (*Cambridge*)

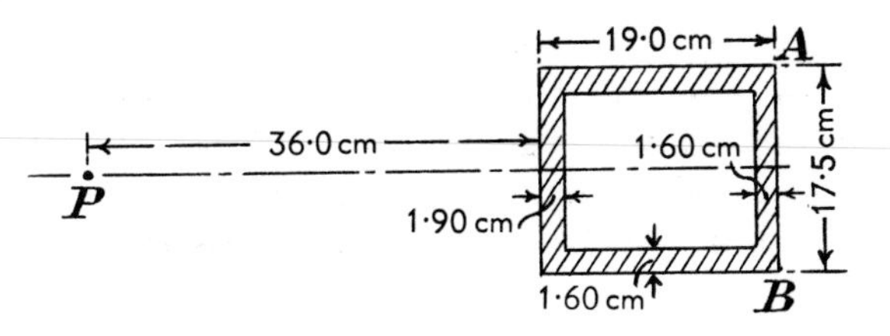

12.5 A vertical masonry chimney has an internal diameter d_1 and an external diameter d_0. The base of the chimney is given a horizontal acceleration a m/s², and the whole chimney moves horizontally with this acceleration. Show that at a section at depth h below the top of the chimney, the resultant normal force acts at a distance $ah/2g$ from the centre of the section. If the chimney behaves as an elastic solid, show that at a depth $g(d_0^2 + d_1^2)/4ad_0$ below the top, tensile stress will be developed in the material. (*Cambridge*)

12.6 A link of a valve gear has to be curved in one plane, for the sake of clearance. Estimate the maximum tensile and compressive stress in the link if the thrust is 2500 N. (*Cambridge*)

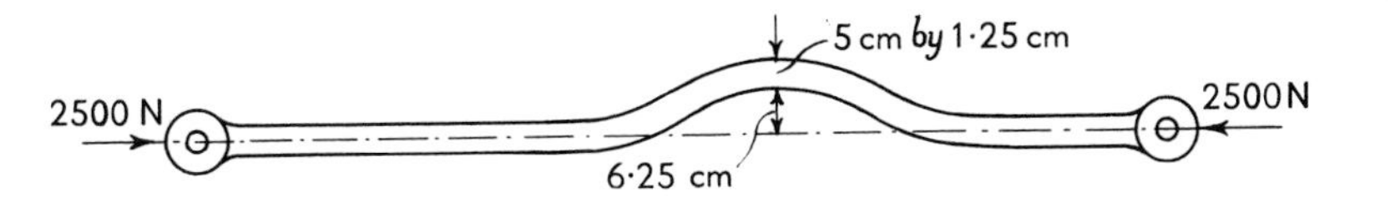

12.7 A cast-iron crank has a section on the line AB of the form shown. Show how to determine the greatest compressive and tensile stresses at AB, normal to the section, due to the thrust P of the connecting rod at the angle ϕ shown.

If the stresses at the section must not exceed 75 MN/m², either in tension or compression, find the maximum value of the thrust P. (*Cambridge*)

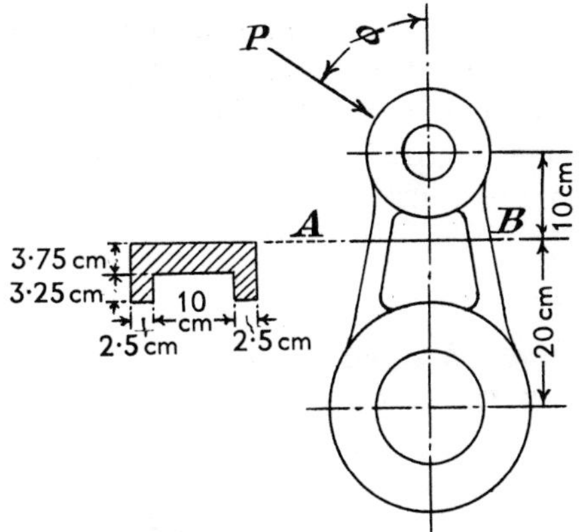

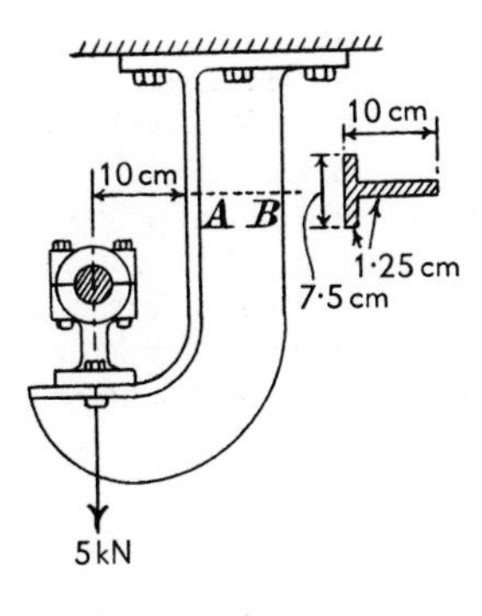

12.8 The load on the bearing of a cast-iron bracket is 5 kN. The form of the section AB is given. Calculate the greatest tensile stress across the section AB and the distance of the neutral axis of the section from the centre of gravity of the section. (*Cambridge*)

13 Deflections of beams

13.1 Introduction

In Chapter 7 we showed that the loading actions at any section of a simply-supported beam or cantilever can be resolved into a bending moment and a shearing force. Subsequently, in Chapters 9 and 10, we discussed ways of estimating the stresses due to these bending moments and shearing forces. There is, however, another aspect of the problem of bending which remains to be treated, namely, the calculation of the *stiffness* of a beam. In most practical cases, it is necessary that a beam should be not only strong enough for its purpose, but also that it should have the requisite stiffness, that is, it should not deflect from its original position by more than a certain amount. Again, there are certain types of beams, such as those carried by more than two supports and beams with their ends held in such a way that they must keep their original directions, for which we cannot calculate bending moments and shearing forces without studying the deformations of the axis of the beam; these problems are statically indeterminate, in fact.

In this chapter we consider methods of finding the deflected form of a beam under a given system of external loads and having known conditions of support.

13.2 Elastic bending of straight beams

It was shown in §9.2 that a straight beam of uniform cross-section, when subjected to end couples M applied about a principal axis, bends into a circular arc of radius R, given by

$$\frac{1}{R} = \frac{M}{EI} \tag{13.1}$$

where EI, which is the product of Young's modulus E and the second moment of area I about the relevant principal axis, is the flexural stiffness of the beam; equation (13.1) holds only for *elastic* bending.

Where a beam is subjected to shearing forces, as well as bending moments, the axis of the beam is no longer bent into a circular arc. To deal with this type of problem, we assume that equation (13.1) still defines the radius of curvature at any point of the beam where the bending moment is M. This implies that where the bending moment varies from one section of a beam to another, the radius of curvature also varies from section to section, in accordance with equation (13.1).

In the unstrained condition of the beam, Cz is the longitudinal centroidal axis, Fig. 13.1, and Cx, Cy are the principal axes in the cross-section. The coordinate axes Cx, Cy are so arranged that the y-axis is vertically downwards. This is convenient since most practical loading conditions give rise to vertically downwards deflections. Suppose bending moments are applied about axes parallel to Cx, so that bending is restricted to the yz plane, since Cx and Cy are principal axes.

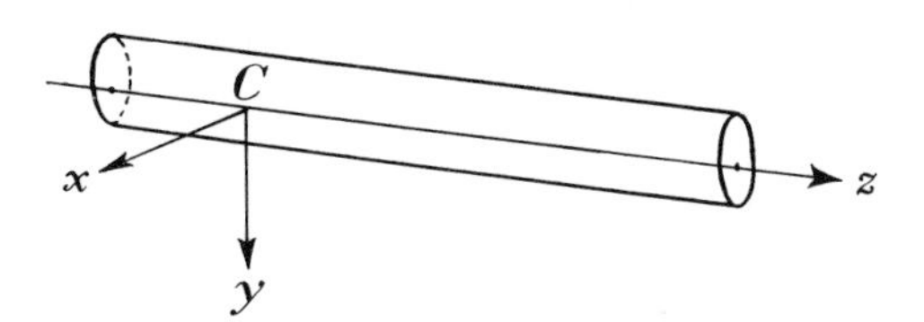

Fig. 13.1 Longitudinal and principal centroidal axes for a straight beam.

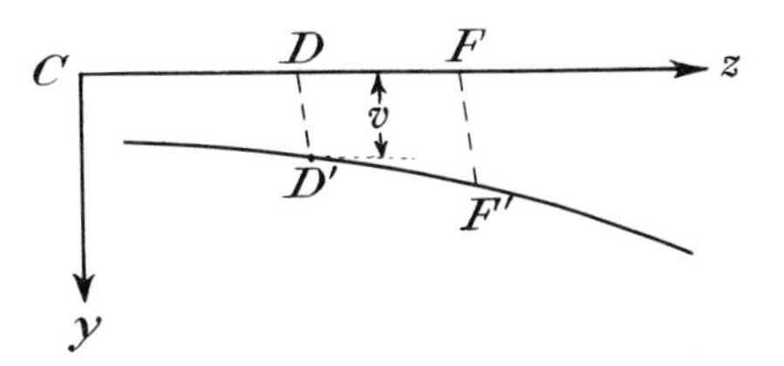

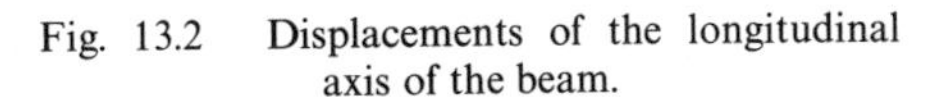

Fig. 13.2 Displacements of the longitudinal axis of the beam.

Now consider a short length of the unstrained beam, corresponding with DF on the axis Cz, Fig. 13.2. In the strained condition D and F are displaced to D' and F', respectively, which lie in the yz-plane. Any point such as D on the axis Cz is displaced by an amount v parallel to Cy; it is also displaced a small, but negligible, amount parallel to Cz.

The radius of curvature R at any section of the beam is then given by

$$\frac{1}{R} = \frac{\dfrac{d^2v}{dz^2}}{\pm\left[1 + \left(\dfrac{dv}{dz}\right)^2\right]^{\frac{3}{2}}} \tag{13.2}$$

We are concerned generally with only small deflections, in which v is small; this implies that (dv/dz) is small, and that $(dv/dz)^2$ is negligible compared with unity. Then, with sufficient accuracy, we may write

$$\frac{1}{R} = \pm \frac{d^2v}{dz^2} \tag{13.3}$$

Then equations (13.1) and (13.3) give

$$\pm EI \frac{d^2v}{dz^2} = M \tag{13.4}$$

We must now consider whether the positive or negative sign is relevant in this equation; we have already adopted the convention in §7.4 that sagging bending moments are positive. When a length of the beam is subjected to sagging bending moments, as in Fig. 13.3, the value of (dv/dz) along the length diminishes as z increases; hence a sagging moment implies that the curvature is negative. Then

$$EI\frac{d^2v}{dz^2} = -M \tag{13.5}$$

where M is the *sagging* bending moment.

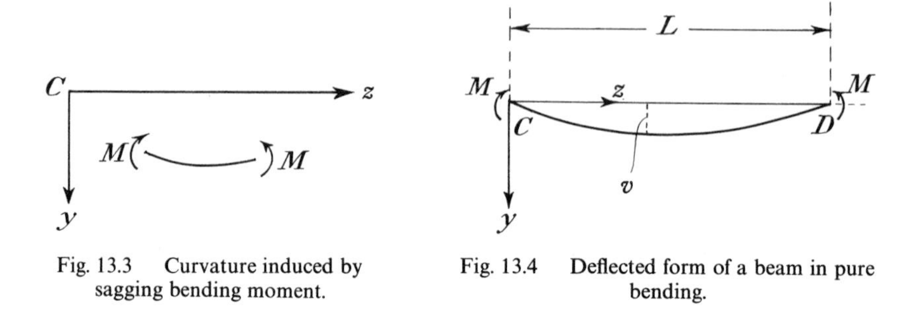

Fig. 13.3 Curvature induced by sagging bending moment.

Fig. 13.4 Deflected form of a beam in pure bending.

Where the beam is loaded on its axis of shear centres, so that no twisting occurs, M may be written in terms of the shearing force F and intensity w of vertical loading at any section. From equation (7.9) we have

$$\frac{d^2M}{dz^2} = \frac{dF}{dz} = -w \tag{7.9}$$

On substituting for M from equation (13.5), we have

$$\frac{d^2}{dz^2}\left[-EI\frac{d^2v}{dz^2}\right] = \frac{dF}{dz} = -w \tag{13.6}$$

This relation is true if EI varies from one section of a beam to another. Where EI is constant along the length of a beam,

$$-EI\frac{d^4v}{dz^4} = \frac{dF}{dz} = -w \tag{13.7}$$

As an example of the use of equation (13.4), consider the case of a uniform beam carrying couples M at its ends, Fig. 13.4. The bending moment at any section is M, so the beam is under a constant bending moment. Equation (13.5) gives

$$EI\frac{d^2v}{dz^2} = -M$$

On integrating once, we have

$$EI\frac{dv}{dz} = -Mz + A \tag{13.8}$$

where A is a constant. On integrating once more

$$EIv = -\tfrac{1}{2}Mz^2 + Az + B \tag{13.9}$$

where B is another constant. If we measure v relative to a line CD joining the ends of the beam, v is zero at each end. Then, $v = 0$, for $z = 0$ and $z = L$

On substituting these two conditions into equations (13.8) and (13.9), we have

$$B = 0 \quad \text{and} \quad A = \tfrac{1}{2}ML$$

Then equation (13.9) may be written

$$EIv = \tfrac{1}{2}Mz(L - z) \tag{13.10}$$

At the mid-length, $z = \frac{1}{2}L$, and

$$v = \frac{ML^2}{8EI} \tag{13.11}$$

which is the greatest deflection. At the ends $z = 0$ and $z = L/2$,

$$\frac{dv}{dz} = \frac{ML}{2EI} \text{ at } C; \quad \frac{dv}{dz} = -\frac{ML}{2EI} \text{ at } D \tag{13.12}$$

It is important to appreciate that equation (13.3), expressing the radius of curvature R in terms of v, is only true if the displacement v is small. We can study more accurately the pure bending of a beam by considering it to be deformed into the arc of a circle, Fig. 13.5; since the bending moment M is constant at all sections of the beam, the

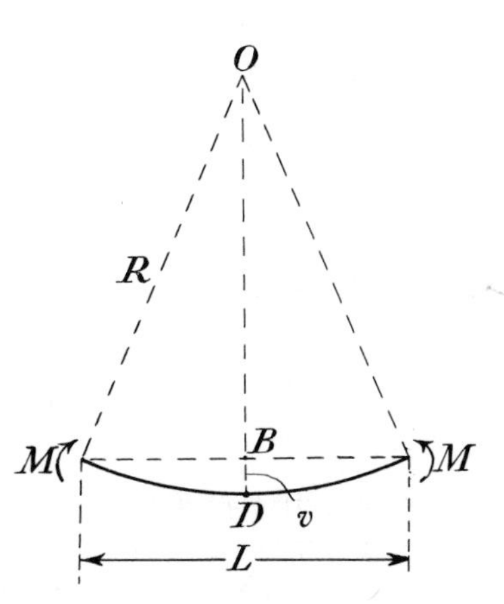

Fig. 13.5 Distortions of a beam in pure bending.

radius of curvature R is the same for all sections. If L be the length between the ends, Fig. 13.5, and D is the mid-point,

$$OB = \sqrt{R^2 - (L^2/4)}$$

Thus the central deflection v, is

$$v = BD = R - \sqrt{R^2 - (L^2/4)}$$

Then

$$v = R\left[1 - \sqrt{1 - \frac{L^2}{4R^2}}\right]$$

Suppose L/R is considerably less than unity; then

$$v = R\left[\frac{1}{2}\left(\frac{L^2}{4R^2}\right) + \frac{1}{8}\left(\frac{L^2}{4R^2}\right)^2 + \dots\right]$$

which can be written

$$v = \frac{L^2}{8R}\left[1 + \frac{L^2}{4R^2} + \dots\right]$$

But

$$\frac{1}{R} = \frac{M}{EI}$$

and so

$$v = \frac{ML^2}{8EI}\left[1 + \frac{M^2L^2}{4(EI)^2} + \dots\right] \tag{13.13}$$

Clearly, if $(L^2/4R^2)$ is negligible compared with unity, we have, approximately,

$$v = \frac{ML^2}{8EI}$$

which agrees with equation (13.11). The more accurate equation (13.13) shows that, when $(L^2/4R^2)$ is not negligible, the relation between v and M is non-linear; for all practical purposes this refinement is unimportant, and we find simple linear relations of the type of equation (13.11) are sufficiently accurate for engineering purposes.

13.3 Simply-supported beam carrying a uniformly distributed load

A beam of uniform flexural stiffness EI and span L is simply-supported at its ends, Fig. 13.6; it carries a uniformly distributed lateral load of w per unit length, which induces bending in the yz plane only. Then the reactions at the ends are each equal to $\frac{1}{2}wL$; if z is measured from the end C, the bending moment at a distance z from C is

$$M = \tfrac{1}{2}wLz - \tfrac{1}{2}wz^2$$

Then from equation (13.5),

$$EI\frac{d^2v}{dz^2} = -M = -\tfrac{1}{2}wLz + \tfrac{1}{2}wz^2$$

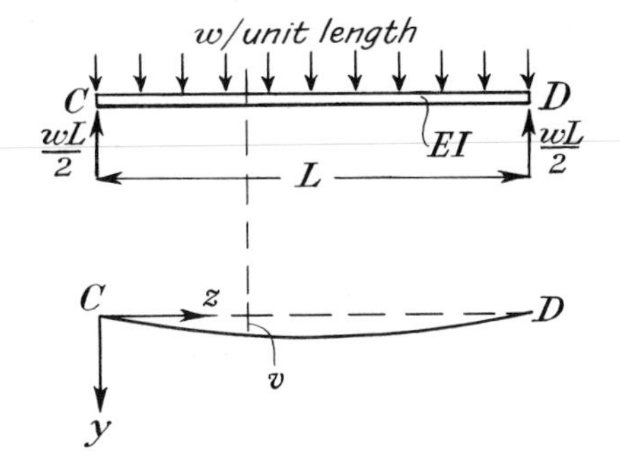

Fig. 13.6 Simply-supported beam carrying a uniformly supported load.

On integrating,

$$EI\frac{dv}{dz} = -\frac{wLz^2}{4} + \frac{wz^3}{6} + A$$

and

$$EIv = -\frac{wLz^3}{12} + \frac{wz^4}{24} + Az + B \tag{13.14}$$

Suppose $v = 0$ at the ends $z = 0$ and $z = L$; then

$$B = 0, \quad \text{and} \quad A = wL^3/24$$

Then equation (13.14) becomes

$$EIv = \frac{wz}{24}[L^3 - 2Lz^2 + z^3] \tag{13.15}$$

The deflection at the mid-length, $z = \frac{1}{2}L$, is

$$v = \frac{5wL^4}{384EI} \tag{13.16}$$

13.4 Cantilever with a concentrated load

A uniform cantilever of flexural stiffness EI and length L carries a vertical concentrated load W at the free end, Fig. 13.7. The bending moment a distance z from the built-in end is

$$M = -W(L - z)$$

Hence equation (13.5) gives

$$EI\frac{d^2v}{dz^2} = W(L - z)$$

Then

$$EI\frac{dv}{dz} = W(Lz - \tfrac{1}{2}z^2) + A \tag{13.17}$$

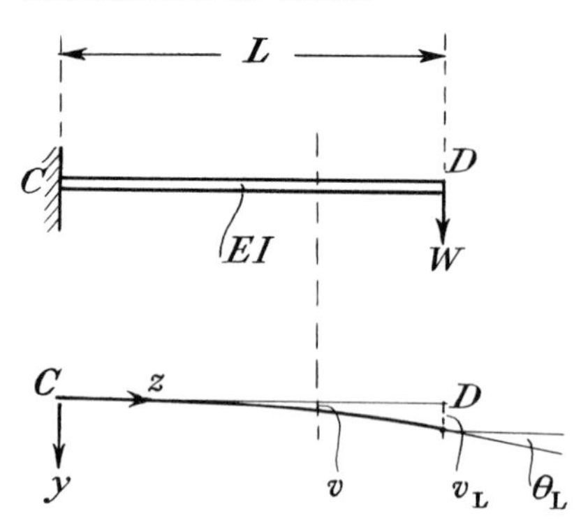

Fig. 13.7 Cantilever carrying a vertical load at the remote end.

and

$$EIv = W(\tfrac{1}{2}Lz^2 - \tfrac{1}{6}z^3) + Az + B$$

At the end $z = 0$, there is zero slope in the deflected form, so that $dv/dz = 0$; then equation (13.17) gives $A = 0$. Furthermore, at $z = 0$ there is also no deflection, so that $B = 0$. Then

$$EIv = \frac{Wz^2}{6}(3L - z)$$

At the free end, $z = L$,

$$v_L = \frac{WL^3}{3EI} \tag{13.18}$$

The slope of the beam at the free end is

$$\theta_L = \left(\frac{dv}{dz}\right)_{z=L} = \frac{WL^2}{2EI} \tag{13.19}$$

When the cantilever is loaded at some point between the ends, at a distance a, say, from the built-in support, Fig. 13.8, the beam between G and D carries no bending moments and therefore remains straight. The deflection at G can be deduced from equation (13.18); for $z = a$,

$$v_a = \frac{Wa^3}{3EI} \tag{13.20}$$

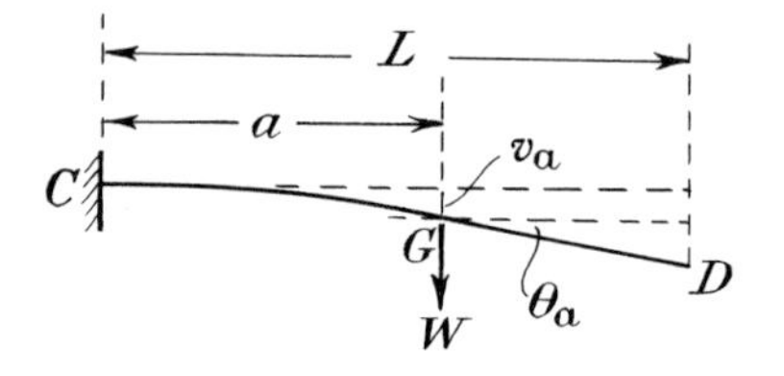

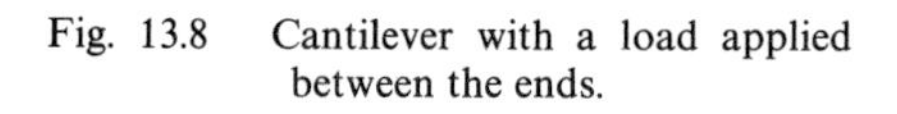
Fig. 13.8 Cantilever with a load applied between the ends.

and the slope at $z = a$ is

$$\theta_a = \frac{Wa^2}{2EI} \tag{13.21}$$

Then the deflection at the free end D of the cantilever is

$$v_L = \frac{Wa^3}{3EI} + (L - a)\frac{Wa^2}{2EI}$$

$$= \frac{Wa^2}{6EI}(3L - a) \tag{13.22}$$

13.5 Cantilever with a uniformly distributed load

A uniform cantilever, Fig. 13.9, carries a uniformly distributed load of w per unit length over the whole of its length. The bending moment at a distance z from C is

$$M = -\tfrac{1}{2}w(L - z)^2$$

Then, from equation (13.5),

$$EI\frac{d^2v}{dz^2} = \tfrac{1}{2}w(L - z)^2 = \tfrac{1}{2}w(L^2 - 2Lz + z^2)$$

Thus

$$EI\frac{dv}{dz} = \tfrac{1}{2}w(L^2z - Lz^2 + \tfrac{1}{3}z^3) + A$$

and

$$EIv = \tfrac{1}{2}w(\tfrac{1}{2}L^2z^2 - \tfrac{1}{3}Lz^3 + \tfrac{1}{12}z^4) + Az + B$$

At the built-in end, $z = 0$, and we have

$$\frac{dv}{dz} = 0 \quad \text{and} \quad v = 0$$

Thus $A = B = 0$. Then

$$EIv = \tfrac{1}{24}w(6L^2z^2 - 4Lz^3 + z^4)$$

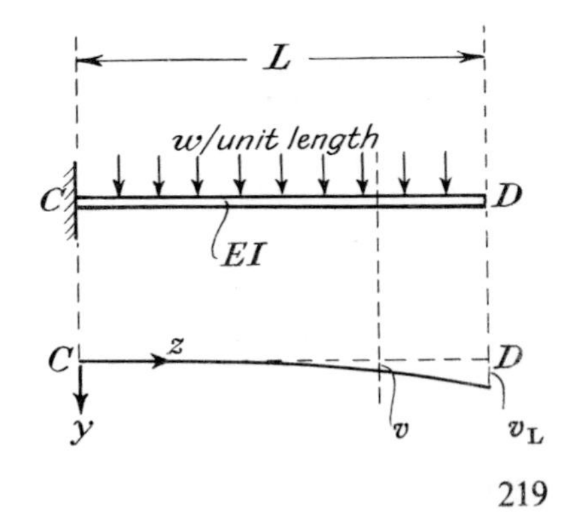

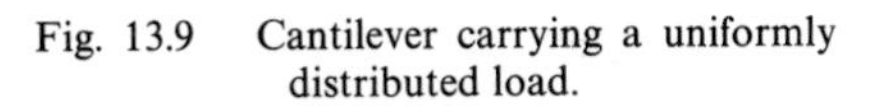
Fig. 13.9 Cantilever carrying a uniformly distributed load.

At the free end, D, the vertical deflection is

$$v_L = \frac{wL^4}{8EI} \tag{13.23}$$

13.6 Propped cantilever with distributed load

The uniform cantilever of Fig. 13.10(i) carries a uniformly distributed load w and is supported on a rigid knife-edge at the end D. Suppose P is the force on the support at

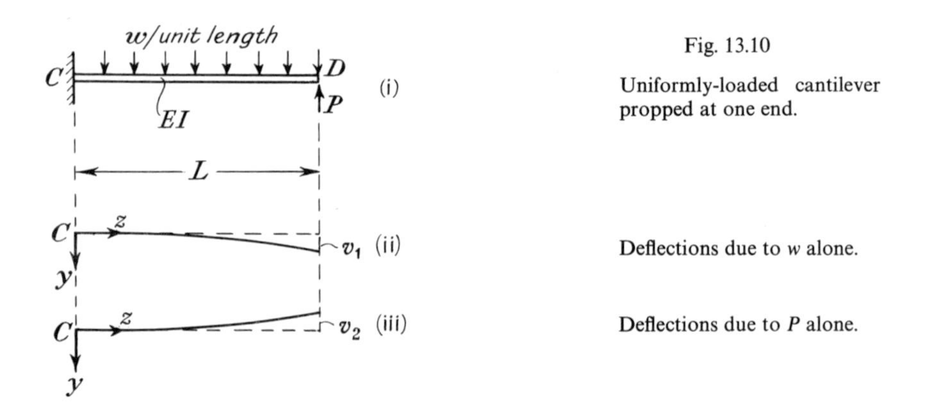

Fig. 13.10

(i) Uniformly-loaded cantilever propped at one end.

(ii) Deflections due to w alone.

(iii) Deflections due to P alone.

D. Then we regard Fig. 13.10(i) as the superposition of the effects of P and w acting separately.

If w acts alone, the deflection at D is given by equation (13.23), and has the value

$$v_1 = \frac{wL^4}{8EI}$$

If the reaction P acted alone, there would be an upwards deflection

$$v_2 = \frac{PL^3}{3EI}$$

at D. If the support maintains zero deflection at D,

$$v_1 - v_2 = 0$$

This gives

$$\frac{PL^3}{3EI} = \frac{WL^4}{8EI}$$

or

$$P = \frac{3wL}{8} \tag{13.24}$$

Problem 13.1: A steel rod 5 cm diameter protrudes 2 m horizontally from a wall. (i) Calculate the deflection due to a load of 1 kN hung on the end of the rod. The weight of the rod may be neglected. (ii) If a vertical steel wire 3 m long, 0·25 cm diameter, supports the end of the cantilever, being taut but unstressed before the load is applied, calculate the end deflection on application of the load. Take $E = 200\ \text{GN/m}^2$. (*R.N.E.C.*)

Solution

(i) The second moment of area of the cross-section is

$$I_x = \frac{\pi}{64}(0{\cdot}050)^4 = 0{\cdot}307 \times 10^{-6}\ \text{m}^4$$

The deflection at the end is then

$$v = \frac{PL^3}{3EI} = \frac{(1000)(2)^3}{3(200 \times 10^9)(0{\cdot}307 \times 10^{-6})} = 0{\cdot}0434\ \text{m}$$

(ii) Let T = tension in the wire; the area of cross-section of the wire is $4{\cdot}90 \times 10^{-6}\ \text{m}^2$. The elongation of the wire is then

$$e = \frac{Tl}{EA} = \frac{T(3)}{(200 \times 10^9)(4{\cdot}90 \times 10^{-6})}$$

The load on the end of the cantilever is then $(1000 - T)$, and this produces a deflection of

$$v = \frac{(1000 - T)(2)^3}{3(200 \times 10^9)(0{\cdot}307 \times 10^{-6})}$$

If this equals the stretching of the wire, then

$$\frac{(1000 - T)(2)^3}{3(200 \times 10^9)(0{\cdot}307 \times 10^{-6})} = \frac{T(3)}{(200 \times 10^9)(4{\cdot}90 \times 10^{-6})}$$

This gives $T = 934$ N, and the deflection of the cantilever becomes

$$v = \frac{(66)(2)^3}{3(200 \times 10^9)(0{\cdot}307 \times 10^{-6})} = 0{\cdot}00276\ \text{m}$$

Problem 13.2: A platform carrying a uniformly distributed load rests on two cantilevers projecting a distance l m from a wall. The distance between the two cantilevers is $\frac{1}{2}l$. In what ratio might the load on the

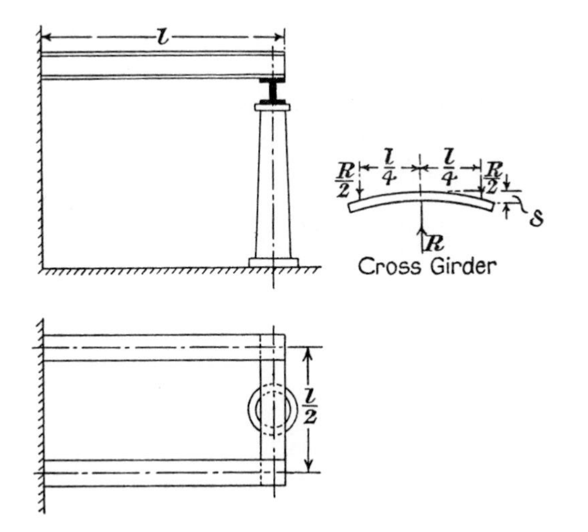

platform be increased if the ends were supported by a cross girder of the same section as the cantilevers, resting on a rigid column in the centre, as shown? It may be assumed that when there is no load on the platform the cantilevers just touch the cross girder without pressure. (*Cambridge*)

Solution

Let w_1 = the safe load per unit length on each cantilever when unsupported. Then the maximum bending moment $= \frac{1}{2}w_1 l^2$.

Let w_2 = the safe load when supported,

δ = the deflection of the end of each cantilever,

$\frac{1}{2}R$ = the pressure between each cantilever and the cross girder. Then the pressure is

$$\frac{R}{2} = \frac{3}{8}w_2 l - \frac{3EI\delta}{l^3}$$

We see from the figure above that

$$\delta = \frac{(R/2)(l/4)^3}{3EI} = \frac{Rl^3}{384EI}$$

I having the same value for the cantilevers and the cross girder. Substituting this value of δ

$$\frac{R}{2} = \frac{3w_2 l}{8} - \frac{R}{128}$$

or

$$R = \frac{48}{65}w_2 l$$

The upward pressure on the end of each cantilever is $\frac{1}{2}R = 24w_2 l/65$, giving a bending moment at the wall equal to $24w_2 l^2/65$. The bending moment of opposite sign due to the distributed load is $\frac{1}{2}w_2 l^2$. Hence it is clear that the maximum bending moment due to both acting together must occur at the wall and is equal to $(\frac{1}{2} - \frac{24}{65})w_2 l^2 = \frac{17}{130}w_2 l^2$. If this is to be equal to $\frac{1}{2}w_1 l^2$, we must have $w_2 = \frac{65}{17}w_1$; in other words, the load on the platform can be increased in the ratio 65/17, or nearly 4/1. The bending moment at the centre of the cross girder is $6w_2 l^2/65$, which is less than that at the wall.

13.7 Simply-supported beam carrying a concentrated lateral load

Consider a beam of uniform flexural stiffness EI and length L, which is simply-supported at its ends C and G, Fig. 13.11. The beam carries a concentrated lateral load W at a distance a from C. Then the reactions at C and G are

$$V_c = \frac{W}{L}(L - a) \qquad V_g = \frac{Wa}{L}$$

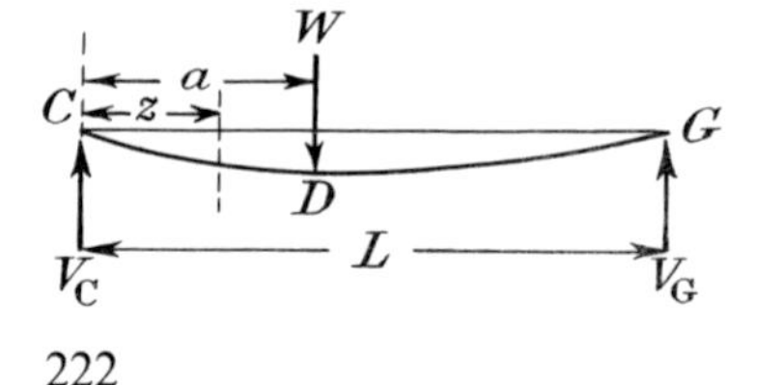

Fig. 13.11 Deflections of a simply-supported beam carrying a concentrated lateral bond.

Now consider a section of the beam a distance z from C; if $z < a$, the bending moment at the section is

$$M = V_c z$$

and if $z > a$,

$$M = V_c z - W(z - a)$$

Then

$$EI\frac{d^2v}{dz^2} = -V_c z \quad \text{for} \quad z < a$$

and

$$EI\frac{d^2v}{dz^2} = -V_c z + W(z - a) \quad \text{for} \quad z > a$$

On integrating these equations, we have

$$EI\frac{dv}{dz} = -\tfrac{1}{2}V_c z^2 + A \quad \text{for} \quad z < a \tag{13.25}$$

$$EI\frac{dv}{dz} = -\tfrac{1}{2}V_c z^2 + W(\tfrac{1}{2}z^2 - az) + A' \quad \text{for} \quad z > a \tag{13.26}$$

and

$$EIv = -\tfrac{1}{6}V_c z^2 + Az + B \quad \text{for} \quad z < a \tag{13.27}$$

$$EIv = -\tfrac{1}{6}V_c z^3 + W(\tfrac{1}{6}z^3 - \tfrac{1}{2}az^2) + A'z + B' \quad \text{for} \quad z > a \tag{13.28}$$

In these equations A, B, A' and B' are arbitrary constants. Now for $z = a$ the values of v given by equations (13.27) and (13.28) are equal, and the slopes given by equations (13.25) and (13.26) are also equal, since there is continuity of the deflected form of the beam through the point D. Then

$$-\tfrac{1}{6}V_c a^3 + Aa + B = -\tfrac{1}{6}V_c a^3 + W(\tfrac{1}{6}a^3 - \tfrac{1}{2}a^3) + A'a + B'$$

and

$$-\tfrac{1}{2}V_c a^2 + A = -\tfrac{1}{2}V_c a^2 + W(\tfrac{1}{2}a^2 - a^2) + A'$$

These two equations give

$$\begin{aligned} A' &= A + \tfrac{1}{2}Wa^2 \\ B' &= B - \tfrac{1}{6}Wa^3 \end{aligned} \tag{13.29}$$

At the extreme ends of the beam $v = 0$, so that when $z = 0$ equation (13.27) gives $B = 0$, and when $z = L$, equation (13.28) gives

$$-\tfrac{1}{6}V_c L^3 + W(\tfrac{1}{6}L^3 - \tfrac{1}{2}aL^2) + A'L + B' = 0$$

We have finally,

$$
\begin{aligned}
A &= \tfrac{1}{6}V_c L^2 - \frac{W}{6L}(L - a)^3 \\
B &= 0 \\
A' &= \tfrac{1}{6}V_c L^2 - \frac{W}{6L}(L - a)^3 + \tfrac{1}{2}Wa^2 \\
B' &= -\tfrac{1}{6}Wa^3
\end{aligned}
\qquad (13.30)
$$

But $V_c = W(L - a)/L$, so that equations (13.30) become

$$
\begin{aligned}
A &= \frac{Wa}{6L}(L - a)(2L - a) \\
B &= 0 \\
A' &= \frac{Wa}{6L}(2L^2 + a^2) \\
B' &= -\tfrac{1}{6}Wa^3
\end{aligned}
\qquad (13.31)
$$

Then equations (13.27) and (13.28) may be written

$$EIv = -\frac{W}{6L}(L - a)z^3 + \frac{Wa}{6L}(2L^2 - 3aL + a^2)z \quad \text{for} \quad z < a \qquad (13.32)$$

$$EIv = -\frac{W}{6L}(L - a)z^3 + \frac{W}{6}(z^3 - 3az^2) + \frac{Wa}{6L}(2L^2 + a^2)z - \frac{Wa^3}{6} \quad \text{for } z > a \qquad (13.33)$$

The second relation, for $z > a$, may be written

$$EIv = -\frac{W}{6L}(L - a)z^3 + \frac{Wa}{6L}(2L^2 - 3aL + a^2)z + \frac{W}{6}(z - a)^3 \qquad (13.34)$$

Then equations (13.32) and (13.33) differ only by the last term of equation (13.34); if the last term of equation (13.34) is discarded when $z < a$, then equation (13.34) may be used to define the deflected form in all parts of the beam.

On putting $z = a$, the deflection at the loaded point D is

$$v_D = \frac{Wa^2(L - a)^2}{3EIL} \qquad (13.35)$$

When W is at the centre of the beam, $a = \frac{1}{2}L$, and

$$v_D = \frac{WL^3}{48EI} \qquad (13.36)$$

This is the maximum deflection of the beam only when $a = \frac{1}{2}L$.

13.8 Use of step-functions

The analysis of the preceding section may be simplified by using step-functions during the process of integration. If $z < a$, the bending moment at any section is

$$M = V_c z$$

and if $z > a$,

$$M = V_c z - W(z - a)$$

Now, suppose we introduce a function Φ of z, which is zero for the range $0 < z < a$, and equal to unity for the range $a < z < L$; then Φ has the 'stepped' form shown in

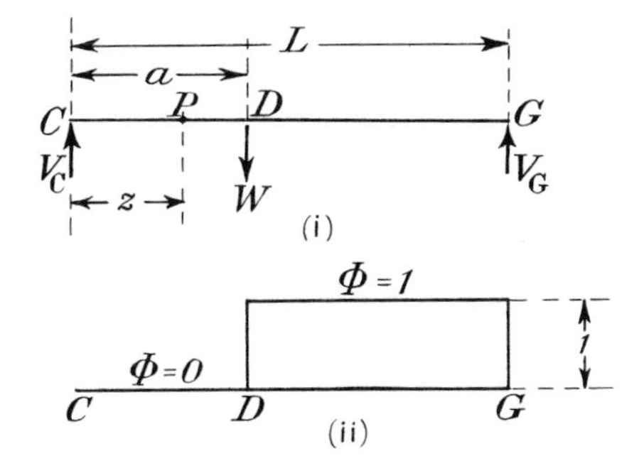

Fig. 13.12 Form of step-function used in deflection analysis of a beam.

Fig. 13.12(ii), and is called a *step-function.* Then we may write the bending moment for all values of z in the form

$$M = V_c z - \Phi[W(z - a)]$$

Since $\Phi = 0$ for $z < a$, the second term vanishes and $M = V_c z$. But for $z > a$, $\Phi = 1$, and the second term is retained. The curved form of the beam is then given by

$$EI \frac{d^2 v}{dz^2} = -V_c z + \Phi[W(z - a)] \tag{13.37}$$

On integrating once, we have

$$EI \frac{dv}{dz} = -\tfrac{1}{2} V_c z^2 + \int_0^z \Phi[W(z - a)]\, dz + A \tag{13.38}$$

where A is a constant. Now, if $z < a$,

$$\int_0^z \Phi[W(z - a)]\, dz = 0$$

since $\Phi = 0$ over this range; and if $z > a$,

$$\int_0^z \Phi[W(z - a)]\, dz = \int_0^a \Phi[W(z - a)]\, dz + \int_a^z \Phi[W(z - a)]\, dz \tag{13.39}$$

But

$$\int_0^a \Phi[W(z - a)]\, dz = 0$$

since $\Phi = 0$ for $z < a$, and we may write

$$\int_a^z \Phi[W(z - a)]\, dz = \Phi \int_0^{z-a} W(z - a)\, d(z - a)$$

Then equation (13.39) becomes

$$\int_0^z \Phi[W(z - a)]\, dz = \Phi \int_0^{z-a} W(z - a)\, d(z - a)$$

and we may now write equation (13.38) in the form

$$EI\frac{dv}{dz} = -\tfrac{1}{2}V_c z^2 + \Phi \int_0^{z-a} W(z - a)\, d(z - a) + A \qquad (13.40)$$

We see that the effect of the step function Φ on the integration is to introduce a new variable $(z - a)$. On integrating equation (13.37), we take Φ as unity and integrate the term $W(z - a)$ with respect to $(z - a)$ and not with respect to z. From equation (13.40), we have finally

$$EI\frac{dv}{dz} = -\tfrac{1}{2}V_c z^2 + \Phi\left[\frac{(z - a)^2}{2}\right] + A \qquad (13.41)$$

The presence of Φ, which is still equal to zero for $z < a$ and to unity for $z > a$, indicates that the term

$$\Phi\left[\frac{W}{2}(z - a)^2\right]$$

is zero for $z < a$, and equal to

$$\frac{W}{2}(z - a)^2$$

for $z > a$.

On repeating the integrating process, we have

$$EIv = \tfrac{1}{6}V_c z^3 + \Phi\left[\frac{W}{6}(z - a)^3\right] + Az + B \qquad (13.42)$$

where the second term on the right-hand side of equation (13.41) is integrated with respect to $(z - a)$, and not z. It remains to find the value of the constants A and B in equation (13.42); at $z = 0$ we have $v = 0$, so that $B = 0$, since $\Phi = 0$ for $z < a$; at $z = L$, again $v = 0$, giving

$$0 = -\tfrac{1}{6}V_c L^3 + \frac{W}{6}(L - a)^3 + AL$$

since $\Phi = 1$ for $z > a$. Then

$$A = \tfrac{1}{6}V_c L^2 - \frac{W}{6L}(L - a)^3$$

Finally,

$$EIv = -\tfrac{1}{6}V_c z^3 + \Phi\left[\frac{W}{6}(z - a)^3\right] + \left[\tfrac{1}{6}V_c L^2 - \frac{W}{6L}(L - a)^3\right]z \qquad (13.43)$$

If we put $V_c = \frac{W}{L}(L - a)$, then

$$EIv = -\frac{W}{6L}(L - a)z^3 + \frac{Wa}{6L}(2L^2 - 3aL + a^2)z + \Phi\left[\frac{W}{6}(z - a)^3\right] \qquad (13.44)$$

This is exactly the same as equation (13.34), established in §13.7, except for the introduction of Φ.

Clearly, the advantage in using a step-function, such as Φ, is that only two constants of integration are introduced. We need not consider continuity of the deflected form of the beam at a point of application of a concentrated load; continuity is ensured automatically by the step-function.

13.9 Simply-supported beam with distributed load over a portion of the span

Suppose firstly that the load is w per unit length over the portion DG, Fig. 13.13; the reactions at the ends of the beam are

$$V_c = \frac{w}{2L}(L - a)^2$$

$$V_g = \frac{w}{2L}(L^2 - a^2)$$

The bending moment at a distance z from C is

$$M = \frac{w}{2L}(L - a)^2 z - \Phi\left[\frac{w}{2}(z - a)^2\right]$$

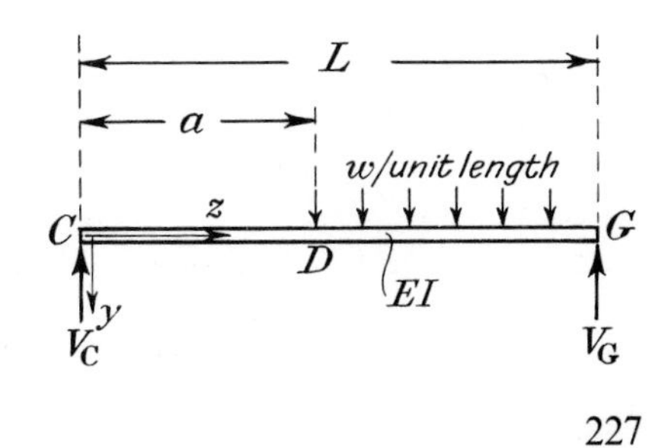

Fig. 13.13 Load extending to one support.

where Φ is a step-function having the values

$$\Phi = 0, \quad \text{for} \quad 0 < z < a$$

$$\Phi = 1, \quad \text{for} \quad a < z < L$$

Then

$$EI\frac{d^2v}{dz^2} = -\frac{w}{2L}(L-a)^2z + \Phi\left[\frac{w}{2}(z-a)^2\right]$$

On integrating twice

$$EIv = -\frac{w}{12L}(L-a)^2z^3 + \Phi\left[\frac{w}{24}(z-a)^3\right] + Az + B$$

The end conditions for the elimination of A and B are $v = 0$ for $z = 0$ and $z = L$. These give $B = 0$, and

$$A = \frac{w}{24L}(L-a)^2(L^2 + 2La - a^2)$$

The equation of the deflection curve is then

$$EIv = -\frac{w}{12L}(L-a)^2z^3 + \Phi\left[\frac{w}{24}(z^3-a)^3\right] + \frac{w}{24L}(L-a)^2(L^2 + 2La - a^2)z \quad (13.45)$$

When the load does not extend to either support, Fig. 13.14(i), the result of equation (13.45) may be used by superposing an upwards distributed load of w per unit length over the length GH on a downwards distributed load of w per unit length over DH, Fig. 13.14(ii). Due to the downwards distributed load alone

$$EIv = -\frac{w}{2L}(L-a)^2z^3 + \Phi_1\left[\frac{w}{24}(z-a)^3\right] + \frac{w}{24L}(L-a)^2(L^2 + 2La - a^2)z$$

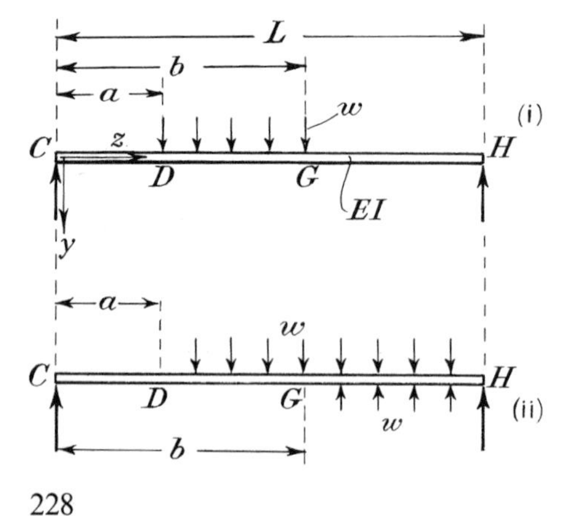

Fig. 13.14 Load not extending to either support.

where

$$\Phi_1 = 0 \quad \text{for} \quad 0 < z < a$$

$$\Phi_1 = 1 \quad \text{for} \quad a < z < L$$

Due to the upwards distributed load over GH acting alone

$$EIv = +\frac{w}{2L}(L - b)^2 z^3 + \Phi_2\left[-\frac{w}{24}(z - b)^3\right] - \frac{w}{24L}(L - b)^2(L^2 + 2Lb - b^2)z$$

where

$$\Phi_2 = 0 \quad \text{for} \quad 0 < z < b$$

$$\Phi_2 = 1 \quad \text{for} \quad b < z < L$$

On superposing the two deflected forms, the resultant deflection is given by

$$EIv = -\frac{wz^3}{2L}(b - a)(2L - a - b) + \Phi_1\left[\frac{w}{24}(z - a)^3\right]$$
$$- \Phi_2\left[\frac{w}{24}(z - b)^3\right] + \frac{wz}{24L}\{(L - a)^2(L^2 + 2La - a^2)$$
$$- (L - b)^2(L^2 + 2Lb - b^2)\} \tag{13.46}$$

13.10 Simply-supported beam with a couple applied at an intermediate point

The simply-supported beam of Fig. 13.15 carries a couple M_a applied to the beam at a point a distance a from C. The vertical reactions at each end are (M_a/L). The bending moment a distance z from C is

$$M = -\frac{M_a z}{L} + \Phi[M_a].$$

where

$$\Phi = 0 \quad \text{for} \quad 0 < z < a$$

$$\Phi = 1 \quad \text{for} \quad a < z < L$$

Then

$$EI\frac{d^2v}{dz^2} = \frac{M_a z}{L} - \Phi[M_a]$$

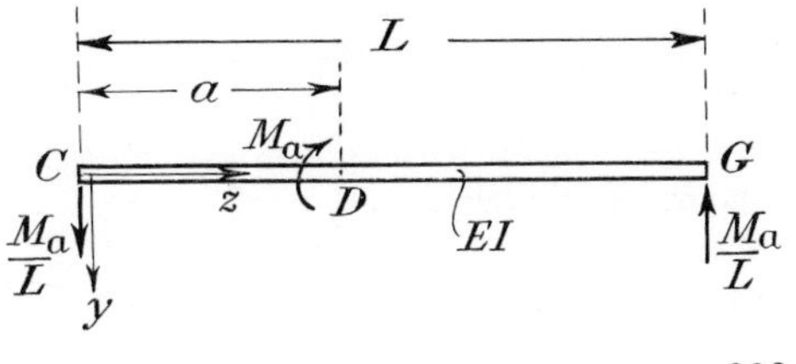

Fig. 13.15 Beam with a couple applied at a point in the span.

As before, the step-function Φ introduces a new variable $(z - a)$, and we have

$$EI \frac{dv}{dz} = \frac{M_a z^2}{2L} - \Phi[M_a(z - a)] + A$$

$$EIv = \frac{M_a z^3}{6L} - \Phi\left[\frac{M_a}{2}(z - a)^2\right] + Az + B$$

The constants A and B are eliminated from the conditions that $v = 0$ at $z = 0$ and $z = L$. These give $B = 0$, and

$$A = \tfrac{1}{6}M(2L^2 - 6La + 3a^2)$$

Then

$$EIv = \frac{M_a z^3}{6L} - \Phi\left[\frac{M_a}{2}(z - a)^2\right] + \frac{M_a z}{6}(2L^2 - 6La + 3a^2)$$

The deflection at D, where $z = a$, is

$$v_D = \frac{M_a a}{3EIL}(L - a)(L - 2a) \qquad (13.47)$$

Problem 13.3: A steel beam rests on two supports 6 m apart, and carries a uniformly distributed load of 100 kN per metre run. The second moment of area of the cross-section is 1×10^{-3} m^4 and $E = 200$ GN/m^2. Estimate the maximum deflection.

Solution

The greatest deflection occurs at mid-length and has the value given by equation (13.16):

$$v = \frac{5wL^4}{384EI} = \frac{5(100 \times 10^3)(6)^4}{384(200 \times 10^9)(1 \times 10^{-3})} = 0{\cdot}00844 \text{ m}$$

Problem 13.4: A uniform, simply-supported beam of span L carries a uniformly distributed lateral load of w per unit length. It is propped on a knife-edge support at a distance a from one end. Estimate the vertical force on the prop.

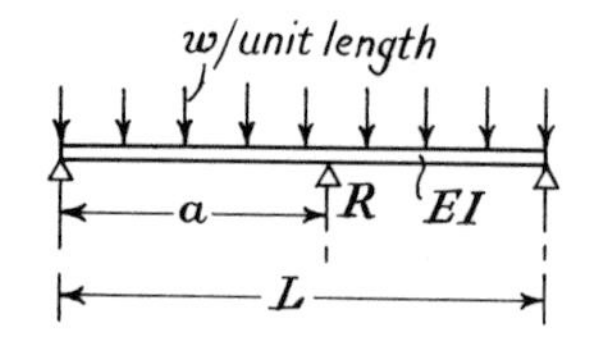

Solution

If the beam is unpropped, then, from equation (13.15), the downwards vertical deflection at the position of the prop is

$$(v)_{z=a} = \frac{wa}{24EI}(L^3 - 2La^2 + a^3)$$

If R is the reaction on the prop, then under the action of R alone the upwards vertical deflection at the prop is, from equation (13.35),

$$(v)_{z=a} = \frac{Ra^2(L - a)^2}{3EIL}$$

If there is no resultant deflection at the prop, we have

$$\frac{Ra^2(L - a)^2}{3EIL} = \frac{wa}{24EI}(L^3 - 2La^2 + a^3)$$

Thus, the reaction on the prop is

$$R = \frac{wL}{8}\left[\frac{1 - 2\left(\frac{a}{L}\right)^2 + \left(\frac{a}{L}\right)^3}{\frac{a}{L}\left(1 - \frac{a}{L}\right)^2}\right]$$

The propping force is least when the prop is at mid-span; in this case, $a/L = 0{\cdot}5$ and $R = 5\,wL/8$.

Problem 13.5: A simply-supported, uniform beam, of span L and flexural stiffness EI, carries a vertical lateral load W at a distance a from one end. Calculate the greatest lateral deflection in the beam.

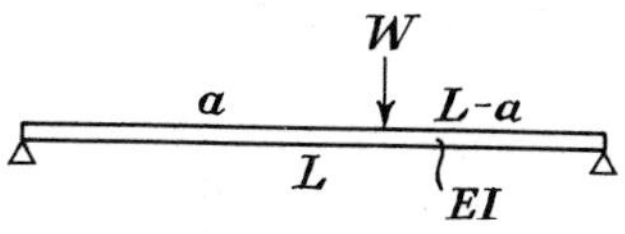

Solution

From section 13.7, the lateral deflection at any point is given by

$$EIv = -\frac{W}{6L}(L - a)z^3 + \frac{Wa}{6L}(2L^2 - 3aL + a^2)z \quad \text{for} \quad z < a$$

$$EIv = -\frac{W}{6L}(L - a)z^3 + \frac{Wz^2}{6}(z - 3a) + \frac{Wa}{6L}(2L^2 + a^2)z - \frac{Wa^3}{6} \quad \text{for} \quad z > a$$

Let us suppose first that $a > \frac{1}{2}L$, when we would expect the greatest deflection to occur in the range $z < a$; over this range

$$EI\frac{dv}{dz} = -\frac{W}{2L}(L - a)z^2 + \frac{Wa}{6L}(2L^2 - 3aL + a^2)$$

This is zero when

$$-\frac{W}{2L}(L - a)z^2 + \frac{Wa}{6L}(2L^2 - 3aL + a^2) = 0$$

i.e. when

$$(L - a)z^2 = \tfrac{1}{3}a(2L^2 - 3aL + a^2)$$

or when

$$z = \sqrt{\frac{a}{3}(2L - a)}$$

If this gives a root in the range $z < a$, then

$$\sqrt{\frac{a}{3}(2L - a)} < a$$

and $2L - a < 3a$, or $a > \frac{1}{2}L$. This is compatible with our earlier suppositions. Then, with $a > \frac{1}{2}L$, the greatest deflection occurs at the point $z = [(a/3)(2L - a)]^{\frac{1}{2}}$, and has the value

$$v_{\max} = \frac{Wa}{9LEI}(2L - a)(L - a)\sqrt{\frac{a}{3}(2L - a)}$$

If $a < \frac{1}{2}L$, the greatest deflection occurs in the range $z > a$; in this case we replace a by $(L - a)$, whence the greatest deflection occurs at the point $z = \sqrt{\frac{1}{3}(L^2 - a^2)}$, and has the value

$$v_{\max} = \frac{Wa}{9LEI}(L^2 - a^2)\sqrt{\frac{a}{3}(2L - a)}$$

13.11 Beam with end couples and distributed load

Suppose the ends of a beam CD, Fig. 13.16, rest on knife-edges, and carry couples M_C and M_D. If, in addition, the beam carries a uniformly distributed lateral load w per unit length, the bending moment a distance z from C is

$$M = \frac{M_C}{L}(L - z) + M_D\frac{z}{L} + \frac{1}{2}wz(L - z)$$

The equation of the deflection curve is then given by

$$EI\frac{d^2v}{dz^2} = -\frac{M_C}{L}(L - z) - M_D\frac{z}{L} - \frac{1}{2}wz(L - z)$$

Then

$$EI\frac{dv}{dz} = -\frac{M_C}{L}(Lz - \tfrac{1}{2}z^2) - \frac{M_D}{L}\left(\frac{z}{2}\right) - \frac{1}{2}w\left(\frac{Lz^2}{2} - \frac{z^3}{3}\right) + A$$

and

$$EIv = -\frac{M_C}{L}\left(\frac{Lz^2}{2} - \frac{z^3}{6}\right) - \frac{M_D}{L}\left(\frac{z^3}{6}\right) - \frac{1}{2}w\left(\frac{Lz^3}{6} - \frac{z^4}{12}\right) + Az + B \qquad (13.48)$$

If the ends of the beam remain at the same level, $v = 0$ for $z = 0$ and $z = L$. Then $B = 0$, and

$$AL = \tfrac{1}{3}M_CL^2 + \tfrac{1}{6}M_DL^2 + \tfrac{1}{24}wL^4$$

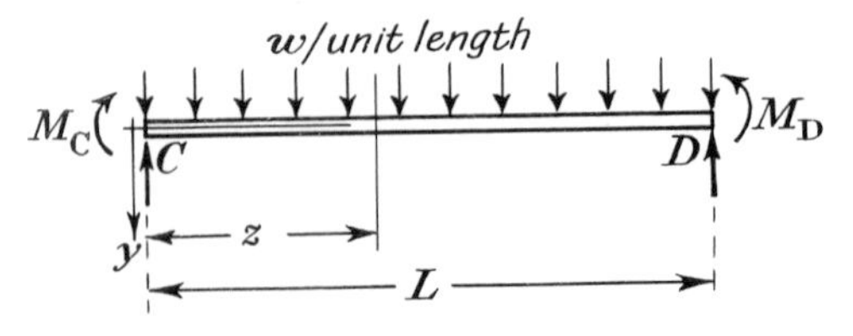

Fig. 13.16 Simply-supported beam carrying end couples and distributed load.

Then

$$EIv = -\frac{M_C}{L}\left(\frac{Lz^2}{2} - \frac{z^3}{6}\right) - \frac{M_D}{L}\left(\frac{z^3}{6}\right) - \tfrac{1}{2}w\left(\frac{Lz^3}{6} - \frac{z^4}{12}\right) + z\left(\frac{M_C L}{3} + \frac{M_D L}{6} + \frac{wL^3}{24}\right)$$

The slopes at the ends are

$$\left(\frac{dv}{dz}\right)_{z=0} = \frac{L}{24EI}(8M_C + 4M_D + wL^2)$$

$$\left(\frac{dv}{dz}\right)_{z=L} = -\frac{L}{24EI}(4M_C + 8M_D + wL^2)$$

Suppose that the end D of the beam now sinks an amount δ downwards relative to C. Then at $v = L$ we have $v = \delta$, instead of $v = 0$. In equation (13.48), A is then given by

$$AL = EI\,\delta + \tfrac{1}{3}M_C L^2 + \tfrac{1}{6}M_D L^2 + \tfrac{1}{24}wL^4$$

For the slopes at the ends we have

$$\left.\begin{aligned} \left(\frac{dv}{dz}\right)_{z=0} &= \frac{L}{24EI}(8M_C + 4M_D + wL^2) + \frac{\delta}{L} \\ \left(\frac{dv}{dz}\right)_{z=L} &= -\frac{L}{24EI}(4M_C + 8M_D + wL^2) + \frac{\delta}{L} \end{aligned}\right\} \qquad (13.49)$$

13.12 Beams with non-uniformly distributed load

When a beam carries a load which is not uniformly distributed the methods of the previous articles can still be employed if M and $\int M\,dz$ are both integrable functions of z, for we have in all cases

$$-EI\frac{d^2v}{dz^2} = M$$

which can be written in the form

$$\frac{d}{dz}\left(\frac{dv}{dz}\right) = -\frac{M}{EI}$$

If I is uniform along the beam the first integral of this is

$$\frac{dv}{dz} = A - \frac{1}{EI}\int M\,dz \qquad (13.50)$$

where A is a constant. The second integral is

$$v = Az + B - \frac{1}{EI}\iint M\,dz\,dz \tag{13.51}$$

If M and $\int M\,dz$ are integrable functions of z the process of finding v can be continued analytically, the constants A and B being found from the terminal conditions. Failing this the integrations must be performed graphically. This is most readily done by plotting the bending-moment curve, and from that deducing a curve of areas representing $\int M\,dz$. From this curve a third is deduced representing $\iint M\,dz\,dz$.

Problem 13.6. A uniform, simply-supported beam carries a distributed lateral load varying in intensity from w_0 at one end to $2w_0$ at the other. Calculate the greatest lateral deflection in the beam.

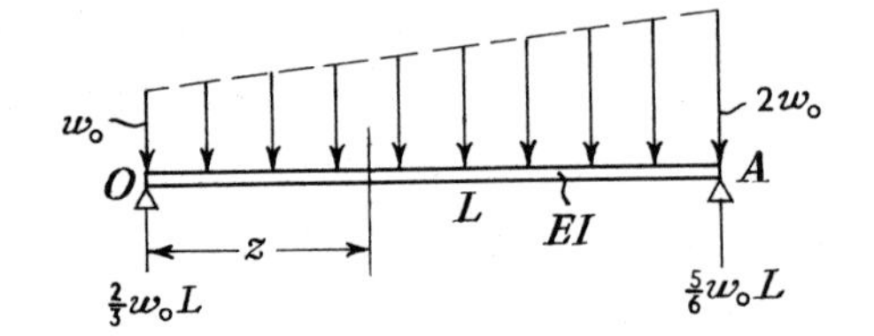

Solution

The vertical reactions at O and A are $\frac{2}{3}w_0L$ and $\frac{5}{6}w_0L$. The bending moment at any section a distance z from O is then

$$M = \tfrac{2}{3}w_0Lz - \tfrac{1}{2}w_0z^2 - \frac{w_0z^3}{6L}$$

Then

$$EI\frac{d^2v}{dz^2} = -\left[\tfrac{2}{3}w_0Lz - \tfrac{1}{2}w_0z^2 - \frac{w_0z^3}{6L}\right]$$

On integrating once,

$$EI\frac{dv}{dz} = -\left[\frac{w_0Lz^2}{3} - \frac{w_0z^3}{6} - \frac{w_0z^4}{24L} + C_1\right]$$

where C_1 is a constant. On integrating further,

$$EIv = -\left[\frac{w_0Lz^3}{9} - \frac{w_0z^5}{24} - \frac{w_0z^5}{120L} + C_1z + C_2\right]$$

where C_2 is a further constant. If $v = 0$ at $z = 0$ and $z = L$, we have

$$C_1 = -\tfrac{11}{180}w_0L^3 \quad\text{and}\quad C_2 = 0$$

Then

$$EIv = \tfrac{11}{180}w_0L^3z - \frac{w_0Lz^3}{9} - \frac{w_0z^4}{24} + \frac{w_0z^5}{120L}$$

The greatest deflection occurs at $dv/dz = 0$, i.e. when

$$\tfrac{11}{180}w_0L^3 - \frac{w_0Lz^2}{3} + \frac{w_0z^3}{6} + \frac{w_0z^4}{24L} = 0$$

or when

$$15\left(\frac{z}{L}\right)^4 + 60\left(\frac{z}{L}\right)^3 - 120\left(\frac{z}{L}\right)^2 + 22 = 0$$

The relevant root of this equation is $z/L \doteqdot 0{\cdot}506$ which gives the point of maximum deflection near to the mid-length. The maximum deflection is

$$v_{max} \doteqdot \frac{7{\cdot}03}{360}\frac{w_0L^4}{EI} = 0{\cdot}0195\frac{w_0L^4}{EI}$$

This is negligibly different from the deflection at mid-span, which is

$$(v)_{z=\frac{1}{2}L} = \frac{5w_0L^4}{256EI}$$

13.13 Cantilever with irregular loading

In Fig. 13.17(i) a cantilever is free at D and built-in to a rigid wall at C. The bending moment curve is DM of Fig. 13.17(ii); the bending moments are assumed to be hogging,

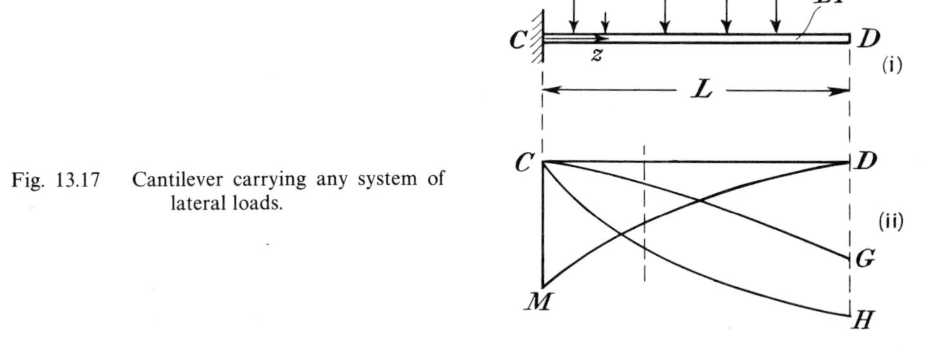

Fig. 13.17 Cantilever carrying any system of lateral loads.

and are therefore negative. The curve CH represents $\int_0^z M\,dz$, and its ordinates are drawn downwards since M is negative. The curve CG is then constructed from CH by finding

$$\int_0^z \int_0^z M\,dz\,dz$$

In equation (13.51), the constants A and B are both zero since $v = 0$ and $dv/dz = 0$ at $z = 0$. Then CD is the base line for both curves.

13.14 Beams of varying section

When the second moment of area of a beam varies from one section to another, equations (13.50) and (13.51) take the forms

$$\frac{dv}{dz} = A - \frac{1}{E}\int\frac{M\,dz}{I}$$

and

$$v = Az + B - \frac{1}{E} \iint \frac{M}{I} \, dz \, dz$$

The general method of procedure follows the same lines as before. If (M/I) and $\int (M/I) \, dz$ are integrable functions of z, then (dv/dz) and v may be evaluated analytically; otherwise graphical methods must be employed, when a curve of (M/I) must be taken as the starting point instead of a curve of M.

Problem 13.7: A cantilever strip has a length L, a constant breadth b and thickness t varying in such a way that when the cantilever carries a lateral end load W, the centre line of the strip is bent into a circular arc. Find the form of variation of the thickness t.

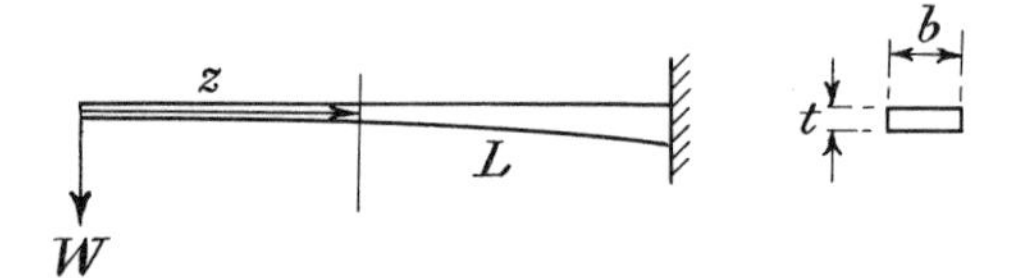

Solution

The second moment of area, I, at any section is

$$I = \tfrac{1}{12} bt^3$$

The bending moment at any section is $(-Wz)$, so that

$$EI \frac{d^2 v}{dz^2} = Wz$$

Then

$$\frac{d^2 v}{dz^2} = \frac{Wz}{EI}$$

If the cantilever is bent into a circular arc, then d^2v/dz^2 is constant, and we must have

$$\frac{Wz}{EI} = \text{constant}$$

This requires that

$$\frac{z}{I} = \text{constant}$$

or

$$I \propto z$$

Thus,

$$\tfrac{1}{12} bt^3 \propto z$$

or

$$t \propto z^{\frac{1}{3}}$$

Any variation of the form

$$t = t_0\left(\frac{z}{L}\right)^{\frac{1}{3}}$$

where t_0 is the thickness at the built-in end will lead to bending in the form of a circular arc.

Problem 13.8: The curve M, below, represents the bending moment at any section of a timber cantilever of variable bending stiffness. The second moments of area are given in the table below. Taking $E = 11\ \text{GN/m}^2$, deduce the deflection curve.

z (from supported end) (m)	0	0·1	0·2	0·3	0·5	0·7	0·9	1·1	1·3	1·5	1·6	1·7
I (m⁴)	50·8	27·4	17·4	12·25	5·65	3·23	1·69	0·783	0·278	0·074	0·0298	0 × 10⁻⁶

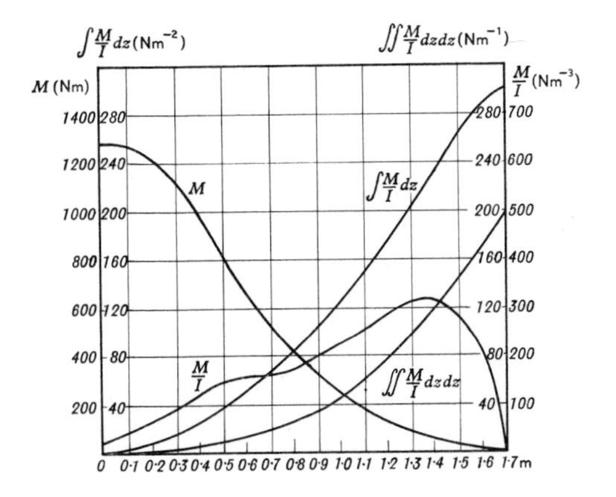

Solution

The first step is to calculate M/I at each section and to plot the M/I curve. We next plot the area under this curve at any section to give the curve

$$\int_0^z \frac{M}{I}\,dz$$

From this, the curve

$$\iint_0^z \frac{M}{I}\,dz\,dz$$

is plotted to give the deflected form

$$v = \frac{1}{E}\iint_0^z \frac{M}{I}\,dz\,dz$$

The maximum deflection at the free end of the cantilever is

$$v \doteqdot \frac{1}{E}(300 \times 10^6) = \frac{300 \times 10^6}{11 \times 10^9} = 0{\cdot}0272\ \text{m}$$

13.15 Non-uniformly distributed load and terminal couples; the method of 'moment-areas'

Consider a simply-supported beam carrying end moments M_C and M_D, as in Fig. 13.16, and a distributed load of varying intensity w. Suppose M_0 is the bending moment at any section due to the load w acting alone on the beam. Then

$$M = M_0 + \frac{M_C}{L}(L - z) + \frac{M_D}{L} z$$

The differential equation for the deflection curve is

$$EI \frac{d^2 v}{dz^2} = -M_0 - \frac{M_C}{L}(L - z) - \frac{M_D}{L} z \tag{13.52}$$

The integral between the limits $z = 0$ and $z = L$ is

$$EI\left[\left(\frac{dv}{dz}\right)_{z=L} - \left(\frac{dv}{dz}\right)_{z=0}\right] = -\tfrac{1}{2}M_C L - \tfrac{1}{2}M_D L - \int_0^L M_0 \, dz \tag{13.53}$$

Again, on multiplying equation (13.52) by z, we have

$$EI\left(z \frac{d^2 v}{dz^2}\right) = -M_0 z - \frac{M_C z}{L}(L - z) - \frac{M_D}{L} z^2 \tag{13.54}$$

But

$$EI\left(z \frac{d^2 v}{dz^2}\right) = EI \frac{d}{dz}\left(z \frac{dv}{dz} - v\right)$$

Thus, on integrating equation (13.54),

$$EI\left[z \frac{dv}{dz} - v\right]_0^L = -\frac{M_D}{L}\left[\frac{z^3}{3}\right]_0^L - \frac{M_C}{L}\left[\frac{Lz^2}{2} - \frac{z^3}{3}\right]_0^L - \int_0^L M_0 z \, dz \tag{13.55}$$

But if $v = 0$ when $z = 0$ and $z = L$, then equation (13.55) becomes

$$EI\left[L\left(\frac{dv}{dz}\right)_{z=L}\right] = -\tfrac{1}{3}M_D L^2 - \tfrac{1}{6}M_C L^2 - \int_0^L M_0 z \, dz$$

Then

$$\left(\frac{dv}{dz}\right)_{z=L} = -\frac{M_D L}{3EI} - \frac{M_C L}{6EI} - \frac{1}{EIL}\int_0^L M_0 z \, dz \tag{13.56}$$

On substituting this value of $(dv/dz)_{z=L}$ into equation (13.53),

$$\left(\frac{dv}{dz}\right)_{z=0} = \frac{M_C L}{3EI} + \frac{M_D L}{6EI} + \frac{1}{EI}\int_0^L M_0 \, dz - \frac{1}{EIL}\int_0^L M_0 z \, dz \tag{13.57}$$

The integral $\int_0^L M_0 \, dz$ is the area of the bending moment curve due to the load w alone; $\int_0^L M_0 z \, dz$ is the moment of this area about the end $z = 0$ of the beam. If A is the area of the bending moment diagram due to the lateral loads only, and $\bar{z}$ is the distance of its centroid from $z = 0$, then

$$A = \int_0^L M_0 \, dz, \qquad \bar{z} = \frac{1}{A}\int_0^L M_0 z \, dz$$

and equations (13.57) and (13.56) may be written

$$\left(\frac{dv}{dz}\right)_{z=0} = \frac{M_C L}{3EI} + \frac{M_D L}{6EI} + \frac{A(L - \bar{z})}{EIL} \tag{13.58}$$

$$\left(\frac{dv}{dz}\right)_{z=L} = -\frac{M_C L}{6EI} - \frac{M_D L}{3EI} - \frac{A\bar{z}}{EIL} \tag{13.59}$$

This method of analysis, making use of A and $\bar{z}$, is known as the *method of moment-areas*; it can be extended to deal with most problems of beam deflections.

When the section of the beam is not constant, equation (13.52) becomes

$$E\frac{d^2v}{dz^2} = -\frac{M_0}{I} - \frac{M_C}{I} + \frac{M_C - M_D}{L}\left(\frac{z}{I}\right)$$

The slopes at the ends of the beam are then given by

$$E\left[\left(\frac{dv}{dz}\right)_{z=L} - \left(\frac{dv}{dz}\right)_{z=0}\right] = -\int_0^L M_0 \frac{dz}{I} - M_C\int_0^L \frac{dz}{I} + \frac{1}{L}(M_C - M_D)\int_0^L \frac{z\,dz}{I}$$

and

$$E\left[\left(\frac{dv}{dz}\right)_{z=L}\right] = -\frac{1}{L}\int_0^L M_0 \frac{z\,dz}{I} - \frac{M_C}{L}\int_0^L \frac{z\,dz}{I} + \frac{1}{L^2}(M_C - M_D)\int_0^L \frac{z^2\,dz}{I}$$

It is necessary to plot five curves of (M_0/I), $(1/I)$, (z/I), (z^2/I), $(M_0 z/I)$, and to find their areas.

As an example of the use of equations (13.58) and (13.59), consider the beam of Fig. 13.18(i), which carries end couples, M_C and M_D, and a concentrated load W at a distance a from C.

The bending moment diagram for W acting alone is the triangle CBD, Fig. 13.18(ii). The area of this triangle is

$$A = \tfrac{1}{2}L\left(\frac{Wa}{L}\right)(L - a) = \frac{Wa}{2}(L - a)$$

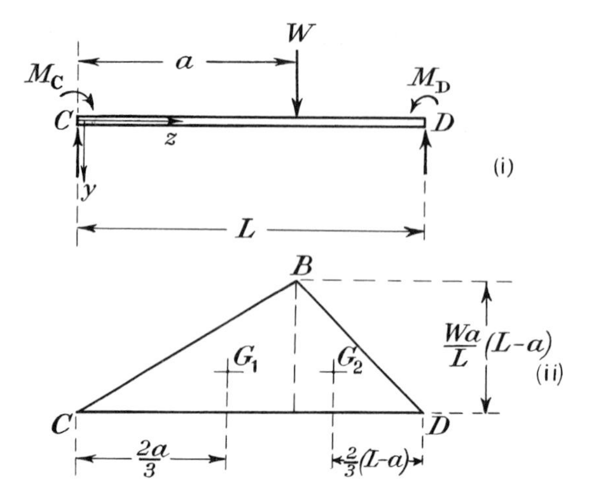

Fig. 13.18 Moment-area solution of a beam carrying end couples and a concentrated load.

To evaluate its first moment about C, divide the triangle into two right-angled triangles, having centroids at G_1 and G_2, respectively. Then

$$
\begin{aligned}
A\bar{z} &= \tfrac{1}{2}a\left[\frac{Wa}{L}(L - a)\right]\frac{2a}{3} + \tfrac{1}{2}[L - a]\left[\frac{Wa}{L}(L - a)\right]\left[\tfrac{1}{3}(L + 2a)\right] \\
&= \tfrac{1}{6}Wa(L^2 - a^2).
\end{aligned}
$$

Then equations (13.58) and (13.59) give

$$
\left(\frac{dv}{dz}\right)_{z=0} = \frac{M_C L}{3EI} + \frac{M_D L}{6EI} + \frac{Wa}{6EIL}(a^2 - 3aL + 2L^2)
$$

$$
\left(\frac{dv}{dz}\right)_{z=L} = -\frac{M_C L}{6EI} - \frac{M_D L}{3EI} - \frac{Wa}{6EIL}(L^2 - a^2)
$$

13.16 Use of Fourier series

The following method of estimating deflections enables us to deal with any load distribution on a simply-supported beam. The load distribution, which may vary in intensity along the length of the beam, can always be expressed as a Fourier series of sines or cosines; this, as we shall see, leads to a similar representation of the deflected form.

Consider first the case of a beam of length L, simply-supported at each end, and carrying a lateral load of intensity

$$
w = w_0 \sin \frac{n\pi z}{L} \qquad (13.60)
$$

per unit length, z being measured from one end, and n being an integer. If the beam has a uniform flexural stiffness EI, we have from equation (13.7),

$$EI\frac{d^4v}{dz^4} = w = w_0 \sin\frac{n\pi z}{L}$$

On integrating we have,

$$EI\frac{d^3v}{dz^3} = -\frac{w_0 L}{n\pi}\cos\frac{n\pi z}{L} + A$$

$$EI\frac{d^2v}{dz^2} = -w_0\left(\frac{L}{n\pi}\right)^2\sin\frac{n\pi z}{L} + Az + B$$

If the beam is simply-supported at each end, $d^2v/dz^2 = 0$ at $z = 0$ and $z = L$; then $A = B = 0$, and

$$EI\frac{d^2v}{dz^2} = -w_0\left(\frac{L}{n\pi}\right)^2\sin\frac{n\pi z}{L} \tag{13.61}$$

But $EI\, d^2v/dz^2 = -M$, so the variation of bending moment due to the distributed load of equation (13.60) is also sinusoidal. On integrating equation (13.61) further, we have

$$EI\frac{dv}{dz} = w_0\left(\frac{L}{n\pi}\right)^3\cos\frac{n\pi z}{L} + C$$

$$EIv = w_0\left(\frac{L}{n\pi}\right)^4\sin\frac{n\pi z}{L} + Cz + D$$

If there is no deflection of the beam at $z = 0$ and $z = L$, we have $C = D = 0$. Then, finally,

$$EIv = w_0\left(\frac{L}{n\pi}\right)^4\sin\frac{n\pi z}{L} \tag{13.62}$$

Thus the sinusoidally distributed load gives rise to a sinusoidal variation of deflection.

More generally the distributed load may have the value

$$w = w_1 \sin\frac{\pi z}{L} + w_2 \sin\frac{2\pi z}{L} + w_3 \sin\frac{3\pi z}{L} + \dots$$

which is written conveniently in the form

$$w = \sum_{n=1}^{\infty} w_n \sin\frac{n\pi z}{L} \tag{13.63}$$

We superpose the effects of $w_1 \sin(\pi z/L)$, $w_2 \sin(2\pi z/L)$, etc., acting separately. The resulting deflected form is given by

$$EIv = \sum_{n=1}^{\infty} w_n\left(\frac{L}{n\pi}\right)^4\sin\frac{n\pi z}{L} \tag{13.64}$$

Clearly, for any given distributed load we have to evaluate the coefficients of the Fourier series of equation (13.63). It is easily shown that, in equation (13.63), w_n has the value

$$w_n = \frac{2}{L}\int_0^L w \sin\frac{n\pi z}{L}\,dz \tag{13.65}$$

When a simply-supported beam carries a uniformly distributed load of intensity w_0, equation (13.65) gives

$$w_n = \frac{2}{L}\int_0^L w_0 \sin\frac{n\pi z}{L}\,dz = \frac{2w_0}{L}\left[-\frac{L}{n\pi}\cos\frac{n\pi z}{L}\right]_0^L$$

Then

$$w_n = \frac{2w_0}{n\pi}[1 - \cos n\pi]$$

When n is even, $\cos n\pi = +1$, and $w_n = 0$; when n is odd, $\cos n\pi = -1$, and

$$w_n = \frac{4w_0}{n\pi}$$

Thus

$$EIv = \frac{4w_0}{\pi}\left(\frac{L}{\pi}\right)^4\left[\sin\frac{\pi z}{L} + \frac{1}{3^5}\sin\frac{3\pi z}{L} + \frac{1}{5^5}\sin\frac{5\pi z}{L} + \ldots\right] \tag{13.66}$$

At the mid-length, $z = \frac{1}{2}L$, and

$$v = \frac{4w_0L^4}{\pi^5 EI}\left[1 - \frac{1}{3^5} + \frac{1}{5^5} - \frac{1}{7^5} + \ldots\right]$$

Clearly, all terms other than the first in this infinite series are very small compared with unity, and we may write approximately,

$$v \doteqdot \frac{4w_0L^4}{\pi^5 EI} \doteqdot \frac{w_0L^4}{76{\cdot}6EI} \doteqdot \frac{5{\cdot}01}{384}\frac{w_0L^4}{EI} \tag{13.67}$$

On comparing this result with equation (13.16), we see that the error involved in taking only the first term of the series of equation (13.66) is about 1 in 500.

When a simply-supported beam carries a single concentrated load W at a distance $z = a$ from one support, the equivalent distributed load has the form

$$w = \frac{2W}{L}\left[\sin\frac{\pi a}{L}\sin\frac{\pi z}{L} + \sin\frac{2\pi a}{L}\sin\frac{2\pi z}{L} + \ldots\right]$$

Equation (13.64) then gives

$$EIv = \frac{2WL^3}{\pi^4}\left[\sin\frac{\pi a}{L}\sin\frac{\pi z}{L} + \frac{1}{2^4}\sin\frac{2\pi a}{L}\sin\frac{2\pi z}{L} + \ldots\right]$$

At the mid-length, $z = \frac{1}{2}L$, and

$$v = \frac{2WL^3}{\pi^4 EI}\left[\sin\frac{\pi a}{L} - \frac{1}{3^4}\sin\frac{2\pi a}{L} + \ldots\right]$$

If the load is at the mid-length, $a = \frac{1}{2}L$, and

$$v = \frac{2WL^3}{\pi^4 EI}\left[1 + \frac{1}{3^4} + \ldots\right]$$

Taking only the first term in the series, we have

$$v \doteqdot \frac{2WL^3}{\pi^4 EI} \doteqdot 0{\cdot}985\,\frac{WL^3}{48EI}$$

On comparing this result with equation (13.36), we see that the error involved is only about 1·5 in 100.

13.17 The funicular analogue of beam deflections

It has been shown in §13.2 that the relation between the deflection v of a beam, the bending moment M, and the flexural stiffness EI, is

$$\frac{d^2v}{dz^2} = -\frac{M}{EI} \tag{13.5}$$

It has also been shown, in §7.3, that the relation between the bending moment M, and the intensity w of lateral load, is

$$\frac{d^2M}{dz^2} = -w \tag{7.9}$$

These differential equations are of the same form, and show that v is related to M/EI in the same way as M is related to w. Consider, first, a simply-supported beam of uniform

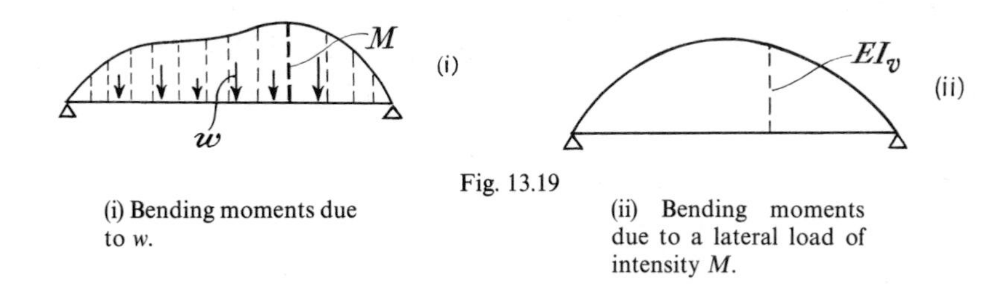

Fig. 13.19

(i) Bending moments due to w.

(ii) Bending moments due to a lateral load of intensity M.

flexural rigidity EI; the lateral loads of intensity w give rise to bending moments M; now consider the beam under a distributed lateral load of intensity M, Fig. 13.19(i).

The bending moments due to the load of intensity M are equal to EIv since

$$\frac{d^2}{dz^2}(EIv) = -M$$

The beam of Fig. 13.19(ii) is the *funicular analogue* of the beam of Fig. 13.19(i).

Problem 13.9: Using a funicular analogue, find the deflected form of a simply-supported beam carrying a uniformly distributed lateral load of intensity w.

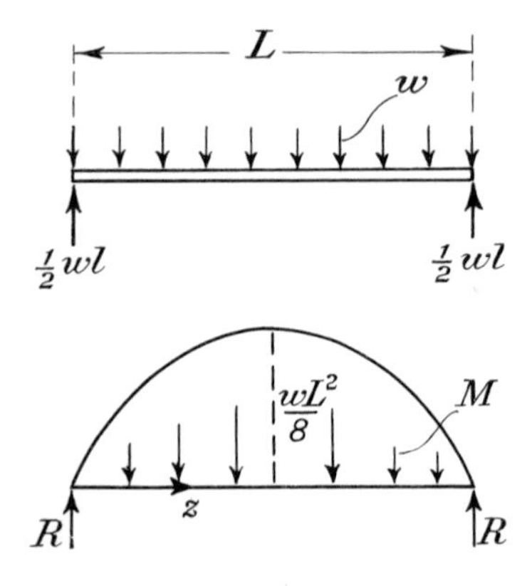

Solution

Due to the loads w, the bending moment a distance z from one support is

$$M = \tfrac{1}{2}wLz - \tfrac{1}{2}wz^2 = \tfrac{1}{2}wz(L - z)$$

Now suppose the beam carries a lateral load of intensity M; the bending moment at any section is

$$EIv = Rz \int_0^z M(z - z')\, dz'$$

Now R is half the total vertical load, so that

$$R = \tfrac{1}{2}\int_0^L M\, dz = \tfrac{1}{4}w\int_0^L z(L - z)\, dz$$

$$= \frac{wL^3}{4}\int_0^1 \left(\frac{z}{L}\right)\left(1 - \frac{z}{L}\right) d\left(\frac{z}{L}\right) = \frac{wL^3}{4}\left[\frac{1}{2}\left(\frac{z}{L}\right)^2 - \frac{1}{3}\left(\frac{z}{L}\right)^3\right]_0^1 = \frac{wL^3}{24}$$

Then

$$EIv = \frac{wL^3z}{24} - \frac{w}{2}\int_0^z z'(L - z')(z - z')\, dz'$$

$$= \tfrac{1}{24}wL^3z - \tfrac{1}{2}w\int_0^z [z'^3 - (L + z)z'^2 + zLz']\, dz'$$

$$= \tfrac{1}{24}wL^3z - \tfrac{1}{2}w[\tfrac{1}{4}z^4 - \tfrac{1}{3}z^3(L + z) + \tfrac{1}{2}zLz^2]$$

$$= \frac{w}{24}[L^3z - 2Lz^3 + z^4]$$

which agrees with the result already established in §13.3.

13.18 Deflections of beams due to shear

In our simple theory of bending of beams, we assumed that plane sections remain plane during bending. The effect of shearing forces in a beam is to distort plane cross-sections into curved planes. In the cantilever of Fig. 13.20, the cross-section DH warps as the force F is applied, due to the shearing strains in the fibres of the beam. We assume

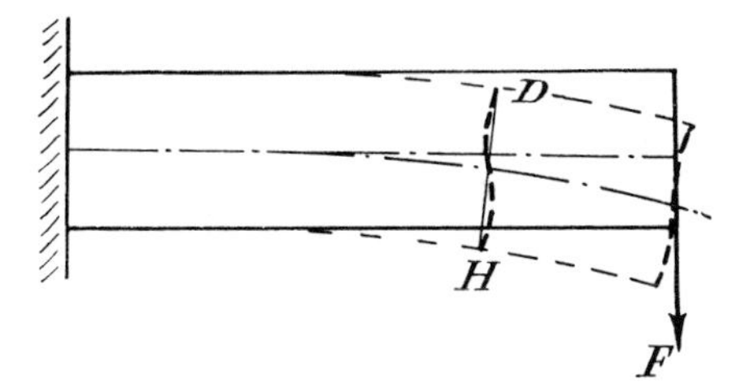

Fig. 13.20 Shearing distortions in a cantilever.

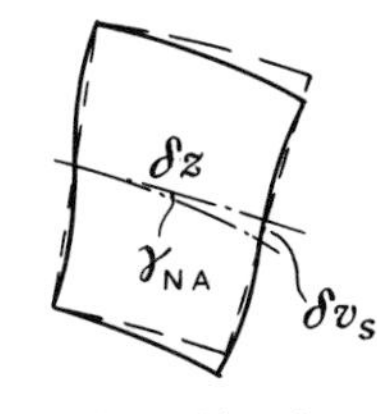

Fig. 13.21 Shearing deflection at the neutral axis of a beam.

that the shearing stresses set up by F are distributed in the manner already discussed in Chapter 10. This is not true strictly, since shearing distortions no longer allow sections to remain plane; however, we assume these shearing effects are secondary, and we are justified therefore in estimating them on our original theory.

Suppose the shearing stress at the neutral axis of the beam is τ_{NA}, then the shearing strain at the neutral axis is

$$\gamma_{NA} = \frac{\tau_{NA}}{G} \tag{13.68}$$

where G is the shearing modulus. The additional deflection arising from shearing of the cross-section is then

$$\delta v_s = \gamma_{NA} \,.\, \delta z = \frac{\tau_{NA}}{G} \,.\, \delta z$$

Then

$$\frac{dv_s}{dz} = \frac{\tau_{NA}}{G} \tag{13.69}$$

For a cantilever of thin rectangular cross-section, §10.2,

$$\tau_{NA} = \frac{3F}{2ht} \tag{13.70}$$

where h is the depth of the cross-section, and t is the thickness. Then

$$\frac{dv_s}{dz} = \frac{3F}{2Ght}$$

Then

$$v_s = \frac{3Fz}{2Ght} + A \tag{13.71}$$

At $z = 0$, there is no shearing deflection, so $A = 0$. At the end $z = L$,

$$(v_s)_L = \frac{3FL}{2Ght} \tag{13.72}$$

The bending deflection at the free end, $z = L$, is

$$(v)_L = \frac{FL^3}{3EI} = \frac{4FL^3}{Eh^3t} \tag{13.73}$$

Then the total end deflection is

$$v_L = \frac{4FL^3}{Eh^3t} + \frac{3FL}{2Ght}$$

$$= \frac{4FL^3}{Eh^3t}\left[1 + \frac{3E}{8G}\left(\frac{h}{L}\right)^2\right] \tag{13.74}$$

For most materials $(3E/8G)$ is of order unity, so the contribution of the shear to the total deflection is equal approximately to $(h/L)^2$. Clearly, the shearing deflection is important only for deep beams.

Problem 13.10: A 1·5 m length of the beam of Problem 11·2 is simply-supported at each end, and carries a concentrated lateral load of 10 kN at the mid-span. Compare the central deflections due to bending and shearing.

Solution

From Problem 11.2, the second moment of area of the equivalent steel I-beam is $12{\cdot}1 \times 10^{-6}$ m^4. The central deflection due to bending is, therefore,

$$v_B = \frac{Wl^3}{48E_sI_s} = \frac{(10 \times 10^3)(1{\cdot}5)^3}{48(200 \times 10^9)(12{\cdot}1 \times 10^{-6})} = 0{\cdot}290 \times 10^{-3} \text{ m}$$

The average shearing stress in the timber is

$$\frac{5 \times 10^3}{(0{\cdot}15)(0{\cdot}075)} = 0{\cdot}445 \text{ MN/m}^2$$

If the shearing modulus for timber is

$$4 \times 10^9 \text{ N/m}^2$$

the shearing strain in the timber is

$$\gamma = \frac{0{\cdot}445 \times 10^6}{4 \times 10^9} = 0{\cdot}111 \times 10^{-3}$$

The resulting central deflection due to shearing is

$$v_s = \gamma \times 0{\cdot}75 = (0{\cdot}111 \times 10^{-3})(0{\cdot}75) = 0{\cdot}0833 \times 10^{-3} \text{ m}$$

Thus, the shearing deflection is nearly 30 per cent of the bending deflection. The estimated total central deflection is

$$v = v_B + v_s = 0{\cdot}373 \times 10^{-3}\ \text{m}$$

FURTHER PROBLEMS
(answers on page 398)

13.11 A straight girder of uniform section and length L rests on supports at the ends, and is propped up by a third support in the middle. The weight of the girder and its load is w per unit length. If the central support does not yield, prove that it takes a load equal to $\frac{5}{8}wL$.

13.12 A horizontal steel girder of uniform section, 15 m long, is supported at its extremities and carries loads of 120 kN and 80 kN concentrated at points 3 m and 5 m from the two ends respectively. I for the section of the girder is $1{\cdot}67 \times 10^{-3}\ \text{m}^4$ and $E = 200\ \text{GN/m}^2$. Calculate the deflections of the girder at points under the two loads. (*Cambridge*)

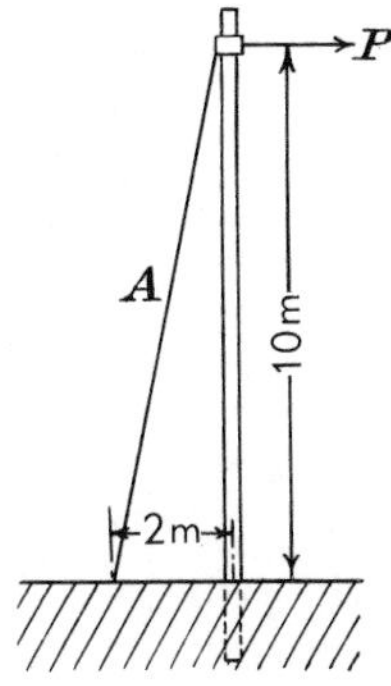

13.13 A wooden mast, with a uniform diameter of 30 cm, is built into a concrete block, and is subjected to a horizontal pull at a point 10 m from the ground. The wire guy A is to be adjusted so that it becomes taut and begins to take part of the load when the mast is loaded to a maximum stress of $7\ \text{MN/m}^2$. Estimate the slack in the guy when the mast is unloaded. Take E for timber $= 10\ \text{GN/m}^2$ (*Cambridge*)

13.14 A bridge across a river has a span $2l$, and is constructed with beams resting on the banks and supported at the middle on a pontoon. When the bridge is unloaded the three supports are all at the same level, and the pontoon is such that the vertical displacement is equal to the load on it multiplied by a constant λ. Show that the load on the pontoon, due to a concentrated load W, placed one-quarter of the way along the bridge, is given by

$$\frac{11W}{16\left(1 + \dfrac{6EI\lambda}{l^3}\right)}$$

where I is the second moment of area of the section of the beams. (*Cambridge*)

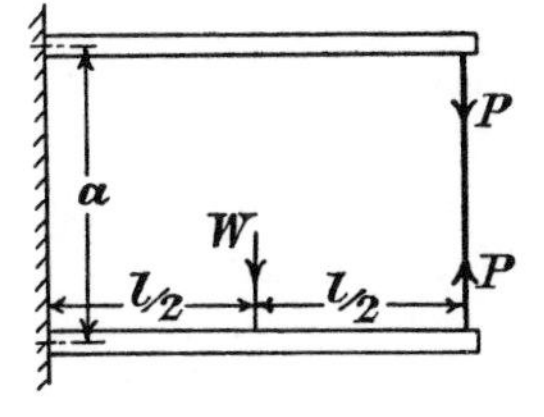

13.15 Two equal steel beams are built-in at one end and connected by a steel rod as shown. Show that the pull in the tie rod is

$$P = \frac{5Wl^3}{32\left(\dfrac{6aI}{\pi d^2} + l^3\right)}$$

where d is the diameter of the rod, and I is the second moment of area of the section of each beam about its neutral axis. (*Cambridge*)

13.16 A beam carries a load, w per unit length, which extends over a distance a from one end. Measuring z from that end show that the deflection can be expressed in the form

$$v = \frac{2wL^4}{\pi^5 EI}\sum_{k=1}^{k=\infty}\frac{1}{k^5}\left(1 - \cos\frac{k\pi a}{L}\right)\sin\frac{k\pi z}{L}$$

where L is the length of the beam, which is freely supported at each end.

14 Built-in and continuous beams

14.1 Introduction

In all our investigations of the stresses and deflections of beams having two supports, we have supposed that the supports exercise no constraint on bending of the beam, i.e. the axis of the beam has been assumed free to take up any inclination to the line of supports. This has been necessary because, without knowing how to deal with the deformation of the axis of the beam, we were not in a position to find the bending moments on a beam when the supports constrain the direction of the axis. We shall now investigate this problem. When the ends of a beam are fixed in direction so that the axis of the beam has to retain its original direction at the points of support, the beam is said to be built-in.

Consider a straight beam resting on two supports A and B (Fig. 14.1) and carrying vertical loads. If there is no constraint on the axis of the beam, it will become curved in the manner shown by broken lines, the extremities of the beam rising off the supports.

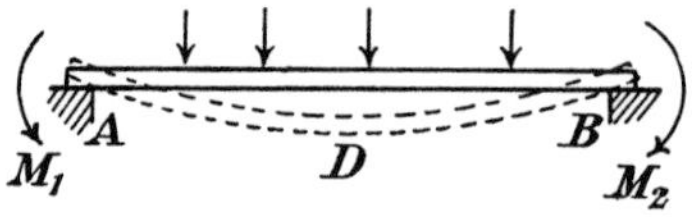

Fig. 14.1 Beam with end couples.

In order to make the ends of the beam lie flat on the horizontal supports, we shall have to apply couples as shown by M_1 and M_2. If the beam is firmly built into two walls, or bolted down to two piers, or in any other way held so that the axis cannot tip up at the ends in the manner indicated, the couples such as M_1 and M_2 are supplied by the resistance of the supports to deformation. These couples are termed *fixed-end moments*, and the main problem of the built-in beam is the determination of these couples; when we have found these we can draw the bending moment diagram and calculate the stresses in the usual way. The couples M_1 and M_2 in Fig. 14.1 must be such as to produce curvature in the opposite direction to that caused by the loads.

14.2 Built-in beam with a single concentrated load

We may deduce the bending moments in a built-in beam under any conditions of lateral loading from the case of a beam under a single concentrated lateral load.

Consider a uniform beam, of flexural stiffness EI, and length L, which is built-in to end supports C and G, Fig. 14.2. Suppose a concentrated vertical load W is applied to

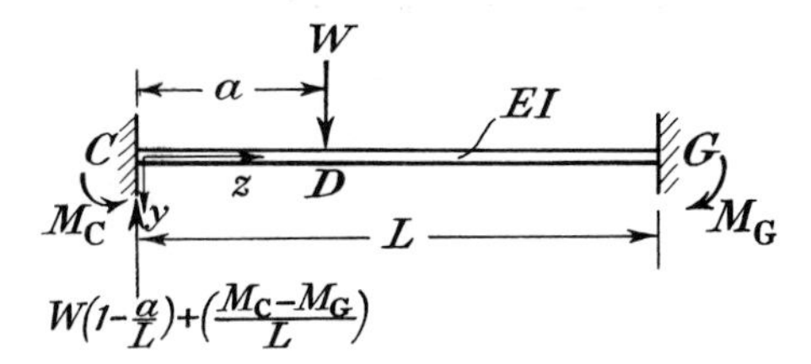

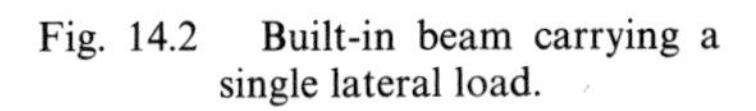
Fig. 14.2 Built-in beam carrying a single lateral load.

the beam at a distance a from C. If M_C and M_G are the restraining moments at the supports, then the vertical reaction at C is

$$W\left(1 - \frac{a}{L}\right) + \frac{1}{L}(M_C - M_G)$$

The bending moment in the beam at a distance z from C is therefore

$$M = \left[W\left(1 - \frac{a}{L}\right) + \frac{1}{L}(M_C - M_G)\right]z - M_C - \Phi[W(z - a)]$$

where Φ is a step-function of z, having the values

$$\Phi = 0 \quad \text{for} \quad z < a$$

$$\Phi = 1 \quad \text{for} \quad z > a$$

Then, for the deflected form of the beam, the displacement is given by

$$EI\frac{d^2v}{dz} = -\left[W\left(1 - \frac{a}{L}\right) + \frac{1}{L}(M_C - M_G)\right]z + M_C + \Phi[W(z - a)] \qquad (14.1)$$

where v is the deflection of the beam parallel to Cy. On integrating equation (14.1) we have

$$EI\frac{dv}{dz} = -\left[W\left(1 - \frac{a}{L}\right) + \frac{1}{L}(M_C - M_G)\right]\frac{z^2}{2} + M_C z + \Phi\left[\frac{W}{2}(z - a)^2\right] + A \qquad (14.2)$$

$$EIv = -\left[W\left(1 - \frac{a}{L}\right) + \frac{1}{L}(M_C - M_G)\right]\frac{z^3}{6} + \frac{M_C z^2}{2} + \Phi\left[\frac{W}{6}(z - a)^3\right] + Az + B \qquad (14.3)$$

At the end $z = 0$, we have $v = 0$ and $dv/dz = 0$; there are two similar conditions for the end $z = L$. The four conditions give $A = B = 0$, and

$$-\left[W\left(1 - \frac{a}{L}\right) + \frac{1}{L}(M_C - M_G)\right]\frac{L^2}{2} + M_C L + \frac{W}{2}(L - a)^2 = 0$$

$$-\left[W\left(1 - \frac{a}{L}\right) + \frac{1}{L}(M_C - M_G)\right]\frac{L^3}{6} + \frac{M_C L^2}{6} + \frac{W}{6}(L - a)^3 = 0$$

These equations give

$$M_C = Wa\left(\frac{L - a}{L}\right)^2 \tag{14.4}$$

$$M_G = W(L - a)\left(\frac{a}{L}\right)^2 \tag{14.5}$$

M_C and M_G are referred to as the *fixed-end moments* of the beam; M_C is measured anticlockwise, and M_G clockwise.

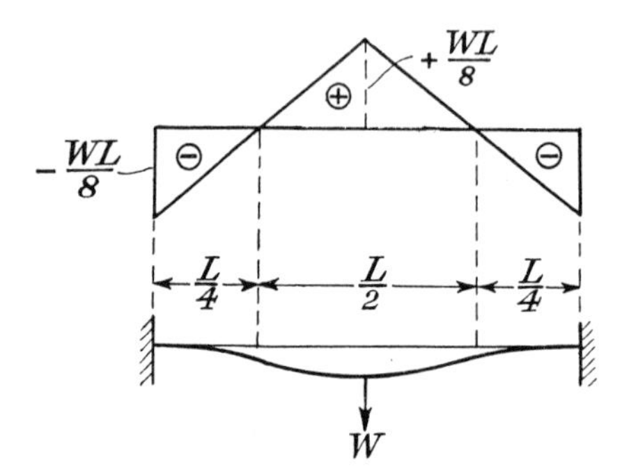

Fig. 14.3 Variation in bending moment in a built-in beam carrying a concentrated load at mid-length.

In the particular case when the load W is applied at the mid-length, $a = \frac{1}{2}L$, and

$$M_C = M_G = \frac{WL}{8}$$

The bending moments in the beam vary linearly from hogging moments of $WL/8$ at each end to a sagging moment of $WL/8$ at the mid-length, Fig. 14.3. There are points of contraflexure, or zero bending moment, at distances $L/4$ from each end.

14.3 Fixed-end moments for other loading conditions

The built-in beam of Fig. 14.4 carries a uniformly-distributed load of w per unit length over the section of the beam from $z = a$ to $z = b$. Consider the loading on an elemental

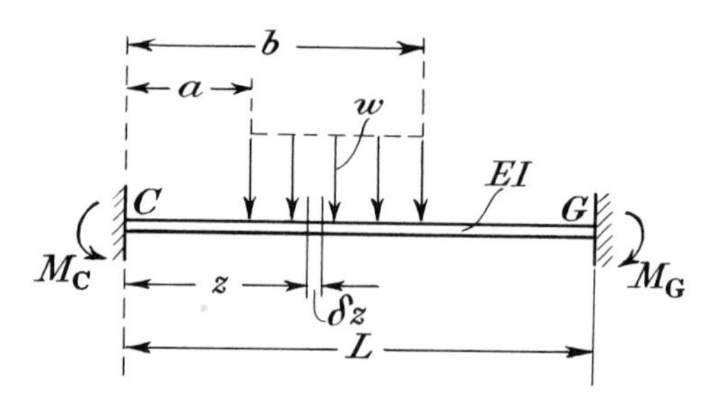

Fig. 14.4 Distributed load over part of the span of a built-in beam.

length δz of the beam; the vertical load on the element is $w\,\delta z$, and this induces a restraining moment at C of amount

$$\delta M_C = w\,\delta z\,\frac{z(L - z)^2}{L^2}$$

from equation (14.4). The total moment at C due to all such loads is

$$M_C = \int_a^b \frac{w}{L^2}\,z(L - z)^2\,dz$$

which gives

$$M_C = \frac{w}{L^2}\left[\frac{L^2}{2}(b^2 - a^2) - \frac{2L}{3}(b^3 - a^3) + \tfrac{1}{4}(b^4 - a^4)\right] \tag{14.6}$$

M_G may be found similarly. When the load covers the whole of the span, $a = 0$ and $b = L$, and equation (14.6) reduces to

$$M_C = \frac{wL^2}{12} \tag{14.7}$$

In this particular case, $M_G = M_C$; the variation of bending moment is parabolic, and of the form shown in Fig. 14.5; the bending moment at the mid-length is $wL^2/24$, so

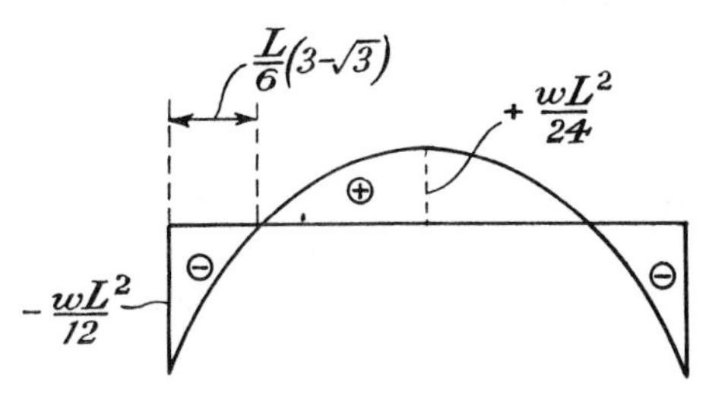

Fig. 14.5 Variation of bending moment in a built-in beam carrying a uniformly distributed load over the whole span.

the fixed-end moments are also the greatest bending moments in the beam. The points of contraflexure, or points of zero bending moment, occur at a distance

$$\frac{L}{6}(3 - \sqrt{3}) \tag{14.8}$$

from each end of the beam.

When a built-in beam carries a number of concentrated lateral loads, W_1, W_2, W_3, Fig. 14.6, the fixed-end moments are found by adding together the fixed-end moments due to the loads acting separately. For example,

$$M_C = \sum_{r=1,2,3} W_r a_r \left(\frac{L - a_r}{L}\right)^2 \tag{14.9}$$

for the case shown in Fig. 14.6

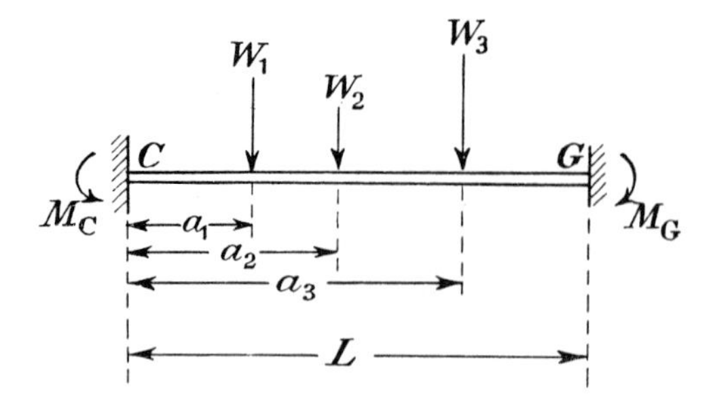

Fig. 14.6 Built-in beam carrying a number of concentrated loads.

We may treat the case of a concentrated couple M_0, applied a distance a from the end C, Fig. 14.7, as a limiting case of two equal and opposite loads W a small distance δa apart. The fixed-end moment at C is

$$M_C = -\frac{Wa}{L^2}(L-a)^2 + \frac{W(a+\delta a)}{L^2}(L-a-\delta a)^2$$

If δa is small,

$$M_C \doteqdot -\frac{Wa}{L^2}(L-a)^2 + \frac{W}{L^2}[a(L-a)^2 + \delta a(L-a)(L-3a)]$$

which gives

$$M_C = \frac{W\,\delta a}{L^2}(L-a)(L-3a)$$

But if δa is small, M_0 is statically equivalent to the couple $W\,\delta a$, and

$$M_C = \frac{M_0}{L^2}(L-a)(L-3a) \tag{14.10}$$

Similarly,

$$M_G = \frac{M_0}{L^2}a(2L-3a) \tag{14.11}$$

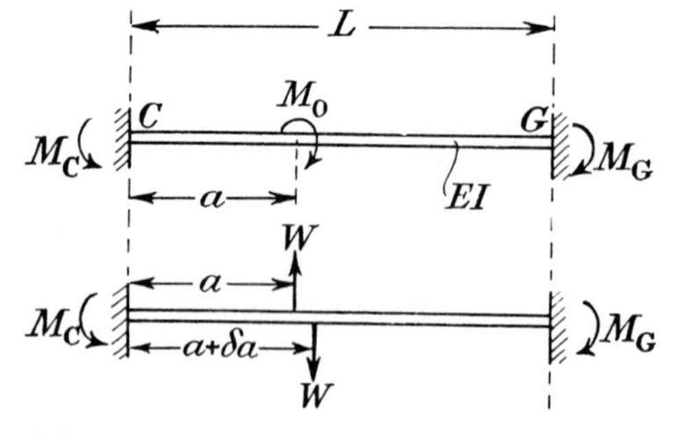

Fig. 14.7 Built-in beam carrying a concentrated couple.

14.4 Disadvantages of built-in beams

The results we have obtained above show that a beam which has its ends firmly fixed in direction is both stronger and stiffer than the same beam with its ends simply-supported. On this account it might be supposed that beams would always have their ends built-in whenever possible; in practice it is not often done. There are several objections to built-in beams: in the first place a small subsidence of one of the supports will tend to set up large stresses, and, in erection, the supports must be aligned with the utmost accuracy; changes of temperature also tend to set up large stresses. Again, in the case of live loads passing over bridges, the frequent fluctuations of bending moment, and vibrations, would quickly tend to make the degree of fixing at the ends extremely uncertain.

Most of these objections can be obviated by employing the double cantilever construction. Since the bending moments at the ends of a built-in beam are of opposite sign to those in the central part of the beam, there must be points of inflexion, i.e., points where the bending moment is zero. At these points a hinged joint might be made in the beam, the axis of the hinge being parallel to the bending axis, since there is no bending moment to resist. If this is done at each point of inflexion, the beam will appear as a central girder freely supported by two end cantilevers; the bending moment curve and deflection curve will be exactly the same as if the beam were solid and built-in. With this construction the beam is able to adjust itself to changes of temperature or subsidence of the supports.

14.5 Effect of sinking of supports

When the ends of a beam are prevented from rotating but allowed to deflect with respect to each other, bending moments are set up in the beam. The uniform beam of Fig. 14.8 is displaced so that no rotations occur at the ends but the remote end is displaced downwards an amount δ relative to C. The end reactions consist of equal couples M_C and equal and opposite shearing forces $2M_C/L$, since the system is anti-symmetric about the mid-point of the beam. The half-length of the beam behaves as a cantilever

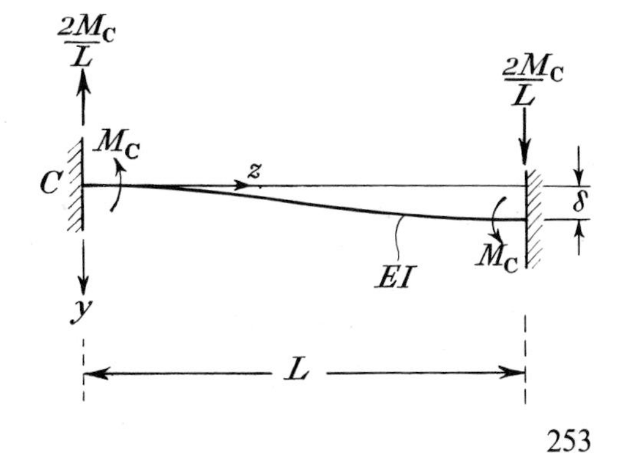

Fig. 14.8 End moments induced by the sinking of the supports of a built-in beam.

carrying an end load $2M_C/L$; then, from equation (13.18),

$$\tfrac{1}{2}\delta = \frac{(2M_C/L)(L/2)^2}{3EI} = \frac{M_C L^2}{12EI}$$

Therefore

$$M_C = \frac{6EI\,\delta}{L^2} \tag{14.12}$$

For a downwards deflection δ, the induced end moments are both anticlockwise; these moments must be superimposed on the fixed-end moments due to any external lateral loads on the beam.

Problem 14.1: A horizontal beam 6 m long is built-in at each end. The elastic section modulus is $0{\cdot}933 \times 10^{-3}$ m³. Estimate the uniformly-distributed load over the whole span causing an elastic bending stress of 150 MN/m².

Solution

The maximum bending moments occur at the built-in ends, and have value

$$M_{\max} = \frac{wL^2}{12}$$

If the bending stress is 150 MN/m²,

$$M_{\max} = \frac{\sigma I}{y} = \sigma Z_e = (150 \times 10^6)(0{\cdot}933 \times 10^{-3}) = 140 \text{ kNm}$$

Then

$$w = \frac{12}{L^2}(M_{\max}) = 46{\cdot}7 \text{ kN/m}$$

14.6 Continuous beams

When the same beam runs across three or more supports it is spoken of as a *continuous beam*. Suppose we have three spans, as in Fig. 14.9, each bridged by a separate beam; the beams will bend independently in the manner shown. In order to make the axes of the three beams form a single continuous curve across the supports B and C, we shall have to apply to each beam couples acting as shown by the arrows. When the beam is one continuous girder these couples, on any bay such as BC, are supplied by the action of the adjacent bays. Thus AB and CD, bending downwards under their own loads, try to bend BC upwards, as shown by the broken curve, thus applying the couples M_B and M_C to the bay BC. This upward bending is of course opposed by the down

Fig. 14.9 Bending moments at the supports of a continuous beam.

load on BC, and the general result is that the beam takes up a sinuous form, being, in general, concave upwards over the middle portion of each bay and convex upwards over the supports.

In order to draw the bending moment diagram for a continuous beam we must first find the couples such as M_B and M_C. In some cases there may also be external couples applied to the beam, at the supports, by the action of other members of the structure.

When the bending moments at the supports have been found, the bending moment and shearing force diagrams can be drawn for each bay according to the methods discussed in Chapter 7.

14.7 Slope-deflection equations for a single beam

In dealing with continuous beams we can make frequent use of the end slope and deflection properties of a single beam under any conditions of lateral loading. The uniform beam of Fig. 14.10(i) carries any system of lateral loads; the ends are supported

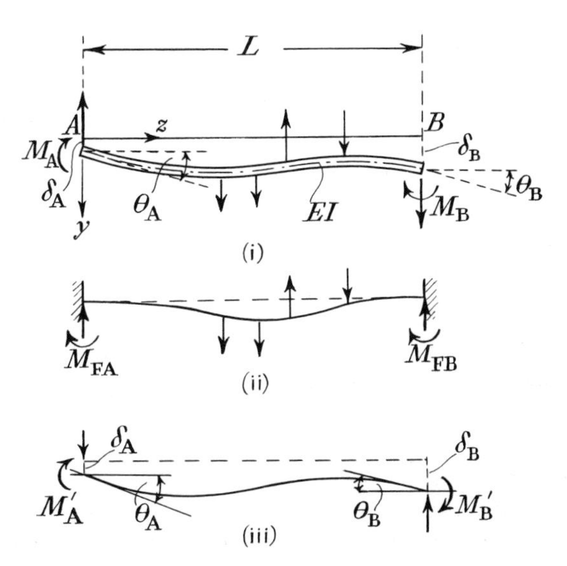

Fig. 14.10 The single beam under any conditions of lateral load and end support shown in (i) can be regarded as the superposition of the built-in end beam of (ii) and the beam with end couples and end deformations of (iii).

in an arbitrary fashion, the displacements and moments being as shown in the figure. In addition there are lateral forces at the supports. The rotations at the supports are θ_A and θ_B, respectively, reckoned positive if clockwise; M_A and M_B are also taken positive clockwise for our present purposes. The displacements δ_A and δ_B are taken positive downwards.

The loaded beam of Fig. 14.10(i) may be regarded as the superposition of the loading conditions of Figs. 14.10(ii) and (iii). In Fig. 14.10(ii) the beam is built-in at each end; the moments at each end are easily calculable from the methods discussed in §§14.2

and 14.3. The fixed-end moments for this condition will be denoted by M_{FA} and M_{FB}. In Fig. 14.10(iii) the beam carries no external loads between its ends, but end displacements and rotations are the same as those in Fig. 14.10(i); the end couples for this condition are M_A' and M_B'. The superposition of Figs. 14.10(ii) and (iii) gives the external loading and end conditions of Fig. 14.10(i). We must find then the end couples in Fig. 14.10(iii); from equations (13.49), putting $w = 0$, we have

$$\theta_A = \frac{M_A'L}{3EI} - \frac{M_B'L}{6EI} + \frac{1}{L}(\delta_B - \delta_A)$$

$$\theta_B = -\frac{M_A'L}{6EI} + \frac{M_B'L}{3EI} + \frac{1}{L}(\delta_B - \delta_A)$$

Then

$$\theta_A + \frac{1}{L}(\delta_A - \delta_B) = \frac{1}{6EI}(2M_A' - M_B')$$

$$\theta_B + \frac{1}{L}(\delta_A - \delta_B) = \frac{L}{6EI}(2M_B' - M_A')$$

But for the superposition we have

$$M_A' = M_A - M_{FA} \qquad M_B' = M_B - M_{FB}$$

Thus

$$\theta_A + \frac{1}{L}(\delta_A - \delta_B) = \frac{L}{6EI}[2(M_A - M_{FA}) - (M_B - M_{FB})] \tag{14.13}$$

$$\theta_B + \frac{1}{L}(\delta_A - \delta_B) = \frac{L}{6EI}[2(M_B - M_{FB}) - (M_A - M_{FA})] \tag{14.14}$$

These are known as the *slope-deflection equations*; they give the values of the unknown moments, M_A and M_B.

14.8 The three-moment equation

Consider a continuous beam which is supported at A, B and C, Fig. 14.11. Suppose EI_1 and EI_2 are the bending stiffnesses of the lengths L_1 and L_2, respectively. The supports

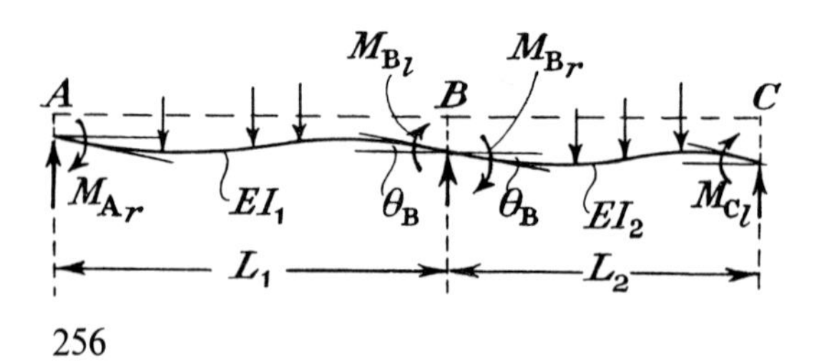

Fig. 14.11 Continuous beam of two spans.

A, B and C sink amounts δ_A, δ_B, δ_C, respectively, under the external loads; M_{Ar} is the bending moment in AB just to the right of A, and M_{Bl} is the bending moment in AB just to the left of B; M_{Br} and M_{Cl} are defined similarly. If no external moments are applied at B,

$$M_{Bl} + M_{Br} = 0 \quad \text{or} \quad M_{Br} = -M_{Bl}$$

The fixed-end moments required to prevent both vertical displacements and rotations in the beams at A, B and C are M_{FAr} in AB to the right of A, and M_{FBl} in AB to the left of B; M_{FBr} and M_{FCl} are defined similarly. All clockwise fixed-end moments are reckoned positive. Then for the rotation of beam AB at B, we have, from equation (14.14),

$$\theta_B = \frac{L_1}{6EI_1}[2(M_{Bl} - M_{FBl}) - (M_{Ar} - M_{FAr})] + \frac{1}{L_1}[\delta_B - \delta_A] \tag{14.15}$$

But beam BC rotates by the same amount at B, and from equation (14.13), we have

$$\theta_B = \frac{L_2}{6EI_2}[2(M_{Br} - M_{FBr}) - (M_{Cl} - M_{FCl})] + \frac{1}{L_2}[\delta_C - \delta_B] \tag{14.16}$$

On eliminating θ_B between equations (14.15) and (14.16), we have

$$M_{Ar}\left(\frac{L_1}{EI_1}\right) - 2M_{Bl}\left(\frac{L_1}{EI_1} + \frac{L_2}{EI_2}\right) - M_{Cl}\left(\frac{L_2}{EI_2}\right) + \frac{6}{L_1}(\delta_A - \delta_B)$$

$$+ \frac{6}{L_2}(\delta_C - \delta_B) + \frac{L_1}{EI_1}[2M_{FBl} - M_{FAr}] + \frac{L_2}{EI_2}[M_{FCl} - 2M_{FBr}] = 0 \tag{14.17}$$

This equation defines a relation between the three moments M_{Ar}, M_{Bl} $(= -M_{Br})$, and M_{Cl} in the three spans. Continued application of this equation to all groups of two adjacent beams of a continuous beam leads to a solution of all the bending moments. Equation (14.17) is the *three-moment equation.*

Problem 14.2: A continuous uniform beam $ABCD$ has a weight w per unit length, and rests on four knife-edge supports at A, B, C and D. The middle span carries a load W as shown. Find the reactions on the supports at A and D. (*Cambridge*)

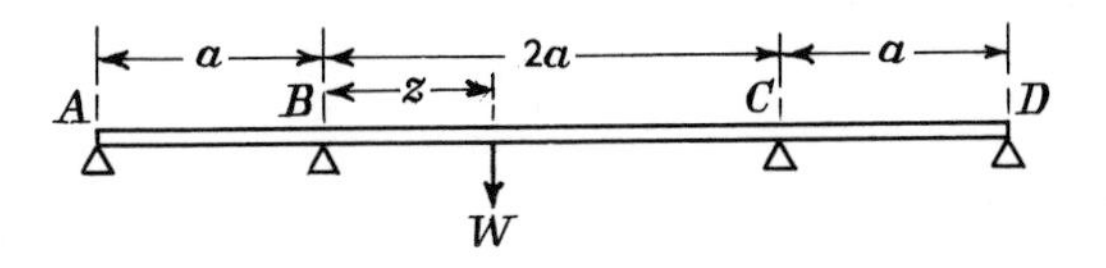

Solution

The fixed-end moments for AB are

$$M_{FA} = -\frac{wa^2}{12} \qquad M_{FB} = +\frac{wa^2}{12}$$

For BC, they are

$$M_{FB} = -\frac{4wa^2}{12} - \frac{W}{(2a)^2}(2a - z)^2 z$$

$$= -\tfrac{1}{3}wa^2 - Wz\left(1 - \frac{z}{2a}\right)^2$$

$$M_{FC} = +\tfrac{1}{3}wa^2 + Wz^2\left(1 - \frac{z}{2a}\right)$$

And for CD,

$$M_{FC} = -\frac{wa^2}{12} \qquad M_{FD} = +\frac{wa^2}{12}$$

The three-moment equation (14.17), applied to AB, gives

$$-2M_B(3a) - M_C(2a) + a\left(\frac{wa^2}{6} + \frac{wa^2}{12}\right) + 2a\left[\tfrac{1}{3}wa^2 + Wz^2\left(1 - \frac{z}{2a}\right)\right.$$

$$\left. + \tfrac{2}{3}wa^2 + 2Wz\left(1 - \frac{z}{2a}\right)^2\right] = 0$$

This reduces to

$$6M_B + 2M_C = \tfrac{9}{4}wa^2 + 2Wz\left(1 - \frac{z}{2a}\right)\left(1 - \frac{z}{2a}\right)$$

Similarly, on applying the three-moment equation to BCD we have

$$6M_C + 2M_B = \tfrac{9}{4}wa^2 + \frac{2W}{4a^2}(2a - z)(z)(2a + z)$$

Then

$$6M_C + 2M_B = \tfrac{9}{4}wa^2 + 2Wz\left(1 - \frac{z}{2a}\right)\left(1 + \frac{z}{2a}\right)$$

The two simultaneous equations in M_B and M_C give

$$M_B = \tfrac{9}{32}wa^2 + \frac{Wz}{16a^2}(2a - z)(5a - 2z)$$

$$M_C = \tfrac{9}{32}wa^2 + \frac{Wz}{16a^2}(2a - z)(a + 2z)$$

The reactions at A and D are then

$$R_A = \tfrac{1}{2}wa - \frac{M_B}{a} = \tfrac{7}{32}wa - \frac{Wz}{16a}\left(2 - \frac{z}{a}\right)\left(5 - \frac{2z}{a}\right)$$

$$R_D = \tfrac{1}{2}wa - \frac{M_C}{a} = \tfrac{7}{32}wa - \frac{Wz}{16a}\left(2 - \frac{z}{a}\right)\left(1 + \frac{2z}{a}\right)$$

Problem 14.3: A railway line for a single railway is constructed with two main girders, continuous over four supports, forming two side spans each 50 m long and a centre span 60 m long. The four supports are at the same level, the girders are of uniform section, and the ends are free from bending moment. The dead load on each main girder is 70 kN per metre run, and the live load 10 kN per metre run. Taking the case when a side span alone carries live load, determine the bending moments over the supports. Sketch the bending moment diagram and determine the pressures on the four supports. *(Cambridge)*

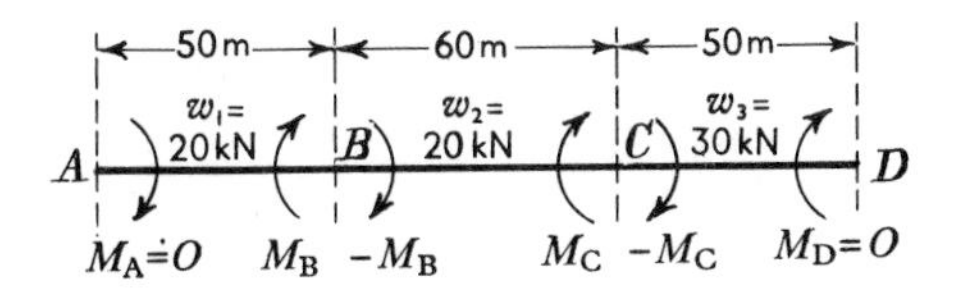

Solution

Let the right-hand span carry live load; the equations for finding the bending moments at B and C are given by the three-moment equation. Applying the equation first to the three supports A, B, C, we have

$$50M_A - 220M_B - 60M_C = -1705 \times 10^6 \text{ Nm}^2$$

and, since $M_A = 0$, this reduces to

$$M_B + 0{\cdot}273M_C = 7{\cdot}75 \times 10^6$$

Again, applying the equation to B, C, D, we have, remembering that $M_D = 0$

$$-60M_B - 220M_C = -2017{\cdot}5 \times 10^6 \text{ Nm}^2$$

or

$$M_B + 3{\cdot}66M_C = 33{\cdot}6 \times 10^6$$

Solving the simultaneous equations for M_B and M_C

$$M_B = 5{\cdot}67 \times 10^6 \text{ Nm}$$

$$M_C = 7{\cdot}61 \times 10^6 \text{ Nm}$$

The bending moment diagram for each bay, on account of the distributed load only, is a parabola of height $wl^2/8$, which gives 6·25, 9·00, 9·37 $\times$ 10^6 Nm for AB, BC, CD respectively. These parabolas are shown dotted. The bending moments due to the fixing moments are given by the straight lines AB', $B'C'$, $C'D$. The complete bending moment diagram is then drawn by replotting the parabolas on the lines $AB'C'D$, as shown by the full curve. If we only require to sketch the bending moment diagram it is sufficient to find the principal points on the curve.

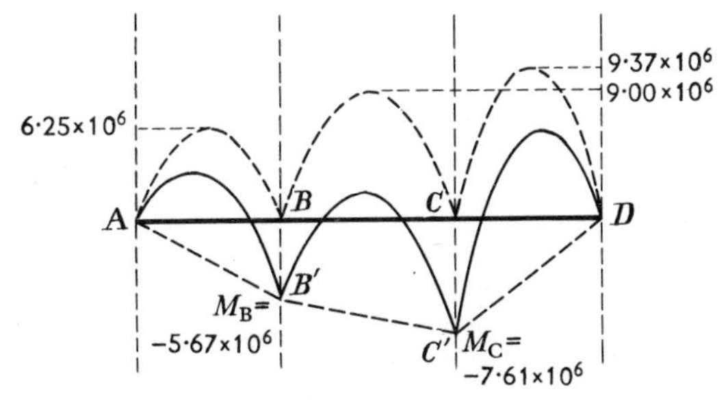

If R_{Bl} and R_{Br} denote the reactions to the left and right of B, and so on, then

$$R_{Ar} = 500 \times 10^3 - (5{\cdot}67 \times 10^6/50) = (500 - 113)10^3 = 387 \times 10^3 \qquad R_A = 387 \text{ kN}$$

$$R_{Bl} = (500 + 113)10^3 = 613 \times 10^3$$

$$R_{Br} = 600 \times 10^3 - (1{\cdot}94 \times 10^6/60) = (600 - 324)10^3 = 276 \times 10^3 \qquad R_B = 889 \text{ kN}$$

$$R_{Cl} = (600 + 324)10^3 = 924 \times 10^3$$

$$R_{Cr} = 750 \times 10^3 + (7{\cdot}61 \times 10^6/50) = (750 + 152)10^3 = 902 \times 10^3 \qquad R_C = 1826 \text{ kN}$$

$$R_{Dl} = (750 - 152)10^3 = 598 \times 10^3 \qquad R_D = 598 \text{ kN}$$

Problem 14.4: A wooden beam, 30 cm by 30 cm cross-section and 4 m long, is supported at its ends A and C and at its mid-length, B, on three girders. Each of these girders is such that a load of 25 kN applied at the points A, B or C deflects the corresponding girder 1 cm. Loads of 50 kN are applied to the beam ABC at points distant 0·5 m and 1·5 m to the right and left of B. Find the magnitudes of the reactions at A, B and C, taking E for the beam as 9 GN/m^2. (*Cambridge*)

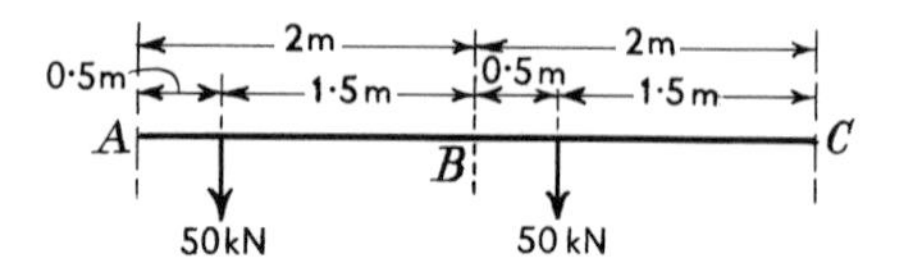

Solution

Let R_A, R_B, R_C denote the reactions at the supports, and $\delta_A, \delta_B, \delta_C$ the deflections of the supporting girders. Then

$$\delta_A = (0{\cdot}4 \times 10^{-6})R_A$$

$$\delta_B = (0{\cdot}4 \times 10^{-6})R_B$$

$$\delta_C = (0{\cdot}4 \times 10^{-6})R_C$$

Again, if M_B is the clockwise moment at B in AB,

$$R_A = (50 \times 10^3)(1{\cdot}5)/2 - M_B/2 = (75 \times 10^3 - M_B)/2$$

$$R_C = (50 \times 10^3)(0{\cdot}5)/2 - M_B/2 = (25 \times 10^3 - M_B)/2$$

$$R_B = (100 \times 10^3) - (R_A + R_C) = (50 \times 10^3) + M_B$$

Thus,

$$\delta_A = (0{\cdot}2 \times 10^{-6})(75 \times 10^3 - M_B)$$

$$\delta_C = (0{\cdot}2 \times 10^{-6})(25 \times 10^3 - M_B)$$

$$\delta_B = (0{\cdot}4 \times 10^{-6})(50 \times 10^3 + M_B)$$

Therefore,

$$\delta_B - \delta_A = (0{\cdot}2 \times 10^{-6})(25 \times 10^3 + 3M_B)$$

$$\delta_B - \delta_C = (0{\cdot}2 \times 10^{-6})(75 \times 10^3 + 3M_B)$$

For the beam,

$$I = \tfrac{1}{12}(0{\cdot}3)(0{\cdot}3)^3 = 0{\cdot}675 \times 10^{-3} \text{ m}^4$$

and therefore

$$EI = (9 \times 10^9)(0{\cdot}675 \times 10^{-3}) = 6{\cdot}08 \times 10^6 \text{ Nm}^2$$

With $M_A = M_C = 0$, the equation of three moments is

$$-4M_B + \frac{6EI}{L^2}[(\delta_A - \delta_B) + (\delta_C - \delta_B)]$$

$$+ (2 \times 0{\cdot}469 + 1{\cdot}406)10^4$$

$$+ (0{\cdot}469 + 2 \times 1{\cdot}406)10^4 = 0$$

On substituting for δ_A, δ_B, δ_C, this becomes

$$1{\cdot}74M_B - 5{\cdot}97 \times 10^4 = 0$$

Thus

$$M_B = -34{\cdot}3 \text{ kNm}$$

Then

$$R_A = 54{\cdot}7 \text{ kN}$$

$$R_B = 15{\cdot}7 \text{ kN}$$

$$R_C = 29{\cdot}6 \text{ kN}$$

Problem 14.5: One of the wing spars of a certain aeroplane is represented* by the straight line CC'; the supports are at A, B, C, A', B', C'. The loads are completely symmetrical about the centre line of the machine, and the supports are assumed not to move. Couples are applied to the spar as shown; the clockwise bending moment to the left of C and the anti-clockwise bending moment to the right of C' are 117·5 Nm. It is required to find the bending moments at the supports.

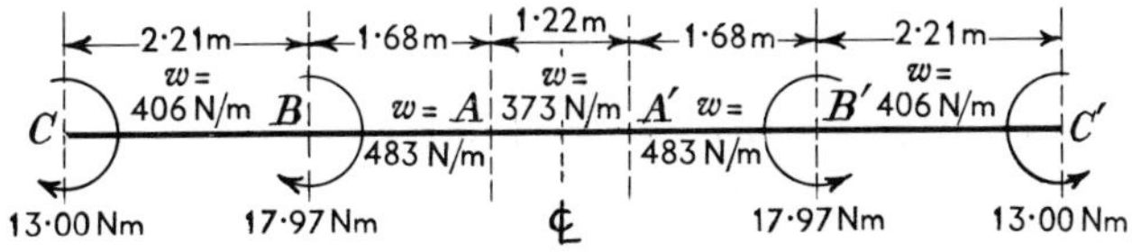

Solution

We have by symmetry

$$M_{Cl} = -M_{C'r} = 117{\cdot}5;\ M_{Bl} = -M_{B'r};\ M_{Br} = -M_{B'l};\ M_A = -M_A$$

Also

$$M_{Cr} = -M_{Cl} - 13{\cdot}00$$

$$M_{Br} = -M_{Bl} - 17{\cdot}97$$

For CBA the three-moment equation is,

$$2{\cdot}21M_{Cr} - 4{\cdot}42M_{Bl} + 3{\cdot}36M_{Br} - 1{\cdot}68M_A + \tfrac{1}{4}(406 \times 2{\cdot}21^3 + 483 \times 1{\cdot}68^3) = 0$$

For BAA'

$$1{\cdot}68M_{Br} - 6{\cdot}00M_A - 1{\cdot}22M_{A'} + \tfrac{1}{4}(483 \times 1{\cdot}68^3 + 373 \times 1{\cdot}22^3) = 0$$

Then

$$M_{Cr} = -104{\cdot}5 \text{ Nm} = M_{C'l}$$

We have then

$$M_{Bl} = 1773 \text{ Nm} = M_{B'r}$$

$$M_A = 67{\cdot}5 \text{ Nm}$$

$$M_{Br} = -159{\cdot}5 \text{ Nm}$$

We have thus found all the bending moments at the supports. The bending moment diagram for the whole beam can then be drawn.

* In reality the loading along the spar acts upwards and the couples applied at the supports are counter-clockwise; the spar is drawn upside down to agree with our conventions of sign.

FURTHER PROBLEMS
(answers on page 398)

14.6 A beam 8 m span is built-in at the ends, and carries a load of 60 kN at the centre, and loads of 30 kN, 2 m from each end. Calculate the maximum bending moment and the positions of the points of inflexion.

14.7 A girder of span 7 m is built-in at each end and carries two loads of 80 kN and 120 kN respectively placed at 2 m and 4 m from the left end. Find the bending moments at the ends and centre, and the points of contraflexure. (*Birmingham*)

14.8 A continuous girder of uniform section rests on four supports at the same level, forming three equal spans of 35 m. The girder carries a load of 65 kN per metre uniformly distributed. Draw to scale the bending moment diagram for the whole girder, and calculate the loads carried by each support. (*Cambridge*)

14.9 A continuous girder of two equal spans, each 25 m long, is carried on three supports, all at the same level. Find the maximum bending moments and shearing forces—(*a*) when one span is loaded with 15 kN per metre; (*b*) when both spans are loaded with the same intensity of load. The beam cannot rise above any of the supports.

Draw the bending moment and shearing force diagrams in each case.

14.10 A thin steel strip is laced in and out through a row of stiff pegs. The pegs are all 1·25 cm diameter, and their centres are spaced 50 cm apart in a straight line. There are five pegs in all, and the section of the strip is a rectangle 2·5 cm by 0·16 cm. Find the thrusts on the several pegs, taking $E = 200$ GN/m^2. (*Cambridge*)

14.11 A straight elastic beam of uniform section rests on four similar elastic supports which are spaced a distance L apart. The supports are such that they compress δ for each unit load upon them. Show that, when a uniformly distributed load of total amount W comes on the beam, the reactions at the centre props are each

$$\frac{\dfrac{11}{6} + \dfrac{3EI\delta}{L^3}}{5 + \dfrac{12EI\delta}{L^3}}$$

(*Cambridge*)

15 Plastic bending of mild-steel beams

15.1 Introduction

We have seen that in the bending of a beam the greatest direct stresses occur in the extreme longitudinal fibres; when these stresses attain the yield-point values, or exceed the limit of proportionality, the distribution of stresses over the depth of the beam no longer remains linear, as in the case of elastic bending.

The general problem of the plastic bending of beams is complicated; plastic bending of a beam is governed by the forms of the stress-strain curves of the material in tension and compression. Mild steel, which is used extensively as a structural material, has tensile and compressive properties which lend themselves to a relatively simple treatment of the plastic bending of beams of this material. The tensile and compressive stress-strain curves for an annealed mild steel have the forms shown in Fig. 15.1; in the elastic range

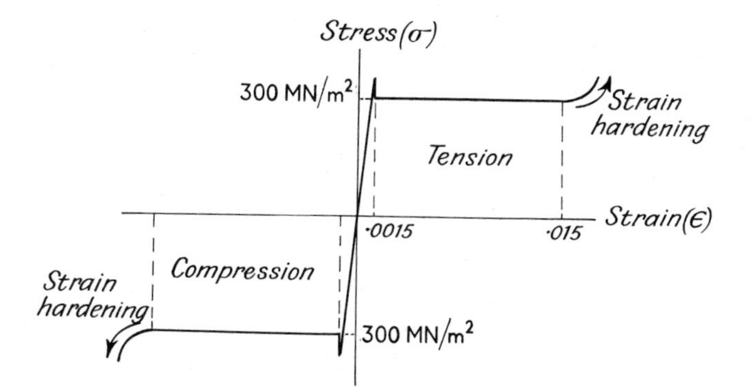

Fig. 15.1 Tensile and compressive stress-strain curves of an annealed mild steel.

Young's modulus is the same for tension and compression; for an annealed mild steel the yield stresses are the same for tension and compression, and of the order of 300 MN/m². The yield point corresponds to a strain of the order 0·0015. When the strain corresponding with the upper yield point is exceeded straining takes place continuously at a constant lower yield stress until a strain of about 0·015 is attained; at this stage further straining is accompanied by an increase in stress, and the material is said to *strain-harden.* This region of strain-hardening begins at strains about ten times larger than the strains at the yield point of the material.

In applying these stress-strain curves to the plastic bending of mild-steel beams we simplify the problem by ignoring the upper yield point of the material; we assume

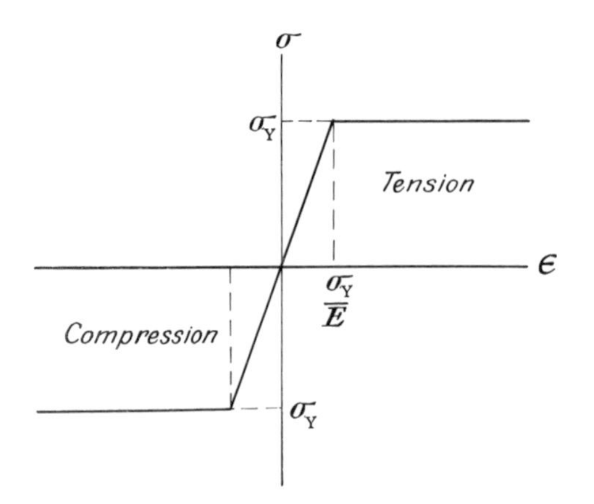

Fig. 15.2 Idealized tensile and compressive stress-strain curves of annealed mild steel.

the material is elastic, with a Young's modulus E, up to a yield stress σ_Y; in the plastic range of the material continuous yielding takes place at this constant stress σ_Y, Fig. 15.2. We assume that the yield stress, σ_Y, and Young's modulus, E, are the same for tension and compression. These idealized stress-strain curves for tension and compression are then similar in form.

15.2 Beam of rectangular cross-section

As an example of the application of these idealized stress-strain curves for mild steel, consider the uniform bending of a beam of rectangular cross-section; b is the breadth of the cross-section and h its depth, Fig. 15.3(i). Equal and opposite moments M are applied to the ends of a length of the beam. We found that in the elastic bending of a rectangular beam there is a linear distribution of direct stresses over a cross-section of the beam; an axis at the mid-depth of the cross-section is unstrained and therefore a neutral axis. The stresses are greatest in the extreme fibres of the beam; the yield stress, σ_Y, is attained in the extreme fibres, Fig. 15.3(ii), when

$$M = \frac{2\sigma_Y I}{h} = M_Y \text{ (say)}$$

where I is the second moment of area of the cross-section about the axis of bending. But $I = \frac{1}{12}bh^3$, and so

$$M_Y = \tfrac{1}{6}bh^2\sigma_Y \qquad (15.1)$$

As the beam is bent beyond this initial yielding condition experiment shows that plane cross-sections of the beam remain nearly plane as in the case of elastic bending. The centroidal axis remains a neutral axis during inelastic bending, and the greatest strains occur in the extreme tension and compression fibres. But the stresses in these extreme fibres cannot exceed σ_Y, the yield stress; at an intermediate stage in the bending of the

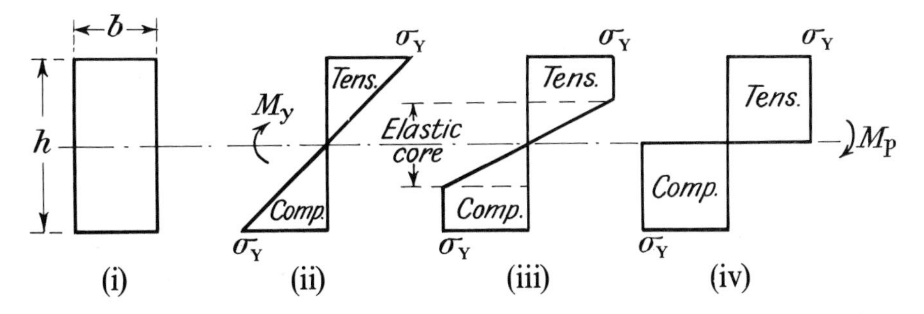

Fig. 15.3 Stages in the elastic and plastic bending of a rectangular mild-steel beam.

beam a central core is still elastic, but the extreme fibres have yielded and become plastic, Fig. 15.3(iii). If the curvature of the beam be increased the elastic core is diminished in depth; finally a condition is reached where the elastic core is reduced to negligible proportions, and the beam is more or less wholly plastic, Fig. 15.3(iv); in this final condition there is still a central unstrained, or neutral, axis; fibres above the neutral axis are stressed to the yield point in tension, while fibres below the neutral axis to the yield point in compression. In the ultimate fully-plastic condition the resultant longitudinal tension in the upper half-depth of the beam is

$$\tfrac{1}{2}bh\sigma_Y$$

There is an equal resultant compression in the lower half-depth. There is, therefore, no resultant longitudinal thrust in the beam; the bending moment for this fully-plastic condition is

$$M_P = (\tfrac{1}{2}bh\sigma_Y)(\tfrac{1}{2}h) = \tfrac{1}{4}bh^2\sigma_Y \tag{15.2}$$

This ultimate moment is usually called the *fully-plastic moment* of the beam; comparing equations (15.1) and (15.2) we get

$$M_P = \tfrac{3}{2}M_Y \tag{15.3}$$

Thus plastic collapse of a rectangular beam occurs at a moment 50% greater than the bending moment at initial yielding of the beam.

15.3 Elastic-plastic bending of a rectangular mild-steel beam

In §15.2 we introduced the concept of a fully-plastic moment, M_P, of a mild steel beam; this moment is attained when all longitudinal fibres of the beam are stressed into the plastic range of the material. Between the stage at which the yield stress is first exceeded and the ultimate stage at which the fully-plastic moment is attained, some fibres at the centre of the beam are elastic and those remote from the centre are plastic. At an intermediate stage the bending is elastic-plastic.

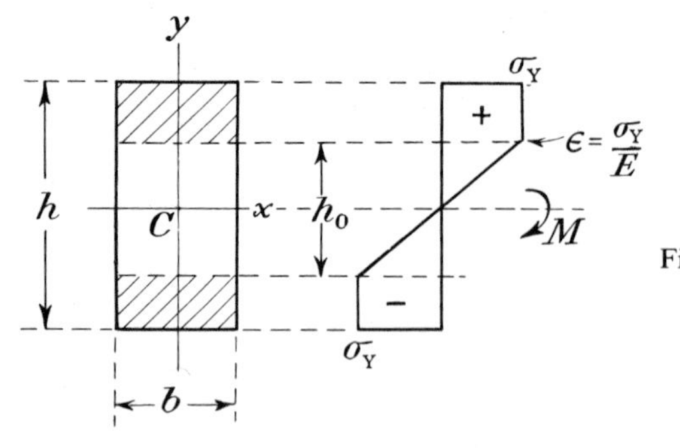

Fig. 15.4 Elastic-plastic bending of a rectangular section beam.

Consider again a mild-steel beam of rectangular cross-section, Fig. 15.4, which is bent about the centroidal axis Cx. In the elastic-plastic range, a central region of depth h_0 remains elastic; the yield stress σ_Y is attained in fibres beyond this central elastic core. If the central region of depth h_0 behaves as an elastic beam, the radius of curvature, R, is given by

$$\frac{2\sigma_Y}{h_0} = \frac{E}{R} \tag{15.4}$$

where E is Young's modulus in the elastic range of the material. Then

$$h_0 = \frac{2R\sigma_Y}{E} \tag{15.5}$$

Now, the bending moment carried by the elastic core of the beam is

$$M_1 = \sigma_Y \frac{bh_0^2}{6} \tag{15.6}$$

and the moment due to the stresses in the extreme plastic regions is

$$M_2 = \sigma_Y \left[\frac{bh^2}{4} - \frac{bh_0^2}{4}\right] \tag{15.7}$$

The total moment is, therefore,

$$M = M_1 + M_2 = \sigma_Y \frac{bh^2}{4} + \sigma_Y \left[\frac{bh_0^2}{6} - \frac{bh_0^2}{4}\right]$$

which gives

$$M = \sigma_Y \frac{bh^2}{4}\left[1 - \frac{h_0^2}{3h^2}\right] \tag{15.8}$$

But the fully-plastic moment, M_P, of the beam is

$$M_P = \sigma_Y \frac{bh^2}{4}$$

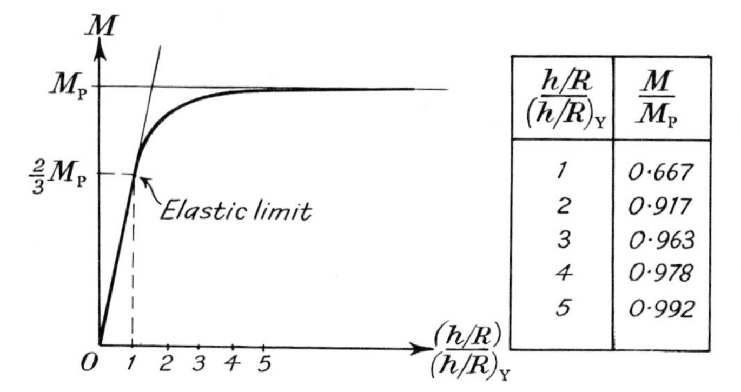

$\frac{h/R}{(h/R)_Y}$	$\frac{M}{M_P}$
1	0·667
2	0·917
3	0·963
4	0·978
5	0·992

Fig. 15.5 Moment-curvature relation for the elastic-plastic bending of a rectangular mild-steel beam.

Thus equation (15.8) may be written

$$M = M_P\left[1 - \frac{h_0^2}{3h^2}\right] \tag{15.9}$$

On substituting for h_0 from equation (15.5),

$$\frac{M}{M_P} = 1 - \frac{4}{3}\left(\frac{\sigma_Y}{E}\right)^2\left(\frac{R}{h}\right)^2 \tag{15.10}$$

At the onset of plasticity in the beam,

$$\frac{h}{R} = \frac{2\sigma_Y}{E} = \left(\frac{h}{R}\right)_Y \quad \text{(say)} \tag{15.11}$$

Then equation (15.10) may be written

$$\frac{M}{M_P} = 1 - \frac{1}{3}\frac{(h/R)_Y{}^2}{(h/R)^2} \tag{15.12}$$

Values of (M/M_P) for different values of $(h/R)/(h/R)_Y$ are given in Fig. 15.5; the elastic limit of the beam is reached when

$$M = \tfrac{2}{3}M_P = M_Y \text{ (say)}$$

As M is increased beyond M_Y, the fully-plastic moment M_P is approached rapidly with increase of curvature $(1/R)$ of the beam; M is greater than 99% of the fully-plastic moment when the curvature is only 5 times as large as the curvature at the onset of plasticity.

15.4 Fully-plastic moment of an I-section; shape factor

The cross-sectional dimensions of an I-section are shown in Fig. 15.6; in the fully-plastic condition, the centroidal axis Cx is a neutral axis of bending. The tensile fibres

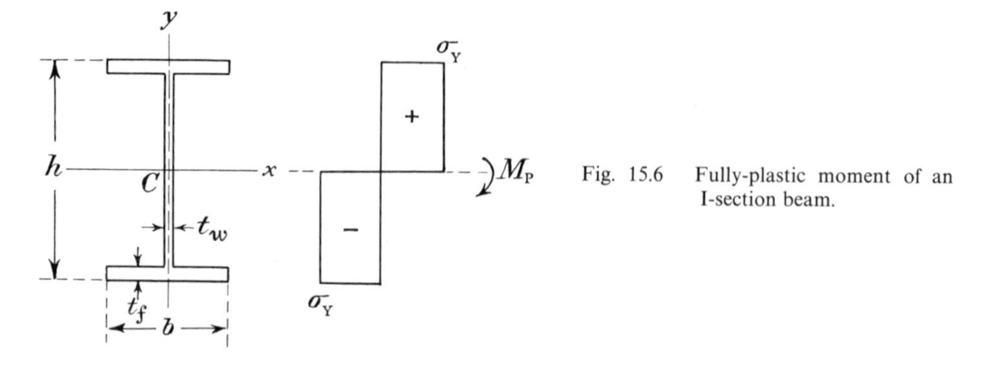

Fig. 15.6 Fully-plastic moment of an I-section beam.

of the beam all carry the same stress σ_Y; the total longitudinal force in the upper flange is

$$\sigma_Y b t_f$$

and its moment about Cx is

$$\sigma_Y b t_f(\tfrac{1}{2}h - \tfrac{1}{2}t_f) = \tfrac{1}{2}\sigma_Y b t_f(h - t_f)$$

Similarly, the total force in the tensile side of the web is

$$\sigma_Y(\tfrac{1}{2}h - t_f)t_w$$

and its moment about Cx is

$$\tfrac{1}{2}\sigma_Y(\tfrac{1}{2}h - t_f)^2 t_w = \tfrac{1}{8}\sigma_Y t_w(h - 2t_f)^2$$

The compressed longitudinal fibres contribute moments of the same magnitudes. The total moment carried by the beam is therefore

$$M_P = \sigma_Y[bt_f(h - t_f) + \tfrac{1}{4}t_w(h - 2t_f)^2] \tag{15.13}$$

In the case of elastic bending we defined the elastic section modulus, Z_e, as a geometrical property, which when multiplied by the allowable bending stress gives the allowable bending moment on the beam. In equation (15.13) suppose

$$Z_P = bt_f(h - t_f) + \tfrac{1}{4}t_w(h - 2t_f)^2 \tag{15.14}$$

Then Z_P is the *plastic section modulus* of the I-beam, and

$$M_P = \sigma_Y Z_p \tag{15.15}$$

As a particular case consider an I-section having dimensions:

$$h = 20 \text{ cm}, \qquad t_w = 0{\cdot}70 \text{ cm}$$
$$b = 10 \text{ cm}, \qquad t_f = 1{\cdot}00 \text{ cm}$$

Then

$$Z_P = (0{\cdot}1)(0{\cdot}010)(0{\cdot}2 - 0{\cdot}010) + \tfrac{1}{4}(0{\cdot}007)(0{\cdot}2 - 0{\cdot}020)^2$$
$$= 0{\cdot}247 \times 10^{-3}\ \text{m}^3$$

The elastic section modulus is approximately

$$Z_e = 0{\cdot}225 \times 10^{-3}\ \text{m}^3$$

If M_Y is the bending moment at which the yield stress σ_Y is first reached in the extreme fibres of the beam, then

$$\frac{M_P}{M_Y} = \frac{Z_P}{Z_e} = \frac{0{\cdot}247}{0{\cdot}225} = 1{\cdot}10 \tag{15.16}$$

Thus, in this case, the fully-plastic moment is only 10% greater than the moment at initial yielding. The ratio (Z_P/Z_e) is sometimes called the *shape factor*.

15.5 More general case of plastic bending

In the case of the rectangular and I-section beams treated so far, the neutral axis of bending coincided with an axis of symmetry of the cross-section. For a section which is unsymmetrical about the axis of bending, the position of the neutral axis must be found first. The beam in Fig. 15.7 has one axis of symmetry, Oy; the beam is bent into the fully-plastic condition about Ox, which is perpendicular to Oy. The axis Ox is the neutral axis of bending; the total longitudinal force on the fibres above Ox is $A_1\sigma_Y$, where A_1 is the area of the cross-section of the beam above Ox. If A_2 is the area of the cross-section below Ox, then the total longitudinal force on the fibres below Ox is $A_2\sigma_Y$. If there is no resultant longitudinal thrust in the beam, then

$$A_1\sigma_Y = A_2\sigma_Y$$

that is,

$$A_1 = A_2 \tag{15.17}$$

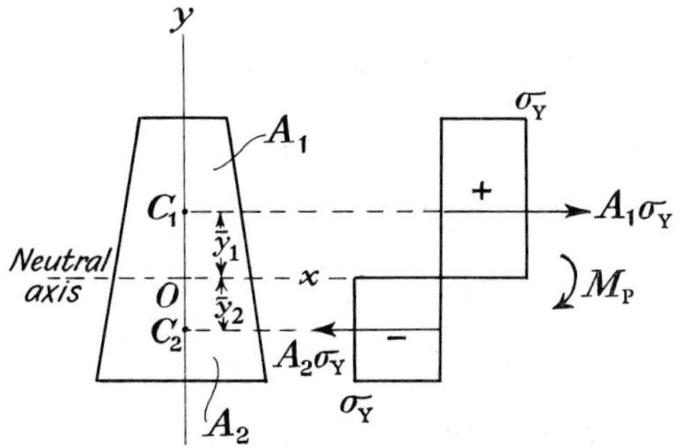

Fig. 15.7 Plastic bending of a beam having one axis of symmetry in the cross-section, but unsymmetrical about the axis of bending.

The neutral axis Ox divides the beam cross-section into equal areas therefore. If the total area of cross-section is A, then

$$A_1 = A_2 = \tfrac{1}{2}A$$

Then

$$A_1\sigma_Y = A_2\sigma_Y = \tfrac{1}{2}A\sigma_Y$$

Suppose C_1 is the centroid of the area A_1, and C_2 the centroid of A_2; if the centroids C_1 and C_2 are distances $\bar{y}_1$ and $\bar{y}_2$, respectively, from the neutral axis Ox, then

$$M_P = \tfrac{1}{2}A\sigma_Y(\bar{y}_1 + \bar{y}_2) \tag{15.18}$$

The plastic section modulus is

$$Z_p = \frac{M_P}{\sigma_Y} = \tfrac{1}{2}A(\bar{y}_1 + \bar{y}_2) \tag{15.19}$$

Problem 15.1: A 10 cm by 10 cm T-section is of uniform thickness 1·25 cm. Estimate the plastic section modulus for bending about an axis perpendicular to the web.

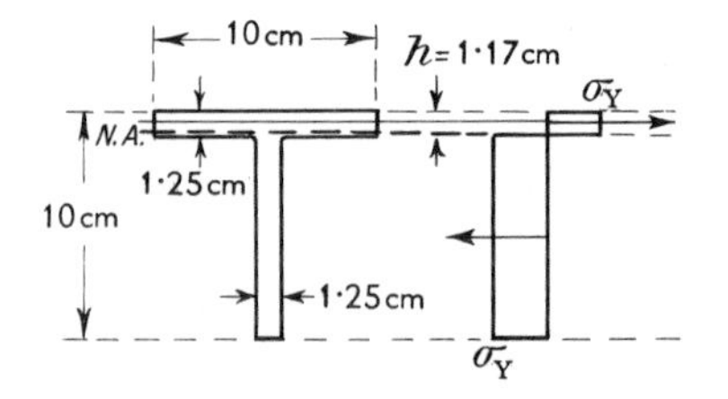

Solution

The neutral axis of plastic bending divides the section into equal areas. If the neutral axis is a distance h below the extreme edge of the flange,

$$(0{\cdot}1)h = (0{\cdot}0875)(0{\cdot}0125) + (0{\cdot}1)(0{\cdot}0125 - h)$$

Then

$$h = 0{\cdot}0117 \text{ m}$$

Then

$$M_P = \tfrac{1}{2}(0{\cdot}1)(0{\cdot}0117)^2\sigma_Y + \tfrac{1}{2}(0{\cdot}0875)(0{\cdot}0008)^2\sigma_Y + \tfrac{1}{2}(0{\cdot}0883)^2(0{\cdot}0125)\sigma_Y$$
$$= (0{\cdot}0557 \times 10^{-3})\sigma_Y$$

The plastic section modulus is then

$$Z_P = \frac{M_P}{\sigma_Y} = 0{\cdot}0557 \times 10^{-3} \text{ m}^3$$

The elastic section modulus is

$$Z_e = 0{\cdot}0311 \times 10^{-3} \text{ m}^3$$

Then

$$\frac{M_P}{M_Y} = \frac{Z_p}{Z_e} = \frac{0{\cdot}0557}{0{\cdot}0311} = 1{\cdot}79$$

15.6 Comparison of elastic and plastic section moduli

For bending of a beam about a centroidal axis Cx, the elastic section modulus is

$$Z_e = \frac{I}{y_{max}} \tag{15.20}$$

where I is the second moment of area of the cross-section about the axis of bending, and y_{max} is the distance of the extreme fibre from the axis of bending.

From equation (15.19) the plastic section modulus of a beam is

$$Z_p = \tfrac{1}{2}A(\bar{y}_1 + \bar{y}_2) \tag{15.21}$$

Values of Z_e and Z_p for some simple cross-sectional forms are shown in Table 15.1. In the solid rectangular and circular sections Z_p is considerably greater than Z_e; the difference between Z_p and Z_e is less marked in the case of thin-walled sections.

15.7 Regions of plasticity in a simply-supported beam

The mild-steel beam shown in Fig. 15.8 has a rectangular cross-section; it is simply-supported at each end, and carries a central lateral load W. The variation of bending moment has the form shown in Fig. 15.8(ii); the greatest bending moment occurs under

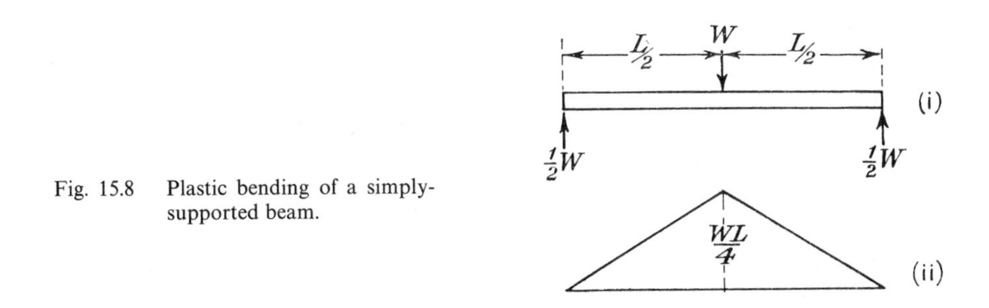

Fig. 15.8 Plastic bending of a simply-supported beam.

the central load and has the value $WL/4$. From the preceding analysis we see that a section may take an increasing bending moment until the fully-plastic moment M_P of the section is reached. The ultimate strength of the beam is reached therefore when

$$M_P = \frac{WL}{4} \tag{15.22}$$

Table 15.1 COMPARISON OF ELASTIC AND PLASTIC SECTION MODULI FOR SOME SIMPLE CROSS-SECTIONAL FORMS.

Cross-sectional form	Elastic section modulus, Z_e	Plastic section modulus, Z_p	Shape factor, $\frac{Z_p}{Z_e}$
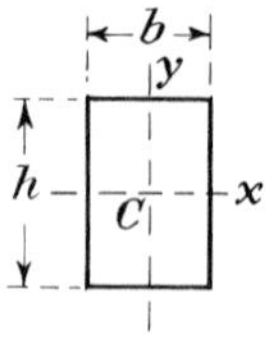 Solid rectangular section	Axis Cy: $\frac{1}{6}b^2h$ Axis Cx: $\frac{1}{6}bh^2$	Axis Cy: $\frac{1}{4}b^2h$ Axis Cx: $\frac{1}{4}bh^2$	1·5 1·5
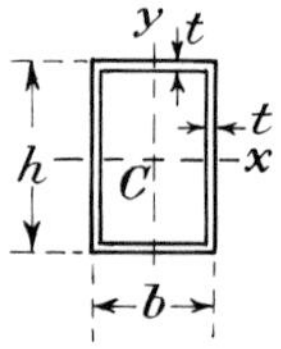 Thin-walled rectangular box of uniform wall-thickness, t	$t \ll h$; $t \ll b$ Axis Cy: $bt(h + \frac{1}{3}b)$ Axis Cx: $ht(b + \frac{1}{3}h)$	Axis Cy: $bt(h + \frac{1}{2}b)$ Axis Cx: $ht(b + \frac{1}{2}h)$	$\frac{h + \frac{1}{2}b}{h + \frac{1}{3}b}$ $\frac{b + \frac{1}{2}h}{b + \frac{1}{3}h}$
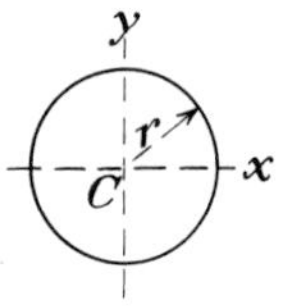 Solid circular section	Axis Cy or Cx: $\frac{\pi r^3}{4}$	Axis Cy or Cx: $\frac{4r^3}{3}$	$\frac{16}{3\pi}$
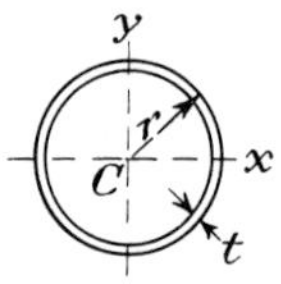 Thin-walled circular tube	$t \ll r$ Axis Cy or Cx: $\pi r^2 t$	$4r^2t$	$\frac{4}{\pi}$
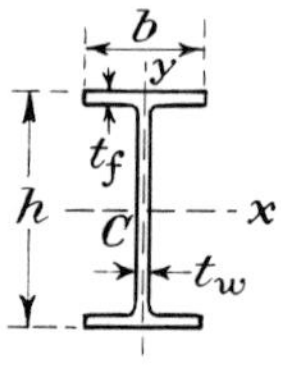 Thin-walled I-section	$t_f \ll b$; $t_w \ll h$ Axis Cy: $\frac{1}{3}b^2t$ Axis Cx: $h[bt_f + \frac{1}{6}ht_w]$	Axis Cy: $\frac{1}{2}b^2t$ Axis Cx: $h[bt_f + \frac{1}{4}ht_w]$	1·5 $\frac{bt_f + \frac{1}{4}ht_w}{bt_f + \frac{1}{6}ht_w}$

If b is the breadth and h the depth of the rectangular cross-section, the bending moment, M_Y, at which the yield stress, σ_Y, is first attained in the extreme fibres is

$$M_Y = \sigma_Y \frac{bh^2}{6} = \tfrac{2}{3} M_P$$

At the ultimate strength of the beam

$$W = \frac{4M_P}{L} = \frac{4}{L}\left[\sigma_Y \frac{bh^2}{4}\right] \tag{15.23}$$

The beam is wholly elastic for a distance of

$$\frac{2}{3}\left(\frac{L}{2}\right) = \tfrac{1}{3}L \tag{15.24}$$

from each end support, Fig. 15.9, since the bending moments in these regions are not greater than M_Y. The middle-third length of the beam is in an elastic-plastic state;

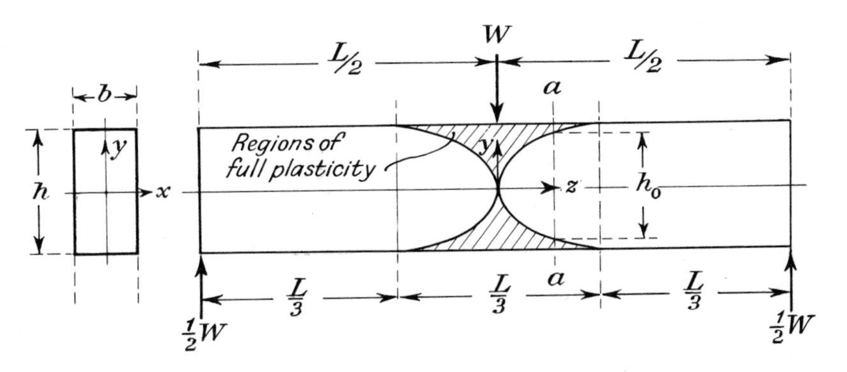

Fig. 15.9 Regions of plasticity in a simply-supported beam carrying a distributed load; in the figure, the depth of the beam is exaggerated.

in this central region consider a transverse section $a - a$ of the beam, a distance z from the mid-length. The bending moment at this section is

$$M = \tfrac{1}{2}W(\tfrac{1}{2}L - z) \tag{15.25}$$

If W has attained its ultimate value given by equation (15.22),

$$M = \frac{2M_P}{L}(\tfrac{1}{2}L - z) \tag{15.26}$$

Suppose the depth of the elastic core of the beam at this section is h_0, Fig. 15.9; then from equation (15.9),

$$M = M_P\left(1 - \frac{h_0^2}{3h^2}\right)$$

On substituting this value of M into equation (15.26), we have

$$1 - \frac{h_0^2}{3h^2} = 1 - \frac{2z}{L} \tag{15.27}$$

and thus

$$h_0^2 = \frac{6h^2}{L} z \tag{15.28}$$

The total depth, h_0, of the elastic core varies parabolically with z, therefore; from equation (15.28), $h_0 = h$ when $z = \frac{1}{6}L$. The regions of full plasticity are wedge-shaped; the shapes of the regions developed in an actual mild steel beam may be affected by, first, the stress-concentrations under the central load W, and, second, the presence of shearing stresses on sections such as $a - a$, Fig. 15.9; equation (15.28) is true strictly for conditions of pure bending only.

For a simply-supported rectangular beam carrying a total uniformly distributed load W, Fig. 15.10, the bending moment at the mid-length is

$$M_P = \frac{WL}{8}$$

at the ultimate load-carrying capacity of the beam. At a transverse section $a - a$, a distance z from the mid-length, the moment is

$$M = \frac{W}{8L}(L^2 - 4z^2) = \frac{M_P}{L^2}(L^2 - 4z^2)$$

$$= M_P\left[1 - 4\left(\frac{z}{L}\right)^2\right] \tag{15.29}$$

From equation (15.9), the depth h_0 of the elastic core at the section $a - a$ is given by

$$M = M_P\left[1 - \frac{h_0^2}{3h^2}\right]$$

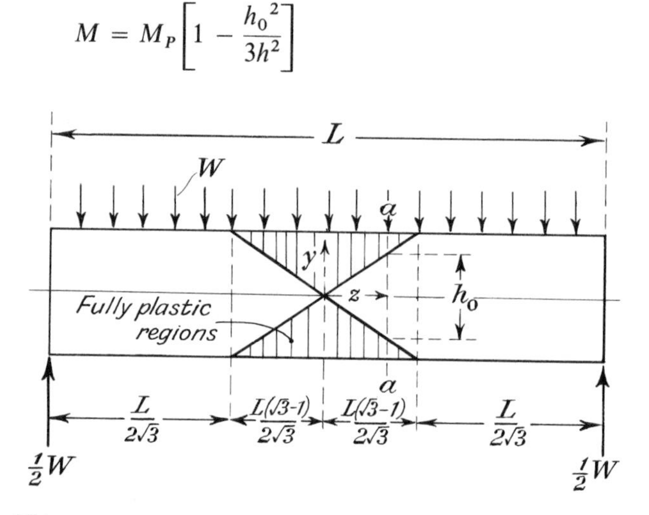

Fig. 15.10 Regions of plasticity in a simply-supported beam carrying a distributed load; in the figure, the depth of the beam is exaggerated.

Then

$$h_0^2 = 12h^2\left(\frac{z}{L}\right)^2$$

or

$$h_0 = 2\sqrt{3}h\left(\frac{z}{L}\right) \qquad (15.30)$$

The limit of the wholly elastic length of the beam is given by $h = h_0$, or $z = L/(2\sqrt{3})$. The regions of plasticity near the mid-section are triangular-shaped, Fig. 15.10.

15.8 Plastic collapse of a built-in beam

A uniform beam of length L is built-in at each end to rigid walls, and carries a uniformly distributed load w per unit length, Fig. 15.11. If the material remains elastic, the bending moment at each end is $wL^2/12$, and at the mid-length $wL^2/24$. The bending moment is therefore greatest at the end supports; if yielding occurs first at a bending moment M_Y, then the lateral load at this stage is given by

$$M_Y = \frac{wL^2}{12} \qquad (15.31)$$

or

$$wL = \frac{12M_Y}{L} \qquad (15.32)$$

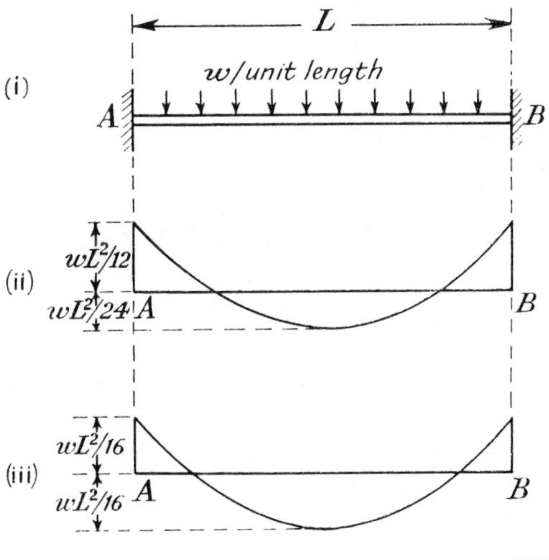

Fig. 15.11 Plastic regions of a uniformly loaded built-in beam.

If the load w be increased beyond the limit of elasticity, plastic hinges first develop at the remote ends. The beam only becomes a mechanism when a third plastic hinge develops at the mid-length. On considering the statical equilibrium of a half-span of the beam we find that the moments at the ends and the mid-length, for plastic hinges at these sections, are

$$M_P = \frac{wL^2}{16} \tag{15.33}$$

or

$$wL = \frac{16M_P}{L} \tag{15.34}$$

Clearly, the load causing complete collapse is at least $\frac{1}{3}$rd greater than that at which initial yielding begins because M_P is greater than M_Y.

FURTHER PROBLEMS

(answers on page 398)

15.2 A uniform mild steel beam AB is 4 m long; it is built-in at A and simply-supported at B. It carries a single concentrated load at a point 1·5 m from A. If the plastic section modulus of the beam is $0{\cdot}433 \times 10^{-3}$ m^3, and the yield stress of the material is 235 MN/m^2, estimate the value of the concentrated load causing plastic collapse.

15.3 A uniform mild steel beam is supported on four knife edges equally spaced a distance 8 m apart. Estimate the intensity of uniformly distributed lateral load over the whole length causing collapse, if the plastic section modulus of the beam is $1{\cdot}690 \times 10^{-3}$ m^3, and the yield stress of the material is 235 MN/m^2.

15.4 A uniform beam rests on three supports A, B, C, with two spans each 5 m long. The collapse load is to be 100 kN per metre, and $\sigma_Y = 235$ MN/m^2. What will be a suitable mild steel section using a shape factor 1·15?

15.5 If, in Problem 15.4, AB is 8 m, BC is 7 m, and the collapse loads are to be 100 kN/m on AB, 50 kN/m on BC, find a suitable mild steel section I-beam, with $\sigma_Y = 235$ MN/m^2.

15.6 A continuous beam $ABCD$ has spans each 8 m long; it is 45 cm by 15 cm, with flanges 2·5 cm thick and web 1 cm thick. Find the collapse load if the whole beam carries a uniformly-distributed load. Which spans collapse? $\sigma_Y = 235$ MN/m^2.

15.7 A mild steel beam 5 cm square section is subjected to a thrust of 200 kN acting in the plane of one of the principal axes, but may be eccentric. What eccentricity will cause the whole section to become plastic if $\sigma_Y = 235$ MN/m^2?

16 Torsion of circular shafts and thin-walled tubes

16.1 Introduction

In Chapter 3 we introduced the concepts of shearing stress and shearing strain; these have an important application in torsion problems. Such problems arise in shafts transmitting heavy torques, in eccentrically-loaded beams, in aircraft wings and fuselages, and many other instances. These problems are very complex in general, and at this elementary stage we can go no further than studying uniform torsion of circular shafts, thin-walled tubes, and thin-walled open sections.

16.2 Torsion of a thin circular tube

The simplest torsion problem is that of the twisting of a uniform thin circular tube; the tube shown in Fig. 16.1 is of thickness t, and the mean radius of the wall is r; L is the length of the tube. Shearing stresses τ are applied around the circumference of the

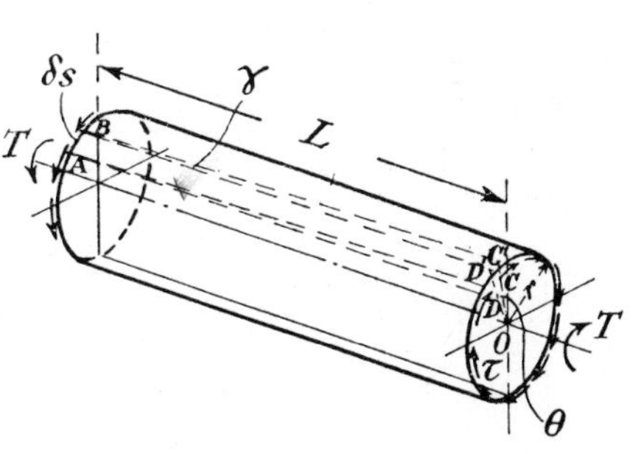

Fig. 16.1 Torsion of a thin-walled circular tube.

tube at each end, and in opposite directions. If the stresses τ are uniform around the boundary, the total torque T at each end of the tube is

$$T = (2\pi r t)\tau r = 2\pi r^2 t \,.\, \tau \tag{16.1}$$

Thus the shearing stress around the circumference due to an applied torque T is

$$\tau = \frac{T}{2\pi r^2 t} \tag{16.2}$$

We consider next the strains caused by these shearing stresses. We note firstly that complementary shearing stresses are set up in the wall parallel to the longitudinal axis of the tube. If δs is a small length of the circumference then an element of the wall $ABCD$, Fig. 16.1, is in a state of pure shearing stress. If the remote end of the tube is assumed not to twist, then the longitudinal element $ABCD$ is distorted into the parallelogram $ABC'D'$, Fig. 16.1, the angle of shearing strain being

$$\gamma = \frac{\tau}{G} \tag{16.3}$$

if the material is elastic, and has a shearing modulus G. But if θ is the angle of twist of the near end of the tube we have

$$\gamma L = r\theta \tag{16.4}$$

Hence

$$\theta = \frac{\gamma L}{r} = \frac{\tau L}{Gr} \tag{16.5}$$

It is sometimes more convenient to define the twist of the tube as the rate of change of twist per unit length; this is given by (θ/L), and from equation (16.5) this is equal to

$$\frac{\theta}{L} = \frac{\tau}{Gr} \tag{16.6}$$

16.3 Torsion of solid circular shafts

The torsion of a thin circular tube is a relatively simple problem since the shearing stress may be assumed constant throughout the wall thickness. The case of a solid circular shaft is more complex since the shearing stresses are variable over the cross-section of the shaft. The solid circular shaft of Fig. 16.2 has a length L and radius a

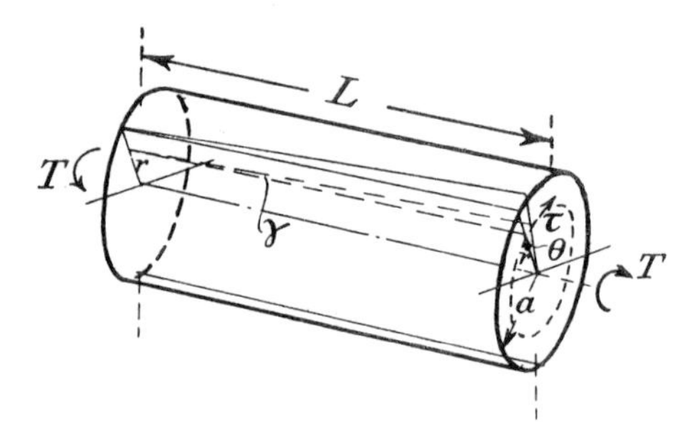

Fig. 16.2 Torsion of a solid circular shaft.

in the cross-section. When equal and opposite torques T are applied at each end about a longitudinal axis we assume that

(i) the twisting is uniform along the shaft, that is, all normal cross-sections the same distance apart suffer equal relative rotation,
(ii) cross-sections remain plane during twisting, and
(iii) radii remain straight during twisting.

If θ is the relative angle of twist of the two ends of the shaft, then the shearing strain γ of an elemental tube of thickness δr and at radius r is

$$\gamma = \frac{r\theta}{L} \tag{16.7}$$

If the material is elastic, and has a shearing modulus G, (§3.4), then the circumferential shearing stress on this elemental tube is

$$\tau = G\gamma = \frac{Gr\theta}{L} \tag{16.8}$$

The thickness of the elemental tube is δr, so the total torque on this tube is

$$(2\pi r\ \delta r)\tau r = 2\pi r^2 \tau\ \delta r$$

The total torque on the shaft is then

$$T = \int_0^a 2\pi r^2 \tau\ dr$$

On substituting for τ from equation (16.8), we have

$$T = 2\pi\left(\frac{G\theta}{L}\right)\int_0^a r^3\ dr \tag{16.9}$$

Now

$$2\pi \int_0^a r^3\ dr = \frac{\pi a^4}{2} \tag{16.10}$$

This is the polar second moment of area of the cross-section about an axis through the centre, and is usually denoted by J. Then equation (16.9) may be written

$$T = \frac{GJ\theta}{L} \tag{16.11}$$

We may combine equations (16.8) and (16.11) in the form

$$\frac{T}{J} = \frac{\tau}{r} = \frac{G\theta}{L} \tag{16.12}$$

We see from equation (16.8) that τ increases linearly with r, from zero at the centre of the shaft to $Ga\theta/L$ at the circumference. Along any radius of the cross-section, the shearing stresses are normal to the radius and in the plane of the cross-section, Fig. 16.3.

16.4 Torsion of a hollow circular shaft

It frequently arises that a torque is transmitted by a hollow circular shaft. Suppose a_1 and a_2 are the internal and external radii, respectively, of such a shaft, Fig. 16.4.

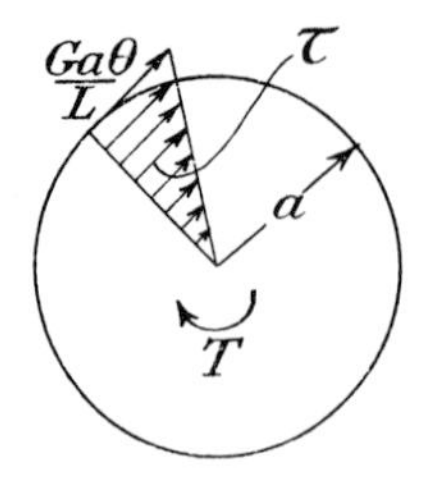

Fig. 16.3 Variation of shearing stresses over the cross-section for elastic torsion of a solid circular bar.

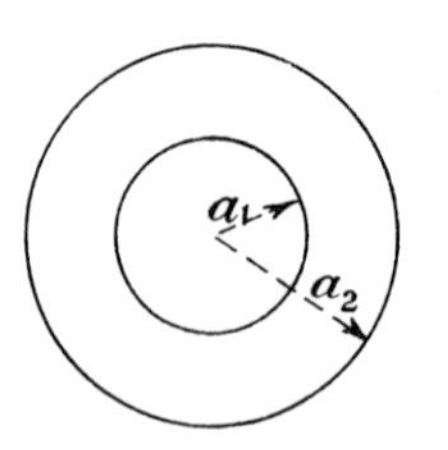

Fig. 16.4 Cross-section of a hollow circular shaft.

We make the same general assumptions as in the torsion of a solid circular shaft. If τ is the shearing stress at radius r, the total torque on the shaft is

$$T = \int_{a_1}^{a_2} 2\pi r^2 \tau \, dr \tag{16.13}$$

If we assume, as before, that radii remain straight during twisting, and that the material is elastic, we have

$$\tau = \frac{Gr\theta}{L}$$

Then equation (16.13) becomes

$$T = \int_{a_1}^{a_2} \left(\frac{G\theta}{L}\right) 2\pi r^3 \, dr = \frac{GJ\theta}{L} \tag{16.14}$$

where

$$J = \int_{a_1}^{a_2} 2\pi r^3 \, dr \tag{16.15}$$

J, here, is the polar second moment of area or, more generally, the *torsion constant* of the cross-section about an axis through the centre; J has the value

$$J = \int_{a_1}^{a_2} 2\pi r^3 \, dr = \frac{\pi}{2}(a_2{}^4 - a_1{}^4) \tag{16.16}$$

Thus, for both hollow and solid shafts, we have the relation

$$\frac{T}{J} = \frac{\tau}{r} = \frac{G\theta}{L}$$

Problem 16.1: What torque, applied to a hollow circular shaft of 25 cm outside diameter and 17·5 cm inside diameter will produce a maximum shearing stress of 75 MN/m^2 in the material? (*Cambridge*)

Solution

We have

$$r_1 = 12{\cdot}5 \text{ cm}, \qquad r_2 = 8{\cdot}75 \text{ cm}$$

Then

$$J = \frac{\pi}{2}[(0{\cdot}125)^4 - (0{\cdot}0875)^4] = 0{\cdot}292 \times 10^{-3} \text{ m}^4$$

If the shearing stress is limited to 75 MN/m^2, the torque is

$$T = \frac{J\tau}{r_1} = \frac{(0{\cdot}292 \times 10^{-3})(75 \times 10^6)}{(0{\cdot}125)} = 175{\cdot}5 \text{ kNm}$$

Problem 16.2: A ship's propeller shaft has external and internal diameters of 25 cm and 15 cm. What power can be transmitted at 110 rev/minute with a maximum shearing stress of 75 MN/m^2, and what will then be the twist in degrees of a 10 m length of the shaft? $G = 80$ GN/m^2. (*Cambridge*)

Solution

In this case

$$r_1 = 0{\cdot}125 \text{ m}, \qquad r_2 = 0{\cdot}075 \text{ m}, \qquad l = 10 \text{ m}$$

$$J = \frac{\pi}{2}[(0{\cdot}125)^4 - (0{\cdot}075)^4] = 0{\cdot}335 \times 10^{-3} \text{ m}^4$$

and

$$\tau = 75 \text{ MN/m}^2$$

Then

$$T = \frac{J\tau}{r_1} = \frac{(0{\cdot}335 \times 10^{-3})(75 \times 10^6)}{0{\cdot}125}$$

$$= 201 \text{ kNm}$$

At 110 rev/min the power generated is

$$(201 \times 10^3)(2\pi \times \tfrac{110}{60}) = 2{\cdot}32 \times 10^6 \text{ Nm/s}$$

The angle of twist is

$$\theta = \frac{TL}{GJ} = \frac{(201 \times 10^3)(10)}{(80 \times 10^9)(0{\cdot}335 \times 10^{-3})} = 0{\cdot}075 \text{ radians} = 4{\cdot}3^\circ$$

Problem 16.3: A solid circular shaft of 25 cm diameter is to be replaced by a hollow shaft, the ratio of the external to internal diameters being 2 to 1. Find the size of the hollow shaft if the maximum shearing stress is to be the same as for the solid shaft. What percentage economy in mass will this change effect? (*Cambridge*)

Solution

Let r = the inside radius of the new shaft, then $2r$ = the outside radius of the new shaft,

$$J \text{ for the new shaft} = \frac{\pi}{2}(16r^4 - r^4) = 7{\cdot}5\pi r^4$$

$$J \text{ for the old shaft} = \frac{\pi}{2} \times (0{\cdot}125)^4 = 0{\cdot}384 \times 10^{-3} \text{ m}^4$$

If T be the applied torque, the maximum shearing stress for the old shaft is

$$\frac{T(0{\cdot}125)}{0{\cdot}384 \times 10^{-3}}$$

and that for the new one is

$$\frac{T(2r)}{7{\cdot}5\pi r^4}$$

If these are equal,

$$\frac{T(0{\cdot}125)}{0{\cdot}384 \times 10^{-3}} = \frac{T(2r)}{7{\cdot}5\pi r^4}$$

Then

$$r^3 = 0{\cdot}261 \times 10^{-3}\ \mathrm{m}^3$$

or

$$r = 0{\cdot}0640\ \mathrm{m}$$

Hence the internal diameter will be 0·128 m and the external diameter 0·256 m. To find the saving in mass, we have

$$\frac{\text{area of new cross-section}}{\text{area of old cross-section}} = \frac{(0{\cdot}128)^2 - (0{\cdot}064)^2}{(0{\cdot}125)^2} = 0{\cdot}785$$

Thus, the saving in mass is about 21 %.

Problem 16.4: A ship's propeller shaft transmits $7{\cdot}5 \times 10^6$ W at 240 rev/min. The shaft has an internal diameter of 15 cm. Calculate the minimum permissible external diameter if the shearing stress in the shaft is to be limited to 150 MN/m^2. (*Cambridge*)

Solution

If T is the torque on the shaft, then

$$T\left(\frac{2\pi \times 240}{60}\right) = 7{\cdot}5 \times 10^6$$

Thus

$$T = 298\ \mathrm{kNm}$$

If d_1 is the outside diameter of the shaft, then

$$J = \frac{\pi}{32}(d_1{}^4 - 0{\cdot}150^4)\ \mathrm{m}^4$$

If the shearing stress is limited to 150 MN/m^2, then

$$\frac{Td_1}{2J} = 150 \times 10^6$$

Thus,

$$Td_1 = (300 \times 10^6)J$$

On substituting for J and T

$$(298 \times 10^3)d_1 = (300 \times 10^6)\left(\frac{\pi}{32}\right)(d_1{}^4 - 0{\cdot}150^4)$$

This gives

$$\left(\frac{d_1}{0{\cdot}150}\right)^4 - 3\left(\frac{d_1}{0{\cdot}150}\right) - 1 = 0$$

On solving this by trial-and-error, we get

$$d_1 = 1{\cdot}54(0{\cdot}150) = 0{\cdot}231 \text{ m}$$

or

$$d_1 = 23{\cdot}1 \text{ cm}$$

16.5 Principal stresses in a twisted shaft

It is important to appreciate that uniform torsion of circular shafts, of the form discussed in §16.3, involves no shearing between concentric elemental tubes of the shaft. Shearing stresses τ occur in a cross-section of the shaft, and complementary shearing stresses parallel to the longitudinal axis, Fig. 16.5. An element $ABCD$ in

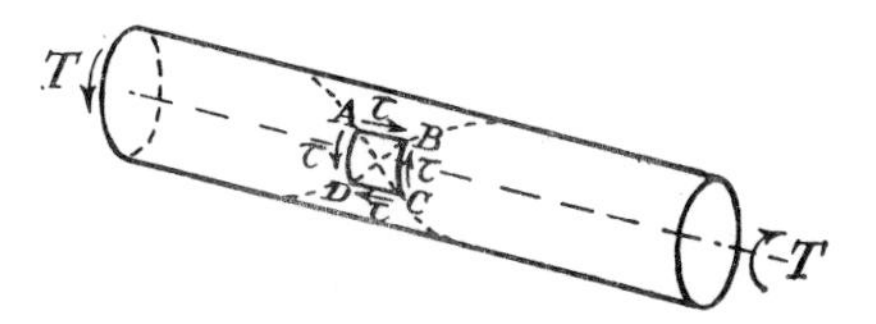

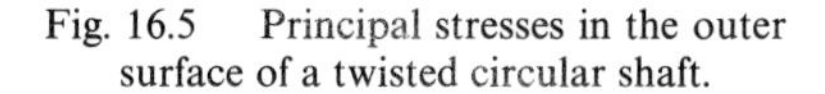
Fig. 16.5 Principal stresses in the outer surface of a twisted circular shaft.

the surface of the shaft is in a state of pure shear. The principal planes make angles of 45° with the axis of the shaft, therefore, and the principal stresses are $\pm\tau$. If the element $ABCD$ is square, then the principal planes are AC and BD. The direct stress on AC is tensile and of magnitude τ; the direct stress on BD is compressive and of the same magnitude. Principal planes such as AC cut the surface of the shaft in a helix; for a brittle material, weak in tension, we should expect breakdown in a torsion test to occur by tensile fracture along planes such as AC. The failure of a twisted bar of a brittle material is shown in Fig. 16.6.

Fig. 16.6 Failure in torsion of a circular bar of brittle cast iron showing a tendency to tensile fracture across a helix on the surface of the specimen.

The torsional failure of ductile materials occurs when the shearing stresses attain the yield stress of the material. The greatest shearing stresses in a circular shaft occur in

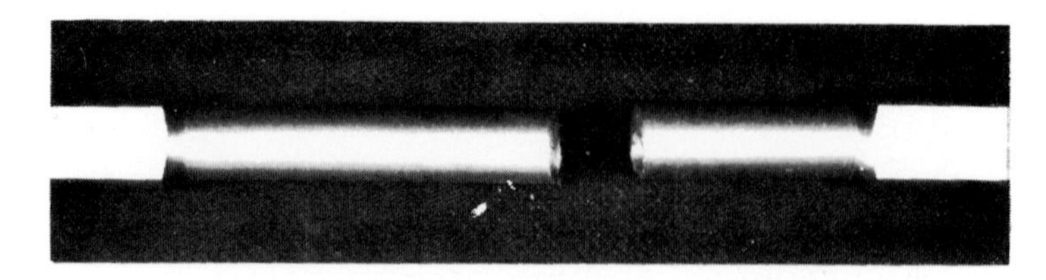

Fig. 16.7 Failure in torsion of a circular bar of ductile cast iron showing a shearing failure over a normal cross-section of the bar.

a cross-section and along the length of the shaft. A circular bar of a ductile material usually fails by breaking off over a normal cross-section, as shown in Fig. 16.7.

16.6 Torsion combined with thrust or tension

When a circular shaft is subjected to longitudinal thrust, or tension, as well as twisting, the direct stresses due to the longitudinal load must be combined with the shearing stresses due to torsion in order to evaluate the principal stresses in the shaft. Suppose the shaft is axially-loaded in tension so that there is a longitudinal direct stress σ at all points of the shaft. If τ is the shearing stress at any point, then we are interested in the

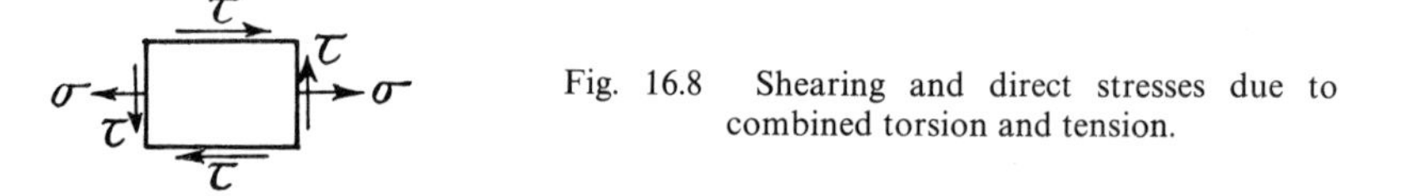

Fig. 16.8 Shearing and direct stresses due to combined torsion and tension.

principal stresses of the system shown in Fig. 16.8; for this system the principal stresses, from equations (5.12), have the values

$$\tfrac{1}{2}\sigma \pm \tfrac{1}{2}\sqrt{\sigma^2 + 4\tau^2} \tag{16.17}$$

and the maximum shearing stress, from equation (5.14), is

$$\tau_{max} = \tfrac{1}{2}\sqrt{\sigma^2 + 4\tau^2} \tag{16.18}$$

Problem 16.5: A steel shaft, 20 cm external diameter and 7·5 cm internal, is subjected to a twisting moment of 30 kNm, and a thrust of 50 kN. Find the shearing stress due to the torque alone and the percentage increase when the thrust is taken into account. (*RNC*)

Solution

For this case, we have

$$r_1 = 0{\cdot}100\text{ m}, \qquad r_2 = 0{\cdot}0375\text{ m}$$
$$A = \pi(r_1{}^2 - r_2{}^2) = 0{\cdot}0270\text{ m}^2$$

The compressive stress is

$$\sigma = -\frac{P}{A} = -\frac{50 \times 10^3}{0{\cdot}0270} = -1{\cdot}85\text{ MN/m}^2$$

Now

$$J = \frac{\pi}{2}(r_1{}^4 - r_2{}^4) = 0{\cdot}00247 \text{ m}^4$$

The shearing stress due to torque alone is

$$\tau = \frac{Tr_1}{J} = \frac{(30 \times 10^3)(0{\cdot}100)}{0{\cdot}00247} = 1{\cdot}22 \text{ MN/m}^2$$

The maximum shearing stress due to the combined loading is

$$\tau_{max} = \tfrac{1}{2}[\sigma^2 + 4\tau^2]^{\frac{1}{2}} = 1{\cdot}53 \text{ MN/m}^2$$

Problem 16.6: A thin steel tube of 2·5 cm diameter and 0·16 cm thickness has an axial pull of 10 kN, and an axial torque of 23·5 Nm applied to it. Find the magnitude and direction of the principal stresses at any point. (*Cambridge*)

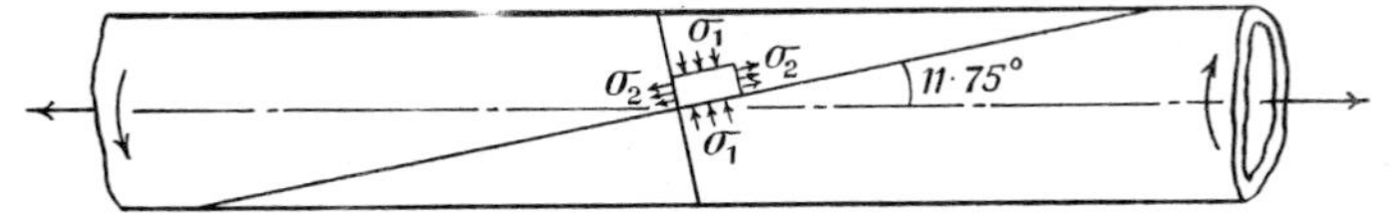

Solution

It will be easier, and sufficiently accurate, to neglect the variation in the shearing stress from the inside to the outside of the tube. Let

τ = the mean shearing stress due to torsion,

r = the mean radius = 0·0109 m

t = the thickness = 0·016 m

Then the moment of the total resistance to shear

$$= 2\pi r^2 \tau t = (1{\cdot}19 \times 10^{-6})\tau \text{ Nm}$$

If this is equal to 23·5 Nm, then

$$\tau = 19{\cdot}75 \text{ MN/m}^2$$

The area of the cross-section, approximately

$$= 2\pi r t = 0{\cdot}1098 \times 10^{-3} \text{ m}^2$$

Hence the tensile stress is

$$\sigma = \frac{10 \times 10^3}{0{\cdot}1098 \times 10^{-3}} = 91{\cdot}1 \text{ MN/m}^2$$

The principal stresses are

$$\tfrac{1}{2}\{\sigma \pm \sqrt{\sigma^2 + 4\tau^2}\} = \tfrac{1}{2}(91{\cdot}1 \pm 99{\cdot}3) \text{ MN/m}^2$$

Then

$$\sigma_1 = -4{\cdot}1 \text{ MN/m}^2, \qquad \sigma_2 = +95{\cdot}2 \text{ MN/m}^2$$

the positive sign denoting tension. The planes across which they act make angles θ and $(\theta + \pi/2)$ with the axis, where

$$\tan 2\theta = \frac{2\tau}{\sigma} = \frac{39{\cdot}5}{91{\cdot}1} = 0{\cdot}434$$

whence $\theta = 11{\cdot}75°$.

16.7 Strain energy of elastic torsion

In §16.3 we found that the torque-twist relationship for a circular shaft has the form

$$T = \frac{GJ\theta}{L} \tag{16.11}$$

This shows that the angle of twist, θ, of one end relative to the other, increases linearly with T. If one end of the shaft is assumed to be fixed, then the work done in twisting

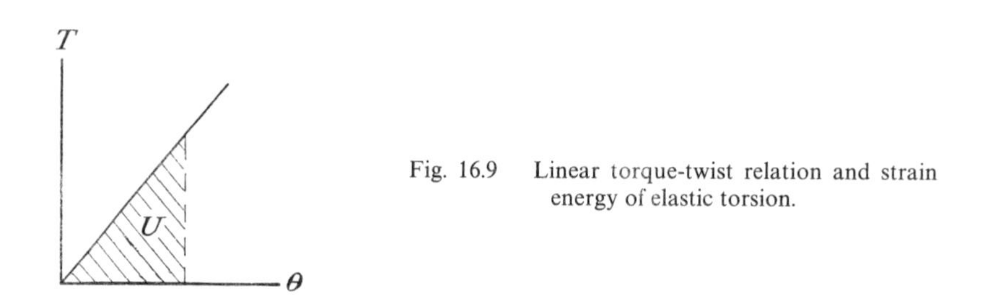

Fig. 16.9 Linear torque-twist relation and strain energy of elastic torsion.

the other end through an angle θ is the area under the $T - \theta$ relationship, Fig. 16.9. This work is conserved in the shaft as strain energy, which has the value

$$U = \tfrac{1}{2}T\theta \tag{16.19}$$

On using equation (16.11) we may eliminate either θ or T, and we have

$$U = \left(\frac{L}{2GJ}\right)T^2 = \left(\frac{GJ}{2L}\right)\theta^2 \tag{16.20}$$

16.8 Plastic torsion of a circular shaft

When a circular shaft is twisted the shearing stresses are greatest in the surface of the shaft. If the limit of proportionality of the material in shear is at a stress τ_Y, then this stress is first attained in the surface of the shaft at a torque

$$T = \frac{J\tau_Y}{a} \tag{16.21}$$

where J is the polar second moment of area, and a is the radius of the cross-section.

Suppose the material has the idealized shearing stress-strain curve shown in Fig. 16.10; behaviour is elastic up to a shearing stress τ_Y, the shearing modulus being G. Beyond the limit of proportionality shearing proceeds at a constant stress τ_Y. This behaviour is nearly true of mild steels with a well-defined yield point.

If we are dealing with a solid circular shaft, then after the onset of plasticity in the surface fibres the shearing stresses vary radially in the form shown in Fig. 16.11. The

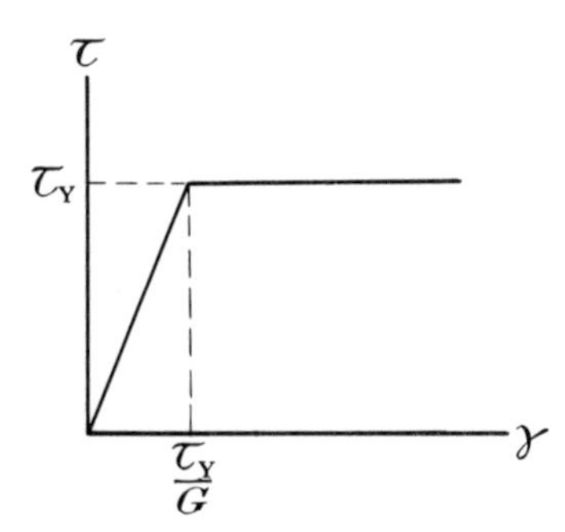

Fig. 16.10 Idealized shearing stress-strain curve of mild steel.

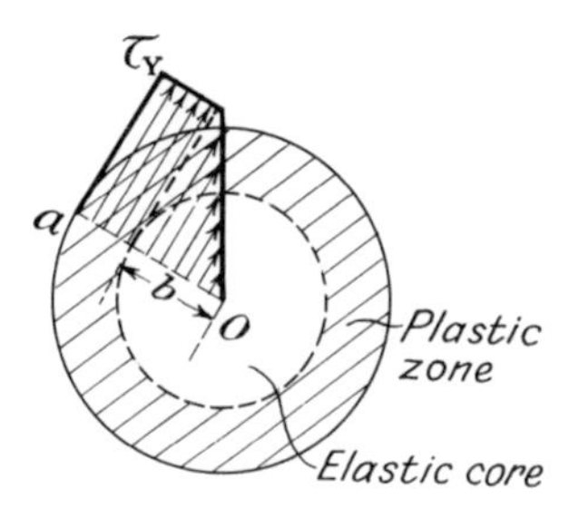

Fig. 16.11 Elastic-plastic torsion of a solid circular shaft.

material within a radius b is still elastic; the material beyond a radius b is plastic and is everywhere stressed to the yield stress τ_Y.

The torque sustained by the elastic core is

$$T_1 = \frac{J_1 \tau_Y}{b} = \frac{\pi}{2} b^3 \tau_Y \tag{16.22}$$

where subscripts 1 refer to the elastic core. The torque sustained by the outer plastic zone is

$$T_2 = \int_b^a 2\pi r^2 \tau_Y \, dr = \frac{2\pi}{3} \tau_Y [a^3 - b^3] \tag{16.23}$$

The total torque on the shaft is

$$\begin{aligned} T &= T_1 + T_2 = \pi \tau_Y (\tfrac{2}{3} a^3 - \tfrac{1}{6} b^3), \\ &= \frac{2\pi a^3}{3} \tau_Y \left[1 - \frac{b^3}{4a^3} \right] \end{aligned} \tag{16.24}$$

The angle of twist of the elastic core is

$$\theta = \frac{\tau_Y L}{Gb} \tag{16.25}$$

where L is the length of the shaft. We assume that the outer plastic region suffers the same angle of twist; this is tantamount to assuming that radii remain straight during plastic torsion of the shaft. Equation (16.25) gives

$$b = \frac{\tau_Y L}{G\theta} \tag{16.26}$$

Then the torque becomes

$$T = \tfrac{2}{3} \pi a^3 \tau_Y \left[1 - \frac{(\tau_Y/G)^3}{4(a/L)^3} \left(\frac{1}{\theta^3} \right) \right] \tag{16.27}$$

At the onset of plasticity

$$\theta = \frac{\tau_Y L}{Ga} = \theta_Y \quad \text{(say)} \tag{16.28}$$

Then, for any other condition of torsion,

$$\theta = \frac{\tau_Y L}{Gb} = \theta_Y \left(\frac{a}{b}\right) \tag{16.29}$$

which gives

$$\left(\frac{b}{a}\right) = \frac{\theta_Y}{\theta} \tag{16.30}$$

and equation (16.27) becomes

$$T = \frac{2\pi a^3}{3} \tau_Y \left[1 - \frac{1}{4}\left(\frac{\theta_Y}{\theta}\right)^3\right] \tag{16.31}$$

When θ becomes very large, T approaches the value

$$\frac{2\pi a^3}{3} \tau_Y = T_P, \quad \text{(say)} \tag{16.32}$$

which is the torque on the shaft when it is fully plastic. For smaller values of θ, we have then

$$\frac{T}{T_P} = 1 - \frac{1}{4}\left(\frac{\theta_Y}{\theta}\right)^3 \tag{16.33}$$

This relation, which is plotted in Fig. 16.12 for values of θ/θ_P up to 5, shows that the fully plastic torque T_Y is approached rapidly after the elastic limit is exceeded. The

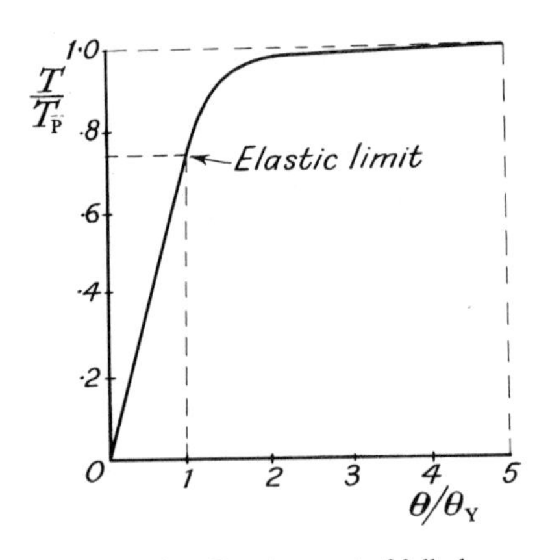

Fig. 16.12 Development of full plasticity in the torsion of a solid circular shaft.

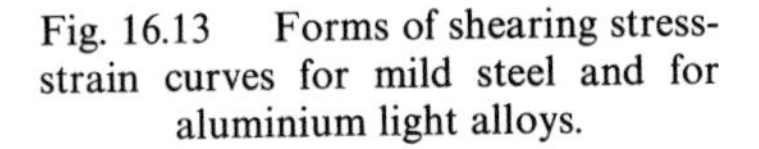

Fig. 16.13 Forms of shearing stress-strain curves for mild steel and for aluminium light alloys.

torque T_Y at the elastic limit is

$$T_Y = \tfrac{3}{4}T_P \tag{16.34}$$

If a torsion test is carried out on a thin-walled circular tube of mean radius r and thickness t, the average shearing stress due to a torque T is

$$\tau = \frac{T}{2\pi r^2 t}$$

from equation (16.2). If θ is the angle of twist of a length L of the tube, the shearing strain is

$$\gamma = \frac{r\theta}{L}$$

from equation (16.4). Thus, from a torsion test, in which values of T and θ are measured, the shearing stress τ and strain γ can be deduced. The resulting variation of τ with γ is called the *shearing stress-strain curve* of the material; the forms of these stress-strain curves are similar to tensile and compressive stress-strain curves, as shown in Fig. 16.13. In the elastic range of a material

$$\tau = G\gamma$$

where G is the *shearing modulus* of the material (§3.4). It is important to appreciate that the shearing stress-strain curve cannot be directly deduced from a torsion test of a *solid* circular bar, although the limit of proportionality can be estimated reasonably accurately.

16.9 Torsion of thin tubes of non-circular cross-section

In general the problem of the torsion of a shaft of non-circular cross-section is a complex one; in the particular case when the shaft is a hollow thin tube we can develop, however, a simple theory giving results which are sufficiently accurate for engineering purposes.

Consider a thin-walled closed tube of uniform section throughout its length. The thickness of the wall at any point is t, Fig. 16.14, although this may vary at points around

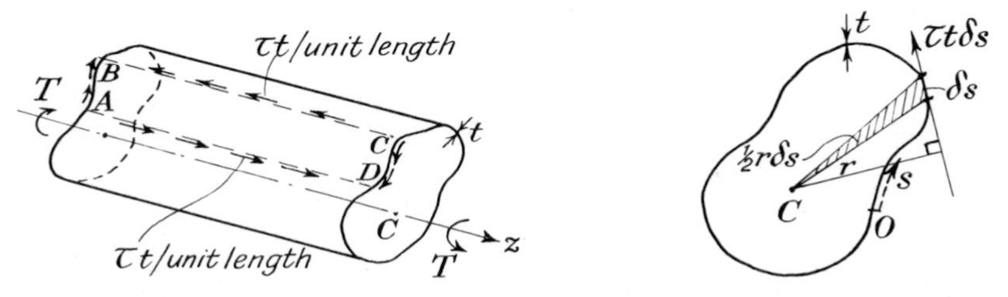

Fig. 16.14 Torsion of a thin-walled tube of any cross-section.

the circumference of the tube. Suppose torques T are applied at each end so that the tube twists about a longitudinal axis Cz. We assume that the torque T is distributed over the end of the tube in the form of shearing stresses which are parallel to the tangent to the wall at any point, Fig. 16.14, and that the ends of the tube are free from axial restraint. If the shearing stress at any point of the circumference is τ, then an equal complementary shearing stress is set up along the length of the tube. Consider the equilibrium of the section $ABCD$ of the wall: if the shearing stress τ at any point is uniform throughout the wall thickness then the shearing force transmitted over the edge BC is τt per unit length. For longitudinal equilibrium of $ABCD$ we must have that τt on BC is equal and opposite to τt on AD; but the section $ABCD$ is an arbitrary one, and we must have that τt is constant for all parts of the tube. Suppose this constant value of τt is

$$\tau t = q \tag{16.35}$$

This is called the *shear flow;* it has the units of a load per unit length of the circumference of the tube.

Suppose we measure a distance s round the tube from some point O on the circumference, Fig. 16.14. The force acting along the tangent to an element of length δs in the cross-section is $\tau t\, \delta s$. Suppose r is the length of the perpendicular from the centre of twist C onto the tangent. Then the moment of the force $\tau t\, \delta s$ about C is

$$\tau t r\, \delta s$$

The total torque on the cross-section of the tube is therefore

$$T = \oint \tau t r\, ds \tag{16.36}$$

where the integration is carried out over the whole of the circumference. But τt is constant and equal to q for all values of s. Then

$$T = \tau t \oint r\, ds = q \oint r\, ds \tag{16.37}$$

Now $\oint r\, ds$ is twice the area, A, enclosed by the centre line of the wall of the tube, and so

$$T = 2Aq \tag{16.38}$$

The shearing stress at any point is then

$$\tau = \frac{q}{t} = \frac{T}{2At} \tag{16.39}$$

To find the angle of twist of the tube we consider the strain energy stored in the tube, and equate this to the work done by the torques T in twisting the tube. When a material is subjected to shearing stresses τ the strain energy stored per unit volume of material is, from equation (3.5),

$$\frac{\tau^2}{2G}$$

where G is the shearing modulus of the material. In the tube the shearing stresses are varying around the circumference but not along the length of the tube. Then the strain energy stored in a longitudinal element of length L, width δs and thickness t is

$$\left(\frac{\tau^2}{2G}\right)Lt\,\delta s$$

The total strain energy stored in the tube is therefore

$$U = \oint \frac{\tau^2}{2G} \,.\, Lt\,ds \tag{16.40}$$

where the integration is carried out over the whole circumference of the tube. But τt is constant, and equal to q, and we may write

$$U = \oint \frac{\tau^2}{2G} \,.\, Lt^2 \,.\, \frac{ds}{t} = \frac{q^2 L}{2G}\oint\frac{ds}{t} \tag{16.41}$$

If the ends of the tube twist relative to each other by an angle θ, then the work done by the torques T is

$$W = \tfrac{1}{2}T\theta \tag{16.42}$$

On equating U and W, we have

$$\theta = \frac{q^2 L}{GT}\oint\frac{ds}{t} \tag{16.43}$$

But from equation (16.38) we have

$$q = \frac{T}{2A} \tag{16.44}$$

Then equation (16.43) may be written

$$\theta = \frac{TL}{4A^2G}\oint\frac{ds}{t} \tag{16.45}$$

For a tube of uniform thickness t,

$$\theta = \frac{TL}{4A^2G}\left(\frac{S}{t}\right) \tag{16.46}$$

where S is the total circumference of the tube.

Equation (16.45) can be written in the form

$$\theta = \frac{TL}{GJ}$$

where

$$J = \frac{4A^2}{\oint\frac{ds}{t}}$$

J is the *torsion constant* for the section; for circular cross-sections J is equal to the polar second moment of area, but this is not true in general.

16.10 Torsion of a flat rectangular strip

A long flat strip of rectangular cross-section has a breadth b, thickness t, and length L. For uniform torsion about the centroid of the cross-section, the strip may be treated as a set of concentric thin hollow tubes, all twisted by the same amount. Consider such an elemental tube which is rectangular in shape the longer sides being a distance y from the central axis of the strip; the thickness of the tube is δy, Fig. 16.15. If δT is

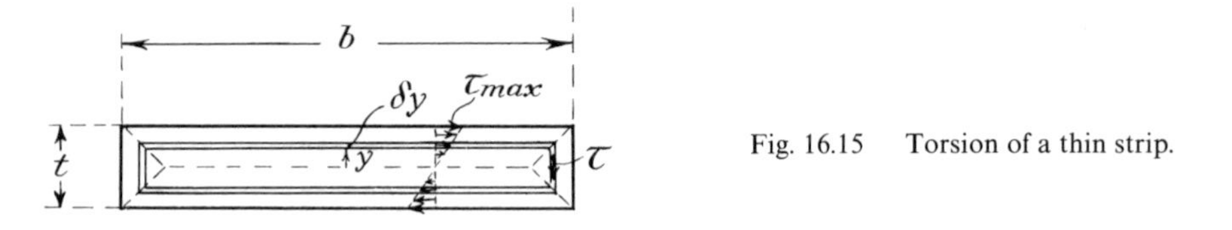

Fig. 16.15 Torsion of a thin strip.

the torque carried by this elemental tube then the shearing stress in the longer sides of the tube is, from equation (16.39),

$$\tau = \frac{\delta T}{4by\,\delta y} \tag{16.47}$$

where b is assumed very much greater than t. This relation gives

$$\frac{dT}{dy} = 4by\tau \tag{16.48}$$

For the angle of twist of the elemental tube we have, from equation (16.46),

$$\theta = \frac{2bL\,\delta T}{16b^2y^2G\,\delta y} \tag{16.49}$$

where L is the length of the strip. This gives the further relation

$$\frac{dT}{dy} = 8by^2G\frac{\theta}{L} \tag{16.50}$$

On comparing equations (16.48) and (16.50), we have

$$\tau = 2yG\left(\frac{\theta}{L}\right) \tag{16.51}$$

This shows that the shearing stress τ varies linearly throughout the thickness of the strip, having a maximum value in the surface of

$$\tau_{max} = Gt\left(\frac{\theta}{L}\right) \tag{16.52}$$

An important feature is that the shearing stresses τ act parallel to the longer side b of the strip, and that their directions reverse over the thickness of the strip. This approximate solution gives an inexact picture of the shearing stresses near the corners of the cross-section. We ought to consider not rectangular elemental tubes but flat tubes with

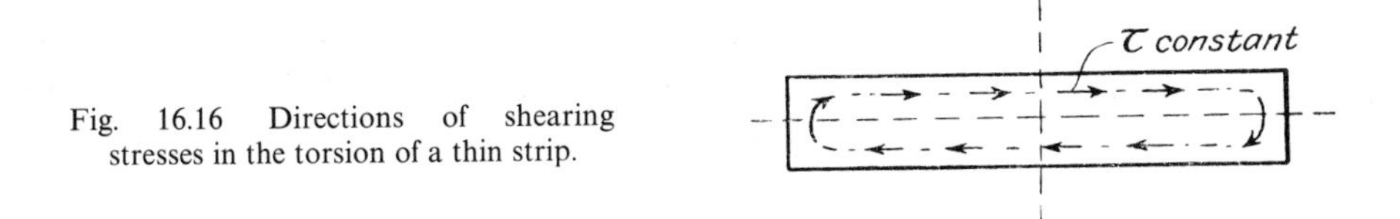

Fig. 16.16 Directions of shearing stresses in the torsion of a thin strip.

curved ends. The contours of constant shearing stress are then continuous curves, Fig. 16.16.

The total torque on the cross-section is

$$T = \int_0^{\frac{1}{2}t} 8by^2 G\left(\frac{\theta}{L}\right) dy = \tfrac{1}{3}bt^3 G \frac{\theta}{L} \tag{16.53}$$

The polar second moment of area of the cross-section about its centre is

$$J = \tfrac{1}{12}(bt^3 + b^3t) \tag{16.54}$$

If b is very much greater than t, then, approximately,

$$J = \tfrac{1}{12}b^3t \tag{16.55}$$

The geometrical constant occurring in equation (16.53) is $\frac{1}{3}bt^3$; thus, in the torsion of a thin strip we cannot use the polar second moment of area for J in the relation

$$\frac{T}{J} = \frac{G\theta}{L} \tag{16.56}$$

Instead we must use

$$J = \tfrac{1}{3}bt^3 \tag{16.57}$$

16.11 Torsion of thin-walled open sections

We may extend the analysis of the preceding section to the uniform torsion of thin-walled open-sections of any cross-sectional form. In the angle section of Fig. 16.17, we take elemental tubes inside the two limbs of the section. If t_1 and t_2 are small compared with b_1 and b_2, the maximum shearing stresses in limbs 1 and 2 are

$$\tau_1 = Gt_1\left(\frac{\theta}{L}\right) \qquad \tau_2 = Gt_2\left(\frac{\theta}{L}\right) \tag{16.58}$$

where the angle of twist per unit length, θ/L, is common to both limbs. The greatest shearing stress occurs then in the surface of the thicker limb of the cross-section. The

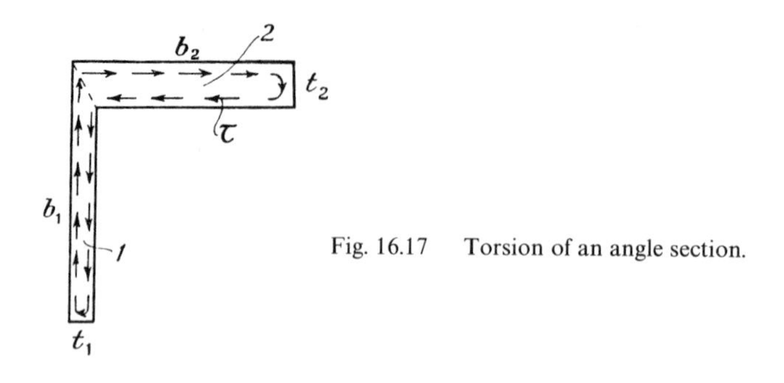

Fig. 16.17 Torsion of an angle section.

total torque is the summation of the torques carried by the two limbs, and has the value

$$T = \tfrac{1}{3}(b_1t_1^3 + b_2t_2^3)G\left(\frac{\theta}{L}\right) \tag{16.59}$$

In general, for a thin-walled open-section of any shape the shearing stress in the surface of a section of thickness t is

$$\tau = Gt\left(\frac{\theta}{L}\right) \tag{16.60}$$

The total torque on the section is

$$T = G\left(\frac{\theta}{L}\right)\sum \tfrac{1}{3}bt^3 \tag{16.61}$$

where the summation is carried out for all limbs of the cross-section.

FURTHER PROBLEMS

(answers on page 398)

16.7 Find the maximum shearing stress in a propeller shaft 40 cm external, and 20 cm internal diameter, when subjected to a torque of 450 kNm. If $G = 80$ GN/m^2, what is the angle of twist in a length of 20 diameters? What diameter would be required for a solid shaft with the same maximum stress and torque? *(RNC)*

16.8 A propeller shaft, 45 m long, transmits 10 MW at 80 rev/min. The external diameter of the shaft is 57 cm, and the internal diameter 24 cm. Assuming that the maximum torque is 1·19 times the mean torque, find the maximum shearing stress produced. Find also the relative angular movement of the ends of the shaft when transmitting the average torque. Take $G = 80$ GN/m^2. *(RNC)*

16.9 A steel tube, 3 m long, 3·75 cm diameter, 0·06 cm thick, is twisted by a couple of 50 Nm. Find the maximum shearing stress, the maximum tensile stress, and the angle through which the tube twists. Take $G = 80$ GN/m^2 *(Cambridge)*

16.10 Compare the mass of a solid shaft with that of a hollow one to transmit a given power at a given speed with a given maximum shearing stress, the inside diameter of the hollow shaft being two-thirds of the outside diameter. *(Cambridge)*

16.11 A 2·5 cm circular steel shaft is provided with enlarged portions A and B. On to this enlarged portion a steel tube 0·125 cm thick is shrunk. While the shrinking processs is going on, the 2·5 cm shaft is held twisted

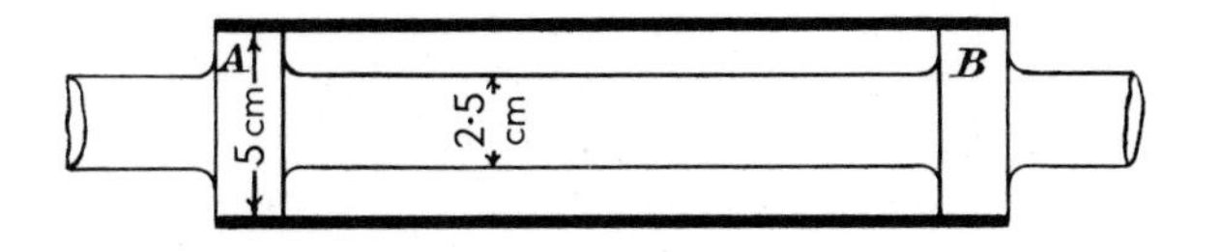

by a couple of magnitude 50 Nm. When the tube is firmly set on the shaft this twisting couple is removed. Calculate what twisting couple is left on the shaft, the shaft and tube being made of the same material.

(*Cambridge*)

16.12 A thin tube of mean diameter 2·5 cm and thickness 0·125 cm is subjected to a pull of 7·5 kN, and an axial twisting moment of 125 Nm. Find the magnitude and direction of the principal stresses. (*Cambridge*)

17 The principle of virtual work and its applications

17.1 Introduction

The analysis of many stress and strain problems can be simplified by making use of a special form of the principle of virtual work. We can also use to advantage methods based on strain energy and complementary energy. We begin with a discussion of pin-jointed frames, and extend the concepts later to problems of beams.

17.2 The principle of virtual work

In its simplest form the *principle of virtual work* is that

> *For a system of forces acting on a particle, the particle is in statical equilibrium if, when it is given any virtual displacement, the net work done by the forces is zero.*

A virtual displacement is any arbitrary displacement of the particle. In the virtual displacement the forces are assumed to remain constant and parallel to their original lines of action. Consider a particle under the action of three forces F_1, F_2, F_3, Fig. 17.1.

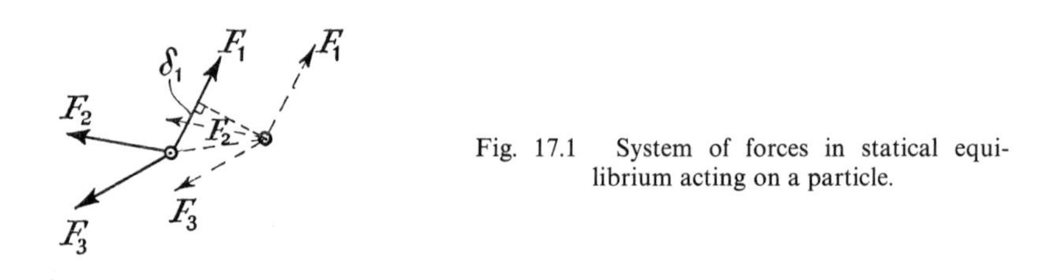

Fig. 17.1 System of forces in statical equilibrium acting on a particle.

Imagine the particle to be given a virtual displacement of any magnitude in any direction. Suppose the displacements of the particle along the lines of action of the forces F_1, F_2, F_3, are δ_1, δ_2, δ_3, respectively; these are known as *corresponding* displacements. Then the forces form a system in statical equilibrium if

$$F_1\delta_1 + F_2\delta_2 + F_3\delta_3 = 0 \tag{17.1}$$

On the basis of the principle of virtual work we can show that the resultant of the forces acting on a particle in statical equilibrium is zero. Suppose the forces F_1, F_2, F_3, acting on the particle of Fig. 17.1, have a resultant of magnitude R in some direction; then by giving the particle a suitable virtual displacement, Δ, say, in the direction of R, the net work done is

$$R\Delta$$

But by the principle of virtual work the net work is zero, so that

$$R\Delta = 0 \tag{17.2}$$

Since Δ can be non-zero, R must be zero. Hence, by adopting the principle of virtual work as a basic concept, we can show that the resultant of a system of forces in statical equilibrium is zero.

17.3 The displacements of a pin-jointed frame

The principle of virtual work can be adapted easily to study the displacements of a statically determinate pin-jointed frame. In Chapter 2 we found the displacements by a graphical method; it is frequently more useful to make a direct analytical approach.

Consider the plane pin-jointed frame of Fig. 17.2, which carries external loads W_1 and W_2 at joints B and C. If there are no gross distortions of the frame, the axial forces

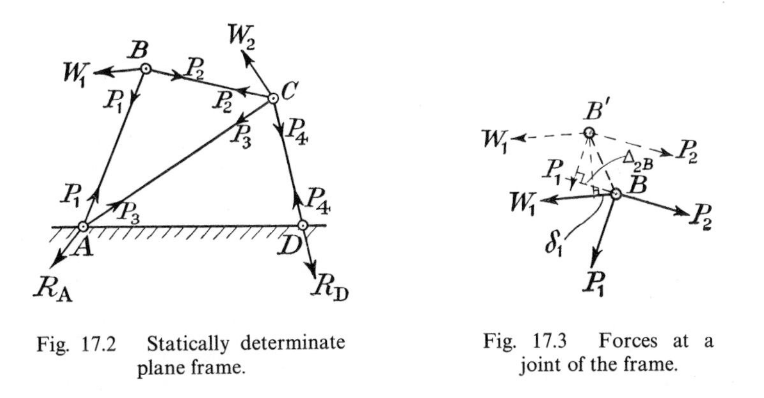

Fig. 17.2 Statically determinate plane frame.

Fig. 17.3 Forces at a joint of the frame.

in the bars are found by considering statical equilibrium at each joint. Suppose the *tensile* forces in the members are given by P_1, P_2, P_3 and P_4. Now consider the internal and external forces acting at joint B; clearly, P_1, P_2 and W_1 form an equilibrium system. Suppose the system is given a virtual displacement from B to B', Fig. 17.3; the displacement of W_1 along its line of action is δ_1, and the displacements of P_1 and P_2 in directions *opposite* to their lines of action are Δ_{1B} and Δ_{2B}, respectively. Then by the principle of virtual work we have

$$W_1\delta_1 - P_1\Delta_{1B} - P_2\Delta_{2B} = 0$$

Then

$$W_1\delta_1 = P_1\Delta_{1B} + P_2\Delta_{2B} \tag{17.3}$$

Now suppose that virtual displacements are also given to all other joints of the frame, including the foundations A and D. For the joint C, we have

$$W_2\delta_2 = P_2\Delta_{2C} + P_3\Delta_{2C} + P_4\Delta_{4C} \tag{17.4}$$

where δ_2 is the corresponding virtual displacement of C, and $\Delta_{2C}, \Delta_{3C}, \Delta_{4C}$ are the corresponding displacements of P_2, P_3, P_4, respectively, in directions opposite to their lines of action. If R_A and R_D are the reactions at A and D, respectively, we have similarly for these joints

$$R_A\delta_A = P_1\Delta_{1A} + P_3\Delta_{3A} \tag{17.5}$$

$$R_D\delta_B = P_4\Delta_{4D} \tag{17.6}$$

On adding equations (17.3), (17.4), (17.5) and (17.6), we have

$$W_1\delta_1 + W_2\delta_2 + R_A\delta_A + R_D\delta_D = P_1(\Delta_{1B} + \Delta_{1A}) + P_2(\Delta_{2B} + \Delta_{2C}) + P_3(\Delta_{3C} + \Delta_{3A}) + P_4(\Delta_{4C} + \Delta_{4D}) \tag{17.7}$$

The left-hand side of this equation represents the work done by the external forces on the frame, including the reactions at the supports A and D, during virtual displacements

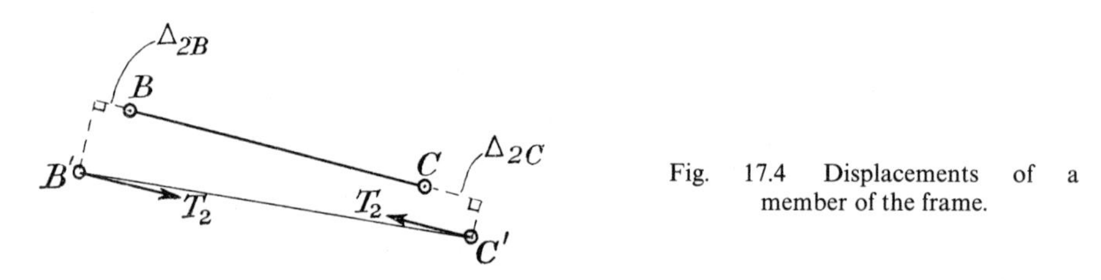

Fig. 17.4 Displacements of a member of the frame.

of the joints. The right-hand side of the equation is the work done against the internal tensions in the frame; if we consider the member BC, Fig. 17.4, we have

$$(\Delta_{2B} + \Delta_{2C})$$

is the extension of the member BC implied by the virtual displacements of B and C, provided these displacements are small. If

$$e_2 = \Delta_{2B} + \Delta_{2C}$$

then equation (17.7) is written in the form

$$\sum_j W\delta = \sum_m Pe \tag{17.8}$$

where the summation on the left-hand side is carried out for all joints, and that on the right-hand side for all members of the frame. In equation (17.8), the external forces, such as W, are in statical equilibrium with the internal tensions, such as P; the virtual extensions of the members, such as e, are compatible with the virtual displacements, such as δ, of the joints.

We have finally, therefore,

> *For a pin-jointed frame, in a condition of statical equilibrium, the work of the external forces in any set of small virtual displacements of the joints is equal to the work done against the internal forces in the frame in a compatible set of extensions of the bars.*

In this form, the principle of virtual work can be used to find the displacements of a structure. Consider first the pin-jointed frame shown in Fig. 17.5; the displacements of the joints of this frame have already been found in Problem 2.2 using a displacement diagram. This frame is statically-determinate; the tensile forces in the bars due to a vertical load W at D are shown in Fig. 17.5(i). Each bar has a cross-sectional area a, and all the bars are made from the same material of Young's modulus E; if the bars

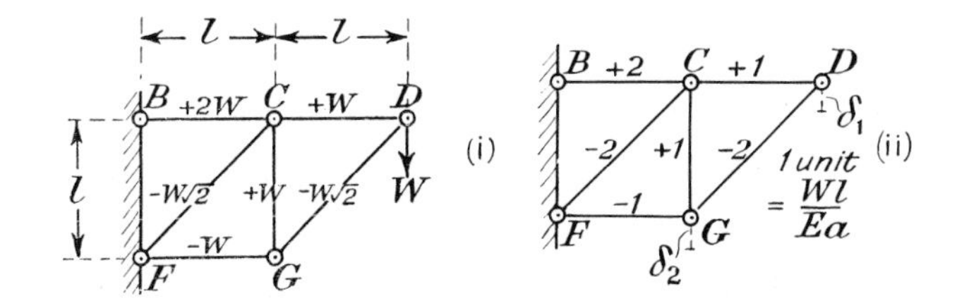

Fig. 17.5 Displacements of a plane frame. (i) A system of external and internal forces in statical equilibrium. (ii) A system of extensions compatible with displacements of the joints.

remain elastic, the extensions due to the load W at D can be found; these extensions are shown in Fig. 17.5(ii). Now the external and internal forces of Fig. 17.5(i) form a system of forces in statical equilibrium, and the extensions of the members in Fig. 17.5(ii) are compatible with a vertical displacement δ_1, say, of joint D. Then the principle of virtual work, in the form of equation (17.8), gives

$$\sum_j W\delta = \sum_m Pe$$

or

$$W\delta_1 = \frac{Wl}{Ea}[(2 \times 2W) + (1 \times W) + (-1 \times -W) + (-2 \times -W\sqrt{2}) + (1 \times W) + (-2 \times -W\sqrt{2})]$$

Then

$$\delta_1 = (7 + 4\sqrt{2})\frac{Wl}{Ea}$$

which is the vertical deflection of the joint D due to the vertical load W at D.

If we wish to find the displacement of an unloaded joint of the frame, we proceed as follows: suppose we wish to find the vertical displacement of the joint G due to a vertical load W at D; we require, therefore, the displacement δ_2 in Fig. 17.5(ii); consider the frame with a unit vertical load at G, as in Fig. 17.6; the internal forces for this

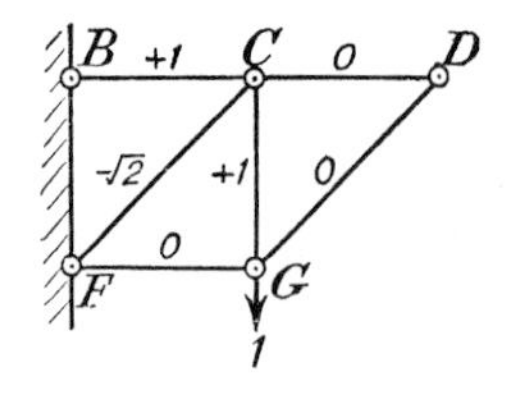

Fig. 17.6 Displacement of an unloaded joint; a system of external and internal forces in statical equilibrium.

condition of loading are shown in Fig. 17.6. The principle of virtual work, applied to the forces of Fig. 17.6 and the extensions and displacements of Fig. 17.5(ii), gives

$$1 \times \delta_2 = \frac{Wl}{Ea}[(2 \times 1) + (-2 \times -\sqrt{2}) + (1 \times 1)]$$

Then

$$\delta_2 = (3 + 2\sqrt{2})\frac{Wl}{Ea}$$

In our discussion of the use of virtual work, no restrictions were imposed on the load-extension characteristics of the bars of a pin-jointed frame; the principle of virtual work can be applied to any type of structure, provided the displacements are small enough

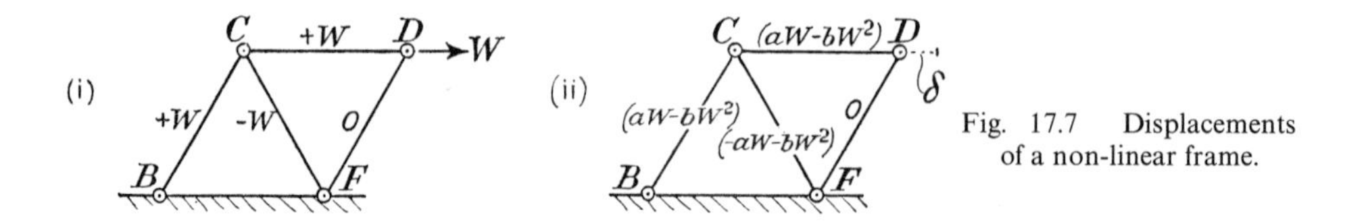

Fig. 17.7 Displacements of a non-linear frame.

to have a negligible effect on the shape of the structure. The frame shown in Fig. 17.7 has already been discussed in Problem 2.3; the members of the frame are non-linear, the extension of a member under a tensile force P being given by

$$e = aP - bP^2$$

The forces in the members due to a horizontal load W at D are shown in Fig. 17.7(i); the extensions of the members can then be found; these are shown in Fig. 17.7(ii). If δ is the horizontal displacement of the joint D, then by the principle of virtual work, we have

$$W\delta = 2W(aW - bW^2) + W(aW + bW^2)$$

Then

$$\delta = 3aW - bW^2$$

Problem 17.1: A pin-jointed frame is hinged at B and supported on a roller at D. Estimate the horizontal deflection of the joint C due to a vertical load W at F. All members of the frame are of the same length l, the same Young's modulus E, and the same cross-sectional area A.

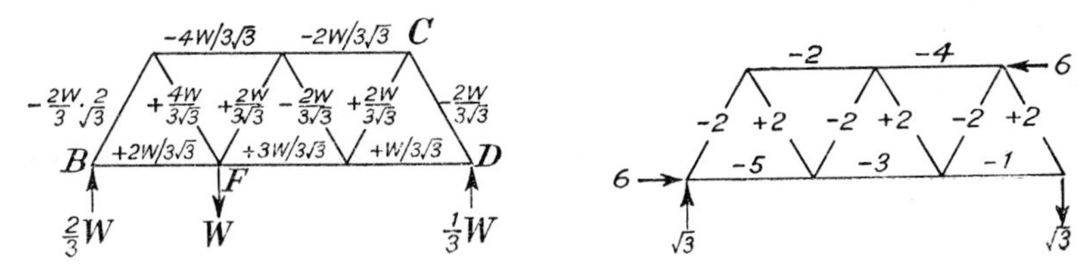

Solution

The loads in the members due to W are proportional to the extensions of the members. A load of 6 units applied horizontally at C gives a convenient system of forces in equilibrium. We then have

$$6 \times \delta = \frac{Wl}{EA}\cdot\frac{1}{3\sqrt{3}} \quad [\text{illegible}]\; 8 - 10 - 9 - 1 + 8 + 8 - 4 - 4 - 4 - 4]$$

Then

$$\delta = -\frac{2Wl}{9EA\sqrt{3}}$$

Problem 17.2: In a pin-jointed frame the four horizontal members are made of a material of coefficient of linear expansion α_1. The inclined and vertical members are all made of the same material of coefficient of linear expansion α_2. Estimate the vertical deflection of the joint D due to a temperature rise θ.

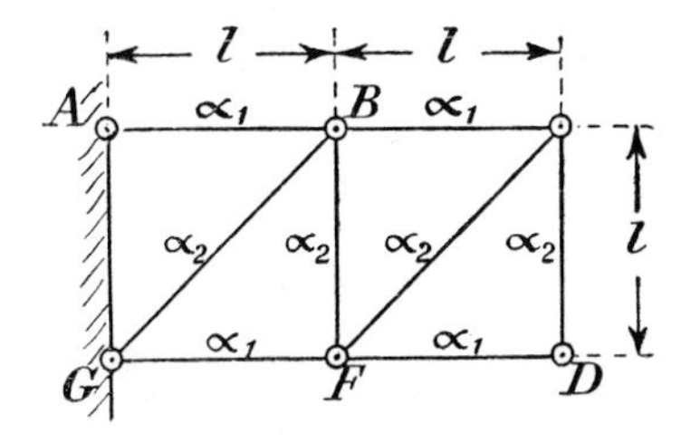

Solution

The system of extensions and displacements arises in this case from the temperature increase θ. We evaluate also the internal forces in the frame due to a unit vertical load at D. By the principle of virtual work

$$1 \times \delta = 2\alpha_1 l\theta - 2\alpha_2 l\theta$$

Then

$$\delta = 2l\theta(\alpha_1 - \alpha_2)$$

There is no vertical deflection of D if $\alpha_1 = \alpha_2$.

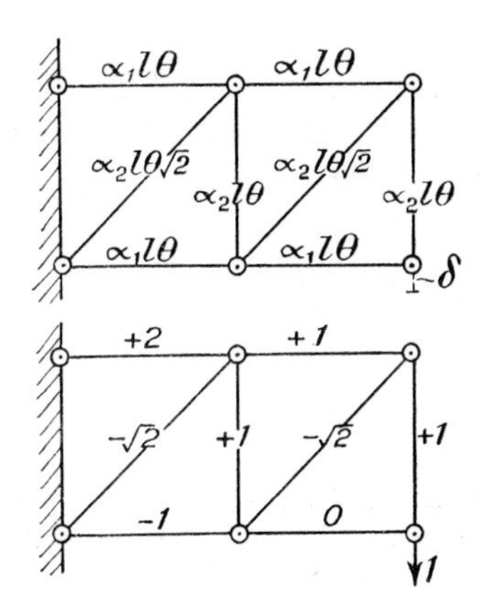

17.4 Statically indeterminate pin-pointed frames

In the modified form developed in §17.3, the principle of virtual work may be used to solve problems of statically indeterminate frames. The frame shown in Fig. 17.8 has one redundant member; any of the members may be treated as redundant, and we consider the member AB. Suppose the tensile force in this member is R, when the frame carries some external loads. In Fig. 17.8(ii) the member AB has been replaced by two equal and opposite forces R; the internal forces in the frame are now calculable in terms of R and the external loads, if the joints suffer no gross distortions. Suppose the tensile force in a typical member, such as CD, is P; the resulting extension of the member CD

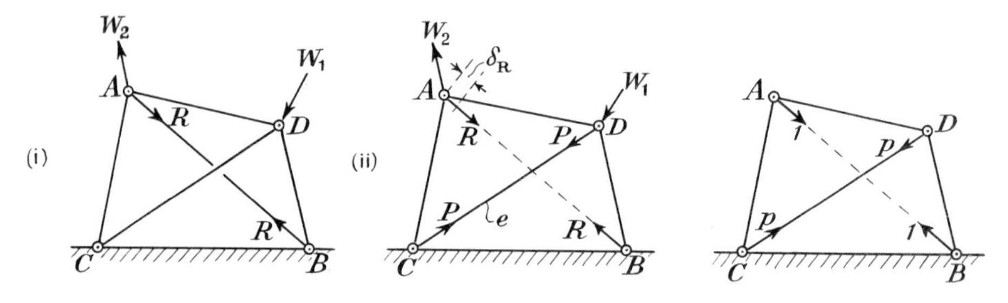

Fig. 17.8 Solution of a statically indeterminate frame.

Fig. 17.9 Internal forces due to equal and opposite unit loads.

can then be found, and is equal to e, say; both P and e are functions of R and the external loads. In Fig. 17.8(ii) δ_R is the displacement of the load R at A along its line of action; δ_R can be found by considering the virtual work of the system of forces in Fig. 17.9 on the extensions and displacements of Fig. 17.8(ii); in Fig. 17.9 unit loads are applied between the joints A and B, and the resulting tension in a typical member, such as CD, is p. Then

$$1 \times \delta_R = \sum_m pe \qquad (17.9)$$

where the summation is carried out for all members of the frame. Now, the tensile force R applied to the member AB produces an extension of e_{AB}, say, of that member. If the members of the frame are all unstressed before the external loads are applied, the strain compatibility condition at joint A is

$$e_{AB} = -\delta_R$$

Then equation (17.9) gives

$$e_{AB} + \sum_m pe = 0 \qquad (17.10)$$

Now e_{AB} and e are functions of R and the external loads, and so equation (17.10) yields the value of R.

If the member AB is not initially of the exact length to fit into the frame, stresses are set up by this initial lack of fit. Suppose AB is a length λ *longer* than it should be for an exact fit in the unloaded frame. Then the strain compatible condition becomes

$$e_{AB} + \lambda = -\delta_R$$

Then

$$(\lambda + e_{AB}) + \sum_m pe = 0 \tag{17.11}$$

Problem 17.3: In a pin-jointed frame all the members remain elastic; all members have the same Young's modulus E and cross-sectional area a. If there are no initial lacks of fit of the members find the force in the member AC.

Solution

Suppose R is the tensile force in the redundant member AC. The axial forces in the members are then calculable in terms of W and R, and the extensions of the members can also be found. Equal and opposite unit loads

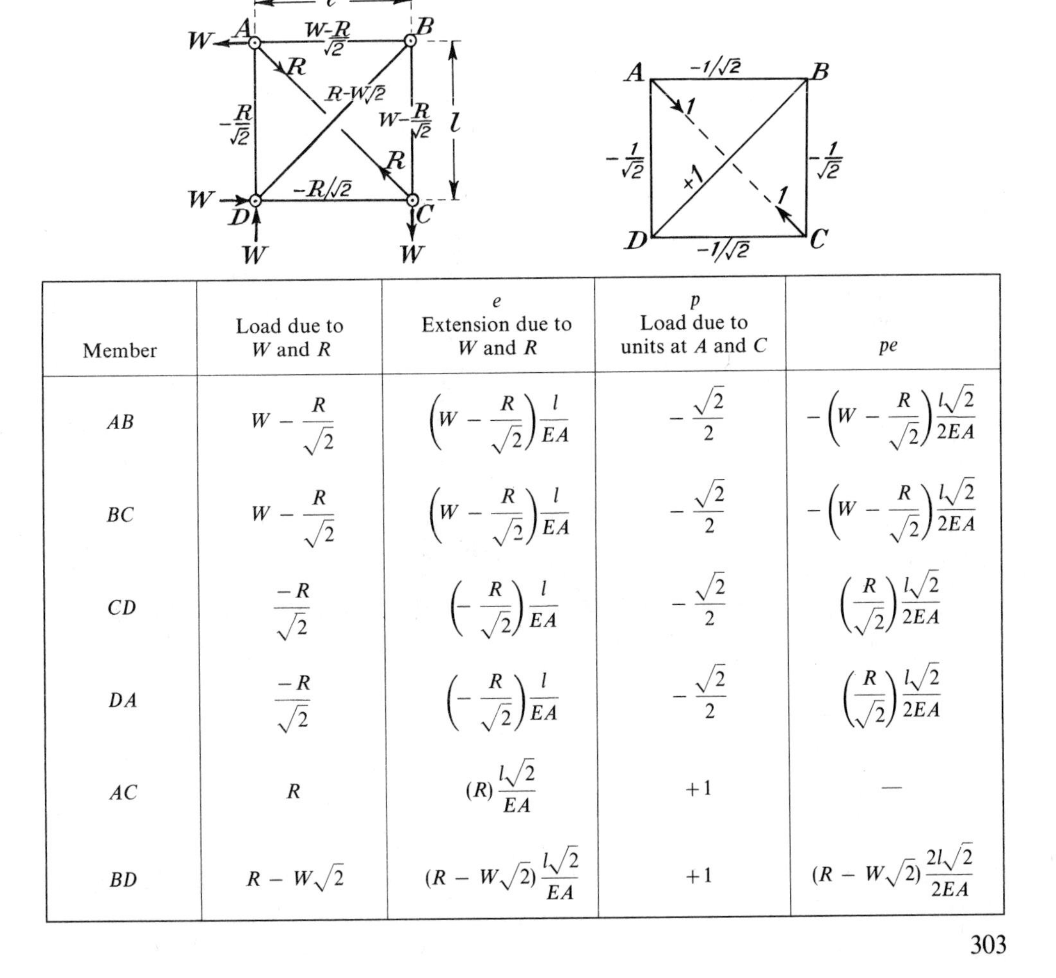

Member	Load due to W and R	e Extension due to W and R	p Load due to units at A and C	pe
AB	$W - \dfrac{R}{\sqrt{2}}$	$\left(W - \dfrac{R}{\sqrt{2}}\right)\dfrac{l}{EA}$	$-\dfrac{\sqrt{2}}{2}$	$-\left(W - \dfrac{R}{\sqrt{2}}\right)\dfrac{l\sqrt{2}}{2EA}$
BC	$W - \dfrac{R}{\sqrt{2}}$	$\left(W - \dfrac{R}{\sqrt{2}}\right)\dfrac{l}{EA}$	$-\dfrac{\sqrt{2}}{2}$	$-\left(W - \dfrac{R}{\sqrt{2}}\right)\dfrac{l\sqrt{2}}{2EA}$
CD	$\dfrac{-R}{\sqrt{2}}$	$\left(-\dfrac{R}{\sqrt{2}}\right)\dfrac{l}{EA}$	$-\dfrac{\sqrt{2}}{2}$	$\left(\dfrac{R}{\sqrt{2}}\right)\dfrac{l\sqrt{2}}{2EA}$
DA	$\dfrac{-R}{\sqrt{2}}$	$\left(-\dfrac{R}{\sqrt{2}}\right)\dfrac{l}{EA}$	$-\dfrac{\sqrt{2}}{2}$	$\left(\dfrac{R}{\sqrt{2}}\right)\dfrac{l\sqrt{2}}{2EA}$
AC	R	$(R)\dfrac{l\sqrt{2}}{EA}$	$+1$	—
BD	$R - W\sqrt{2}$	$(R - W\sqrt{2})\dfrac{l\sqrt{2}}{EA}$	$+1$	$(R - W\sqrt{2})\dfrac{2l\sqrt{2}}{2EA}$

are then applied at A and C. From equation (17.10)

$$R\sqrt{2}\frac{l}{EA} = \frac{l}{2EA}\left[2\sqrt{2}\left(W - \frac{R}{\sqrt{2}}\right) - 2\sqrt{2}\left(\frac{R}{\sqrt{2}}\right) - 2\sqrt{2}(R - W\sqrt{2})\right]$$

Then

$$R = \tfrac{1}{2}\sqrt{2}W$$

Problem 17.4: In Problem 17.3 the member AC is initially too long by an amount $\alpha l/EA$. Estimate the force in the member AC.

Solution

From equation (17.11) we have

$$\frac{\alpha l}{EA} + R\sqrt{2}\,.\frac{l}{EA} = \frac{l}{2EA}[2W(1 + \sqrt{2}) - 2R(2 + \sqrt{2})]$$

Thus

$$R = \frac{W\sqrt{2}}{2} - \frac{\alpha}{2}(\sqrt{2} - 1)$$

Problem 17.5: In Problem 17.3 the members are elastic but behave non-linearly; the extension e of any member is given by

$$e = aP - bP^2$$

If there is no initial lack of fit of the members find the force in the member AC.

Solution

The extensions of the members are no longer proportional to the tensile loads. We draw up a new table:—

Member	Load due to W and R	e Extension due to W and R	p Load due to 1 unit loads at A and C	pe
AB	$W - \frac{R}{\sqrt{2}}$	$a\left(W - \frac{R}{\sqrt{2}}\right) - b\left(W - \frac{R}{\sqrt{2}}\right)^2$	$-\frac{1}{\sqrt{2}}$	$-\frac{1}{\sqrt{2}}\left[a\left(W - \frac{R}{\sqrt{2}}\right) - b\left(W - \frac{R}{\sqrt{2}}\right)^2\right]$
BC	$W - \frac{R}{\sqrt{2}}$	$a\left(W - \frac{R}{\sqrt{2}}\right) - b\left(W - \frac{R}{\sqrt{2}}\right)^2$	$-\frac{1}{\sqrt{2}}$	$-\frac{1}{\sqrt{2}}\left[a\left(W - \frac{R}{\sqrt{2}}\right) - b\left(W - \frac{R}{\sqrt{2}}\right)^2\right]$
CD	$\frac{-R}{\sqrt{2}}$	$a\left(\frac{-R}{\sqrt{2}}\right) - b\left(\frac{-R}{\sqrt{2}}\right)^2$	$-\frac{1}{\sqrt{2}}$	$-\frac{1}{\sqrt{2}}\left[a\left(-\frac{R}{\sqrt{2}}\right) - b\left(-\frac{R}{\sqrt{2}}\right)^2\right]$
DA	$\frac{-R}{\sqrt{2}}$	$a\left(\frac{-R}{\sqrt{2}}\right) - b\left(\frac{-R}{\sqrt{2}}\right)^2$	$-\frac{1}{\sqrt{2}}$	$-\frac{1}{\sqrt{2}}\left[a\left(-\frac{R}{\sqrt{2}}\right) - b\left(-\frac{R}{\sqrt{2}}\right)^2\right]$
AC	R	$aR - bR^2$	1	—
BD	$R - W\sqrt{2}$	$a(R - W\sqrt{2}) - b(R - W\sqrt{2})^2$	1	$a(R - W\sqrt{2}) - b(R - W\sqrt{2})^2$

Equation (17.10) gives

$$aR - bR^2 = \frac{2}{\sqrt{2}}\left[a\left(W - \frac{R}{\sqrt{2}}\right) - b\left(W - \frac{R}{\sqrt{2}}\right)^2\right]$$
$$+ \frac{2}{\sqrt{2}}\left[a\left(-\frac{R}{\sqrt{2}}\right) - b\left(\frac{R}{\sqrt{2}}\right)^2\right] + [a(R - W\sqrt{2}) - b(R - W\sqrt{2})^2]$$

Then R is the solution of the quadratic equation

$$bR^2 + [a\sqrt{2} - Wb(2 + \sqrt{2})]R + \frac{\sqrt{2}}{2}bW^2 = 0$$

17.5 Temperature stresses in redundant frames

In a statically determinate frame small changes in the lengths of the members give rise to no internal stresses in the members of the frame. Any temperature changes, giving rise to changes in length of the members, do not produce internal stresses, therefore.

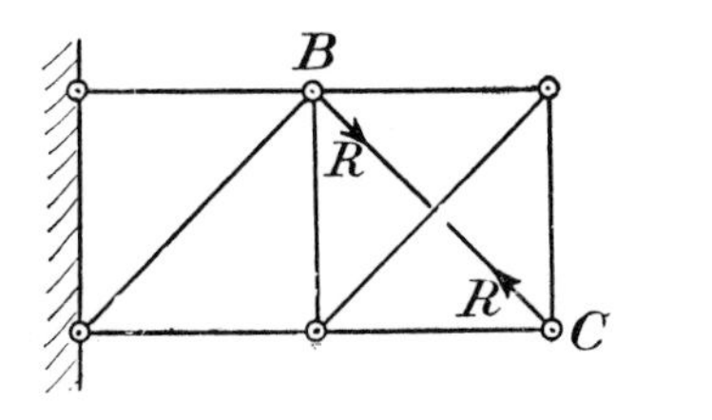

Fig. 17.10 Temperature stresses in a redundant frame.

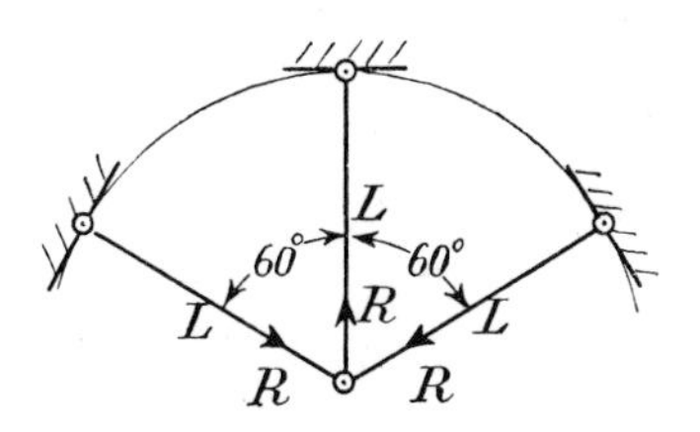

Fig. 17.11 Example of temperature stresses.

This is not true necessarily of redundant frames; consider, for example, the statically indeterminate pin-jointed frame shown in Fig. 17.10, in which the members fit without internal stresses at some initial temperature conditions. Suppose the temperatures of the members are raised, the distance between the pin-joints at the supports remaining unchanged. The tensile load induced in the redundant member BC is R; the tensile load induced in any other member may be written kR, where k is constant for any member, since the frame has only one redundant member. Suppose any member extends an amount e_θ due to temperature change alone, and an amount e due to its tensile load. Then the extensions $(e_\theta + e)$ of the members are compatible with each other, while the internal forces are in equilibrium with each other. Thus, by the principle of virtual work

$$\sum kR(e_\theta + e) + R(e_\theta + e)_{BC} = 0 \qquad (17.12)$$

As an example, consider the pin-jointed frame shown in Fig. 17.11; each member has a cross-sectional area A, Young's modulus E, and coefficient of linear expansion α; initially there are no temperature stresses. If there is a general temperature rise of θ, the expansion of each member due to θ alone is $\alpha L\theta$. If the force set up in the central member is R, its extension due to R alone is RL/EA; the forces set up in the inclined members are $(-R)$, and their resulting extensions are $(-RL/EA)$. By equation (17.12)

$$R\left(\alpha L\theta + \frac{RL}{EA}\right) + 2(-R)\left(\alpha L\theta - \frac{RL}{EA}\right) = 0$$

Hence

$$R = \tfrac{1}{3}EA\alpha\theta$$

17.6 Deflections of beams

In a pin-jointed frame subjected to loads applied to the joints only the tensile load in any member is constant throughout the length of that member. In the case of a beam

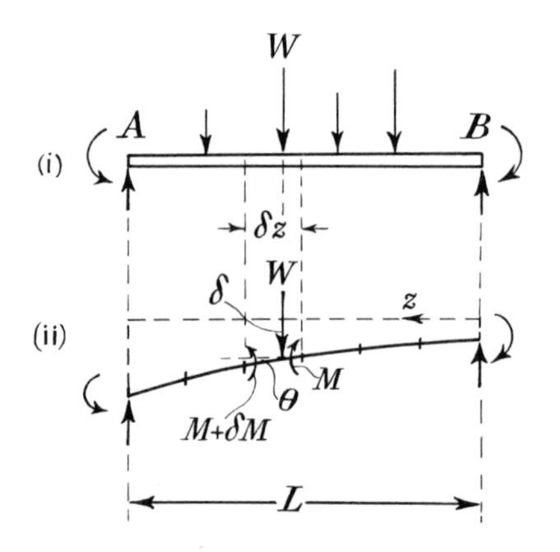

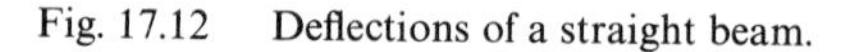
Fig. 17.12 Deflections of a straight beam.

under lateral loads the bending moments and shearing forces may vary from one section to another, so that the state of stress is not uniform along the length of the beam. In applying the principle of virtual work to problems of beams we must consider the loading actions on and the virtual displacements of an elemental length of the beam.

Consider a straight beam AB, Fig. 17.12, which is in statical equilibrium under the action of a system of external forces and couples. The beam is divided into a number of short lengths; the loading actions on a short length such as δz consist of bending moments M and $(M + \delta M)$, an external lateral load W, and lateral shearing forces at the ends of the short length. Now suppose the short lengths of the beam are given small virtual displacements, θ. If the elements remain connected to each other, then for given values of θ the external forces, such as W, suffer certain displacements, such as δ. Then the values of θ and δ form a *compatible system* of rotations and displacements, and the virtual work of any system of forces and couples in statical equilibrium on these rotations and displacements is zero. Then

$$\sum \delta M \times \theta + \sum W \times \delta = 0 \tag{17.13}$$

since the net work of the internal shearing forces is zero. The summation $\sum \delta M \times \theta$ is carried out for all short lengths of the beam, while the summation $\sum W \times \delta$ is carried out for all external loads, including couples and force reactions at points of support. If the virtual rotations θ are small, the virtual displacements δ can be found easily. If the lengths δz of the beam are infinitesimally small,

$$\sum \delta M \times \theta = \int_{z=0}^{z=L} \theta \, dM \tag{17.14}$$

where the integration is carried out over the whole length L of the beam. But

$$\int_{z=0}^{z=L} \theta \, dM = \left[M\theta - \int M \, d\theta \right]_{z=0}^{z=L}$$

Now

$$[M\theta]_{z=0}^{z=L} = [M\theta]_{z=L} - [M\theta]_{z=0}$$

and is the work of the end couples on their respective virtual displacements; this work has already been taken account of in the summation $\sum W \times \delta$, so that equation (17.13) becomes

$$\sum W \times \delta = \int_{z=0}^{z=L} M \, d\theta = \int_0^L M\left(\frac{d\theta}{dz}\right) dz \tag{17.15}$$

Now $(d\theta/dz)$ is the curvature of the beam when it is given the virtual rotations and displacements. If we put

$$\frac{d\theta}{dz} = \frac{1}{R} \tag{17.16}$$

where R is the radius of curvature of the beam, then

$$\sum W \times \delta = \int_0^L M\left(\frac{1}{R}\right) dz \tag{17.17}$$

This relation is similar in form to equation (17.8) which we derived for pin-jointed frames; instead of the summation $\sum P\Delta$ on the right-hand side of equation (17.8), we have an integral over the whole length of the beam.

As an example of the application of equation (17.17), consider the cantilever shown in Fig. 17.13, having a uniform flexural stiffness EI. The cantilever carries a vertical load W at the free end; the bending moment at any section due to W is Wz, so that, if

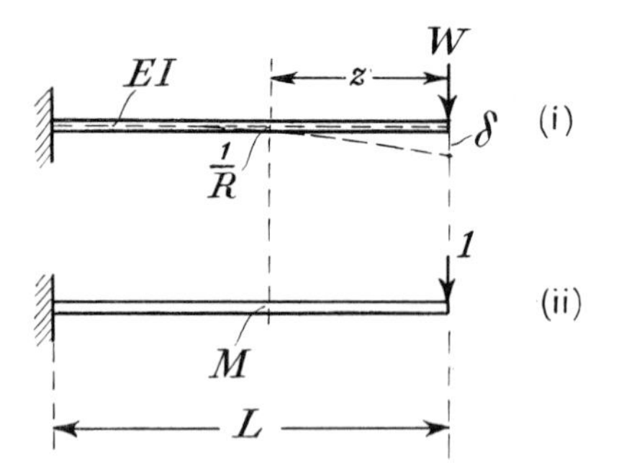

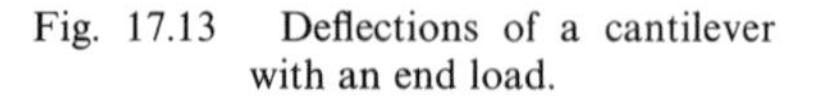

Fig. 17.13 Deflections of a cantilever with an end load.

the beam remains elastic, the corresponding curvature at any section is

$$\frac{1}{R} = \frac{Wz}{EI}$$

Suppose the corresponding deflection of W is δ, Fig. 17.13; then the values of $1/R$ and δ form a system of compatible curvatures and displacements. We derive a simple system of forces and couples in statical equilibrium by applying a unit vertical load at the end of the cantilever; the bending moment at any section due to this unit load is

$$M = 1 \times z = z$$

Then, from equation (17.17),

$$1 \times \delta = \int_0^L M\left(\frac{1}{R}\right) dz = \int_0^L \frac{Wz^2}{EI} dz$$

Then

$$\delta = \frac{WL^3}{3EI}$$

Problem 17.6: A simply-supported beam, of uniform flexural stiffness EI, carries a lateral load W at a distance a from the end A. Estimate the vertical deflection of W.

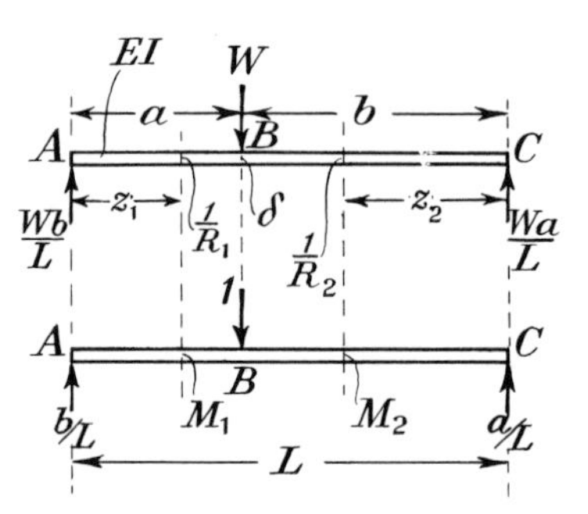

Solution

The bending moment a distance z_1 from A, for the section AB, is

$$\frac{Wbz_1}{L}$$

The curvature for AB is therefore

$$\frac{1}{R_1} = \frac{Wbz_1}{EIL}$$

Similarly, the curvature at any section in BC is

$$\frac{1}{R_2} = \frac{Waz_2}{EIL}$$

Now consider the beam with a unit vertical load at B; the bending moments at sections in AB and BC are, respectively,

$$M_1 = \frac{bz_1}{L}, \qquad M_2 = \frac{az_2}{L}$$

Then, equation (17.17) gives

$$\delta = \int_0^a M_1\left(\frac{1}{R_1}\right) dz_1 + \int_0^b M_2\left(\frac{1}{R_2}\right) dz_2$$

$$= \int_0^a \frac{Wb^2}{EIL^2} z_1{}^2\, dz_1 + \int_0^b \frac{Wa^2}{EIL^2} z_2{}^2\, dz_2$$

Therefore

$$\delta = \frac{Wa^2b^2}{3EIL^2}[a + b) = \frac{Wa^2b^2}{3EIL}$$

Problem 17.7: A cantilever of uniform flexural stiffness EI carries a uniformly-distributed load of intensity w. Estimate the vertical deflection of the free end.

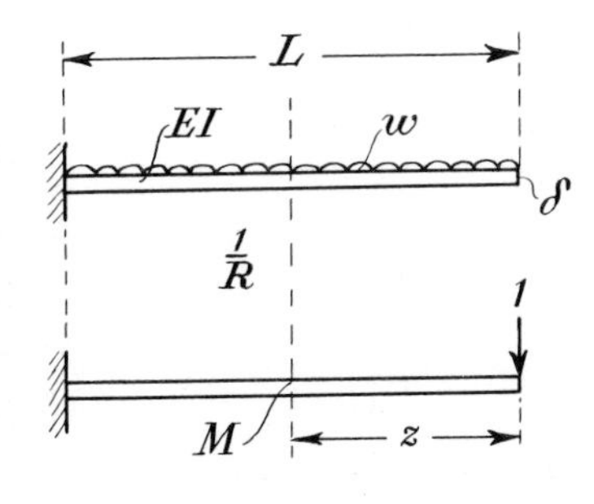

Solution

Due to the distributed load, the curvature at any section is

$$\frac{1}{R} = \frac{wz^2}{2EI}$$

For a unit vertical load at the free end, the bending moment at any section is

$$M = z$$

Then equation (17.17) gives

$$\delta = \int_0^L M\left(\frac{1}{R}\right) dz = \int_0^L \frac{wz^3}{2EI} dz$$

Then

$$\delta = \frac{wL^4}{8EI}$$

Problem 17.8: A semicircular thin ring has a radius r and uniform flexural stiffness EI. The ring carries equal and opposite loads W at the ends. Find the increase in distance between the loaded points.

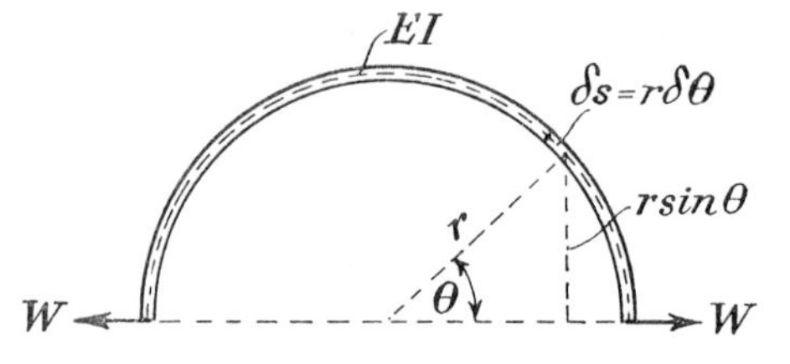

Solution

The bending moment at any angular position θ is

$$M = Wr \sin\theta$$

If the ring is thin, the change of curvature at any section is

$$\frac{1}{R} = \frac{M}{EI}$$

Now consider the virtual work of the forces and couples on their resulting displacements; if δ is the increase in distance between the loaded points

$$W \times \delta = \int_{\theta=0}^{\theta=\pi} M\left(\frac{1}{R}\right) ds = \int_0^\pi \frac{M^2 r}{EI} d\theta = \frac{W^2 r^3}{EI}\int_0^\pi \sin^2\theta \, d\theta$$

Then

$$\delta = \frac{\pi W r^3}{2EI}$$

17.7 Statically indeterminate beam problems

The principle of virtual work may also be used in solving statically indeterminate beam problems. Consider, for example, the beam of Fig. 17.14, which is built-in at A and supported on a roller at B; the beam is of uniform flexural stiffness EI, and carries a uniformly distributed lateral load of intensity w. Suppose the statically indeterminate reaction at B is W; then the bending moment at any section is

$$\tfrac{1}{2}wz^2 - Wz$$

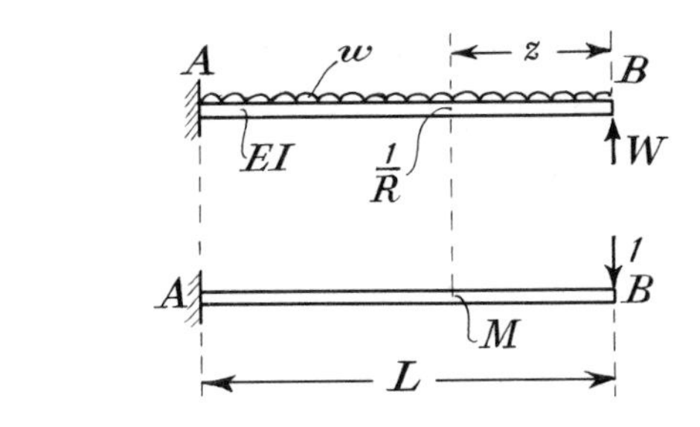

Fig. 17.14 Propped cantilever under uniform lateral loading.

and if the beam remains elastic the resulting curvature at any section is

$$\frac{1}{R} = \frac{1}{EI}(\tfrac{1}{2}wz^2 - Wz)$$

The bending moment at any section due to a unit lateral load at B is

$$M = z$$

Then, for no deflection at B in Fig. 17.14,

$$1 \times 0 = \int_0^L M\left(\frac{1}{R}\right) dz = \int_0^L \frac{z}{EI}\left(\frac{wz^2}{2} - Wz\right) dz$$

Then

$$\int_0^L \tfrac{1}{2}wz^3\, dz = \int_0^L Wz^2\, dz^2$$

Thus

$$W = \frac{3wL}{8}$$

17.8 Plastic bending of mild-steel beams

The principle of virtual work is not limited in its application to linear problems of the type discussed in the preceding problems. It is useful, for example, in solving problems of plastic bending; the uniform mild-steel beam of Fig. 17.15 has a fully-plastic moment M_P. At collapse of the beam plastic hinges develop at A and B. Suppose the point B is now given a virtual displacement δ; if δ is small, AB rotates through an angle (δ/a), and BC through an angle $[\delta/(L - a)]$. The work of the system of forces and couples of Fig. 17.15(ii) on the virtual displacements and rotations of Fig. 17.15(iii) is zero. Then

$$W\delta = M_P\left[\frac{2\delta}{a} + \frac{\delta}{(L - a)}\right]$$

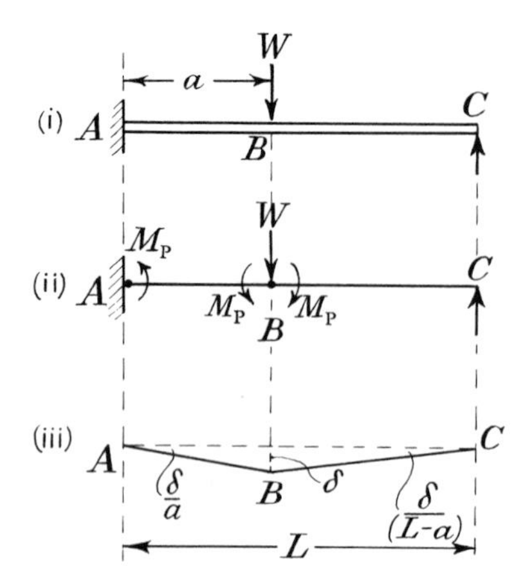

Fig. 17.15 Plastic bending of a mild-steel beam.

Then

$$W = \frac{M_P(2L - a)}{a(L - a)}$$

This is the value of W at plastic collapse of the beam.

Problem 17.9: A uniform mild-steel beam has a fully-plastic moment M_P. Find the intensity of uniformly distributed loading at collapse of the beam.

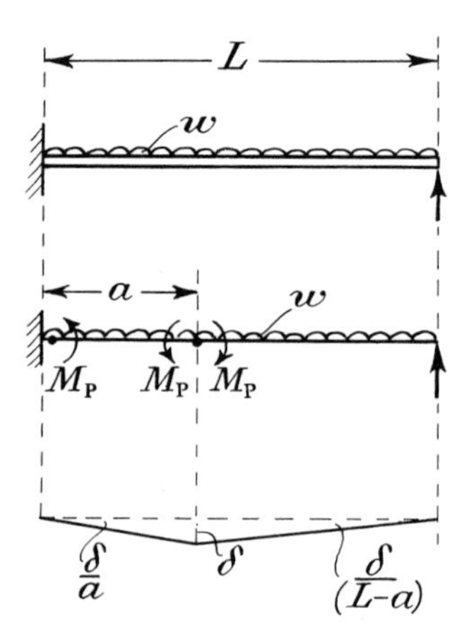

Solution

Suppose that, at plastic collapse, hinges develop at the built-in end, and at a distance a from that end. Then

$$\tfrac{1}{2}wa\delta + \tfrac{1}{2}w(L - a)\delta = M_P\left[\frac{2\delta}{a} + \frac{\delta}{(L - a)}\right].$$

Thus,

$$w = \frac{2\left(2 - \frac{a}{L}\right)}{\left(\frac{a}{L}\right)\left(1 - \frac{a}{L}\right)} \cdot \frac{M_P}{L^2}$$

This is a minimum with respect to (a/L) when

$$\frac{a}{L} = (2 - \sqrt{2})$$

Then the relevant value of w is

$$w = \frac{2M_P}{L^2}(3 + 2\sqrt{2})$$

17.9 Reciprocal characteristics of linear-elastic systems

An interesting property of linear-elastic systems in statical equilibrium may be deduced from the principle of virtual work. Consider a pin-jointed frame, which may be simply-stiff or redundant, Fig. 17.16. Suppose the members of the frame are elastic

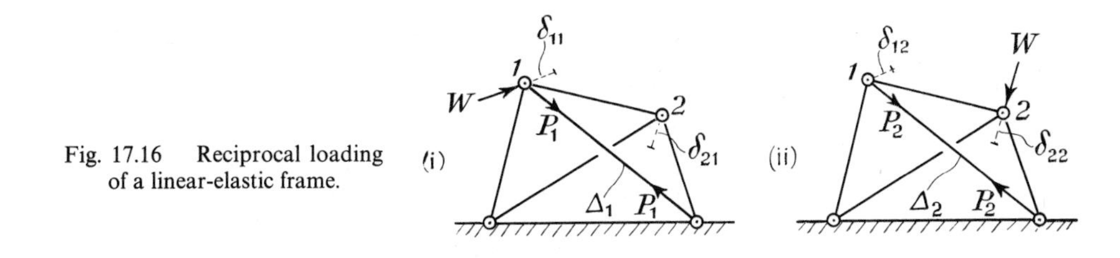

Fig. 17.16 Reciprocal loading of a linear-elastic frame.

and linear; the extension e of any member is proportional therefore to the tensile load P in that member, and we may write

$$e = kP \tag{17.18}$$

where k is a constant for any member of the frame. When an external load W is applied at joint *1*, Fig. 17.16(i), the corresponding displacement of joint *1* is δ_{11}, and the displacement of joint *2* in a given direction is δ_{21}; the tensile load in a typical member is P_1 and its extension is e_1. Now consider the frame under an alternative condition of loading in which W is applied in the given direction at joint *2*, Fig. 17.16(ii); the corresponding displacement of joint *2* is δ_{22}, and the displacement of joint *1*, in the direction of the load W of Fig. 17.16(ii), is δ_{12}; for the load W at joint *2*, the tensile force in a typical member is P_2 and its extension is e_2. On applying the principle of virtual work to the loads of Fig. 17.16(i) and the displacements of Fig. 17.16(ii), we have

$$W\delta_{12} = \sum_m P_1 e_2 \tag{17.19}$$

where the summation is carried out for all members m of the frame. Again, the virtual work of the forces of Fig. 17.16(ii) on the displacements of Fig. 17.16(i) is zero, and so

$$W\delta_{21} = \sum_m P_2 e_1 \tag{17.20}$$

But each member of the frame is elastic and linear, and from equation (17.8),

$$W\delta_{12} = \sum_m P_1 e_2 = \sum_m P_1 k P_2 = \sum_m k P_1 P_2 \tag{17.21}$$

and

$$W\delta_{21} = \sum_m P_2 e_1 = \sum_m P_2 k P_1 = \sum_m k P_2 P_1 \tag{17.22}$$

Thus

$$W\delta_{21} = W\delta_{12}$$

and

$$\delta_{21} = \delta_{12} \tag{17.23}$$

Then the displacement at joint *2* in Fig. 17.16(i) is equal to the displacement at joint *1* in Fig. 17.16(ii); a reciprocal relation holds therefore between the displacements at the joints.

This reciprocal property is useful in finding the influence line of deflection of a beam; consider the beam of Fig. 17.17 which is held rigidly at one end and supported on a roller at the other. The beam may have a non-uniform section along its length, but remains elastic and linear during bending. Suppose the beam carries a unit vertical load at A, Fig. 17.17(i), and that the resulting vertical deflection at B is δ. Then by the reciprocal property the vertical deflection at A due to a unit vertical load at B, Fig. 17.17(ii), is δ also. When the beam carries a unit vertical load at A', the vertical deflection at B is δ', Fig. 17.17(iii); then δ' is also the vertical deflection at A' due to a unit vertical load at B, Fig. 17.17(ii). Thus, the deflected form of the beam due to a unit vertical load at B gives the vertical deflections at B due to a unit vertical load placed anywhere in the

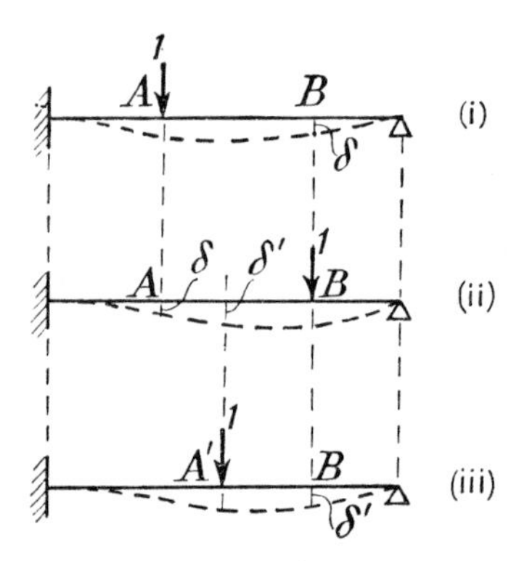

Fig. 17.17 Influence line of deflection of a beam.

span of the beam. The deflected form of Fig. 17.17(ii) is therefore the influence line of vertical deflections at B due to unit vertical loads on the beam.

Problem 17.10: When a linear-elastic cantilever of variable section carries an end load W, the deflected form is

$$v = W\left[a\left(\frac{z}{L}\right)^2 + b\left(\frac{z}{L}\right)^3\right]$$

where a and b are constants. Estimate the deflection of the free end of the cantilever when it carries a uniformly distributed lateral load of intensity w.

Solution

By the reciprocal property of linear-elastic systems,

$$W\delta = \int_0^L W\left[a\left(\frac{z}{L}\right)^2 + b\left(\frac{z}{L}\right)^3\right] dz$$

Then

$$\delta = wL(\tfrac{1}{3}a + \tfrac{1}{4}b)$$

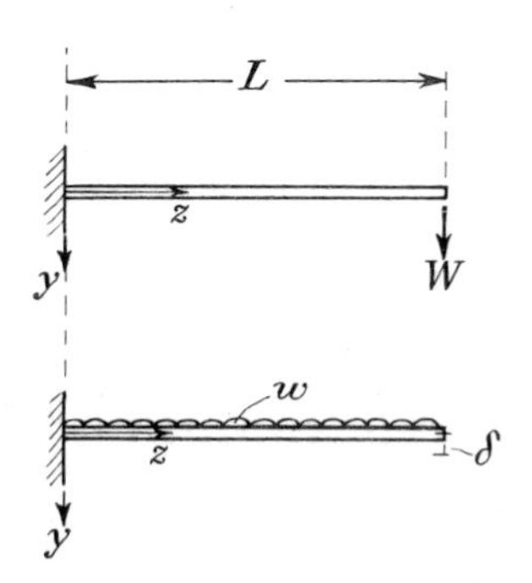

Problem 17.11: If the linear-elastic beam of Problem 17.10 carries a couple M at the free end, estimate the vertical deflection of the free end.

Solution

From Problem 17.10, the deflected form due to W is

$$v = W\left[a\left(\frac{z}{L}\right)^2 + b\left(\frac{z}{L}\right)^3\right]$$

The slope at the free-end is then

$$\theta = \left(\frac{dv}{dz}\right)_{z=L} = \frac{W}{L}(2a + 3b)$$

The reciprocal property then gives

$$W\delta = M\theta = \frac{MW}{L}(2a + 3b)$$

Thus

$$\delta = \frac{M}{L}(2a + 3b)$$

FURTHER PROBLEMS
(answers on page 399)

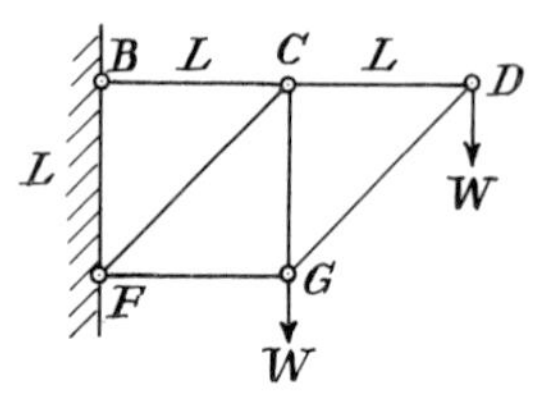

17.12 Find the horizontal displacement of the joint D of the pin-jointed frame shown. Each bar has a cross-sectional area A and Young's modulus E.

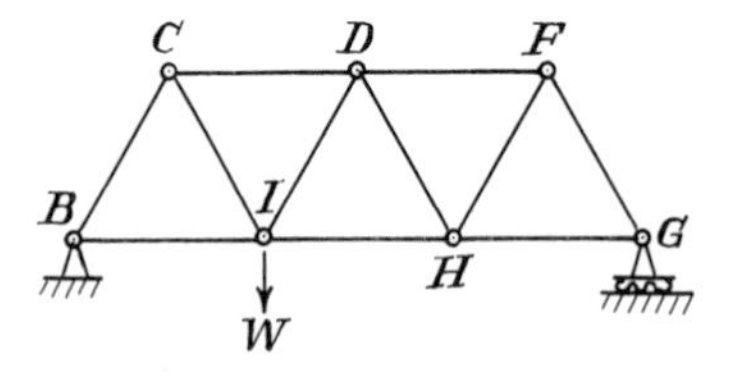

17.13 A pin-jointed frame is hinged at B and supported on a roller at G. Find the vertical deflection of the joint H due to a vertical load W at joint I. Each member of the frame is of length L, area A and Young's modulus E.

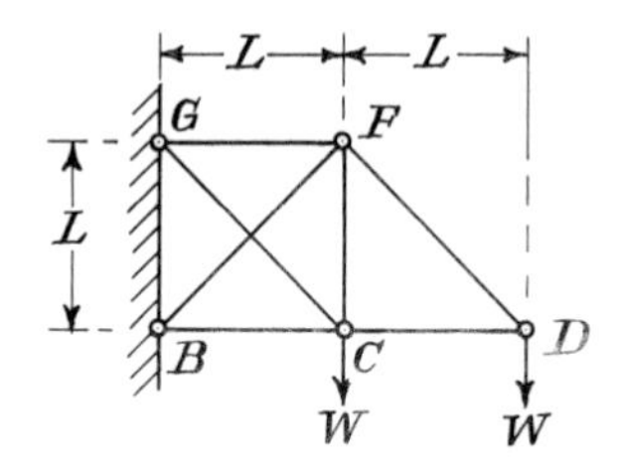

17.14 Find the load in the member FG of the redundant pin-jointed frame shown. All members have the same area A and Young's modulus E.

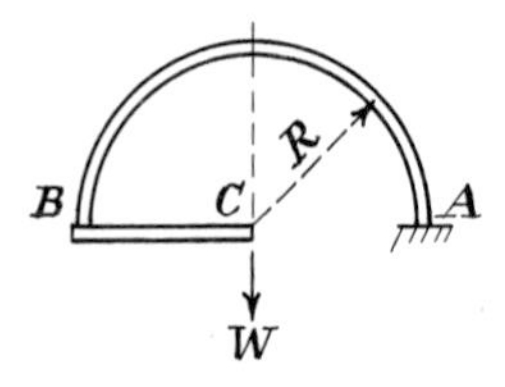

17.15 A thin semicircular bracket, AB, of radius R is built-in at A, and has at B a rigid horizontal arm BC of length R. The arm carries a vertical load W at C. Show that the vertical deflection at C is $\pi WR^3/2EI$, where EI is the flexural rigidity of the strip, and determine the horizontal deflection.

(*Nottingham*)

18 Strain energy and complementary energy

18.1 Properties of the strain energy function

The principle of virtual work may be used to show that the concepts of strain energy and complementary energy are important in stress and strain analysis. We shall discuss these concepts in relation to pin-jointed frames, for the sake of simplicity; the basic ideas may be applied to any type of stress-strain problem.

Consider a plane pin-jointed frame which may be simply-stiff or have redundant members, Fig. 18.1; the frame is pinned to rigid foundations at A and B, and carries external loads W_1 and W_2 at joints C and D, respectively. Suppose that the displacements of the joints C and D in the directions of the loads W_1 and W_2, respectively, are δ_1 and δ_2; then δ_1 and δ_2 are the corresponding displacements of W_1 and W_2. Now suppose that the tensile force in a typical member, such as BC, is P, and that its resulting extension is e. The external forces W_1 and W_2, together with the internal forces, P, etc., form a system of forces in statical equilibrium, while the extensions, e, etc., of the members are compatible with the displacements δ_1 and δ_2. Then by the principle of virtual work

$$W_1\delta_1 + W_2\delta_2 = \sum_m Pe \tag{18.1}$$

where the summation is carried out for all members, m, of the frame. Now consider a system of displacements and extensions in which δ_1 is increased by a small amount $\delta\delta_1$, δ_2 remaining unchanged; suppose the extension e of a typical member increases an amount δe due to the change $\delta\delta_1$ in δ_1. The modified set of extensions, such as $(e + \delta e)$, are compatible with the displacements $(\delta_1 + \delta\delta_1)$ and δ_2, and so by the principle of virtual work

$$W_1(\delta_1 + \delta\delta_1) + W_2\delta_2 = \sum_m P(e + \delta e) \tag{18.2}$$

On subtracting equations (18.1) and (18.2) we have

$$W_1\,\delta\delta_1 = \sum_m P\,\delta e \tag{18.3}$$

where δe is the increase in the extension e of a member due to the increase $\delta\delta_1$ in δ_1. Suppose the members of the frame are elastic but have non-linear load-extension curves of the form shown in Fig. 18.2; the shaded area, which is approximately rectangular and equal to $P\,\delta e$, is the work done in extending the bar a further small amount δe when the tensile load is P. If the member is elastic, this work is conserved in the member as strain energy; if U is the total strain energy of the members of the frame, then

$$\sum_m P\,\delta e = \delta U \tag{18.4}$$

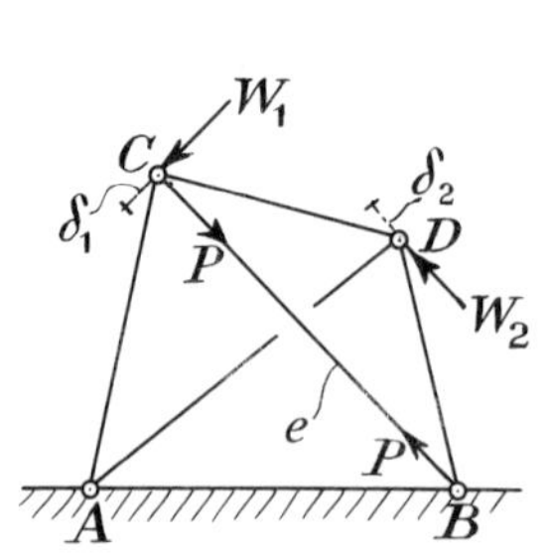

Fig. 18.1 Plane frame under any system of external loads.

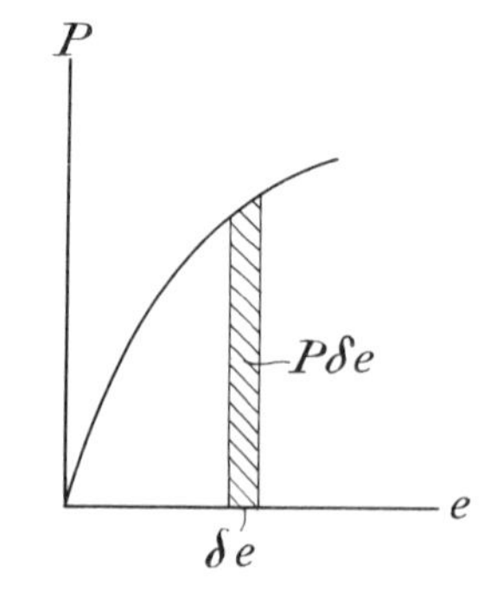

Fig. 18.2 Strain energy of a member of the frame.

where δU is the increase of U resulting from the increase $\delta\delta_1$ of δ_1; then from equation (18.3)

$$\delta U = W_1 \, \delta\delta_1 \tag{18.5}$$

If the change $\delta\delta_1$ is infinitesimally small, we have

$$\frac{\partial U}{\partial \delta_1} = W_1 \tag{18.6}$$

Then the partial derivative of the strain energy function U with respect to the displacement δ_1 of a joint gives the corresponding external load W_1 at that joint. For the load W_2 we have, similarly,

$$\frac{\partial U}{\partial \delta_2} = W_2 \tag{18.7}$$

The use of this property of the strain energy function can be illustrated best by an example; consider the frame shown in Fig. 18.3, consisting of three similar pin-jointed

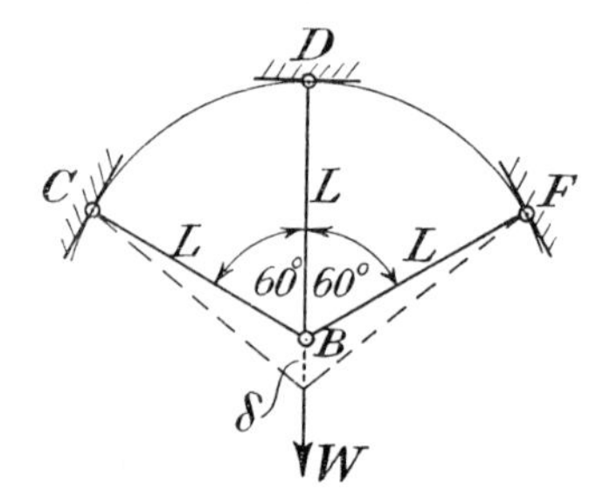

Fig. 18.3 Strain energy as a function of a displacement of the system.

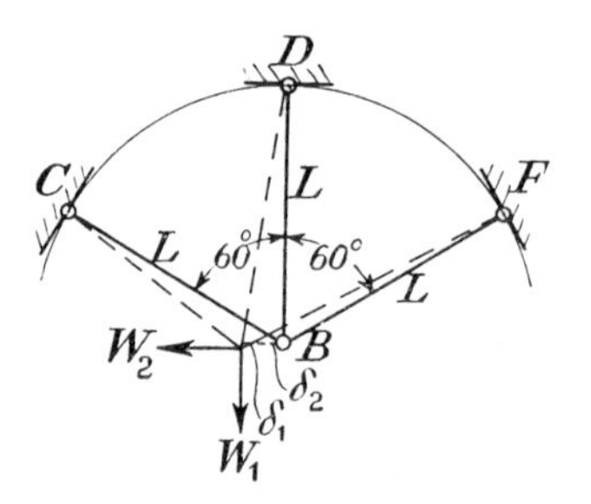

Fig. 18.4 Strain energy as a function of the displacements of a system.

bars, each of length L, cross-sectional area A, and Young's modulus E, and connected together at the joint B. When a vertical load W is applied at B, suppose the joint B is displaced downwards an amount δ. Then the extension of the member BD is δ, and the extensions of the members BC and BF are $\frac{1}{2}\delta$, approximately. If the bars remain elastic, the extension e of any member is

$$e = \frac{PL}{EA}$$

where P is the tensile load in the member. The strain energy stored in extending a bar an amount e is then

$$\int_0^e P\,de = \int_0^e \left(\frac{EA}{L}\right) e\,de = \frac{EAe^2}{2L} \tag{18.8}$$

The total strain energy stored in the three bars of the frame is therefore

$$U = \sum_m \frac{EA}{2L} e^2 = \frac{EA}{2L}\left[(\delta)^2 + (\tfrac{1}{2}\delta)^2 + (\tfrac{1}{2}\delta)^2\right] \tag{18.9}$$

giving

$$U = \frac{3EA\delta^2}{4L}$$

Then by the strain energy property we have

$$\frac{\partial U}{\partial \delta} = \frac{3EA\delta}{2L} = W$$

Thus,

$$\delta = \frac{2WL}{3EA}$$

In Fig. 18.4 the frame of Fig. 18.3 is shown with a horizontal load W_2 at joint B as well as a vertical load W_1. Suppose the corresponding deflections of B in the directions of W_1 and W_2 are δ_1 and δ_2, respectively. Then, in terms of δ_1 and δ_2 the extensions of the bars are

$$e_{BC} = \tfrac{1}{2}\delta_1 - \frac{\sqrt{3}}{2}\delta_2$$

$$e_{BF} = \tfrac{1}{2}\delta_1 + \frac{\sqrt{3}}{2}\delta_2$$

$$e_{BD} = \delta_1$$

If the members behave linearly, as before, then the strain energy stored in the bars is

$$U = \frac{EA}{2L}\left[\left(\tfrac{1}{2}\delta_1 - \frac{\sqrt{3}}{2}\delta_2\right)^2 + \left(\tfrac{1}{2}\delta_1 + \frac{\sqrt{3}}{2}\delta_2\right)^2 + (\delta_1)^2\right].$$

Then

$$W_1 = \frac{\partial U}{\partial \delta_1} = \frac{EA}{2L}\left[\left(\tfrac{1}{2}\delta_1 - \frac{\sqrt{3}}{2}\delta_2\right) + \left(\tfrac{1}{2}\delta_1 + \frac{\sqrt{3}}{2}\delta_2\right) + 2\delta_1\right]$$

and

$$W_2 = \frac{\partial U}{\partial \delta_2} = \frac{EA}{2L}\left[-\sqrt{3}\left(\tfrac{1}{2}\delta_1 - \frac{\sqrt{3}}{2}\delta_2\right) + \sqrt{3}\left(\tfrac{1}{2}\delta_1 + \frac{\sqrt{3}}{2}\delta_2\right)\right]$$

These two equations give

$$\delta_1 = \frac{2W_1L}{3EA}$$

and

$$\delta_2 = \frac{2W_2L}{3EA}$$

We may use the property of the strain energy function in non-linear problems; suppose the bars of the frame of Fig. 18.3 have load-extension relations of the form

$$P = ae + be^2$$

where P is the tensile load in a member, e is the corresponding extension, and a and b are constants. Then the strain energy in a member due to an extension e is

$$\int_0^e P\,de = \int_0^e (ae + be^2)\,de = \tfrac{1}{2}ae^2 + \tfrac{1}{3}be^3$$

The total strain energy of the frame of Fig. 18.3 is then

$$U = \tfrac{1}{2}a[\delta^2 + \tfrac{1}{4}\delta^2 + \tfrac{1}{4}\delta^2] + \tfrac{1}{3}b[\delta^3 + \tfrac{1}{8}\delta^3 + \tfrac{1}{8}\delta^3]$$

which gives

$$U = \frac{3a\delta^2}{4} + \frac{5b\delta^3}{12}$$

Then

$$W = \frac{\partial U}{\partial \delta} = \frac{3a\delta}{2} + \frac{5b\delta^2}{4}$$

In structures suffering inelastic distortions we may still make use of the function U, by considering this to be the work done in deforming the structure. Suppose the bars of the framework of Fig. 18.3 are made of mild steel having a tensile stress-strain curve of the form shown in Fig. 18.5; the material is elastic, having a Young's modulus E, up to a yield stress σ_Y; continuous yielding takes place at the yield stress σ_Y. If the bars are each of length L and cross-sectional area A, then for the elastic range we have

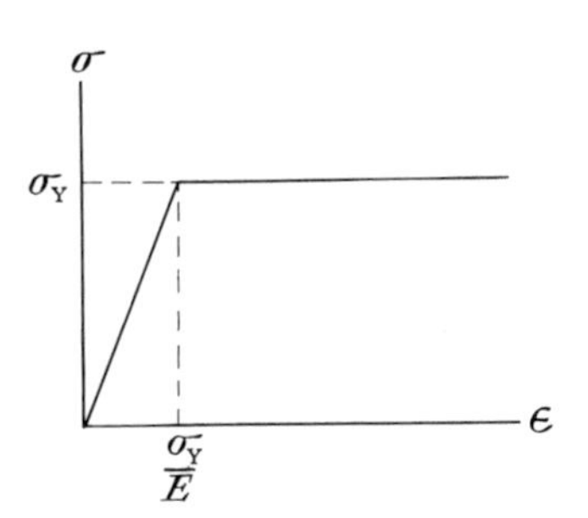

Fig. 18.5 Stress-strain curve for mild-steel members.

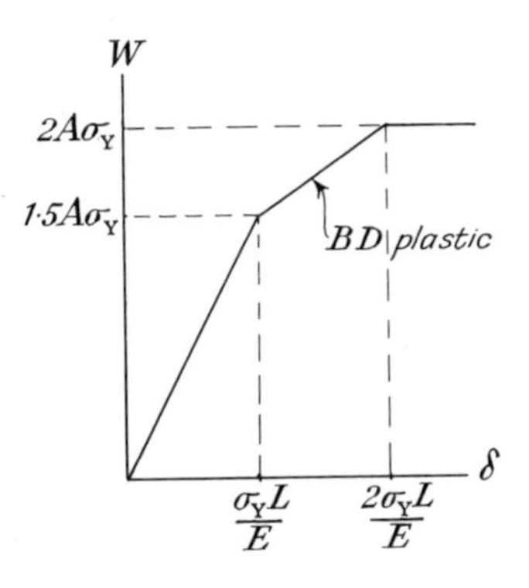

Fig. 18.6 Load-deflection relationship for an inelastic structure.

that the strain energy is

$$U = \frac{3EA\delta^2}{4L}$$

Then

$$\frac{\partial U}{\partial \delta} = W = \frac{3EA\delta}{2L}$$

Now suppose the value of W is increased until the member BD is plastic, the bars BC and BF remaining elastic. Then, if W is increased continuously, the work done in extending the three bars is

$$U = A\sigma_Y\left(\delta - \frac{\sigma_Y L}{2E}\right) + \frac{EA\delta^2}{4L}$$

Then

$$W = \frac{\partial U}{\partial \delta} = A\sigma_Y + \frac{EA\delta}{2L}$$

When all three members are plastic

$$U = A\sigma_Y\left(\delta - \frac{\sigma_Y L}{2E}\right) + 2A\sigma_Y\left(\tfrac{1}{2}\delta - \frac{\sigma_Y L}{2E}\right)$$

Then

$$W = \frac{\partial U}{\partial \delta} = 2A\sigma_Y$$

The W–δ relation for a continuously increasing load is shown in Fig. 18.6; the structure is linear up to $W = 1{\cdot}5A\sigma_Y$. For larger values of W, the member BD becomes plastic. All three members of the frame become plastic when $W = 2A\sigma_Y$. Our analysis of this problem is correct only if W is increased continuously; an inherent difficulty in non-linear problems of this type, in which the work done on the structure is not conserved wholly as strain energy, is that the W–δ relation for unloading of the structure may be different from the W–δ relation for continuously increasing values of W.

18.2 Complementary energy

In §18.1 we have discussed the properties of the strain energy of a structural system; the principle of virtual work leads also to a concept of wider application in stress-strain analysis than that of strain energy; this other property of a structure is known as *complementary energy.*

Consider the statically determinate pin-jointed frame shown in Fig. 18.7; the frame is pinned to a rigid foundation at A and B, and carries external loads W_1 and W_2 at

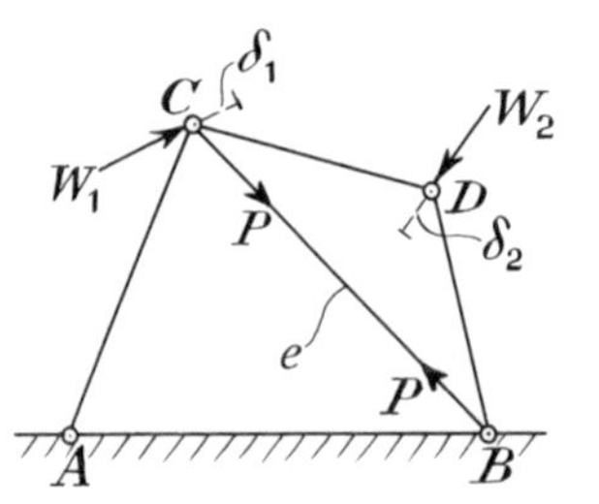

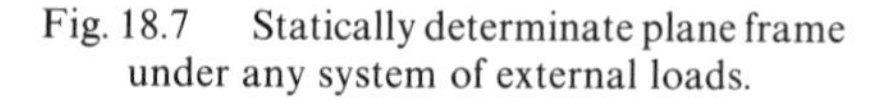
Fig. 18.7 Statically determinate plane frame under any system of external loads.

joints C and D, respectively. Suppose the corresponding displacements of the joints C and D are δ_1, and δ_2, respectively; the tensile force induced in a typical member, such as BC, is P, and its resulting extension is e. The forces W_1, W_2, P, etc. are a system of forces in statical equilibrium, while the extensions e, etc., are compatible with the displacements δ_1, δ_2 of the joints. Thus by the principle of virtual work

$$W_1\delta_1 + W_2\delta_2 = \sum_m Pe \tag{18.10}$$

where the summation is carried out for all members of the frame. Now suppose the external load W_1 is increased in magnitude by a small amount δW_1, the external load W_2 remaining unchanged; due to change in W_1 small changes occur in the forces in the members of the frame, P, for example, increasing to $(P + \delta P)$. Now consider the virtual work of the modified system of forces on the original set of displacements and extensions; we have

$$(W_1 + \delta W_1)\delta_1 + W_2\delta_2 = \sum_m (P + \delta P)e \tag{18.11}$$

On subtracting equations (18.10) and (18.11), we have

$$\delta_1 \times \delta W_1 = \sum_m e\,\delta P \tag{18.12}$$

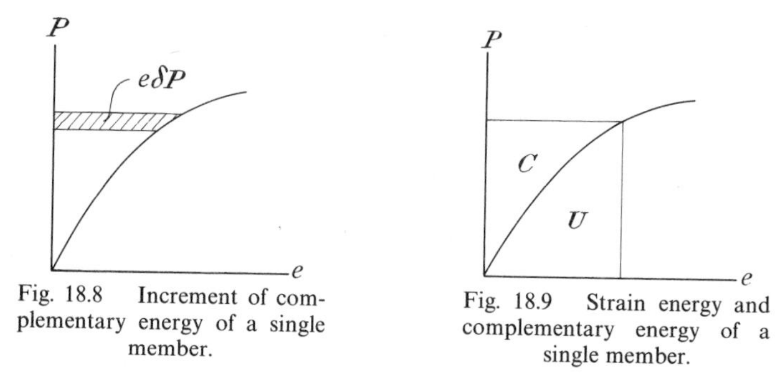

Fig. 18.8 Increment of complementary energy of a single member.

Fig. 18.9 Strain energy and complementary energy of a single member.

The quantity $e\ \delta P$ for a member is the shaded elemental area shown on the load-extension diagram of Fig. 18.8; this is an element of the area C shown in Fig. 18.9. When a bar is extended the work done on the bar is the area below the P–e curve of Fig. 18.9; for a conservative structural member this work is stored as strain energy, which we have already referred to as U. We define the area above the P–e curve of Fig. 18.9 as the *complementary energy*, C, of the member; we have that

$$U + C = Pe \tag{18.13}$$

and

$$\delta C = e\ \delta P \tag{18.14}$$

In equation (18.12) we may write, therefore,

$$\delta_1 \times \delta W_1 = \delta C \tag{18.15}$$

where C is the complementary energy of all members of the frame. If δW_1 is infinitesimally small

$$\frac{\partial C}{\partial W_1} = \delta_1 \tag{18.16}$$

Then the partial derivative of the complementary energy function C with respect to the external load W_1 gives the corresponding displacement δ_1 of that load.

The use of this property of the complementary energy function is illustrated best by a number of examples. Consider first the pin-jointed frame shown in Fig. 18.10; all the members of the frame are of the same material of Young's modulus E, and each

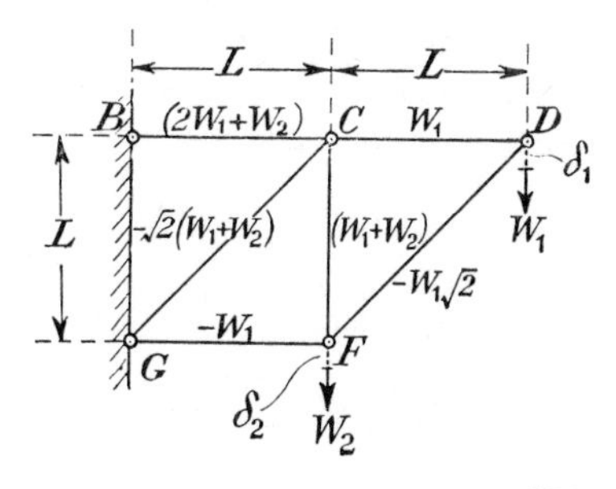

Fig. 18.10 Deflections of a pin-jointed plane frame.

has the same cross-sectional area A. If the members remain elastic the extension e of a member of length l due to a tensile force P is

$$e = \frac{Pl}{EA}$$

The complementary energy of the member due to P is

$$\int_0^P e\,dP = \int_0^P \left(\frac{l}{EA}\right) P\,dP = \frac{P^2 l}{2EA} \tag{18.17}$$

The total complementary energy of the frame is

$$C = \sum_m \frac{P^2 l}{2EA} \tag{18.18}$$

where m is the total number of members. For the frame of Fig. 18.10,

$$C = \frac{L}{2EA}[(2W_1 + W_2)^2 + (W_1)^2 + (W_1 + W_2)^2 + (-W_1)^2$$
$$+ (-W_1\sqrt{2})^2\sqrt{2} + (-\sqrt{2}[W_1 + W_2])^2\sqrt{2}]$$

Then the corresponding displacement of the external load W_1 is

$$\delta_1 = \frac{\partial C}{\partial W_1} = \frac{L}{2EA}[4(2W_1 + W_2) + 2W_1 + 2(W_1 + W_2) + 2W_1$$
$$+ 4\sqrt{2}W_1 + 4\sqrt{2}(W_1 + W_2)]$$

Thus,

$$\delta_1 = \frac{L}{EA}[(7 + 4\sqrt{2})W_1 + (3 + 2\sqrt{2})W_2]$$

Furthermore,

$$\delta_2 = \frac{\partial C}{\partial W_2} = \frac{L}{2EA}[2(2W_1 + W_2) + 2(W_1 + W_2) + 4\sqrt{2}(W_1 + W_2)]$$

Then

$$\delta_2 = \frac{L}{EA}[(3 + 2\sqrt{2})W_1 + 2(1 + \sqrt{2})W_2]$$

If $W_2 = 0$,

$$\delta_2 = \frac{L}{EA}(3 + 2\sqrt{2})W_1$$

Then, by introducing the load W_2, evaluating $\partial C/\partial W_2$ and putting $W_2 = 0$, we can calculate the displacement of an unloaded joint of the frame.

Fig. 18.11 Deflections of a non-linear plane frame.

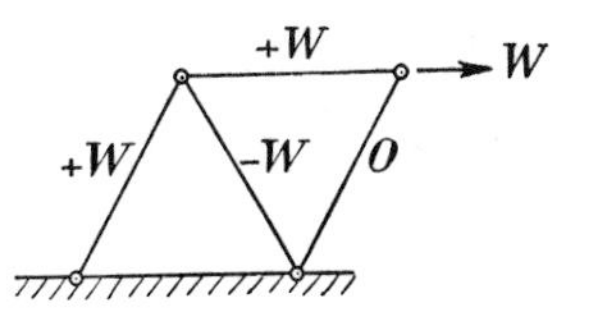

The property of the complementary energy function may be used in studying non-linear frames; in the pin-jointed frame of Fig. 18.11 all the members are of the same length, and the load-extension relation for each member is of the form

$$e = aP - bP^2$$

where a and b are constants; this frame has been studied already in Problem 2.3. The complementary energy of a member due to a tensile force P is

$$\int_0^P e\,dP = \int_0^P (aP - bP^2)\,dP = \tfrac{1}{2}aP^2 - \tfrac{1}{3}bP^3$$

The total complementary energy for the frame is

$$C = \sum_m (\tfrac{1}{2}aP^2 - \tfrac{1}{3}bP^3)$$

Then the corresponding deflection of W is

$$\delta = \frac{\partial C}{\partial W} = \sum_m (aP - bP^2)\frac{\partial P}{\partial W}$$

Thus

$$\delta = 2(aW - bW^2) + (-aW - bW^2)(-1) = 3aW - bW^2$$

18.3 Statically determinate frame carrying two equal and opposite external forces

The important result of §18.2 is that the displacement corresponding to any external force is the partial derivative of the complementary energy function with respect to that force. Now consider a statically determinate frame, Fig. 18.12, under the action of two external forces W_1 and W_2 having opposed lines of action. An important property of a

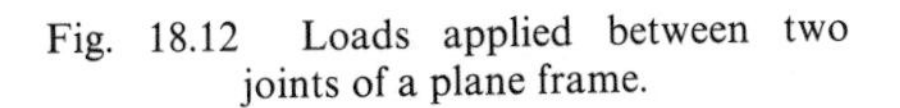
Fig. 18.12 Loads applied between two joints of a plane frame.

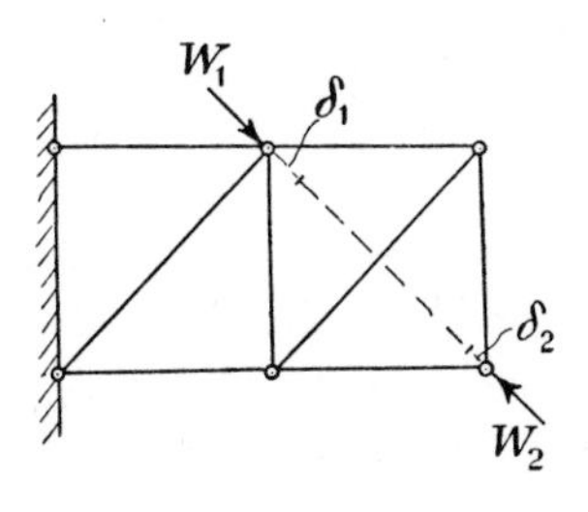

statically determinate frame is that the tensile force P in any member of the frame is related linearly to the external loads, provided the distortions of the frame are small, so we may express the tensile force in any member in the form

$$P = \alpha W_1 + \beta W_2 + \gamma \tag{18.19}$$

where α, β are constants and γ is the force in the member due to external loads other than W_1 and W_2. If C is the total complementary energy in the members of the frame, then the corresponding displacement of P is

$$\delta_1 = \frac{\partial C}{\partial W_1} = \frac{\partial}{\partial W_1} \sum_m \left[\int_0^P e \, dP \right] \tag{18.20}$$

This may be written

$$\delta_1 = \sum_m e \frac{\partial P}{\partial W} \tag{18.21}$$

On substituting for P from equation (18.19),

$$\delta_1 = \sum_m e\alpha \tag{18.22}$$

Similarly, for the corresponding displacement of W_2

$$\delta_2 = \sum_m e\beta \tag{18.23}$$

Then the forces W_1 and W_2 approach each other an amount

$$\delta_1 + \delta_2 = \sum_m e\alpha + \sum_m e\beta = \sum_m e(\alpha + \beta) \tag{18.24}$$

Now, if $W_1 = W_2 = W$(say), we have

$$P = (\alpha + \beta)W + \gamma \tag{18.25}$$

and

$$\delta_1 + \delta_2 = \sum_m e(\alpha + \beta) = \sum_m e \frac{\partial T}{\partial W} = \frac{\partial C}{\partial W} \tag{18.26}$$

Thus, for a pin-jointed frame under the action of equal and opposite forces W, the loaded joints approach an amount $\partial C/\partial W$.

18.4 Solution of statically indeterminate frames using complementary energy

We may apply the concept of complementary energy to the solution of statically indeterminate stress-strain problems. As a particular instance consider the pin-jointed plane frame of Fig. 18.13, which is pinned to a rigid foundation at A and B, and carries external loads W_1 and W_2; the frame has a single redundant member, which we will consider to be CD. Suppose the unknown tensile force in this member is R; we imagine

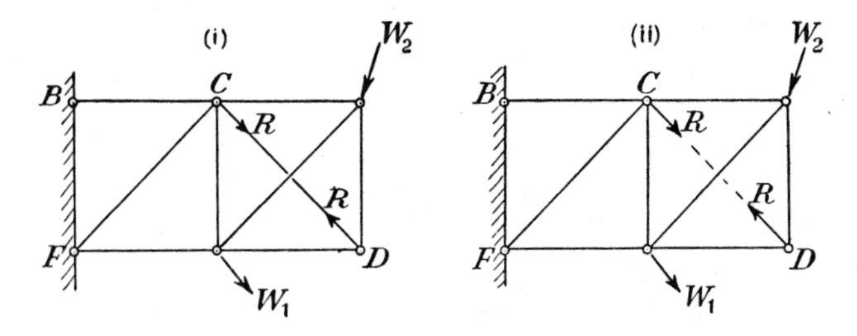

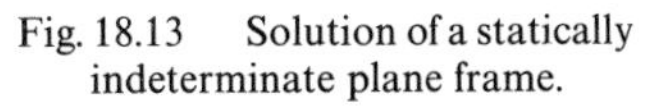
Fig. 18.13 Solution of a statically indeterminate plane frame.

the member CD to be replaced by two equal and opposite forces R at C and D, as in Fig. 18.13(ii). The frame of Fig. 18.13(ii) is statically determinate, so that the joints C and D approach an amount

$$\delta_R = \frac{\partial C_1}{\partial R} \tag{18.27}$$

where C_1 is the complementary energy of all the members of the frame, excluding the member CD. Now the member CD is itself extended an amount

$$e_R = \frac{\partial C_2}{\partial R} \tag{18.28}$$

where C_2 is the complementary energy of the member CD due to the tensile force R. Then the extended member CD fits into the loaded frame of Fig. 18.13(ii) if

$$\delta_R = -e_R \tag{18.29}$$

that is, if

$$\frac{\partial C_1}{\partial R} + \frac{\partial C_2}{\partial R} = 0 \tag{18.30}$$

or, if

$$\frac{\partial}{\partial R}(C_1 + C_2) = 0 \tag{18.31}$$

Now, $(C_1 + C_2)$ is the total complementary energy, C, of all members of the frame, including the member CD. Then for a redundant member, carrying a tensile force R,

$$\frac{\partial C}{\partial R} = 0 \tag{18.32}$$

As an example of the use of this relation, consider the pin-jointed frame shown in Fig. 18.14; all the members have the same cross-sectional A and Young's modulus E. The force in the redundant member BF is R. If the members remain elastic, the complementary energy of all members of the frame is

$$C = \frac{L}{2EA}\left[2\left(W - \frac{R}{\sqrt{2}}\right)^2 + 2\left(-\frac{R}{\sqrt{2}}\right)^2 + R^2\sqrt{2} + (R - W\sqrt{2})^2\sqrt{2}\right]$$

Then

$$\frac{\partial C}{\partial R} = \frac{L}{2EA}\left[-\frac{4}{\sqrt{2}}\left(W - \frac{R}{\sqrt{2}}\right) - \frac{4}{\sqrt{2}}\left(-\frac{R}{\sqrt{2}}\right) + 2\sqrt{2}R + 2\sqrt{2}(R - W\sqrt{2})\right]$$

If $\partial C/\partial R = 0$, then

$$R = \frac{W}{\sqrt{2}}$$

The complementary energy property of redundant frames is also true of non-linear systems; consider the frame shown in Fig. 18.15, which has one redundant member.

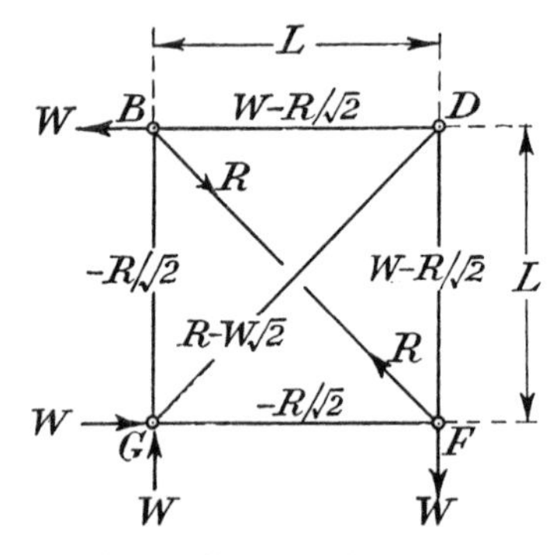

Fig. 18.14 Simple statically indeterminate plane frame.

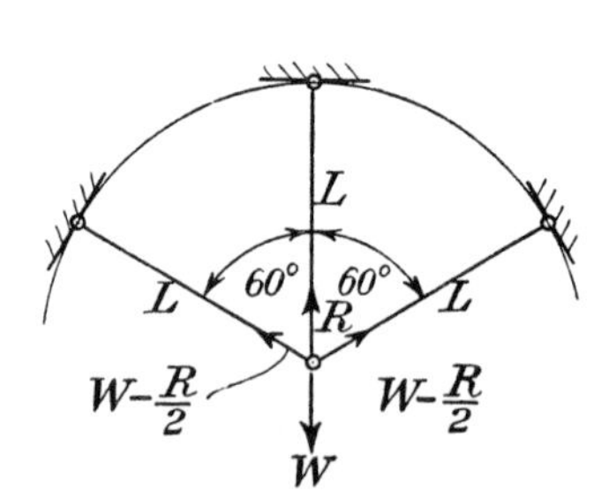

Fig. 18.15 Non-linear statically indeterminate frame.

Suppose the force in the central member is R; the load-extension relation of any member is of the form

$$e = aP - bP^2$$

The complementary energy in any member due to a tensile force P is

$$\int_0^P e\,dP = \int_0^P (aP - bP^2)\,dP = \tfrac{1}{2}aP^2 - \tfrac{1}{3}bP^3$$

The total complementary energy of the frame is

$$C = \sum_m \left(\tfrac{1}{2}aP^2 - \tfrac{1}{3}bP^3\right)$$

Then

$$\frac{\partial C}{\partial R} = \sum_m (aP - bP^2)\frac{\partial P}{\partial R} = 0$$

Thus

$$2\left[a\left(W - \frac{R}{2}\right) - b\left(W - \frac{R}{2}\right)^2\right]\left(-\tfrac{1}{2}\right) + [aR - bR^2] = 0$$

Then R is the root of the quadratic equation

$$\tfrac{3}{4}bR^2 + (bW - \tfrac{3}{2}a)R + (aW - bW^2) = 0$$

18.5 Initial lack of fit of the members of the frame

Suppose the member CD of the frame of Fig. 18.13 is initially too *short* by an amount λ to fit into the unloaded frame. The member CD carrying a tensile load R can be fitted into the loaded frame of Fig. 18.13 if

$$\delta_R = -e_R + \lambda \tag{18.33}$$

Then

$$\delta_R + e_R = \frac{\partial C}{\partial R} = \lambda \tag{18.34}$$

where C is the total complementary energy. As an example, suppose the member BF of the frame of Fig. 18.14 has an initial lack of fit kL/EA; then for the case of linear-elastic members

$$\frac{\partial C}{\partial R} = \frac{L}{2EA}\left[-4W\left(\frac{1}{\sqrt{2}} + 1\right) + 4R(1 + \sqrt{2})\right] = \frac{kL}{EA}$$

Then

$$-\frac{2W}{\sqrt{2}}(1 + \sqrt{2}) + 2R(1 + \sqrt{2}) = k$$

Thus

$$R = \frac{W}{\sqrt{2}} + \frac{k}{2(1 + \sqrt{2})} = \frac{W}{\sqrt{2}} + \frac{k(1 - \sqrt{2})}{2(1 - 2)}$$

and

$$R = \frac{W}{\sqrt{2}} + \frac{k}{2}(\sqrt{2} - 1)$$

18.6 Complementary energy in problems of bending

The complementary energy method may be used to considerable advantage in the solution of problems of bending of straight and thin curved beams. In general we suppose that the moment-curvature relationship for an element of a beam is of the form shown in Fig. 18.16. The complementary energy of bending of an elemental length δs due to a

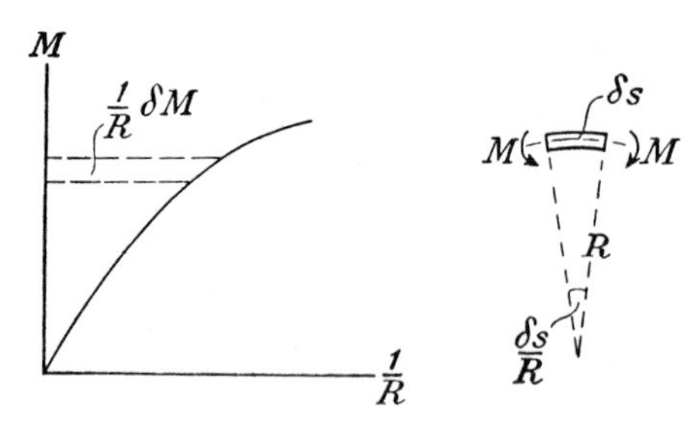

Fig. 18.16 Complementary energy of bending of the element of a beam.

bending moment M is

$$\int_0^M \left(\frac{\delta s}{R}\right) dM$$

For a linear-elastic beam of flexural stiffness EI

$$\frac{1}{R} = \frac{M}{EI}$$

and so the complementary energy is

$$\int_0^M \frac{M}{EI}\, dM\, \delta s = \frac{M^2\, \delta s}{2EI} \tag{18.35}$$

For a length L of the beam, the complementary energy is therefore

$$C = \int_0^L \frac{M^2\, ds}{2EI} \tag{18.36}$$

As in the case of pin-jointed frames, the partial derivative of C with respect to any external load is the corresponding displacement of that load. For statically indeterminate beams, the partial derivative of the complementary energy with respect to a redundant force or couple is zero.

Problem 18.1: Estimate the vertical displacement of the free end of the uniform cantilever shown.

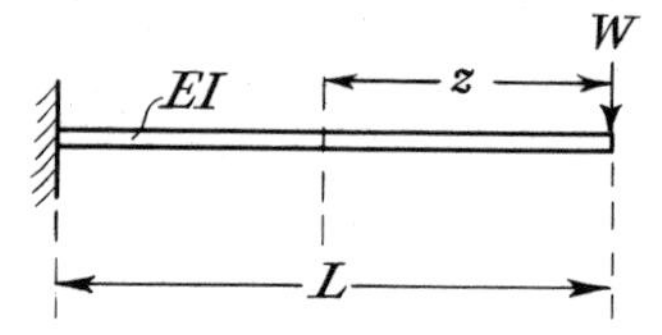

Solution

The complementary energy of bending is

$$C = \int_0^L \frac{M^2\, dz}{2EI} = \int_0^L \frac{W^2 z^2\, dz}{2EI} = \frac{W^2 L^3}{6EI}$$

The corresponding displacement of W is

$$\delta_W = \frac{\partial C}{\partial W} = \frac{WL^3}{3EI}$$

Problem 18.2: A cantilever has a uniform flexural stiffness EI. Estimate the vertical deflection at the free end if the cantilever carries a uniformly distributed lateral load of intensity w.

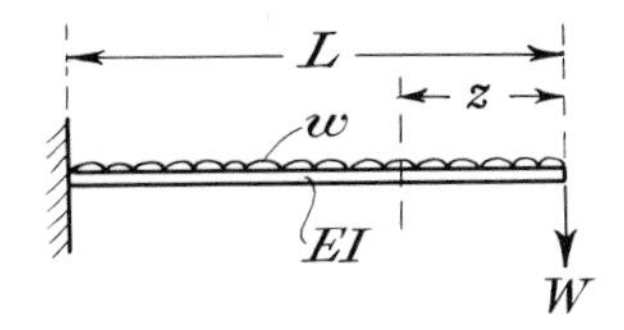

Solution

Introduce a vertical load W at the free end; the bending moment at any section is then

$$M = \tfrac{1}{2}wz^2 + Wz$$

The complementary energy of bending is

$$C = \frac{1}{2EI}\int_0^L (\tfrac{1}{2}wz^2 + Wz)^2\,dz$$

The corresponding displacement of W is

$$\delta_W = \frac{\partial C}{\partial W} = \frac{1}{EI}\int_0^L (\tfrac{1}{2}wz^2 + Wz)z\,dz$$

Now put $W = 0$; then

$$\delta_W = \frac{1}{EI}\int_0^L \tfrac{1}{2}wz^3\,dz = \frac{wL^4}{8EI}$$

Problem 18.3: A cantilever of uniform flexural stiffness EI carries a moment M at the remote end. Estimate the angle of rotation at that end of the beam.

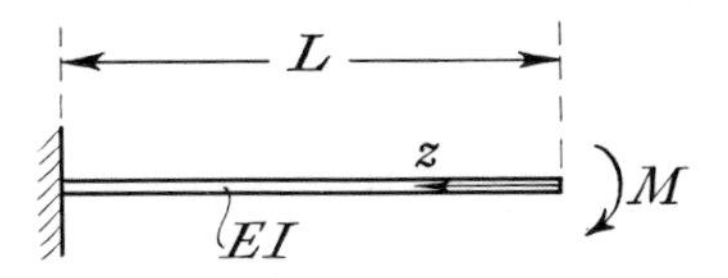

Solution

All sections of the beam carry the same bending moment M, so the complementary energy is

$$C = \int_0^L \frac{M^2\,dz}{2EI} = \frac{M^2L}{2EI}$$

The corresponding displacement of M is

$$\theta_M = \frac{ML}{EI}$$

which is the angle of rotation at the remote end.

Problem 18.4: Solve the problem discussed in §17.7, using complementary energy.

Solution

The bending moment at any section in terms of w and the redundant force W is

$$\tfrac{1}{2}wz^2 - Wz$$

Then

$$C = \int_0^L (\tfrac{1}{2}wz^2 - Wz)^2 \frac{dz}{2EI}$$

The property $\partial C/\partial W = 0$ gives

$$\int_0^L \tfrac{1}{2}wz^3 \, dz = \int_0^L Wz^2 \, dz$$

Then

$$W = \frac{3wL}{8}$$

Problem 18.5: Solve Problem 17.8 using complementary energy.

Solution

The bending moment at any angular position θ is

$$M = Wr \sin \theta$$

Then

$$C = \int_0^\pi \frac{M^2}{2EI} r \, d\theta$$

Thus

$$\delta_W = \frac{\partial C}{\partial W} = \int_0^\pi M \frac{\partial M}{\partial W} \cdot \frac{r \, d\theta}{EI}$$

$$= \int_0^\pi \frac{Wr^3 \sin^2 \theta}{EI} \, . \, d\theta = \frac{\pi Wr^3}{2EI}$$

Problem 18.6: A thin circular ring of radius r and uniform flexural stiffness carries two radial loads W applied along a diameter. Estimate the maximum bending moment in the ring.

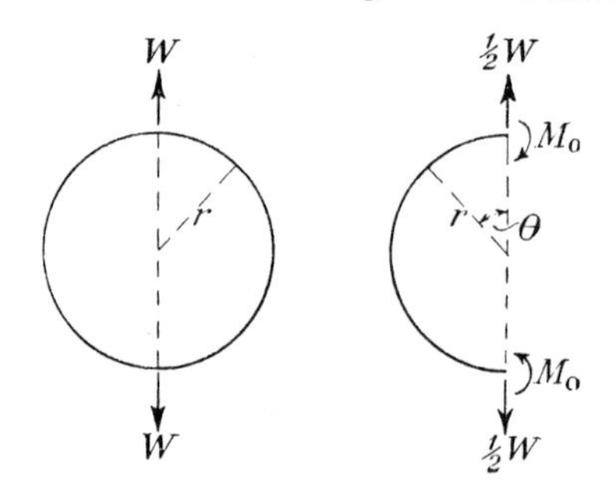

Solution

By symmetry the loading actions on a half-ring are $\frac{1}{2}W$ and M_0. The bending moment at any angular position θ is

$$M = M_0 - \tfrac{1}{2}Wr \sin \theta$$

Then

$$C = \int_0^\pi (M_0 - \tfrac{1}{2}Wr \sin \theta)^2 \frac{r \, d\theta}{2EI}$$

But $\partial C/\partial M_0 = 0$, so that

$$\int_0^{\pi} M_0 \, d\theta = \tfrac{1}{2}Wr \int_0^{\pi} \sin\theta \, d\theta$$

Then

$$M_0 = \frac{Wr}{\pi}$$

FURTHER PROBLEMS
(answers on page 399)

18.7 A Warren girder is supported on a hinge at A and on rollers at F. It carries three equal vertical loads at B, C and D. Each member is of length L, area A and Young's modulus E. Find the vertical deflection of joint D.

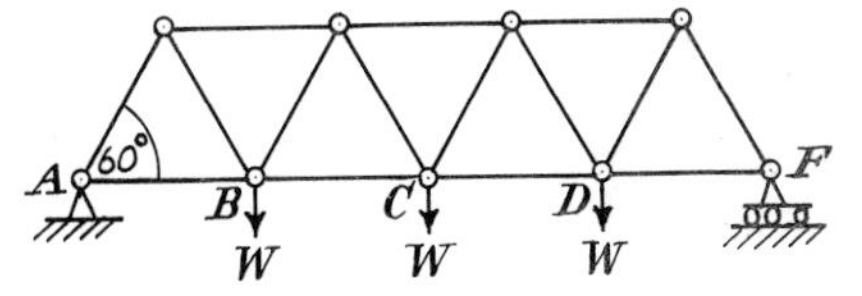

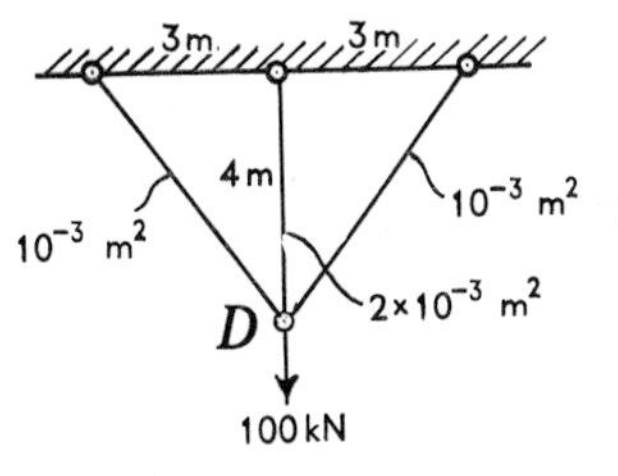

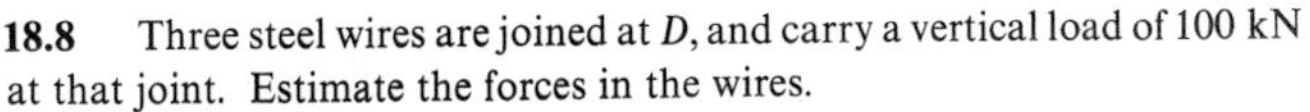
18.8 Three steel wires are joined at D, and carry a vertical load of 100 kN at that joint. Estimate the forces in the wires.

18.9 A beam has a second moment of area of $2I$ over one-half of the span and I over the other half. Find the fixed-end moments when a load of 100 kN is carried at the mid-length.

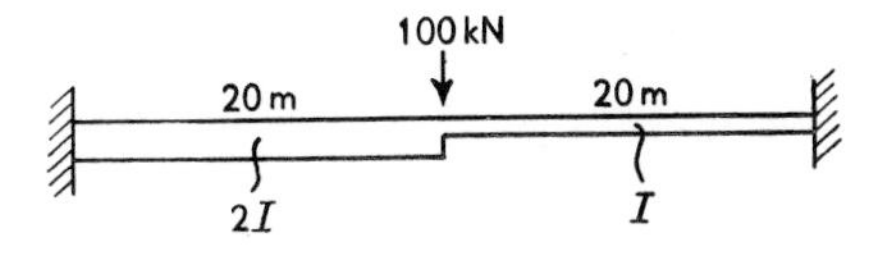

18.10 A ring radius R and uniform cross-section hangs from a single support. Find the position and magnitude of the maximum bending moment due to its own weight. (*London*)

18.11 An 'S' hook follows part of the outline of two equal circles of radius R that just touch. It embraces $\frac{5}{6}$ths of one circle and $\frac{2}{3}$rds of the other. If the ends are pulled apart by a force P, by how much will they be moved if the hook has a constant rigidity EI? (*London*)

19 Springs

19.1 General properties of springs

The common purpose of all kinds of springs is to absorb energy and restore it slowly or rapidly according to the function of the particular spring under consideration. Thus, in the case of clockwork a certain amount of work is done by an external agency in winding up, i.e. deforming, the spring; this work is stored in the form of strain energy and is regained when the spring is allowed to return to its original shape. In clockwork the resumption of the spring's original shape takes place slowly. The other most common use of springs is for absorbing shocks, such as the springs of buffers of railway rolling-stock and the springs of wheels on all manner of vehicles. In such cases some of the kinetic energy of the moving body, the truck, or that due to the vertical motion of the wheels and axles, is converted into strain energy in the spring, the effects of the blow on the truck as a whole being thereby reduced. The springs, in returning to their original shape, give back this energy, tending to reverse the relative motion of the colliding bodies. Springs are also used to provide a means of restoring various mechanisms to their original configurations against the action of some external force, or when an external force is removed.

The properties of a spring which are usually of most interest to the engineer are (i) its capacity for absorbing energy, (ii) the deformation produced by a given load, or vice versa, provided of course that the safe working stress of the material is not exceeded, and sometimes (iii) its natural frequency of vibration.

Springs in practice belong usually to one of two definite families: springs in which a length of rod or wire is made into a coil of some kind, and springs consisting of one or more approximately flat plates.

The *stiffness* of a spring is the load required to produce unit deflection. The *resilience* of a spring is its capacity for storing energy without exceeding a certain stress limit.

19.2 Coiled springs

Coiled springs may be divided into (i) ordinary helical springs, when the axis of the wire has the form of a helix described on a right circular cylinder, (ii) helical springs in which the axis of the wire is a helix described on a right circular cone, and (iii) spiral springs, when the axis forms a plane spiral curve.

19.3 Geometry of helical springs

In Fig. 19.1, CQB is a helix described on a cylinder whose axis is OA, this helix being the central line of the wire forming the spring. FQD is a generator of the cylinder; DS is the tangent at D to the cross-section CDH of the cylinder, and QS is the tangent

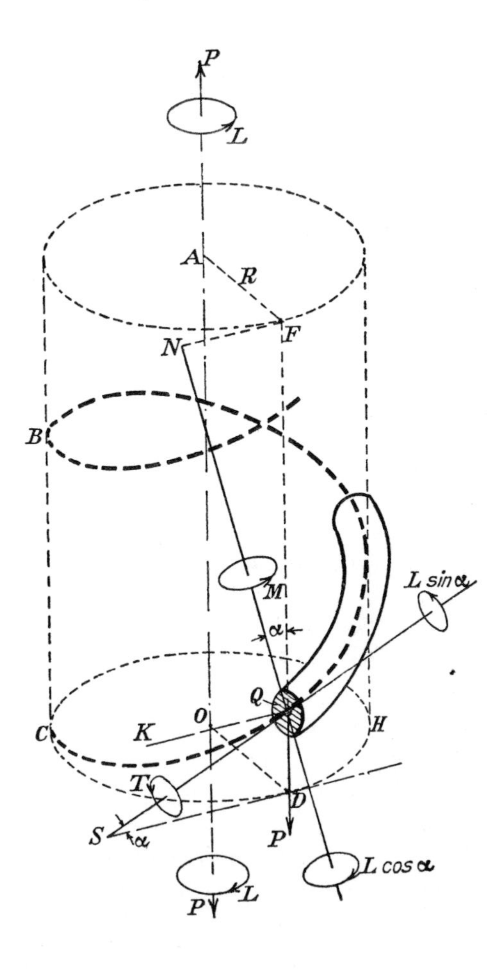

Fig. 19.1 Loading actions at a section of a helical spring.

at Q to the helix. QN is in the tangent plane FDS and perpendicular to QS. QK is parallel to DS.

Let R be the radius of the cylinder, and a the pitch angle of the helix, so that $\angle FQN = \angle QSD = \alpha$. Let l be the length of wire, and n the number of complete convolutions. Then

$$l = 2\pi nR \sec \alpha \tag{19.1}$$

Let the spring be acted on by axial forces P and axial couples L. Now forces P acting along the axis of the spring are equivalent to forces P along DF and couples $P \times R$ about QK.

Consider the forces and couples transmitted by the part CQ of the wire to the part QB; the force P along QD can be resolved into components $P \cos \alpha$ along NQ, i.e. a shearing force in the plane of the normal cross-section of the wire at Q, and $P \sin \alpha$ along QS, i.e. perpendicular to the same cross-section.

The couple PR about QK can be resolved into components $T = PR \cos \alpha$ about QS, i.e., a torsional couple acting on the wire, and $M = PR \sin \alpha$ about QN, i.e., a flexural couple tending to decrease the curvature of the wire.

Similarly a couple L, whose axis is OA, will have a torsional component $L \sin \alpha$ about QS in the same direction as T, as shown in Fig. 19.1, and a flexural component $L \cos \alpha$ about QN tending to increase the curvature of the wire.

19.4 Close-coiled helical spring; axial pull

When the coils of a helical spring are so close together that they can be regarded as practically lying in planes at right angles to the axis of the helix, the angle α is very small, and we speak of the spring as *close-coiled*.

With such springs under an axial tension P only, the bending couple $PR \sin \alpha$ becomes negligible in comparison with the torsional couple; we can also take $l = 2\pi nR$, with sufficient accuracy.

If δ is the axial extension of the spring due to the load P applied gradually, then the work done by P is $\frac{1}{2}P\delta$. The torsional couple at any point on the central line is $T = PR$, approximately. From equation (16.20) the strain energy of the deformed spring is

$$U = \frac{lT^2}{2GJ}$$

where l is the total length of wire, i.e. $2\pi nR$. Hence the energy stored in the spring is

$$U = \frac{\pi n R^3 P^2}{GJ} = \frac{32 n R^3 P^2}{G d^4} \tag{19.2}$$

if the spring is made of round wire of diameter d.

Equating this to the work done by P we obtain an expression for the deflection:

$$\tfrac{1}{2}P\delta = \frac{\pi n R^3 P^2}{GJ}$$

Then

$$\delta = \frac{2\pi n R^3 P}{GJ} = \frac{64 n R^3 P}{G d^4} \tag{19.3}$$

For unit deflection, $\delta = 1$, and the *stiffness* is

$$\frac{P}{\delta} = \frac{GJ}{2\pi nR^3} = \frac{Gd^4}{64nR^3} \tag{19.4}$$

If τ is the maximum permissible shearing stress, the resilience and maximum load can be expressed in terms of τ: from §16.3 we have

$$\tau = \frac{Td}{2J} = \frac{PRd}{2J}$$

Thus

$$P = \frac{2J\tau}{Rd} = \frac{\pi d^3\tau}{16R} \tag{19.5}$$

Substituting in equation (19.2) we obtain a formula for the stored energy, or resilience, in terms of the maximum stress; the *resilience* is given by

$$\frac{4\pi nRJ\tau^2}{Gd^2} = \frac{\pi^2 nRd^2\tau^2}{8G} = \frac{\tau^2}{4G} \times \text{volume of metal} \tag{19.6}$$

It will thus be seen that all such springs have equal mass for equal stress and equal resilience.

The following practical rules are worth noting: A new close-coiled helical spring should be closed up solid *twice* in compression; after this treatment it will take no more permanent set. If P be the load required to close the spring solid, the spring should never, even in tension, carry more than this load P.

19.5 Close-coiled helical spring; axial couple

If a close-coiled spring is acted on by a couple L whose axis is the axis of the helix, we can neglect the torsional couple $L \sin \alpha$, and consider the wire as acted on everywhere by a flexural couple which is approximately equal to L. Let θ be the total angle through which one end of the spring is turned relative to the other. Then the work done by L is $\frac{1}{2}L\theta$, θ being measured in radians. From §9.11 the strain energy of the spring is

$$\frac{L^2 l}{2EI} = \frac{\pi nRL^2}{EI}$$

where EI is the bending stiffness of the wire.

Equating this to the work done by L we have

$$\theta = \frac{2\pi nRL}{EI} \tag{19.7}$$

For a circular section wire of diameter d we have

$$\theta = \frac{128nRL}{Ed^4} \qquad (19.8)$$

The maximum bending stress in the wire is given by

$$\sigma = \frac{Ld}{2I} = \frac{32L}{\pi d^3} \qquad (19.9)$$

Hence the maximum couple for a given stress is

$$L = \frac{\pi d^3 \sigma}{32} \qquad (19.10)$$

For unit rotation, $\theta = 1$, and the stiffness is given by

$$\frac{L}{\theta} = \frac{EI}{2\pi nR} \quad \text{per radian} \qquad (19.11)$$

Problem 19.1: Calculate the number of turns required for 0·6 cm deflection in a spring made of steel 2 cm diameter and forming a cylindrical coil 12·5 cm mean diameter, if the load be 500 N and $G = 80$ GN/m^2

(Cambridge)

Solution

Assuming the coil to be close-coiled, we have from equation (19.3)

$$n = \frac{d^4 G\delta}{64R^3 P}$$

Now,

$d = 0{\cdot}020$ m, $\quad d^4 = 0{\cdot}16 \times 10^{-6}$ m^4, $\quad R = 0{\cdot}0625$ m, $\quad R^3 = 0{\cdot}244 \times 10^{-3}$ m^3,

$P = 500$ N, $\quad \delta = 0{\cdot}006$ m, $\quad G = 80 \times 10^9$ N/m^2

Hence

$$n = 9{\cdot}46$$

19.6 Open-coiled helical spring; axial force

We shall now consider a helical spring where the coils are not so close that the angle α can be treated as small. Let P be the axial load, and let the rest of the notation be as in §19.3. The torsional couple is $PR \cos \alpha$, and the flexural couple $PR \sin \alpha$; hence the strain energy of the spring is

$$\frac{l(PR\cos\alpha)^2}{2GJ} + \frac{l(PR\sin\alpha)^2}{2EI} = \frac{lP^2R^2}{2}\left(\frac{\cos^2\alpha}{GJ} + \frac{\sin^2\alpha}{EI}\right)$$

The work done by P is $\frac{1}{2}P\delta$, and so

$$\delta = lPR^2\left(\frac{\cos^2\alpha}{GJ} + \frac{\sin^2\alpha}{EI}\right) = 2\pi nPR^3 \sec\alpha\left(\frac{\cos^2\alpha}{GJ} + \frac{\sin^2\alpha}{EI}\right) \qquad (19.12)$$

The stiffness is the value of P obtained from this when $\delta =$ unity. For a spring of circular wire of diameter d, equation (19.12) becomes

$$\delta = \frac{64nPR^3 \sec\alpha}{d^4 G}\left(\cos^2\alpha + \frac{2G}{E}\sin^2\alpha\right) \tag{19.13}$$

19.7 Open-coiled helical spring; axial couple

Let the spring be acted on by a couple L whose axis coincides with the axis of the helix. Then the torsion couple at any point is $L \sin\alpha$ and the flexural couple is $L\cos\alpha$. Hence, in this case, the strain energy is

$$\frac{l(L\sin\alpha)^2}{2GJ} + \frac{l(L\cos\alpha)^2}{2EI}$$

whilst the work done by the couple L in twisting one end of the spring through an angle θ relative to the other is $\frac{1}{2}L\theta$, the angle being measured in radians. Equating this to the strain energy we have

$$\theta = lL\left(\frac{\sin^2\alpha}{GJ} + \frac{\cos^2\alpha}{EI}\right) = 2\pi nRL\sec\alpha\left(\frac{\sin^2\alpha}{GJ} + \frac{\cos^2\alpha}{EI}\right) \tag{19.14}$$

For a wire of circular cross-section this gives

$$\theta = \frac{128nRL\sec\alpha}{d^4 E}\left(\frac{E}{2G}\sin^2\alpha + \cos^2\alpha\right) \tag{19.15}$$

In the case of open-coiled springs the stresses due to flexure and torsion must be found separately and the principal stresses calculated.

It should be noted that the formulae of §§19.6 and 19.7 are only approximations as we have treated R and α as constants, whereas really they vary continuously as the load is applied, but the formulae derived here are sufficiently accurate for all practical purposes. The relations between R, α, δ are

$$2\pi Rn = l\cos\alpha, \qquad \delta = (l\cos\alpha)\,\Delta\alpha$$

where $\Delta\alpha$ is the change in α due to straining.

19.8 Plane spiral springs

Figure 19.2 represents a spring whose central line is a plane spiral curve. One end of the spring is anchored to a pin at C, and the inner end is attached to the winding spindle. A couple M is applied to this spindle. Let X and Y be the components of the reaction at C along and perpendicular to the line joining the axis of the spindle to the centre of the pin C. Let (x, y) be the co-ordinates of an element AB ($= \delta s$) referred to the same directions as shown in Fig. 19.2. Then the bending moment on AB is $Yx - Xy$.

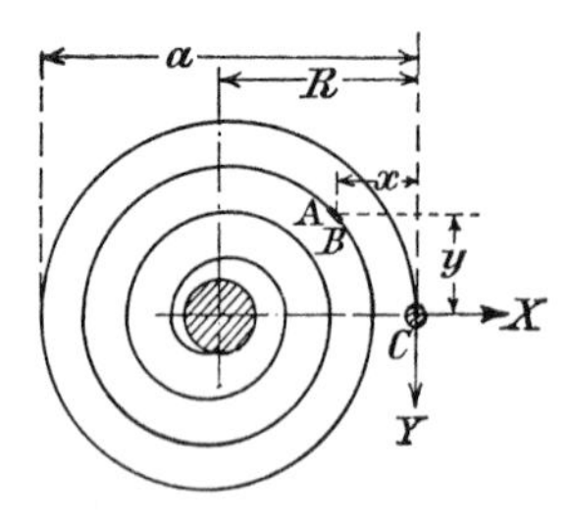

Fig. 19.2 Twisting of a spiral spring.

Let $\delta\theta$ be the change in the angle between the tangents at the ends of the element AB, then

$$\delta\theta = \delta s\left(\frac{1}{r} - \frac{1}{r_0}\right)$$

where r_0 and r denote the radii of curvature at (x, y) before and after straining. But

$$\frac{1}{r} - \frac{1}{r_0} = \frac{Yx - Xy}{EI}$$

where I is the second moment of area of a cross-section of the spring about its neutral axis perpendicular to the plane of the spring. Hence

$$\delta\theta = \frac{Yx - Xy}{EI}\delta s$$

If I is constant this gives, for the total change of angle between the tangents at the extremities of the spring,

$$\delta\theta = \frac{Y}{EI}\int x\,ds - \frac{X}{EI}\int y\,ds$$

Now $\int x\,ds$ and $\int y\,ds$ represent the moment of the whole length of the spring about the directions Y and X respectively. The centroid of the spring will be approximately at the centre of the winding spindle, and if we assume this to be true we shall have

$$\int x\,ds = lR \quad \text{and} \quad \int y\,ds = 0$$

where l is the total length of the spiral. Also $Y \times R = M$, the couple applied to the spindle. Hence

$$\theta = \frac{Ml}{EI} \tag{19.16}$$

The work done in winding, i.e. the energy stored, is

$$\tfrac{1}{2}M\theta = \frac{M^2 l}{2EI}$$

The stress at any section will be given by

$$\sigma = \frac{Yx}{Z_e} = \frac{Mx}{RZ_e}$$

where Z_e is the elastic modulus of the cross-section. The maximum value of x is a (Fig. 19.2), so that the maximum stress is

$$\sigma_{\max} = \frac{Ma}{RZ_e} \tag{19.17}$$

The maximum value of M is

$$M_{\max} = \frac{RZ_e\sigma}{a} \tag{19.18}$$

and the resilience is

$$\frac{R^2 Z_e{}^2 l\sigma_{\max}^2}{2EIa^2} \tag{19.19}$$

When the spring is made of a very thin flat band of metal, anti-clastic curvature arises and the full treatment of the problem becomes more complicated, but is not of great interest to the engineer.

19.9 Close-coiled conical spiral spring

Fig. 19.3 represents a spring whose central line is a spiral curve drawn on a circular cone. Let r_1 be the smallest, and r_2 the largest radius of the spiral, and n the number

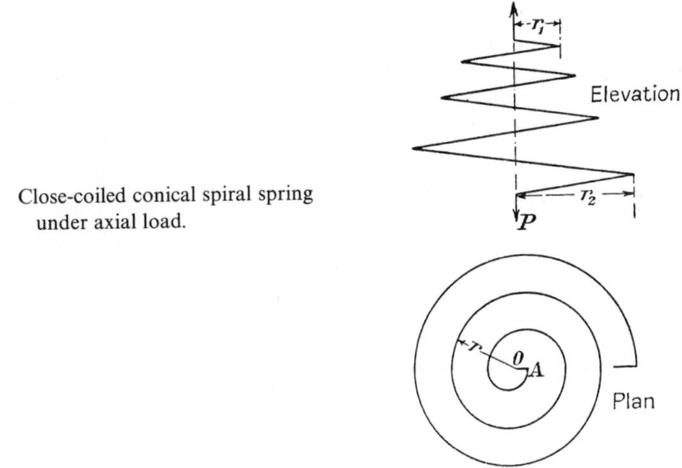

Fig. 19.3 Close-coiled conical spiral spring under axial load.

of complete convolutions of the wire. Let the spiral be such that the polar equation of its projection on a plane at right angles to the axis of the cone is of the form

$$r = a + b\theta$$

Then, if we measure θ from the radius r_1, we have $a = r_1$ and

$$b = (r_2 - r_1)/2\pi n$$

and so

$$r = r_1 + \left[\frac{\theta}{2\pi n}(r_2 - r_1)\right] \tag{19.20}$$

We shall suppose the spring to be close-coiled, i.e. that the convolutions lie very nearly in planes perpendicular to the axis of the cone. Then the element of the spring at radius r is subjected to a torque Pr. The strain energy per unit length is then $P^2r^2/2GJ$. Hence the energy stored in an elemental length δs ($= r\,\delta\theta$) is

$$\delta U = \frac{P^2r^3\,\delta\theta}{2GJ}$$

From equation (19.20) we have

$$\delta\theta = \frac{2\pi n\,.\,\delta r}{r_2 - r_1}$$

Then

$$\delta U = \frac{\pi n P^2 r^3\,\delta r}{GJ(r_2 - r_1)}$$

Hence the total strain energy of the spring is

$$U = \frac{\pi n P^2}{GJ(r_2 - r_1)}\int_{r_1}^{r_2} r^3\,dr = \frac{\pi n P^2}{4GJ}(r_1 + r_2)(r_1{}^2 + r_2{}^2) \tag{19.21}$$

The work done by P is $\frac{1}{2}P\delta$, if δ denotes the total axial extension of the spring; equating this to U we have

$$\delta = \frac{\pi n P}{2GJ}(r_1 + r_2)(r_1{}^2 + r_2{}^2) = \frac{Pl(r_1{}^2 + r_2{}^2)}{2GJ} \tag{19.22}$$

where l denotes the total length of the wire, approximately.

If the spring be acted on by an axial couple L instead of an axial pull, every element of the wire is subjected to the same bending moment, and the change of radius has no direct effect. The formula which we found for a cylindrical helical spring (§19.5) subjected to an axial torque is therefore true also for a close-coiled conical spring, l denoting the total length of wire. It is only in its effect on the value of l that the changing radius alters the formula.

19.10 Approximate theory of leaf springs

We shall now consider the type of spring shown in Fig. 19.4, consisting of n parallel strips of metal of width b. The spring carries a central vertical load W which is balanced by equal end reactions $\frac{1}{2}W$ as shown. We shall assume now that the centre lines of all the plates are initially circular arcs of the same radius R, that each plate has a uniform thickness t, and overlaps the one below it by an amount $a = l/2n$ at each end, and that these overlaps are tapered in width to the triangular shape shown.

Now, since the plates are initially circular arcs of the same radius, each will, when unloaded, touch the one above it at its ends only. If, when the load is applied, the change

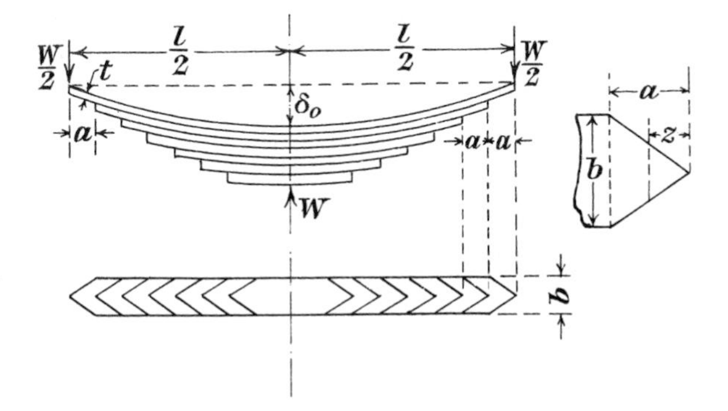

Fig. 19.4 Bending of a leaf spring.

of curvature of each plate is uniform and the same for all the plates, contact will continue to be at the ends only. In these circumstances each plate will bear a downward load $\frac{1}{2}W$ at its ends, and an upward load $\frac{1}{2}W$ at the ends of the plate next below it. Thus the triangular overhanging ends are loaded as cantilevers, whilst the parallel portions are loaded with a uniform bending moment $\frac{1}{2}Wa$. The second moment of area of the cross-section of the parallel portion of each plate is $\frac{1}{12}bt^3$, hence for this portion of the plate the new radius of curvature, r, is given by

$$\frac{1}{R} - \frac{1}{r} = \frac{\text{bending moment}}{EI} = \frac{6Wa}{Ebt^3}$$

Next consider the triangular ends: at a distance z from the point the bending moment is $\frac{1}{2}Wz$, and the moment of inertia is $zbt^3/12a$; hence for this portion we have

$$\frac{1}{R} - \frac{1}{r} = \frac{Wz}{2} \times \frac{12a}{zbt^3E} = \frac{6Wa}{Ebt^3}$$

Thus for the whole length of each plate we have

$$\frac{1}{r} = \frac{1}{R} - \frac{6Wa}{Ebt^3} = \text{constant} \tag{19.23}$$

Thus all the plates are bent into circular arcs of radius r, and contact continues to be at the ends only.

The load W_0 which will straighten out all the plates is obtained by putting $r = \infty$ in equation (19.23); we get

$$W_0 = \frac{Ebt^3}{6aR} = \frac{nEbt^3}{3lR} \tag{19.24}$$

If δ_0 be the initial dip of the top plate, assumed to be small, we have, from the properties of a circle,

$$\delta_0 = \frac{l^2}{8R} \quad \text{or} \quad R = \frac{l^2}{8\delta_0}$$

Substituting for R in equation (19.24) gives

$$W_0 = \frac{8nEbt^3\delta_0}{3l^3} \tag{19.25}$$

Since for each plate the ratio of bending moment to the second moment of area is constant and the same for all the plates, the maximum fibre stress will also be constant for each plate and the same for all. If σ is the maximum fibre stress for a given load W, then

$$\sigma = \frac{\frac{1}{2}Wa(\frac{1}{2}t)}{I} = \frac{3Wa}{bt^2} = \frac{3Wl}{2nbt^2} \tag{19.26}$$

Hence, if σ_0 denotes the stress corresponding with the load W_0 required to straighten the plates,

$$W_0 = \frac{2}{3}\frac{nbt^2\sigma_0}{l} \tag{19.27}$$

The load W_0 is generally known as the *proof* load. From equations (19.25) and (19.27) we have

$$\sigma_0 = \frac{4t\delta_0}{l^2}E \tag{19.28}$$

For a given material, σ_0 and E will be prescribed, so that this equation fixes the proper relationship between the thickness and initial radius of the plates.

Since M/I is constant, the dip δ, for a load W, is given by

$$\delta_0 - \delta = \frac{Ml^2}{8EI} = \frac{1}{8}\left(\frac{Wa}{2}\right)\left(\frac{l^2}{E}\right)\left(\frac{12}{bt^3}\right) = \frac{3Wal^2}{4Ebt^3} = \frac{3Wl^3}{8nEbt^3} \tag{19.29}$$

which is the deflection of the ends of the spring relative to the centre.

Problem 19.2: Find the load required to straighten a carriage spring which has 6 strips, of breadth 7·5 cm and thickness 1 cm, the top strip having a length of 1 m, if the deflection of the top strip when unloaded is

6·00 cm. The overlaps are each equal to half the total length of the bottom strip, and their breadth is uniformly tapered to a point. (*Cambridge*)

Solution

From equation (19.29) we have, since $\delta = 0$,

$$W = \frac{8nEbt^3\delta_0}{3l^3}$$

Now,

$$n = 6, \quad b = 0{\cdot}075 \text{ m}, \quad t = 0{\cdot}010 \text{ m}, \quad \delta_0 = 0{\cdot}060 \text{ m}, \quad l = 1 \text{ m}$$

Hence, taking $E = 200$ GN/m^2,

$$W = 14{\cdot}4 \text{ kN}.$$

FURTHER PROBLEMS
(answers on page 399)

19.3 A helical spring is made of steel wire 0·6 cm in diameter. The coils, 60 in number, are in close formation, and the centre line of the wire lies on a cylinder 7·5 cm diameter. The two ends of the spring are pulled apart by an axial pull of 50 N. What shearing stress is set up in the wire and what elongation is produced? (*Cambridge*)

19.4 A safety valve of 7·5 cm diameter is to blow off at a pressure of 1 MN/m^2 by gauge. It is held by a close-coiled compression spring of circular steel bar. The mean diameter is 15 cm and the initial compression of the spring is 2·5 cm. Find the diameter of the steel and the number of convolutions necessary if the shearing stress allowed is 120 MN/m^2, and $G = 80$ GN/m^2. (*Cambridge*)

19.5 Two shafts in line, which are prevented from moving axially, are connected by a helical spring, the spring fitting loosely on the shafts and having its ends fixed to the shafts. Show that, if the coils of the spring are of circular cross-section, and are inclined at 45° to the axis, the couple per unit angle of twist is given by

$$\frac{r^4}{8\sqrt{2}\,.\,nR}\left(\frac{E}{2} + G\right)$$

where r is the radius of the cross-section and the rest of the notation is as in §19.6. (*Cambridge*)

19.6 A close-coiled helical spring made of 0·6 cm round steel wire has 20 coils, mean diameter 7·5 cm. Find (i) its deflection under a load of 200 N, (ii) the shearing stress in the wire due to this load, (iii) the work done in producing the extension. Take $G = 80$ GN/m^2. (*RNEC*)

19.7 A steel carriage spring is to be 1 m long and to carry a central load of 5 kN. If the plates are 7·5 cm wide and 0·6 cm thick, how many plates will be required if the stress is to be limited to 180 MN/m^2? What will be the deflection of the spring at the centre? To what radius should each piece be curved? Take $E = 200$ GN/m^2. (*RNEC*)

19.8 Prove that, for a given stiffness and given maximum shearing stress, the ratio of the mass of a close-coiled helical spring made of tube to that of one made of solid wire is $k^2/(1 + k^2)$, where k is the ratio of the outside diameter of the tube to the inside diameter.

20 Elastic buckling of columns and beams

20.1 Introduction

In all the problems treated in preceding chapters, we were concerned with the small strains and distortions of a stressed material. In certain types of problems, and especially those involving compressive stresses, we find that a structural member may develop relatively large distortions under certain critical loading conditions. Such structural members are said to *buckle*, or become *unstable*, at these critical loads.

As an example of elastic buckling, we consider firstly the buckling of a slender column under an axial compressive load.

20.2 Flexural buckling of a pin-ended strut

A perfectly straight bar of uniform cross-section has two axes of symmetry Cx, Cy in the cross-section Fig. 20.1. We suppose the bar to be a flat strip of material, Cx being the major axis of the cross-section. End thrusts P are applied along the centroidal axis Cz of the bar, which is hinged about the axis Cx at each end. Suppose L is the length of the bar, and EI its uniform flexural stiffness for bending about Cx.

Now Cx is the weakest axis of bending of the bar, and if bowing of the compressed bar occurs we should expect bending to take place in the yz-plane. Consider the possibility that at some value of P, the end thrust, the strut can buckle laterally in the yz-plane.

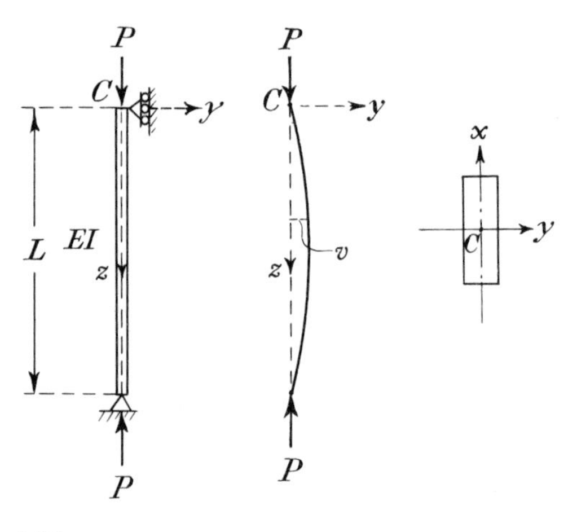

Fig. 20.1 Flexural buckling of a pin-ended strut under axial thrust.

There can be no lateral deflections at the ends of the strut; suppose v is the displacement of the centre line of the bar parallel to Cy at any point. There can be no forces at the hinges parallel to Cy, since these would imply bending moments at the ends of the bar. The only two external forces are the end thrusts P, which are assumed to maintain their original line of action after the onset of buckling. The bending moment at any section of the bar is then

$$M = Pv \tag{20.1}$$

which is a sagging moment in relation to the axes Cz and Cy, in the sense of §13.2. But the moment-curvature relation for the beam at any section is

$$M = -EI\frac{d^2v}{dz^2}$$

provided the deflection v is small. Thus

$$-EI\frac{d^2v}{dz^2} = Pv$$

Then

$$EI\frac{d^2v}{dz^2} + Pv = 0 \tag{20.2}$$

Put

$$\frac{P}{EI} = k^2 \tag{20.3}$$

Then

$$\frac{d^2v}{dz^2} + k^2v = 0 \tag{20.4}$$

The general solution of this differential equation is

$$v = A\cos kz + B\sin kz \tag{20.5}$$

where A and B are arbitrary constants. We have two boundary conditions to satisfy: at the ends $z = 0$ and $z = L$, $v = 0$. Then

$$A = 0 \quad \text{and} \quad B\sin kL = 0$$

Now consider the implications of the equation

$$B\sin kL = 0$$

which is derived from the boundary conditions. If $B = 0$, then both A and B are zero, and obviously the strut is undeflected. If, however, $\sin kL = 0$, B is indeterminate, and the strut may assume the form

$$v = B\sin kz$$

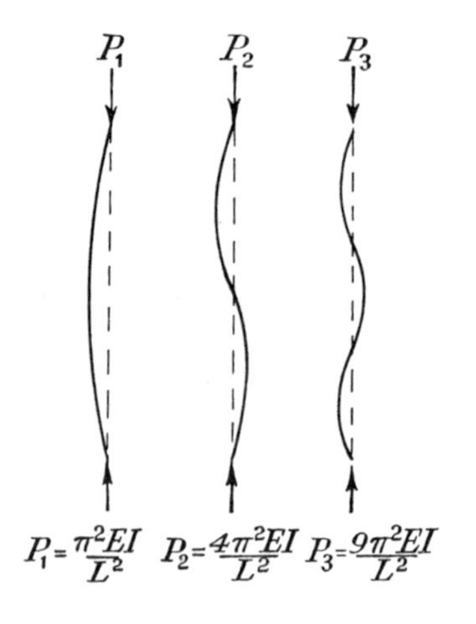

Fig. 20.2 Modes of buckling of a pin-ended strut

This is called a buckled condition of the strut. Obviously B is indeterminate when kL assumes the values,

$$kL = \pi, 2\pi, \dots \text{etc.} \tag{20.6}$$

We need not consider the solution $kL = 0$, which implies $k = 0$, since the solution of the differential equation is not trigonometric in form when $k = 0$. Instability occurs when

$$P = k^2 EI = \frac{\pi^2 EI}{L^2}, \; 4\pi^2 \frac{EI}{L^2} \text{ etc.} \tag{20.7}$$

There are an infinite number of values of P for instability, corresponding with various modes of buckling, Fig. 20.2. The fundamental mode occurs at the lowest critical load

$$P_e = \pi^2 \frac{EI}{L^2} \tag{20.8}$$

This is known as the *Euler formula* and corresponds with buckling in a single longitudinal half-wave. The critical load

$$P = 2^2\pi^2 \frac{EI}{L^2} = 4\pi^2 \frac{EI}{L^2} \tag{20.9}$$

corresponds with buckling in two longitudinal half-waves, and so on for higher modes. In practice the critical load P_e is never exceeded because high stresses develop at this load and collapse of the strut ensues. We are not therefore concerned with buckling loads higher than the lowest buckling load. For all practical purposes the buckling load of a pin-ended strut is given by equation (20.8).

At this load a perfectly straight pin-ended strut is in a state of *neutral equilibrium*; the small deflection

$$v = B \sin kz$$

is indeterminate, since B itself is indeterminate. Theoretically, the strut is in equilibrium at the load $\pi^2 EI/L^2$ for any small value of B, corresponding with a condition of *neutral equilibrium*; at this buckling load we should expect to be able to push the strut into any sinusoidal wave of small amplitude. This can be verified experimentally by compressing a long slender strip of a material which remains elastic during bending.

At values of P less than $\pi^2 EI/L^2$ the strut is in *stable equilibrium;* if the strut is given a small lateral displacement, the displacement is recovered after the removal of the disturbance.

At values of P greater than $\pi^2 EI/L^2$ the strut is in a condition of *unstable equilibrium;* any small lateral disturbance produces motion and finally collapse of the strut. This,

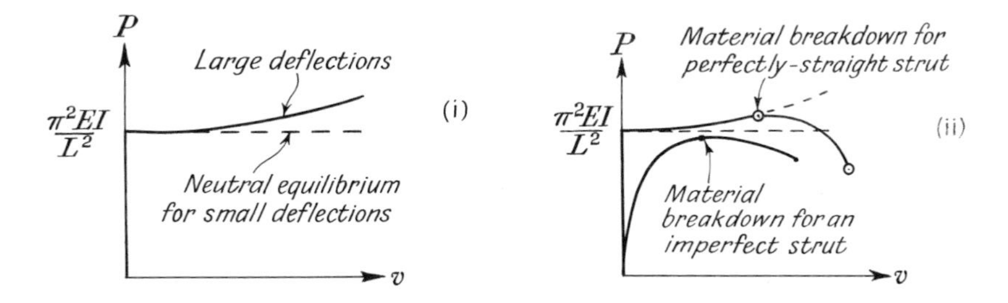

Fig. 20.3 Large deflections and material breakdown of struts.

however, is a hypothetical situation, since in practice the load $\pi^2 EI/L^2$ cannot be exceeded if the loads are static, and not applied suddenly.

The condition of neutral equilibrium at

$$P_e = \pi^2 \frac{EI}{L^2}$$

is only attained for small lateral displacements of the strut. When these displacements become large, the moment-curvature relation

$$M = -EI \frac{d^2 v}{dz^2}$$

is no longer valid; theoretically the problem becomes more difficult. The effect of large lateral displacements is to increase the flexural stiffness of the strut; in this case, provided the material remains elastic, end thrusts greater than $\pi^2 EI/L^2$ are attainable. If the thrust P is plotted against the lateral displacement v at any section, the P–v relation for a perfectly straight strut has the form shown in Fig. 20.3(i), when account is taken of large deflections. Lateral deflections become possible only when

$$P \geqslant \frac{\pi^2 EI}{L^2}$$

This analysis is restricted to the hypothetical case of a perfectly straight strut. When the strut has small imperfections, displacements v are possible for all values of P (Fig. 20.3(ii)), and the hypothetical condition of neutral equilibrium at

$$P = \frac{\pi^2 EI}{L^2}$$

is never attained. All materials have a limit of proportionality; when this is attained the flexural stiffness of the strut usually falls off rapidly. On the P–v diagram for the strut this corresponds with the development of a region of unstable equilibrium.

20.3 Pin-ended strut with eccentric end thrusts

In practice it is difficult, if not impossible, to apply the end thrusts along the longitudinal centroidal axis Cz of a strut. We consider now the effect of an eccentrically applied compressive load P on a uniform strut of flexural stiffness EI and length L.

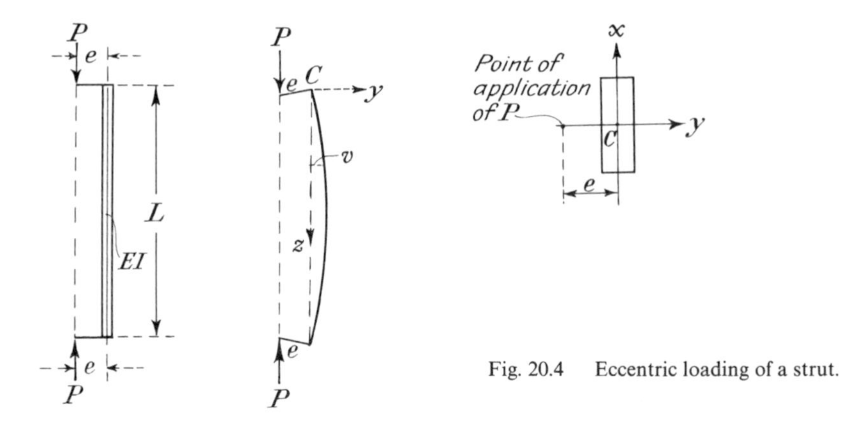

Fig. 20.4 Eccentric loading of a strut.

Suppose the end thrusts are applied at a distance e from the centroid and on the axis Cy of the cross-section. We assume again that the cross-section is that of a flat rectangular strip, Cx being the weaker axis of bending. The end thrusts P are applied to rigid arms attached to the ends of the strut.

An end load P causes the straight strut to bend; suppose v is the displacement of any point on Cz from its original position. The bending moment at that section is

$$M = P(e + v)$$

which is a sagging moment in relation to the axes Cz and Cy. If v is small we have

$$M = -EI \frac{d^2v}{dz^2}$$

Then

$$-EI \frac{d^2v}{dz^2} = P(e + v)$$

Thus

$$EI\frac{d^2v}{dz^2} + Pv = -Pe$$

When $e = 0$, this differential equation reduces to that already derived for an axially loaded strut.

As before, put

$$k^2 = \frac{P}{EI}$$

Then

$$\frac{d^2v}{dz^2} + k^2v = -k^2e$$

The complete solution is

$$v = A\cos kz + B\sin kz - e$$

Now $v = 0$ at $z = 0$ and $z = L$, so that

$$A - e = 0, \quad \text{and} \quad A\cos kL + B\sin kL - e = 0$$

The first of these equations gives $A = e$, and the second gives

$$B = \frac{e(1 - \cos kL)}{\sin kL}$$

Then

$$v = e(\cos kz - 1) + \frac{e(1 - \cos kL)}{\sin kL}\sin kz \tag{20.10}$$

The displacement v at the mid-length, $z = \frac{1}{2}L$, is

$$v_0 = e\left[\left(\cos\frac{kL}{2} - 1\right) + \frac{1 - \cos kL}{\sin kL}\sin\tfrac{1}{2}kL\right]$$

$$= e\left[\frac{2\sin\frac{1}{2}kL(1 - \cos\frac{1}{2}kL)}{\sin kL}\right] \tag{20.11}$$

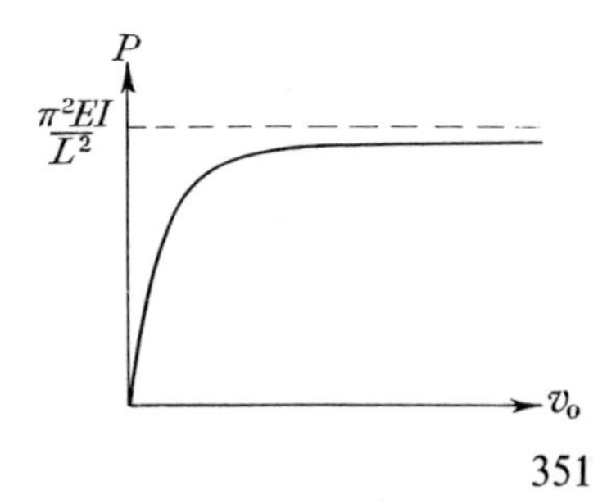

Fig. 20.5 Deflections of an eccentrically loaded strut.

If $\sin \frac{1}{2}kL \neq 0$, we have

$$v_0 = e(\sec \tfrac{1}{2}kL - 1) \tag{20.12}$$

When $P = 0$,

$$\tfrac{1}{2}kL = \frac{L}{2}\sqrt{\frac{P}{EI}} = 0$$

and $v_0 = 0$. As P approaches $\pi^2 EI/L^2$, $\frac{1}{2}kL$ approaches $\pi/2$, and

$$\sec \tfrac{1}{2}kL \to \infty$$

Thus values of v_0 are possible from the onset of loading; the values of v_0 increase non-linearly with increases of P. The value of $P = \pi^2 EI/L^2$ is not attainable since this would imply an infinitely large value of v_0, and material breakdown would occur at some smaller value of P.

It is interesting to evaluate the longitudinal stresses at the mid-length of the strut; the largest lateral deflection occurs at this section, and the greatest bending moment also occurs at this section, therefore. The bending moment is

$$M = P(v_0 + e) = Pe \sec \tfrac{1}{2}kL \tag{20.13}$$

Suppose c is the distance from the centroidal axis Cx to the extreme fibres of the strut. Then the longitudinal bending stress set up by M is

$$\sigma_1 = \frac{Mc}{I} = \frac{Pec \sec \frac{1}{2}kL}{I} \tag{20.14}$$

The average longitudinal compressive stress set up by P is

$$\sigma_2 = \frac{P}{A} \tag{20.15}$$

where A is the cross-sectional area of the strut. Then the maximum longitudinal compressive stress is

$$\sigma_{max} = \sigma_2 + \sigma_1 = \frac{P}{A} + \frac{Pec}{I} \sec \tfrac{1}{2}kL \tag{20.16}$$

Suppose $I = Ar^2$, where r is the radius of gyration of the cross-section about Cx. Then

$$\sigma_{max} = \frac{P}{A}\left[1 + \frac{ec}{r^2} \sec \tfrac{1}{2}kL\right] \tag{20.17}$$

The minimum compressive stress is

$$\sigma_{min} = \frac{P}{A}\left[1 - \frac{ec}{r^2} \sec \tfrac{1}{2}kL\right] \tag{20.18}$$

The value of P giving rise to a maximum compressive stress σ is

$$P = \frac{A\sigma}{1 + \frac{ec}{r^2}\sec\frac{1}{2}kL} \tag{20.19}$$

However, $\frac{1}{2}kL = \frac{L}{2}\sqrt{\frac{P}{EI}}$, and is therefore a function of P, so that the above equations must be solved by trial and error. A good approximation is derived as follows: let $\frac{1}{2}kL = \theta$, then for $0 < \theta < \frac{1}{2}\pi$

$$\sec\theta \doteqdot \frac{1 + 0{\cdot}26\left(\frac{2\theta}{\pi}\right)^2}{1 - \left(\frac{2\theta}{\pi}\right)^2} = \frac{P_e + 0{\cdot}26P}{P_e - P}$$

which leads to the following equation for P:

$$P^2\left(1 - 0{\cdot}26\frac{ec}{r^2}\right) - P\left[P_e\left(1 + \frac{ec}{r^2}\right) + \sigma A\right] + \sigma A P_e = 0$$

If $e = 0$, this has the roots $P = P_e$ or σA.

20.4 Initially-curved pin-ended strut

In practice a strut cannot be made perfectly straight, and our analysis for the flexure of a compressed bar would become more realistic if account could be taken of the slight deviations from straightness of the centroidal axis of a strut.

Consider again a strut consisting of a flat strip of material. Suppose the centroidal longitudinal axis is initially curved, the lateral displacement at any point being v_0 from the axis Oz, Fig. 20.6. Thrusts P are now applied at the ends of the strut and at the

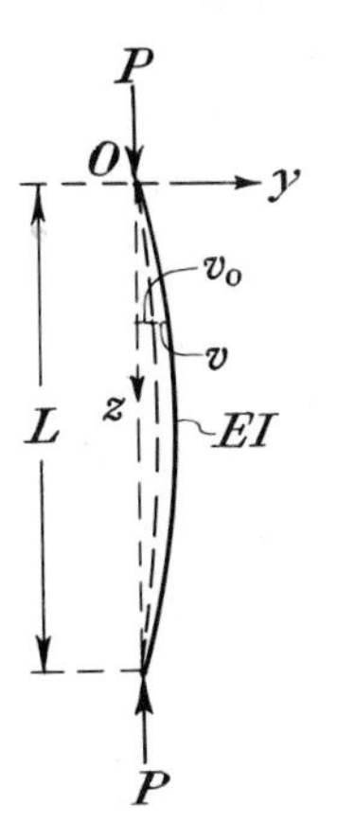

Fig. 20.6 Initially curved strut.

centroids of the end cross-sections. The strut then bends further from its initial unloaded position. Suppose v is the additional lateral displacement at any section due to the application of P. If the ends of the strut are pinned there can be no lateral forces at the ends. The bending moment at any section of the strut is

$$M = P(v + v_0)$$

If the strut is initially unstressed then the bending moment at any section is proportional to the change of curvature at that section. Then

$$M = -EI \frac{d^2 v}{dz^2}$$

since the change of curvature is due only to the additional displacement v of the strut and not the total displacement $(v + v_0)$. Then

$$EI \frac{d^2 v}{dz^2} + P(v + v_0) = 0$$

Put $P/EI = k^2$, as before. Then

$$\frac{d^2 v}{dz^2} + k^2 v = -k^2 v_0$$

Suppose for the sake of simplicity that v_0 is sinusoidal in form; take

$$v_0 = a \sin \frac{\pi z}{L} \tag{20.20}$$

where a is a constant, and is the initial lateral displacement at the centre of the strut. Then

$$\frac{d^2 v}{dz^2} + k^2 v = -k^2 a \sin \frac{\pi z}{L}$$

The general solution is

$$v = A \cos kz + B \sin kz + \frac{k^2 a}{\dfrac{\pi^2}{L^2} - k^2} \sin \frac{\pi z}{L}$$

If the ends are pinned we have

$$v = 0 \quad \text{at} \quad z = 0 \quad \text{and} \quad z = L$$

Then

$$A = 0 \quad \text{and} \quad B \sin kL = 0$$

If k is to assume any non-zero value we must have $B = 0$, so the relation for v reduces to

$$v = \frac{k^2 a}{\dfrac{\pi^2}{L^2} - k^2} \sin \frac{\pi z}{L} \tag{20.21}$$

This may be written

$$v = \frac{a \sin \dfrac{\pi z}{L}}{\dfrac{\pi^2}{k^2 L^2} - 1} \tag{20.22}$$

But $k^2 = P/EI$, so on putting $\pi^2 EI/L^2 = P_e$, we have

$$v = \frac{a \sin \dfrac{\pi z}{L}}{\dfrac{P_e}{P} - 1} = \frac{v_0}{\dfrac{P_e}{P} - 1} \tag{20.23}$$

Now P_e is the buckling load for the perfectly straight strut. The relation for v, which is the additional lateral displacement of the strut, shows that the effect of the end thrust P is to increase v_0 by the factor $1\bigg/\left(\dfrac{P_e}{P} - 1\right)$. Obviously as P approaches P_e, v tends to infinity. The additional displacement at the mid-length of the strut is

$$v_c = \frac{a}{\dfrac{P_e}{P} - 1} \tag{20.24}$$

This relation between P and v_c has the form shown in Fig. 20.7(i); v_c increases rapidly as P approaches P_e. Theoretically, the load P_e can only be attained at an infinitely large deflection. In practice material breakdown would occur before P_e could be

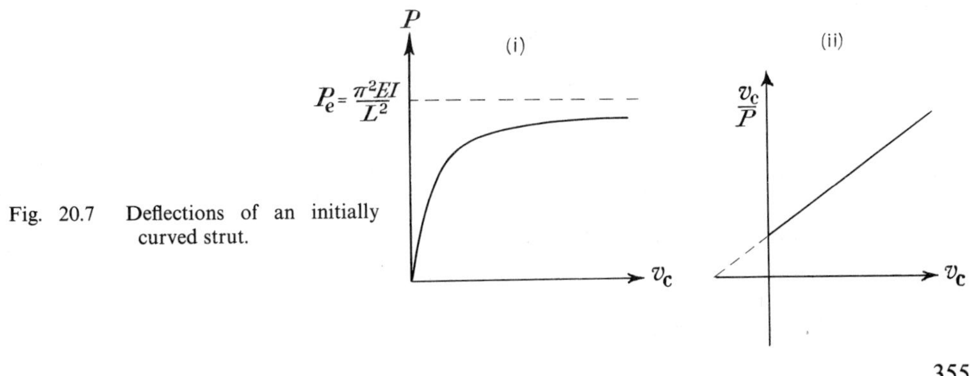

Fig. 20.7 Deflections of an initially curved strut.

attained, and at a finite displacement. We may write the relation for v_c in the form

$$P_e \frac{v_c}{P} - v_c = a \tag{20.25}$$

This gives a linear relation between (v_c/P) and v_c, Fig. 20.7. The negative intercept on the axis of v_c is equal to $(-a)$. If values of (v_c/P) and v_c are plotted in a strut test, it will be found that as the critical condition is approached these variables are related by a straight-line equation of the type discussed above. The slope of this straight line defines P_e, the buckling load for a perfectly-straight strut.

The P–v_c curve is asymptotic to the line $P = P_e$ if the material remains elastic. It is of considerable interest to evaluate the maximum longitudinal compressive stress in the strut. The maximum bending moment occurs at the mid-length, and has the value

$$M = P(a + v_c) = Pa\left[1 + \frac{1}{\dfrac{P_e}{P} - 1}\right] = Pa\left[\frac{P_e}{P_e - P}\right] \tag{20.26}$$

The maximum compressive stress occurs in an extreme fibre, and has the value

$$\sigma_{\max} = \frac{P}{A} + \frac{Pa \, . \, P_e}{P_e - P}\left(\frac{c}{I}\right) = \frac{P}{A}\left[1 + \frac{P_e}{P_e - P}\left(\frac{ac}{r^2}\right)\right] \tag{20.27}$$

where A is the area of the cross-section, c is the distance from the centroidal axis to the extreme fibres, and r is the relevant radius of gyration of the cross-section. Now P/A is the average stress on the strut; if this is equal to σ, then

$$\sigma_{\max} = \sigma\left[1 + \frac{\sigma_e}{\sigma_e - \sigma}\left(\frac{ac}{r^2}\right)\right] \tag{20.28}$$

where

$$\sigma_e = \frac{P_e}{A} = \pi^2 E\left(\frac{r}{L}\right)^2 \tag{20.29}$$

Suppose $\dfrac{ac}{r^2} = \eta$. Then

$$\sigma_{\max} = \sigma\left[1 + \frac{\eta\sigma_e}{\sigma_e - \sigma}\right] \tag{20.30}$$

Then

$$\sigma_{\max}(\sigma_e - \sigma) = \sigma[(1 + \eta)\sigma_e - \sigma]$$

Thus,

$$\sigma^2 - \sigma[\sigma_{\max} + (1 + \eta)\sigma_e] + \sigma_{\max}\sigma_e = 0$$

Then

$$\sigma = \tfrac{1}{2}[\sigma_{max} + (1 + \eta)\sigma_e] - \sqrt{\tfrac{1}{4}[\sigma_{max} + (1 + \eta)\sigma_e]^2 - \sigma_{max}\sigma_e} \tag{20.31}$$

We need not consider the positive square root, since we are only interested in the smaller of the two roots of the equation. This relation gives the value of average stress, σ, at which a maximum compressive stress σ_{max} would be attained for any value of η. If we are interested in the value of σ at which the yield stress σ_Y of a mild-steel strut is attained, we have

$$\sigma = \tfrac{1}{2}[\sigma_Y + (1 + \eta)\sigma_e] - \sqrt{\tfrac{1}{4}[\sigma_Y + (1 + \eta)\sigma_e]^2 - \sigma_Y\sigma_e} \tag{20.32}$$

20.5 Design of pin-ended struts

A commonly used structural material is mild steel. It has been found from tests on mild-steel pin-ended struts that failure of an initially-curved member takes place when the yield stress is first attained in one of the extreme fibres. From a wide range of tests Robertson concluded that the failing loads of mild-steel struts could be estimated if η is taken to be proportional to (L/r)—the slenderness ratio of the strut; Robertson suggests that

$$\eta = 0{\cdot}003\left(\frac{L}{r}\right) \tag{20.33}$$

This value of η gives

$$\sigma = \frac{1}{2}\left[\sigma_Y + \left(1 + 0{\cdot}003\,\frac{L}{r}\right)\sigma_e\right] - \sqrt{\frac{1}{4}\left[\sigma_Y + \left(1 + 0{\cdot}003\,\frac{L}{r}\right)\sigma_e\right]^2 - \sigma_Y\sigma_e} \tag{20.34}$$

This represents a transition curve between yielding of the material for low slenderness ratios, Fig. 20.8, and buckling at high slenderness ratios.

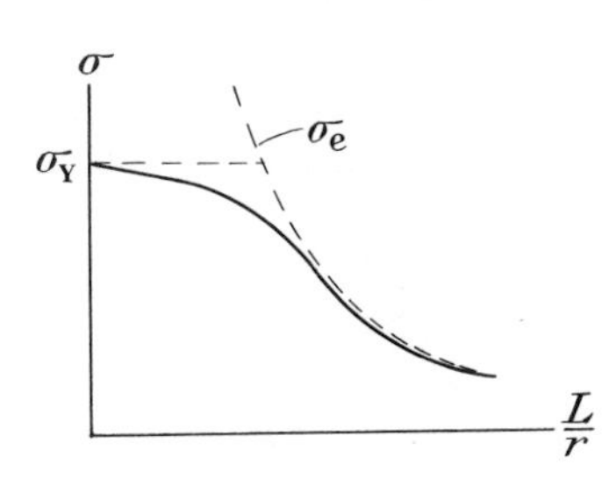

Fig. 20.8 Effect of material breakdown on the buckling of a strut.

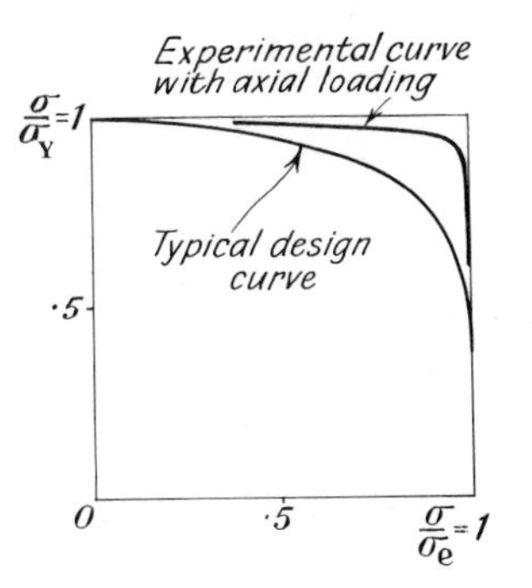

Fig. 20.9 'Interaction' curves for practical struts.

In the case of mild-steel struts under true axial loading buckling occurs either at σ_e—the elastic buckling load—or at σ_Y—the yield stress. If true axial loading could be achieved in practice, all struts would fail at stresses that could be represented either by $\sigma/\sigma_Y = 1$, or $\sigma/\sigma_e = 1$. In a series of strut tests it is found that the test results are usually defined by a curve on the $\sigma/\sigma_Y - \sigma/\sigma_e$ diagram, Fig. 20.9, and not by the two straight lines $\sigma/\sigma_Y = 1$ and $\sigma/\sigma_e = 1$. If the experimental technique is improved to give better axial-loading conditions the curve approaches these two straight lines. Any convenient transition curve on this diagram may be taken as a design curve for practical conditions of axial loading.

20.6 Strut with uniformly distributed lateral loading

In the preceding sections we considered the effects of end eccentricities and initial curvatures on the lateral bending of compressed struts; these produce lateral bending of the strut from the onset of compression.

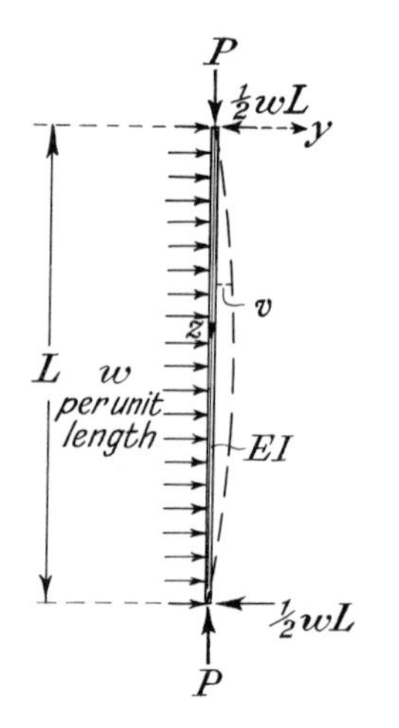

Fig. 20.10 Laterally loaded struts.

A similar problem arises when a compressed strut carries a lateral load. Consider a pin-ended strut of length L and uniform flexural stiffness EI, Fig. 20.10. Suppose the axial thrust on the strut is P, and that there is a lateral load of uniform intensity w per unit length. At the ends of the strut there are lateral shearing forces $\frac{1}{2}wL$.

If v is the lateral deflection at any point of the centroidal axis, then the bending moment at any section is

$$M = -EI\frac{d^2v}{dz^2} = Pv + \tfrac{1}{2}wLz - \tfrac{1}{2}wz^2$$

Then

$$\frac{d^2v}{dz^2} + \frac{Pv}{EI} = -\frac{w}{2EI}(Lz - z^2)$$

If $P/EI = k^2$, then

$$\frac{d^2v}{dz^2} + k^2v = -\frac{wk^2}{2P}(Lz - z^2)$$

The complete solution of this equation is

$$v = A\cos kz + B\sin kz - \frac{w}{2P}\left(Lz - z^2 + \frac{2}{k^2}\right)$$

in which A and B are arbitrary constants. Now, at $z = 0$ and $z = L$ we have $v = 0$, so

$$A - \frac{w}{Pk^2} = 0$$

and

$$A\cos kL + B\sin kL - \frac{w}{Pk^2} = 0$$

Then

$$A = \frac{w}{Pk^2}, \qquad B = \frac{w}{Pk^2}\left[\frac{1 - \cos kL}{\sin kL}\right]$$

Thus

$$v = \frac{w}{Pk^2}\left[\cos kz + \frac{(1 - \cos kL)}{\sin kL}\sin kz - 1 - \tfrac{1}{2}k^2(Lz - z^2)\right] \tag{20.35}$$

The maximum value of v occurs at the mid-length, $z = \frac{1}{2}L$, and is given by

$$v_{max} = \frac{w}{Pk^2}\left[\cos \tfrac{1}{2}kL + \frac{(1 - \cos kL)}{\sin kL}\sin \tfrac{1}{2}kL - 1 - \tfrac{1}{8}k^2L^2\right] \tag{20.36}$$

This may be written

$$v_{max} = \frac{w}{Pk^2}[\sec \tfrac{1}{2}kL - 1 - \tfrac{1}{8}k^2L^2] \tag{20.37}$$

The maximum bending moment also occurs at the mid-length, and has the value

$$M_{max} = Pv_{max} + \tfrac{1}{8}wL^2 \tag{20.38}$$

On substituting for v_{max}, we have

$$M_{max} = \frac{w}{k^2}[\sec \tfrac{1}{2}kL - 1 - \tfrac{1}{8}k^2L^2] + \tfrac{1}{8}wL^2 = \frac{w}{k^2}[\sec \tfrac{1}{2}kL - 1] \tag{20.39}$$

When P is small, k is also small, and

$$\sec \tfrac{1}{2}kL = \frac{1}{\cos \frac{1}{2}kL} \fallingdotseq \left[1 - \tfrac{1}{2}(\tfrac{1}{2}kL)^2 + \frac{1}{24}(\tfrac{1}{2}kL)^4\right]^{-1}$$

Thus, approximately,

$$\sec \tfrac{1}{2}kL \doteqdot 1 + \left[\tfrac{1}{8}(kL)^2 - \frac{1}{384}(kL)^4\right] + \left[\tfrac{1}{8}(kL)^2 - \frac{1}{384}(kL)^4\right]^2$$

$$\doteqdot 1 + \tfrac{1}{8}(kL)^2 + \frac{5}{384}(kL)^4 \qquad (20.40)$$

Then

$$v_{\max} = \frac{w}{Pk^2}\left[\frac{5}{384}k^4L^4\right] = \frac{5}{384}\frac{wL^4}{EI} \qquad (20.41)$$

This agrees with the value of the central deflection of a laterally loaded beam without end thrust. Similarly, when k is small,

$$M_{\max} = \frac{w}{k^2}\left[\tfrac{1}{8}k^2L^2\right] = \tfrac{1}{8}wL^2$$

If we write equation (20.39) in the form

$$M_{\max} = \frac{wL^2}{8}\left[\frac{8(\sec \frac{1}{2}kL - 1)}{k^2L^2}\right] \qquad (20.42)$$

the term in square brackets is the factor by which the bending moment due to w alone must be multiplied to give the correct bending moment.

20.7 Buckling of a strut with built-in ends

We have assumed so far that the ends of the strut are always hinged to some foundation. When the ends are supported so that no rotations can occur, Fig. 20.11, then the relevant mode of instability for the lowest critical load involves points of contraflexure at the quarter points. The buckling load is therefore the same as that of a pin-ended strut of half the length. Then

$$P_{cr} = \frac{\pi^2 EI}{(\frac{1}{2}L)^2} = 4\pi^2\frac{EI}{L^2} \qquad (20.43)$$

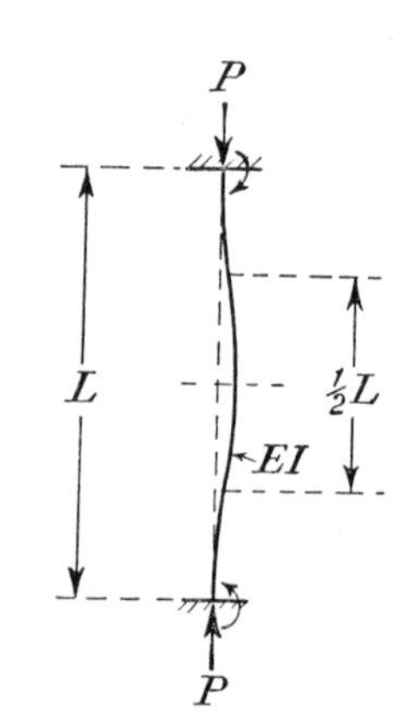

Fig. 20.11 Buckling of a strut with built-in ends.

When the ends of the strut are built-in, no restraining moments are induced at the ends until the strut develops a buckled form.

20.8 Buckling of a strut with one end fixed, and the other end free

When a vertical load P is applied to the free end of a vertical cantilever AB, at the lowest critical load the laterally deflected form of the strut is a sinusoidal wave of length

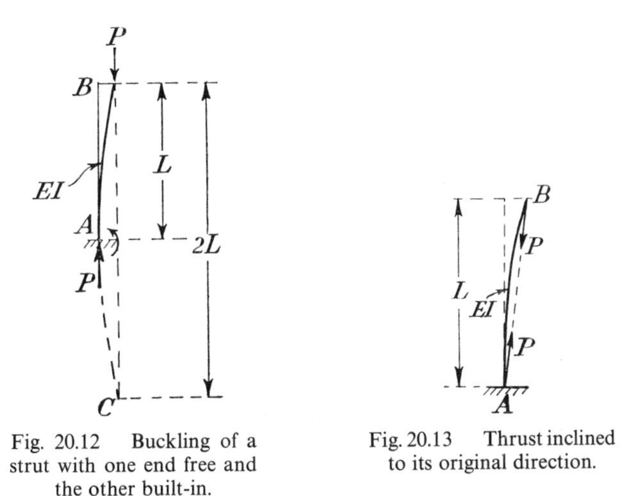

Fig. 20.12 Buckling of a strut with one end free and the other built-in.

Fig. 20.13 Thrust inclined to its original direction.

$2L$. If we consider the reflection of the buckled strut about A, Fig. 20.12, then the strut of length $2L$ behaves as a pin-ended strut. The buckling load is

$$P_{cr} = \frac{\pi^2 EI}{(2L)^2} = \frac{\pi^2 EI}{4L^2} \tag{20.44}$$

An important assumption in the preceding analysis is that the load at the free end of the cantilever is maintained in a vertical direction. If the load is always directed at A, that is, its line of action is BA, Fig. 20.13, in the buckled form, then there is no restraining moment at A, and the cantilever behaves as a pin-ended strut. The buckling load is

$$P_{cr} = \pi^2 \frac{EI}{L^2} \tag{20.45}$$

20.9 Buckling of a strut with one end pinned, and the other end fixed

For other combinations of end conditions we are usually led to more involved calculations. A strut is pinned at its upper end and built-in to a rigid foundation at the lower end, Fig. 20.14. In the buckled form of the strut a lateral shearing force F is induced at the upper end. If v is the deflection of the central axis of the strut parallel to the

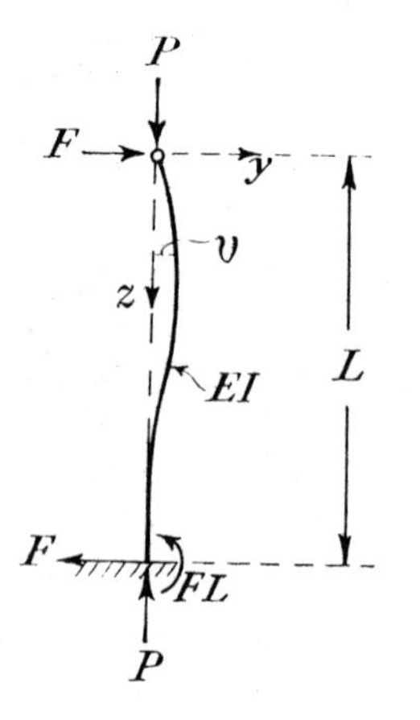

Fig. 20.14 Strut with one end pinned and the other end built-in.

y-axis, the bending moment at any section is

$$M = Pv - Fz$$

But

$$M = -EI\frac{d^2v}{dz^2}$$

Thus

$$-EI\frac{d^2v}{dz^2} = Pv - Fz$$

Put $k^2 = P/EI$. Then

$$\frac{d^2v}{dz^2} + k^2v = \frac{Fk^2z}{P}$$

The general solution is

$$v = A\cos kz + B\sin kz + \frac{F}{P}z$$

where A and B are arbitrary constants; the value of F is also unknown as yet, so there are three unknown constants in this equation. The boundary conditions are

$$v = 0, \quad \text{at} \quad z = 0$$

and

$$v = 0 \quad \text{and} \quad \frac{dv}{dz} = 0, \quad \text{at} \quad z = L$$

These give

$$A = 0$$

$$B \sin kL + \frac{FL}{P} = 0$$

$$Bk \cos kL + \frac{F}{P} = 0$$

The last two of these equations give

$$\frac{B}{F} = -\frac{L}{P \sin kL} = -\frac{1}{Pk \cos kL}$$

Thus

$$kL \cos kL = \sin kL \tag{20.46}$$

This equation gives the values of kL at which B and F are indeterminate, that is, at a condition of neutral equilibrium. The equation may be written

$$kL = \tan kL \tag{20.47}$$

The smallest non-zero value of kL satisfying this equation is approximately equal to 4·49 (see Fig. 20.15). This gives

$$P_{\mathrm{cr}} = k^2EI = 4{\cdot}49^2 \frac{EI}{L^2} = 20{\cdot}2 \frac{EI}{L^2}$$

We may derive an approximate value of kL in the following way: suppose kL is less than $3\pi/2$ by a small amount ϵ, then

$$kL = \frac{3\pi}{2} - \epsilon$$

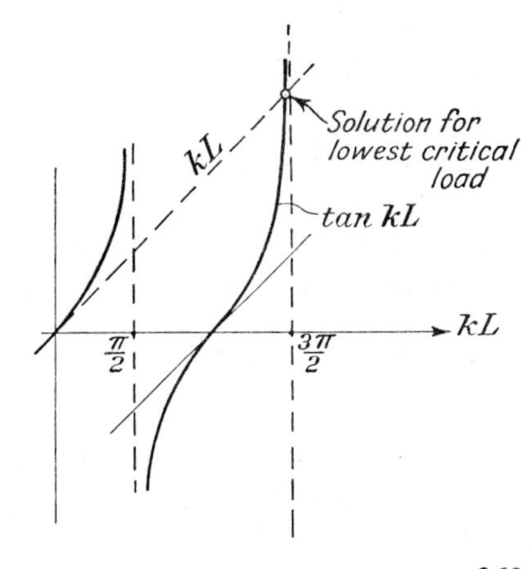

Fig. 20.15 Graphical determination of buckling load.

Then we are interested in the roots of the equation

$$\frac{3\pi}{2} - \epsilon = \tan\left(\frac{3\pi}{2} - \epsilon\right)$$

If ϵ is small, then

$$\frac{3\pi}{2} - \epsilon \doteqdot \cot \epsilon \doteqdot \frac{1}{\epsilon}(1 - \tfrac{1}{3}\epsilon^2)$$

Approximately

$$\frac{3\pi}{2} = \frac{1}{\epsilon}, \quad \text{or} \quad \epsilon = \frac{2}{3\pi}$$

Then

$$kL = \frac{3\pi}{2} - \frac{2}{3\pi} = \frac{9\pi^2 - 4}{6\pi}$$

and

$$P_{cr} = k^2 EI = \left(\frac{9\pi^2 - 4}{6\pi}\right)^2 \frac{EI}{L^2} = 20{\cdot}3\,\frac{EI}{L^2} \tag{20.48}$$

20.10 Flexural buckling of struts with other cross-sectional forms

In §20.2 we considered the strut to be in the form of a flat rectangular strip. We considered buckling to involve bending about the major axis Cx only, Fig. 20.16. In the case of a flat rectangular strip the axis Cx is clearly the weaker axis of bending. In practice, structural sections rarely have this simple cross-sectional form, but frequently have I-, or angle sections, or circular sections.

In general, if the cross-sectional form of a strut has two axes of symmetry, we can consider flexural instability about these two axes independently. If an I-section has two axes of symmetry in the cross-section, Fig. 20.17, flexural instability occurs usually about the axis of smaller bending stiffness, usually Cx. In a rectangular strut, Fig. 20.17, the weaker bending axis is parallel to the longer sides. Circular cross-sectional forms have the property that any two mutually perpendicular diameters are principal centroidal

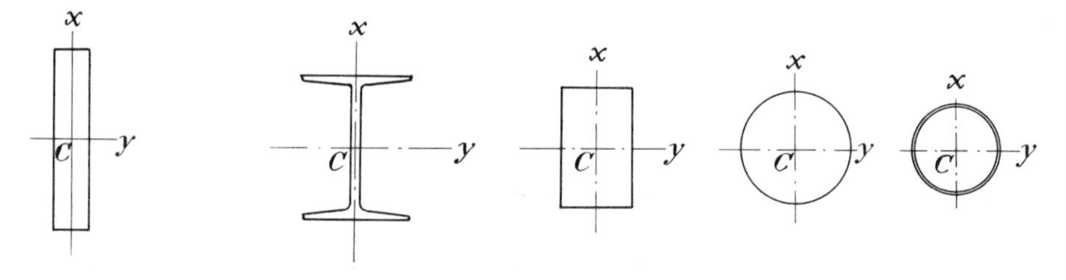

Fig. 20.16 Narrow strip cross-section.

Fig. 20.17 Cross-sections with two axes of symmetry.

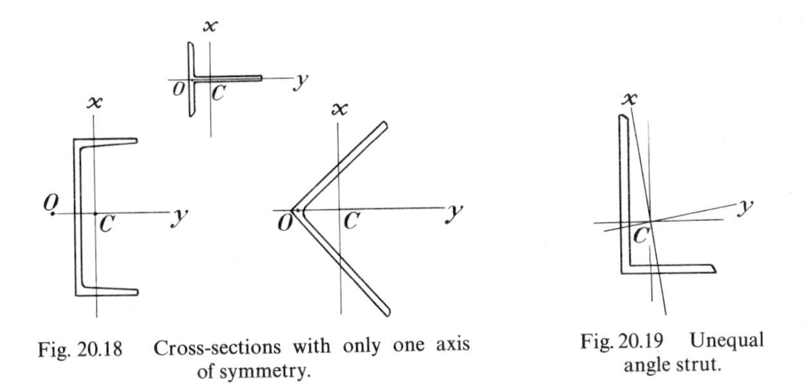

Fig. 20.18 Cross-sections with only one axis of symmetry.

Fig. 20.19 Unequal angle strut.

axes; for these sections flexural instability is equally likely about any principal centroidal axis, Fig. 20.17; when buckling occurs it is usually restricted to one plane. In making these statements we assume the ends of the strut are hinged about both axes Cy and Cx; this can be achieved in practice by loading through ball-ends. When the ends are not supported in the same way about Cy and Cx, then torsional effects may become important in the buckling behaviour.

In the case of cross-sectional forms with only one axis of symmetry, Cy, say, (Fig. 20.18), torsional effects become important if the shear centre is not coincident with the centroid. This is true of channel sections, T-sections, and equal angle sections. Although for certain struts flexural instability occurs about the weaker principal axis Cz, in general twisting also occurs.

In the case of cross-sectional forms with no axes of symmetry, Fig. 20.19, the buckled form always involves torsion, and the flexural buckling load has little meaning. This is true of unequal angle struts.

Problem 20.1: What thrust will a round steel rod take without buckling if it is 1·25 cm diameter, 2 m long, perfectly straight, and pin-jointed at the ends the load being applied exactly along the axis of the rod?

Solution

We have

$$I = \frac{\pi(0{\cdot}0125)^4}{64} = 1{\cdot}20 \times 10^{-9}\ \text{m}^4, \qquad L = 2\ \text{m}$$

Taking $E = 200\ \text{GN/m}^2$, we have

$$P_e = \frac{\pi^2 EI}{L^2} = 591\ \text{N}$$

20.11 Torsional buckling of a cruciform strut

We mentioned above that some struts are prone to torsional buckling effects. A cross-sectional form in which torsional instability occurs independently of any other form of buckling is a symmetrical cruciform section.

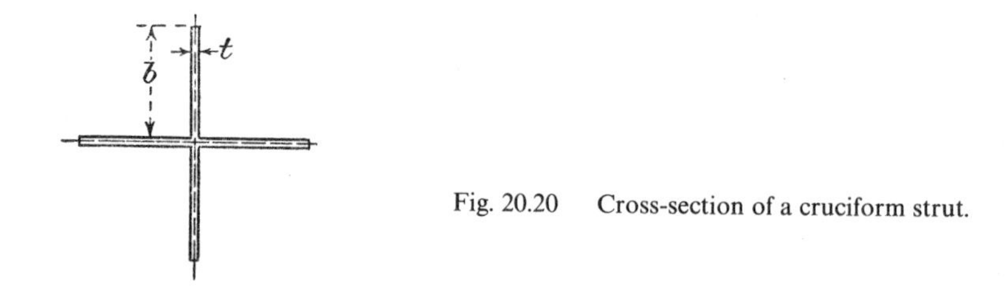

Fig. 20.20 Cross-section of a cruciform strut.

The cruciform has four equally spaced limbs each of breadth b and uniform thickness t, Fig. 20.20. Consider the section under a uniform compressive stress σ, Fig. 20.21(i). We consider the possibility that the section may become unstable by twisting about the longitudinal axis Cz, Fig. 20.21(ii); the stresses σ over the ends remain parallel to Cz during buckling.

Over any cross-section of the cruciform the stress is σ, acting parallel to Cz. Consider an elemental area δA of one limb at a distance x from the axis Cz, Fig. 20.21(iii). If the relative twist between two cross-sections a distance δz apart is $\delta\theta$, then the force

$$\sigma\,\delta A$$

on the elemental area is statically equivalent to a force $\sigma\,\delta A$ acting along the twisted form of the strut and a small force

$$\sigma\,\delta A x\,\frac{d\theta}{dz}$$

acting in the plane of the cross-section. The inclined forces $\sigma\,\delta A$ on the two cross-sections are in equilibrium with each other, but the two forces $\sigma\,\delta Ax(d\theta/dz)$ give rise to a resultant torque at any cross-section. This torque is

$$4\int_0^b \sigma x^2\,\frac{d\theta}{dx}\,dA = 4\sigma\,\frac{d\theta}{dz}\int_0^b x^2\,dA$$

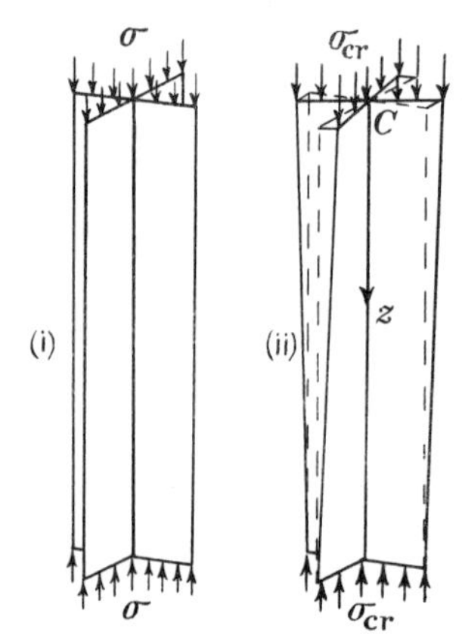

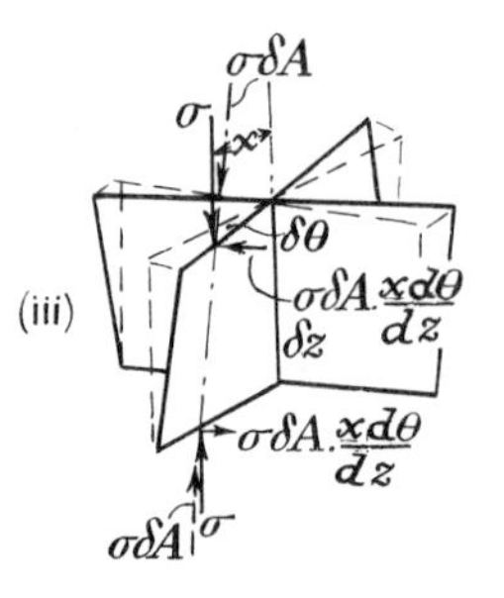

Fig. 20.21 Torsional buckling of a cruciform column.

since there are four limbs. The geometrical quantity $4\int_0^b x^2\, dA$ is the polar second moment of area of the cross-section about Cz. The resultant torque at any cross-section is then

$$\sigma \frac{d\theta}{dz} J_0$$

where

$$J_0 = 4\int_0^b x^2\, dA = 4t\int_0^b x^2\, dz = \tfrac{4}{3}b^3 t$$

Now, we found in Chapter 16 that the torque-twist relation for a cruciform section is

$$\text{Torque} = GJ\frac{d\theta}{dz} = \tfrac{4}{3}Gbt^3\frac{d\theta}{dz}$$

In the case of the compressed cruciform, the twisted form can be maintained if

$$\sigma \frac{d\theta}{dz} J_0 = GJ\frac{d\theta}{dz}$$

Then

$$\sigma = G\left(\frac{J}{J_0}\right) = G\left(\frac{\tfrac{4}{3}bt^3}{\tfrac{4}{3}b^3 t}\right) = G\left(\frac{t}{b}\right)^2 \tag{20.49}$$

20.12 Modes of buckling of a cruciform strut

With a knowledge of the torsional and flexural buckling loads of a cruciform strut, we can estimate the range of struts for which buckling is likely in the two modes.

If b is very much greater than t, and if all the limbs are similar in form, flexural buckling of a pin-ended strut is possible about any axis through the junction of the limbs, since the flexural stiffness is the same for all axes. For flexural instability the critical stress is

$$\sigma_f = \pi^2\frac{EI}{AL^2} \tag{20.50}$$

Now $I = \frac{1}{12}t(2b)^3 = \frac{2}{3}b^3 t$ and $A = 4bt$, and so

$$\sigma_f = \frac{\pi^2}{6}\frac{Eb^2}{L^2} \tag{20.51}$$

Now, as we have seen, the torsional buckling stress is independent of L, and has the value

$$\sigma_t = G\left(\frac{t}{b}\right)^2 \tag{20.52}$$

Then $\sigma_f > \sigma_t$ when

$$\frac{\pi^2}{6}\frac{Eb^2}{L^2} > G\left(\frac{t}{b}\right)^2$$

i.e. when

$$\frac{b^4}{L^2t^2} > \frac{6G}{\pi^2 E} = \frac{6}{2\pi^2(1 + \nu)} = \frac{3}{\pi^2(1 + \nu)} \tag{20.53}$$

If $\nu = 0{\cdot}3$, then

$$\frac{b^4}{L^2t^2} > \frac{3}{1{\cdot}3\pi^2} = 0{\cdot}234 \tag{20.54}$$

Thus torsional buckling takes place when

$$\frac{b^2}{Lt} > \sqrt{0{\cdot}234} = 0{\cdot}484$$

i.e. when

$$\frac{Lt}{b^2} < 2{\cdot}07$$

This condition may be written

$$\left(\frac{L}{b}\right) < 2{\cdot}07\left(\frac{b}{t}\right) \tag{20.55}$$

We can show the domains of flexural and torsional instability by plotting (L/b) against (b/t), Fig. 20.22. For a practical material, yielding or material breakdown occurs when L/b and b/t approach zero; the lower left-hand corner is therefore the yielding domain. Above the straight line

$$\left(\frac{L}{b}\right) = 2{\cdot}07\left(\frac{b}{t}\right)$$

buckling is by flexure, whereas below this line buckling is by torsion.

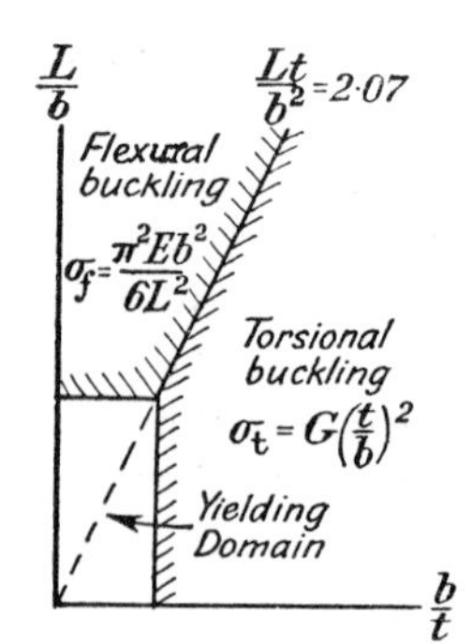

Fig. 20.22 Modes of buckling of a cruciform strut.

20.13 Lateral buckling of a narrow beam

We have seen that the axial compression of a slender strut can lead to a condition of neutral equilibrium, in which at a certain critical compressive load a flexural form of deformation becomes possible. In the case of a cruciform strut we have shown that a form of neutral equilibrium involving torsion is possible under certain conditions.

Problems of structural instability are not restricted entirely to compression members, although there are many problems of this type. In the case of deep beams, for example, lateral buckling may occur, involving torsion and bending perpendicular to the plane

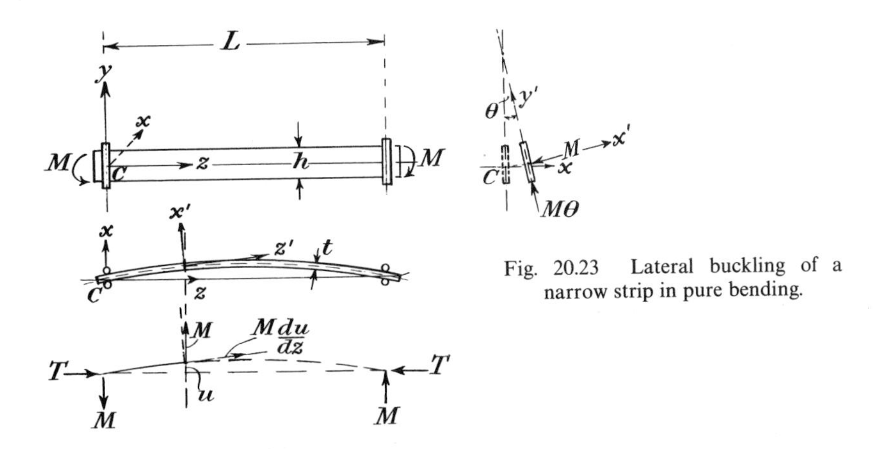

Fig. 20.23 Lateral buckling of a narrow strip in pure bending.

of the depth of the beam. In general this problem is a complex one; however, we can determine some of the factors involved by studying the relatively simple case of the bending of a narrow deep beam.

A long rectangular strip has a depth h and thickness t, which is small compared with h, Fig. 20.23. The principal centroidal axes are Cx, Cy, Cz. At the ends of the beam are vertical rollers which prevent twisting of the beam about a longitudinal axis. The distance between the end supports is L.

The beam is loaded with moments M applied at each end about axes parallel to Cx. Consider the possibility that the beam may become laterally unstable at some critical value of M. If $h \gg t$ then bending displacements in the yz plane may be neglected. Suppose in the buckled form the principal centroidal axes at any cross-section are Cx', Cy', Cz'. The lateral displacement parallel to Cx is u, and θ is the angle of twist about Cz at any cross-section. The moments M are assumed to be maintained along their original lines of action; the only other forces which may be induced at the ends are equal and opposite longitudinal torques T. The bending moment about the axis Cy' is then

$$M\theta$$

and since this gives rise to the curvature of the beam in the xz plane we have

$$M\theta = -EI_y \frac{d^2u}{dz^2}$$

where EI_y is the bending stiffness of the beam about Cy. Again, for twisting about Cz'

$$T + M\frac{du}{dz} = GJ\frac{d\theta}{dz}$$

where GJ is the torsional stiffness about Cx. Differentiation of the second equation gives

$$M\frac{d^2u}{dz^2} = GJ\frac{d^2\theta}{dz^2}$$

Thus, on eliminating u,

$$M\theta = -EI_y \frac{GJ}{M}\frac{d^2\theta}{dz^2}$$

Then

$$\frac{d^2\theta}{dz^2} + \frac{M^2}{GJEI_y}\theta = 0$$

Put

$$k^2 = \frac{M^2}{GJEI_y} \tag{20.56}$$

Then

$$\frac{d^2\theta}{dz^2} + k^2\theta = 0$$

The general solution is

$$\theta = A\cos kz + B\sin kz$$

where A and B are arbitrary constants. If $\theta = 0$ at $z = 0$, then $A = 0$. Further if $\theta = 0$ at $z = L$,

$$B\sin kL = 0$$

If $B = 0$, then both A and B are zero, and no buckling occurs; but B can be non-zero if

$$\sin kL = 0$$

We can disregard the root $kL = 0$, since the general solution is only valid if k is non-zero.

The relevant roots are

$$kL = \pi, 2\pi, 3\pi, \ldots \tag{20.57}$$

The smallest value of critical moment is

$$M_{cr} = k\sqrt{(GJ)(EI_y)} = \frac{\pi}{L}\sqrt{(GJ)(EI_y)}$$

Now, for a beam of rectangular cross-section,

$$GJ = \tfrac{1}{3}Ght^3, \qquad EI_y = \tfrac{1}{12}Eht^3 \tag{20.59}$$

Then

$$M_{cr} = \frac{\pi}{L}\sqrt{\tfrac{1}{36}GEh^2t^6} = \frac{\pi}{L}\frac{ht^3}{6}\sqrt{GE} \tag{20.60}$$

If

$$G = E/2(1 + \nu),$$

then

$$\sqrt{GE} = \sqrt{E^2/2(1 + \nu)} = \frac{E}{\sqrt{2(1 + \nu)}} \tag{20.61}$$

The maximum bending stress at the bending moment M_{cr} is

$$\sigma_{cr} = \frac{M_{cr}}{I_x}\frac{h}{2} = \frac{6M_{cr}}{h^2t} = \frac{\pi E}{\sqrt{2(1 + \nu)}}\frac{t^2}{hL} \tag{20.62}$$

For a strip of given depth h and thickness t, the buckling stress σ_{cr} is proportional to the inverse of (L/t), which is sometimes referred to as the slenderness ratio of the beam.

FURTHER PROBLEMS

(answers on page 399)

20.2 Calculate the buckling load of a pin-jointed strut made of round steel rod 2 cm diameter and 4 m long.

20.3 Find the thickness of a round steel tubular strut 3·75 cm external diameter, 2 m long, pin-jointed at the ends, to withstand an axial load of 10 kN.

20.4 Calculate the buckling load of a strut built-in at both ends, the cross-section being a square 1 cm by 1 cm, and the length 2 m. Take $E = 200$ GN/m^2.

20.5 A steel scaffolding pole acts as a strut, but the load is applied eccentrically—at 7·5 cm distance from the centre line with leverages in the same direction at top and bottom. The pole is tubular, 5 cm external diameter and 0·6 cm thick, 3 m in length between its ends—which are not fixed in direction. If the steel has a yield stress of 300 MN/m^2 and $E = 200$ GN/m^2, estimate approximately the load required to buckle the strut. (*RNEC*)

20.6 Two similar members of the same dimensions are connected together at their ends by two equal rigid links, the links being pin-jointed to the members. At the middle the members are rigidly connected by a

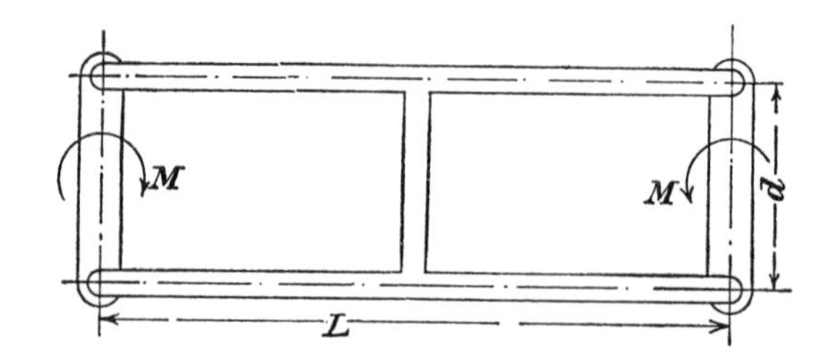

distance piece. Equal couples are applied to the links, the axes of the couples being parallel to the pins of the joints. Show that buckling will occur in the top member if the couples M exceed a value given by the root of the equation

$$\tan \tfrac{1}{2}kL = \tanh \tfrac{1}{2}kL$$

where $k^2 = M/EId$. (*Cambridge*)

21 Vibrations of beams

21.1 Introduction

In the problems discussed in preceding chapters we were concerned with systems of steady external loads. When a structure is subjected to pulsating, or suddenly applied, loads, we must take account of the dynamical behaviour of the components of the structure.

The problem of impact stresses is discussed briefly in Chapter 22; in this chapter we study the problem of a pulsating load applied to a simple beam.

The vibrational behaviour of a beam under the action of a pulsating, or alternating, load is dependent on the frequency of the alternating load and the frequency of free vibrations of the beam. A free vibration is induced in a beam when the beam is distorted in some arbitrary fashion, and then released completely.

21.2 Free vibrations of a mass on a beam

We can simplify the treatment of the free vibrations of a beam by considering its mass to be concentrated at the mid-length. Consider, for example, a uniform simply-supported beam of length L and flexural stiffness EI, Fig. 21.1. Suppose the beam

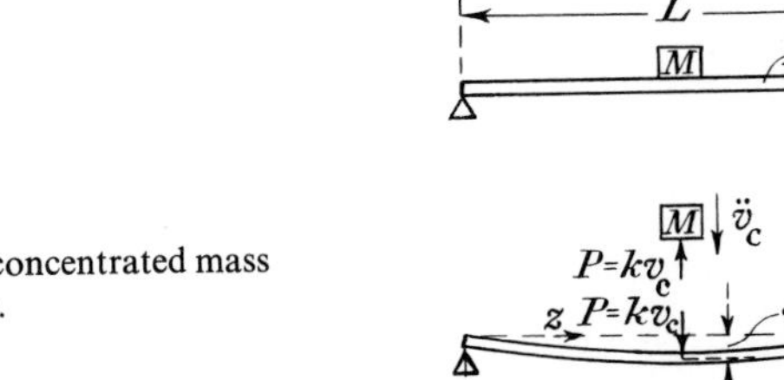

Fig. 21.1 Vibrations of a concentrated mass on a beam.

itself is mass-less, and that a concentrated mass M is held at the mid-span. If we ignore for the moment the effect of the gravitational field, the beam is undeflected when the mass is at rest. Now consider the motion of the mass when the beam is deflected laterally to some position and then released. Suppose v_c is the lateral deflection of the beam at the mid-span at a time t; since the beam is massless the force P on the beam at the mid-span is

$$P = \frac{48EIv_c}{L^3}$$

If $k = 48EI/L^3$, then

$$P = kv_c$$

The massless beam behaves then as a simple elastic spring of stiffness k. In the deflected position there is an equal and opposite reaction P on the mass. The equation of vertical motion of the mass is

$$M\frac{d^2v_c}{dt^2} = -P = -kv_c$$

Thus

$$\frac{d^2v_c}{dt^2} + \frac{kv_c}{M} = 0$$

The general solution of this differential equation is

$$v_c = A\cos\sqrt{\frac{k}{M}}\,t + B\sin\sqrt{\frac{k}{M}}\,t$$

where A and B are arbitrary constants; this may also be written in the form

$$v_c = C\sin\left(\sqrt{\frac{k}{M}}\,t + \epsilon\right)$$

where C and ϵ are also arbitrary constants. Obviously C is the amplitude of a simple-harmonic motion of the beam (Fig. 21.2); v_c first assumes its peak value when

$$\sqrt{\frac{k}{M}}\,t_1 + \epsilon = \frac{\pi}{2}$$

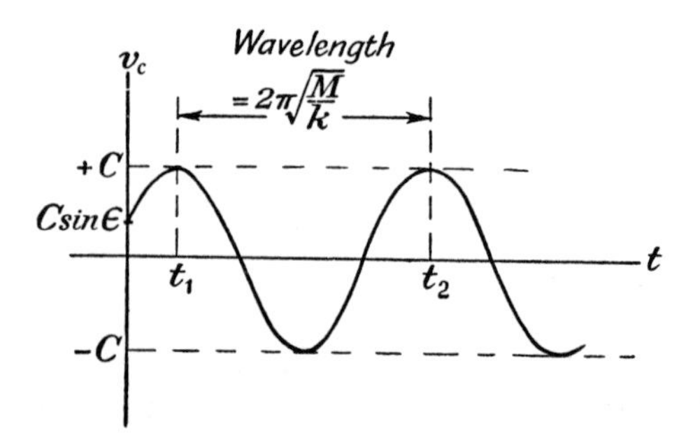

Fig. 21.2 Variation of displacement of beam with time.

and again attains this value when

$$\sqrt{\frac{k}{M}}\,t_2 + \epsilon = \frac{5\pi}{2}$$

The period T of one complete oscillation is then

$$T = t_1 - t_2 = 2\pi\sqrt{\frac{M}{k}} \tag{21.1}$$

The number of complete oscillations occurring in unit time is the frequency of vibrations; this is denoted by n, and is given by

$$n = \frac{1}{T} = \frac{1}{2\pi}\sqrt{\frac{k}{M}} \tag{21.2}$$

The behaviour of the system is therefore directly analogous to that of a simple mass-spring system. On substituting for the value of k we have

$$n = \frac{1}{2\pi}\sqrt{\frac{48EI}{ML^3}} \tag{21.3}$$

Problem 21.1: A steel I-beam, simply supported at each end of a span of 10 m, has a second moment of area of 10^{-4} m^4. It carries a concentrated mass of 500 kg at the mid-span. Estimate the natural frequency of lateral vibrations.

Solution

In this case

$$EI = (200 \times 10^9)(10^{-4}) = 20 \times 10^6 \text{ Nm}^2$$

Then

$$k = \frac{48EI}{L^3} = \frac{48(20 \times 10^6)}{(10)^3} = 960 \times 10^3 \text{ N/m}$$

The natural frequency is

$$n = \frac{1}{2\pi}\sqrt{\frac{k}{M}} = \frac{1}{2\pi}\sqrt{\frac{960 \times 10^3}{500}} = 6{\cdot}97 \text{ cycles/sec}$$

21.3 Free vibrations of a beam with distributed mass

Consider a uniform beam of length L, flexural stiffness EI, and mass m per unit length (Fig. 21.3); suppose the beam is simply-supported at each end, and is vibrating freely in the yz-plane, the displacement at any point parallel to the y-axis being v. We assume

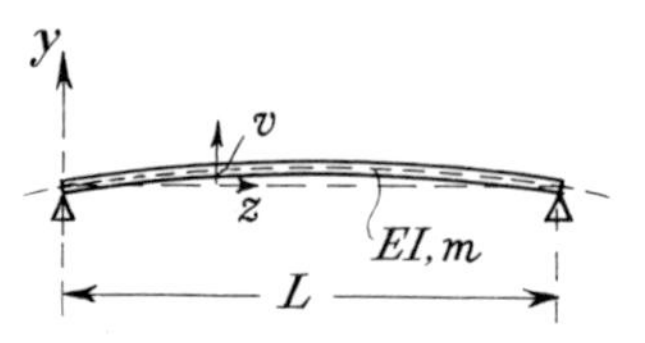

Fig. 21.3 Vibrations of a beam having an intrinsic mass.

firstly that the beam vibrates in a sinusoidal form

$$v = a \sin \frac{\pi z}{L} \sin 2\pi n t \tag{21.4}$$

where a is the lateral displacement, or amplitude, at the mid-length, and n is the frequency of oscillation. The kinetic energy of an elemental length δz of the beam is

$$\tfrac{1}{2} m \, \delta z \left(\frac{dv}{dt} \right)^2 = \tfrac{1}{2} m \, \delta z \left[2\pi n a \sin \frac{\pi z}{L} \cos 2\pi n t \right]^2$$

The bending strain energy in an elemental length is

$$\tfrac{1}{2} EI \left(\frac{d^2 v}{dz^2} \right)^2 \delta z = \tfrac{1}{2} EI \left[\frac{a\pi^2}{L^2} \sin \frac{\pi z}{L} \sin 2\pi n t \right]^2 \delta z$$

The total kinetic energy at any time t is then

$$\tfrac{1}{2} m \left[4\pi^2 n^2 a^2 \cos^2 2\pi n t \int_0^L \sin^2 \frac{\pi z}{L} \, dz \right] \tag{21.5}$$

The total strain energy at time t is

$$\tfrac{1}{2} EI \frac{a^2 \pi^4}{L^4} \sin^2 2\pi n t \int_0^L \sin^2 \frac{\pi z}{L} \, dz \tag{21.6}$$

For the free vibrations we must have the total energy, i.e. the sum of the kinetic and strain energies, is constant and independent of time. This is true if

$$\tfrac{1}{2} m (4\pi^2 n^2 a^2) \cos^2 2\pi n t + \tfrac{1}{2} EI \left(\frac{\pi^4 a^2}{L^4} \right) \sin^2 2\pi n t = \text{constant}$$

For this condition we must have

$$\tfrac{1}{2} m (4\pi^2 n^2 a^2) = \tfrac{1}{2} EI \left(\frac{\pi^4 a^2}{L^4} \right)$$

This gives

$$n^2 = \frac{\pi^2 EI}{4 m L^4} \tag{21.7}$$

Now $mL = M$, say, is the total mass of the beam, so that

$$n = \frac{\pi}{2}\sqrt{\frac{EI}{ML^3}} \tag{21.8}$$

This is the frequency of oscillation of a simply-supported beam in a single sinusoidal half-wave. If we consider the possibility of oscillations in the form

$$v = a \sin \frac{2\pi z}{L} \sin 2\pi n_2 t$$

then proceeding by the same analysis we find that

$$n_2 = 4n_1 = 2\pi\sqrt{\frac{EI}{ML^3}} \tag{21.9}$$

This is the frequency of oscillations of two sinusoidal half-waves along the length of the beam, Fig. 21.4, and corresponds to the second mode of vibration. Other higher modes are found similarly.

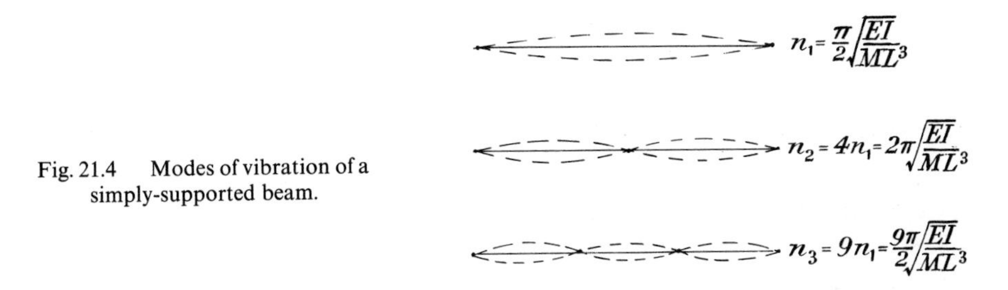

Fig. 21.4 Modes of vibration of a simply-supported beam.

As in the case of the beam with a concentrated mass at the mid-length, we have ignored gravitation effects; when the weight of the beam causes initial deflections of the beam, oscillations take place about this deflected condition; otherwise the effects of gravity may be ignored.

The effect of distributing the mass uniformly along a beam, compared with the whole mass being concentrated at the mid-length, is to increase the frequency of oscillations from

$$\frac{1}{2\pi}\sqrt{\frac{48EI}{ML^3}} \quad \text{to} \quad \frac{\pi}{2}\sqrt{\frac{EI}{ML^3}}$$

If

$$n_1 = \frac{1}{2\pi}\sqrt{\frac{48EI}{ML^3}}, \quad \text{and} \quad n_2 = \frac{\pi}{2}\sqrt{\frac{EI}{ML^3}}$$

then

$$\frac{n_2}{n_1} = \left(\frac{\pi}{2}\right)\frac{2\pi}{\sqrt{48}} = \frac{\pi^2}{4\sqrt{3}} = 1{\cdot}42 \tag{21.10}$$

Problem 21.2: If the steel beam of Problem 21.1 has a mass of 15 kg per metre run, estimate the lowest natural frequency of vibrations of the beam itself.

Solution

The lowest natural frequency of vibrations is

$$n_1 = \frac{\pi}{2}\sqrt{\frac{EI}{ML^3}}$$

Now

$$EI = 20 \times 10^6 \text{ Nm}^2$$

and

$$ML^3 = (15)(10)(10)^3 = 150 \times 10^3 \text{ kgm}^3$$

Then

$$\frac{EI}{ML^3} = \frac{20 \times 10^6}{150 \times 10^3} = 133 \text{ s}^{-2}$$

Thus

$$n_1 = \frac{\pi}{2}\sqrt{133} = 18{\cdot}1 \text{ cycles/sec}$$

21.4 Forced vibrations of a beam carrying a single mass

Consider a light beam, simply-supported at each end and carrying a mass M at mid-span, Fig. 21.5. Suppose the mass is acted upon by an alternating lateral force

$$P \sin 2\pi N t \tag{21.11}$$

which is applied with a frequency N. If v_c be the central deflection of the beam, then the equation of motion of the mass is

$$M\frac{d^2 v_c}{dt^2} + kv_c = P \sin 2\pi N t$$

where $k = 48EI/L^3$. Then

$$\frac{d^2 v_c}{dt^2} + \frac{k}{M}v_c = \frac{P}{M}\sin 2\pi N t$$

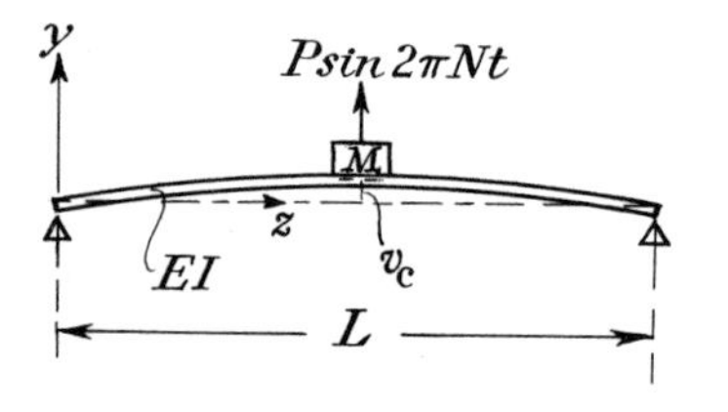

Fig. 21.5 Alternating force applied to a beam.

The general solution is

$$v_c = A\cos\sqrt{\frac{k}{M}}\,t + B\sin\sqrt{\frac{k}{M}}\,t + \frac{\frac{P}{k}\sin 2\pi N t}{1 - 4\pi^2 N^2 \frac{M}{k}} \tag{21.12}$$

in which A and B are arbitrary constants. Suppose initially, i.e. at time $t = 0$, both v_c and dv_c/dt are zero. Then $A = 0$, and

$$B = -\frac{2\pi N \cdot \frac{P}{k}}{1 - 4\pi^2 N^2 \cdot \frac{M}{k}} \frac{1}{\sqrt{\frac{k}{M}}}$$

Then

$$v_c = \frac{P/k}{1 - 4\pi^2 N^2 \cdot \frac{M}{k}}\left[\sin 2\pi N t - 2\pi N\sqrt{\frac{M}{k}}\sin\sqrt{\frac{k}{M}}\,t\right] \tag{21.13}$$

Now, the natural frequency of free vibrations of the system is

$$n = \frac{1}{2\pi}\sqrt{\frac{k}{M}}$$

Then

$$\sqrt{k/M} = 2\pi n$$

and

$$v_c = \frac{P/k}{1 - N^2/n^2}\left[\sin 2\pi N t - \frac{N}{n}\sin 2\pi n t\right] \tag{21.14}$$

Now, the maximum value which

$$\left(\sin 2\pi N t - \frac{N}{n}\sin 2\pi n t\right)$$

may assume is $\left[1 + \frac{N}{n}\right]$, and occurs when $\sin 2\pi N t = -\sin 2\pi n t = 1$. Then

$$v_{c\,\text{max}} = \frac{P/k\left(1 + \frac{N}{n}\right)}{1 - \frac{N^2}{n^2}} = \frac{P/k}{1 - \frac{N}{n}} \tag{21.15}$$

Thus, if $N < n$, $v_{c\,\text{max}}$ is positive and in phase with the alternating load $P \sin 2\pi N t$. As N approaches n, the values of $v_{c\,\text{max}}$ become very large. When $N > n$, $v_{c\,\text{max}}$ is negative and out of phase with $P \sin 2\pi N t$. When $N = n$, the beam is in a condition of resonance.

21.5 Damped free oscillations of a beam

The free oscillations of practical systems are inhibited by damping forces. One of the commonest forms of damping is known as velocity, or *viscous*, damping; the damping

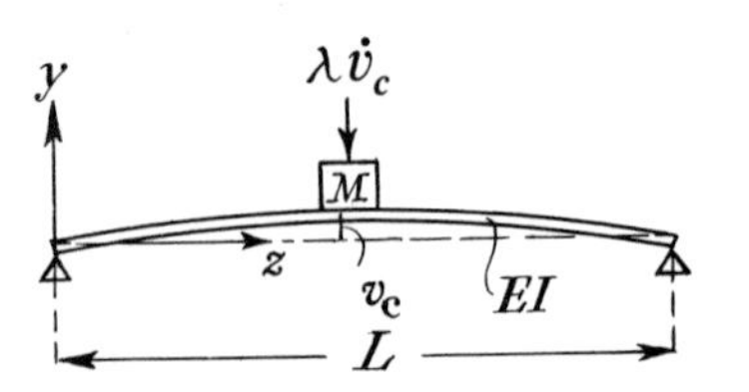

Fig. 21.6 Effect of damping on free vibrations.

force on a particle or mass is proportional to its velocity. Suppose in the beam problem discussed in §21.2 we have as the damping force $\mu(dv_c/dt)$. Then the equation of motion of the mass is

$$M\frac{d^2v_c}{dt^2} = -kv_c - \mu\frac{dv_c}{dt}$$

Thus

$$M\frac{d^2v_c}{dt^2} + \mu\frac{dv_c}{dt} + kv_c = 0$$

Hence

$$\frac{d^2v_c}{dt^2} + \frac{\mu}{M}\frac{dv_c}{dt} + \frac{k}{M}v_c = 0$$

The general solution of this equation is

$$v_c = Ae^{\{-\mu/2M + \sqrt{(\mu/2M)^2 - k/M}\}t} + Be^{\{-\mu/2M - \sqrt{(\mu/2M)^2 - k/M}\}t} \tag{21.16}$$

Now (k/M) is usually very much greater than $(\mu/2M)^2$, and so we may write

$$\begin{aligned} v_c &= Ae^{(-\mu/2M + i\sqrt{k/M})t} + Be^{(-\mu/M - i\sqrt{k/M})t} \\ &= e^{-(\mu/2M)t}[Ae^{i\sqrt{(k/M)}t} + Be^{-i\sqrt{(k/M)}t}] \\ &= e^{-(\mu/2M)t}\left[C\cos\left\{\sqrt{\frac{k}{M}}t + \epsilon\right\}\right] \end{aligned} \tag{21.17}$$

Thus, when damping is present, the free vibrations given by

$$C\cos\left(\sqrt{\frac{k}{M}}t + \epsilon\right)$$

are damped out exponentially, Fig. 21.7. The peak values on the curve of v_c correspond

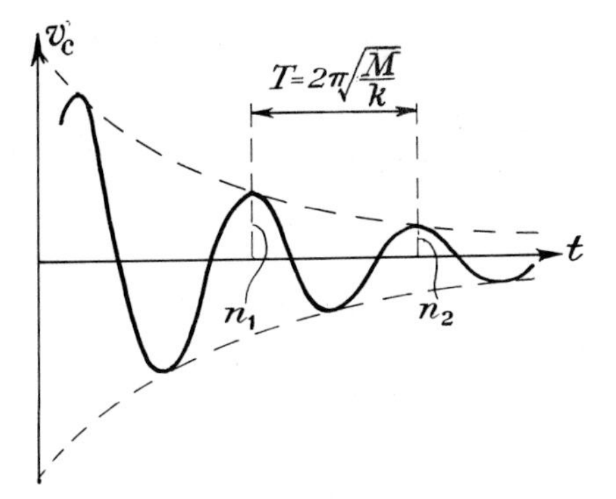

Fig. 21.7 Form of damped oscillation of a beam.

to points of zero velocity. These are given by

$$\frac{dv_c}{dt} = 0$$

or

$$\sqrt{\frac{k}{M}} \sin\left(\sqrt{\frac{k}{M}}\, t + \epsilon\right) - \frac{\mu}{2M} \cos\left(\sqrt{\frac{k}{M}}\, t + \epsilon\right) = 0$$

Obviously the higher peak values are separated in time by an amount

$$T = 2\pi \sqrt{\frac{M}{k}}$$

We note that successive peak values are in the ratio

$$\frac{v_{c1}}{v_{c2}} = \frac{e^{-(\mu/2M)t}[C \cos(\sqrt{(k/M)}t + \epsilon)]}{e^{-(\mu/2M)(t + 2\pi\sqrt{M/k})}[C \cos(\sqrt{(k/M)}t + \epsilon)]} = e^{(\mu/M)\pi\sqrt{M/k}} \tag{21.18}$$

Then

$$\log_e \frac{v_{c1}}{v_{c2}} = \frac{\pi\mu}{M} \sqrt{\frac{M}{k}} \tag{21.19}$$

Now

$$n = \frac{1}{2\pi} \sqrt{\frac{k}{M}}$$

Thus

$$\log_e \frac{v_{c1}}{v_{c2}} = \frac{\mu}{2Mn} \tag{21.20}$$

Hence

$$\mu = 2Mn \log_e \frac{v_{c1}}{v_{c2}} \tag{21.21}$$

21.6 Damped forced oscillations of a beam

We imagine that the mass on the beam discussed in §21.5 is excited by an alternating force $P \sin 2\pi Nt$. The equation of motion becomes

$$M \frac{d^2 v_c}{dt^2} + \mu \frac{dv_c}{dt} + kv_c = P \sin 2\pi Nt$$

The complementary function is the damped free oscillation; since this decreases rapidly in amplitude we may assume it to be negligible after a very long period. Then the particular integral is

$$v_c = \frac{P \sin 2\pi Nt}{MD^2 + \mu D + k}$$

This gives

$$v_c = \frac{P[(k - 4\pi^2 N^2 M) \sin 2\pi Nt - 2\pi N\mu \cos 2\pi Nt]}{(k - 4\pi^2 N^2 M)^2 + 4\pi^2 N^2 \mu^2} \tag{21.22}$$

If we write

$$n = \frac{1}{2\pi} \sqrt{\frac{k}{M}}$$

then

$$v_c = P \left[\frac{k\left(1 - \dfrac{N^2}{n^2}\right) \sin 2\pi Nt - 2\pi N\mu \cos 2\pi Nt}{k^2 \left(1 - \dfrac{N^2}{n^2}\right)^2 + 4\pi^2 N^2 \mu^2} \right] \tag{21.23}$$

The amplitude of this forced oscillation is

$$v_{c\,\max} = \frac{P}{\sqrt{k^2 \left(1 - \dfrac{N^2}{n^2}\right)^2 + 4\pi^2 N^2 \mu^2}} \tag{21.24}$$

21.7 Vibrations of a beam with end thrust

In general, when a beam carries end thrust the period of free undamped vibrations is greater than when the beam carries no end thrust. Consider the uniform beam shown

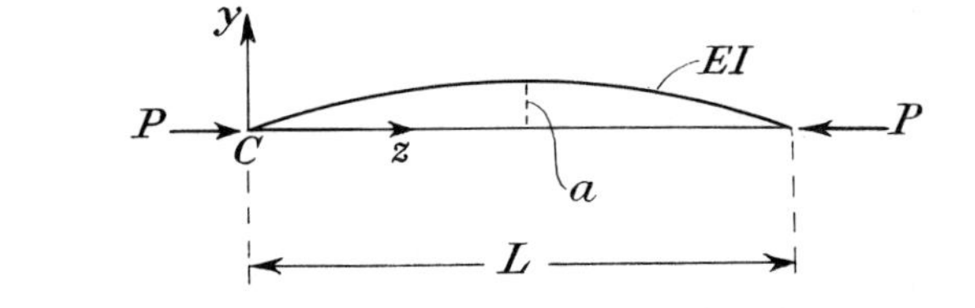

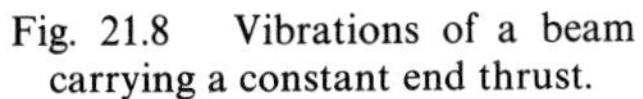
Fig. 21.8 Vibrations of a beam carrying a constant end thrust.

in Fig. 21.8; suppose the beam is vibrating in the fundamental mode so that the lateral displacement at any section is given by

$$v = a \sin \frac{\pi z}{L} \sin 2\pi nt \tag{21.25}$$

If these displacements are small, the shortening of the beam from the straight configuration is approximately

$$\int_0^L \tfrac{1}{2}\left(\frac{dv}{dz}\right)^2 dz = \frac{a^2\pi^2}{4L} \sin^2 2\pi nt \tag{21.26}$$

If m is the mass per unit length of the beam, the total kinetic energy at any instant is

$$\int_0^L \tfrac{1}{2}m\left(2\pi a \sin \frac{\pi z}{L} \cos 2\pi nt\right)^2 dz = m\pi^2 a^2 n^2 L \cos^2 2\pi nt \tag{21.27}$$

The total potential energy of the system is the strain energy stored in the strut together with the potential energy of the external loads; the total potential energy is then

$$\left[\tfrac{1}{4}EIL\left(\frac{a\pi^2}{L^2}\right)^2 - \tfrac{1}{4}P\left(\frac{a^2\pi^2}{L}\right)\right] \sin^2 2\pi nt \tag{21.28}$$

If the total energy of the system is the same at all instants

$$m\pi^2 a^2 n^2 L = \tfrac{1}{4}EIL\left(\frac{a\pi^2}{L^2}\right)^2 - \tfrac{1}{4}P\left(\frac{a^2\pi^2}{L}\right)$$

This gives

$$n^2 = \frac{\pi^2 EI}{4mL^4}\left[1 - \frac{P}{P_e}\right] \tag{21.29}$$

where

$$P_e = \frac{\pi^2 EI}{L^2}$$

and is the Euler load of the column. If we write

$$n_1{}^2 = \frac{\pi^2 EI}{4mL^4} \tag{21.30}$$

then

$$n = n_1 \sqrt{1 - \frac{P}{P_e}} \tag{21.31}$$

Clearly, as P approaches P_e, the natural frequency of the column diminishes and approaches zero.

FURTHER PROBLEMS

(answers on page 399)

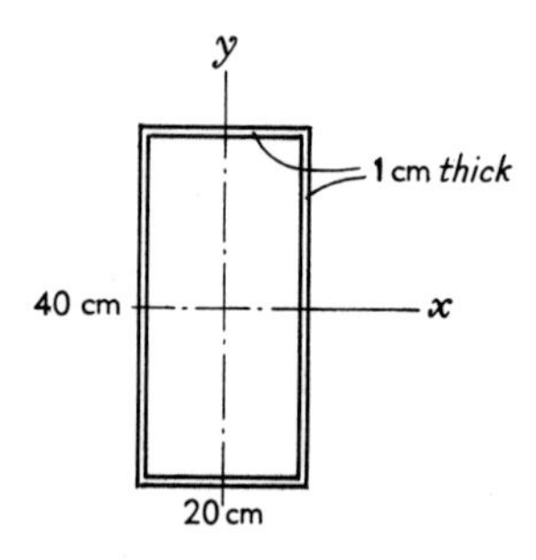

21.3 A doubly-symmetrical beam consists of a hollow rectangular steel section, having the cross-section shown, and of length 10 m. It is simply-supported in bending about both axes Cx, Cy at the ends. Estimate the lowest few natural frequencies of lateral vibrations of the beam about the axes Cx and Cy. Take $E = 200\ \text{GN/m}^2$.

21.4 If the beam of Problem 21.3 carries an axial thrust of 10^3 kN, what is the lowest natural frequency of the beam?

21.5 A light, uniform cantilever, of length L and uniform flexural stiffness EI, carries a mass M at the free end. Estimate the natural frequency of vibrations.

22 Impact stresses in rods

22.1 Introduction

When a structure is struck by a moving body, impact stresses are set up in the structure at the instant of impact. Not all regions of the structure are affected immediately by these impact stresses.

Consider first a long straight rod which is free at one end and held rigidly at the other. Suppose a rigid body strikes the rod axially at the free end. The particles of the rod at this loaded end begin to move under the action of the applied forces resulting from the impact, and we are at once led to consider the effects of the inertias of these particles. As the particles at the loaded end are displaced they build up a wave of compression, which travels eventually to the remote end of the rod. If the reader has ever observed a locomotive starting a long stationary goods train, and a similar train running into fixed buffers, it will be helpful to recall the experience. In the case of starting a train, all the trucks do not begin to move at once, but each starts up the one behind it, so that a 'wave' of tightening up of couplings travels along the train; the last truck of the train is unaffected until the truck in front of it has been moving long enough to tighten the coupling chain. So it is when an axial force is applied suddenly to the end of a rod; *stress waves* travel along the rod, and until the waves reach any particular section we assume there is no stress at, or beyond, this section.

Let us return to the analogy of the train and examine matters a little farther: as each truck is set in motion by the one in front of it, its motion is resisted by the inertia of the truck behind it. But the last truck meets no such resistance, and, as it is acted on by the same pull as the others (neglecting the effects of ground friction), it will tend to move faster than the trucks in front of it; in other words, the last truck will run after the next in front and push it forwards, and so on until the front of the train is reached. Thus a push 'wave' travels up the train, i.e., the wave is reversed in type at the free end of the train. Next, consider the case of the train running into fixed buffers: the engine will be brought to rest first, and gradually each truck in turn will feel the resistance in front and be brought to rest until the last is reached, i.e., a push 'wave' travels down the train from front to rear. But the last truck, having nothing behind it, will rebound and start moving backwards, and so exert a pull on the next truck, and thus a pull 'wave' travels up the train to the engine. Again we see that the type of wave is reversed at the free end. When the pull 'wave' reaches the engine the latter will resist being pulled backwards, and will therefore exert a forward *pull* on the first truck. Thus a pull 'wave' meeting the fixed end of the train is sent back as another pull 'wave,' i.e., the type of wave is not reversed at the fixed end.

So it is with the rod we were considering: a pull suddenly applied to one end travels as a wave of tension to the fixed end, and is reflected from there as another wave of tension; this reflected wave will be reflected from the free end as a wave of compression, and so on.

We remark here two important rules: (i) a wave is reflected from a fixed end without reversal of type: (ii) at a free end the reflected wave is one of tension if the original wave were one of compression—and vice versa.

We shall now apply approximate analytical processes to investigate the phenomena we have described, and derive formulae for the stresses.

22.2 Velocity of propagation of stress in a straight rod

Consider a long straight rod, Fig. 22.1, to which are applied longitudinal impact stresses. The rod is uniform, having a cross-sectional area A, and a Young's modulus E.

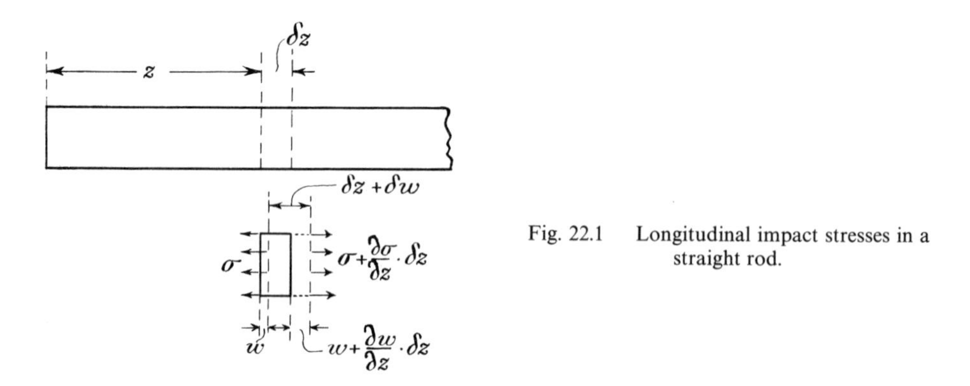

Fig. 22.1 Longitudinal impact stresses in a straight rod.

Consider an elemental length δz of the rod at a distance z from one end; suppose at any time t that the longitudinal displacement of the whole cross-section at a distance z is w parallel to the z axis. On the further edge of the elemental length, the longitudinal displacement is

$$w + \frac{\partial w}{\partial z} \cdot \delta z$$

Thus, the elemental length is stretched an amount

$$\frac{\partial w}{\partial z} \cdot \delta z$$

and the extensional strain of the element is

$$\epsilon_z = \frac{\partial w}{\partial z} \tag{22.1}$$

At the same instant t, suppose the tensile stress in the rod at a distance z is σ; then, on the further edge of the elemental length, the tensile stress is

$$\sigma + \frac{\partial \sigma}{\partial z} \cdot \partial z$$

For longitudinal equilibrium of the element we have

$$A \cdot \frac{\partial \sigma}{\partial z} \cdot \delta z = \rho A \cdot \delta z \cdot \frac{\partial^2 w}{\partial t^2}$$

where ρ is the density of the material. Thus

$$\frac{\partial \sigma}{\partial z} = \rho \frac{\partial^2 w}{\partial z^2} \tag{22.2}$$

If we ignore the effects of lateral strain,

$$\sigma = E\epsilon_z = E \frac{\partial w}{\partial z} \tag{22.3}$$

Then equation (22.2) becomes

$$E \frac{\partial^2 w}{\partial z^2} = \rho \frac{\partial^2 w}{\partial t^2} \tag{22.4}$$

If we put

$$c = \sqrt{\frac{E}{\rho}} \tag{22.5}$$

equation (22.4) may be written

$$\frac{\partial^2 w}{\partial t^2} = c^2 \frac{\partial^2 w}{\partial z^2} \tag{22.6}$$

The general solution of this partial differential equation is

$$w = f(z + ct) + g(z - ct) \tag{22.7}$$

where f is any function of $(z + ct)$, and g is any function of $(z - ct)$. Consider first the function

$$g(z - ct)$$

At time t_1 it has the value

$$g(z - ct_1)$$

which is a function of z, having any form, say of the type shown in Fig. 22.2. At a subsequent time t_2, the function g has the value $g(z - ct_2)$, which may be written

$$g(z - ct_2) = g([z - c(t_2 - t_1)] - ct_1)$$

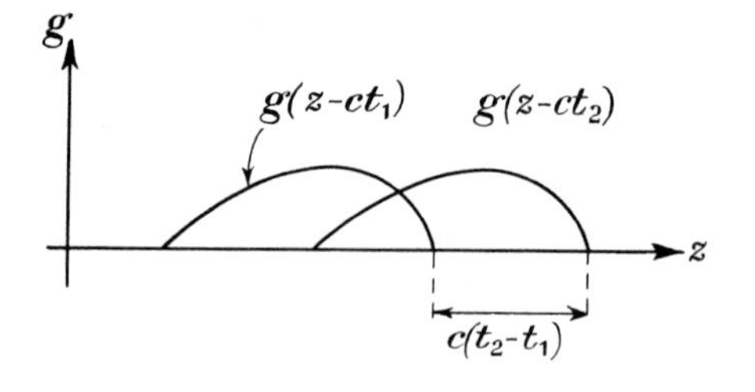

Fig. 22.2 Velocity of propagation of a stress wave.

In other words, $g(z - ct_2)$ has the same *form* as $g(z - ct_1)$ if the variable z be changed to

$$z - c(t_2 - t_1)$$

This means that the function $g(z - ct_2)$ is equivalent to the function $g(z - ct_1)$ displaced an amount

$$c(t_2 - t_1)$$

along the z-axis. In fact $g(z - ct)$ represents a travelling wave of constant form; in a small interval of time δt the wave is displaced an amount $c\,\delta t$, so that c is the wave-velocity along the z-axis. Similarly we can show that $f(z + ct)$ is a wave of constant form travelling with velocity c in the reverse direction, that is, in the negative direction of z. The general solution given by equation (22.7) consists of two waves travelling with equal speeds but in reverse directions.

Table 22.1 VELOCITIES OF STRESS WAVES IN THREE COMMON MATERIALS

Material	$c = \sqrt{\dfrac{E}{\rho}}$
Steel	5100 m/sec
Aluminium alloy	4900 m/sec
Concrete	2400 m/sec

The velocities of propagation of stress waves in steel, aluminium light-alloy, and concrete are given in Table 22.1. In evaluating c it is important to express E in absolute units.

22.3 Constant stress applied at one end of the rod

Consider the particular case for which

$$f(z + ct) = 0, \qquad g(z - ct) = \frac{\sigma_0}{E}(z - ct)$$

for $z < ct$, and

$$f(z + ct) = 0, \qquad g(z + ct) = 0$$

for $z > ct$. Then $w = 0$ for $z > ct$, and for $z < ct$,

$$w = \frac{\sigma_0}{E}(z - ct) \tag{22.8}$$

This represents a wave of the form shown in Fig. 22.3, the triangle *Oab* increasing in size as the wave travels along the rod. Within the distorted region,

$$\sigma = E\frac{\partial w}{\partial z} = \sigma_0 \tag{22.9}$$

so that the assumed value of the function g corresponds to a constant tensile stress σ_0 in the region *Ob* of the rod. Since $Ob = 0$ when $t = 0$, there is no displacement at the

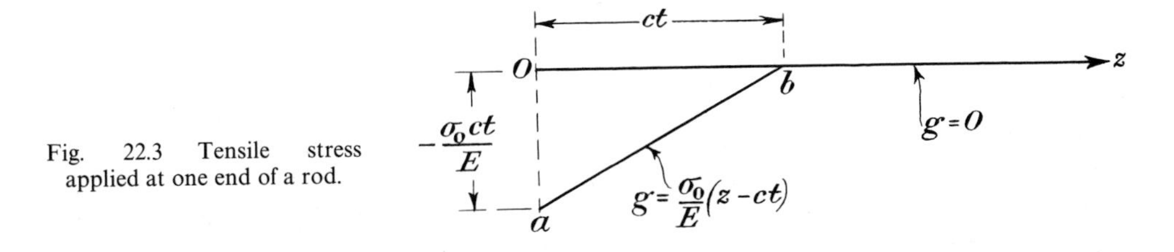

Fig. 22.3 Tensile stress applied at one end of a rod.

end O initially, and the problem corresponds with a rod on which a tensile stress σ_0 is suddenly applied at the end O, and maintained constant in magnitude; all sections of the rod eventually attain the stress σ_0 as the wave *Oab* travels along the rod. The velocity of the particles in the stressed region is

$$\dot{w} = \frac{\partial w}{\partial t} = -\frac{\sigma_0 c}{E}$$

On substituting for c from equation (22.5), we have

$$\dot{w} = -\frac{\sigma_0}{\sqrt{\rho E}} \tag{22.10}$$

Thus the particle velocity, $\dot{w}$, is different, *both in magnitude and direction*, from the wave velocity, c.

Problem 22.1: A vertical wire is being wound onto a drum at a speed of 3 m/sec when the free end of the wire is suddenly held fixed. Show that the instantaneous stress induced is about 120 MN/m^2. For the wire $E = 200$ GN/m^2, and $\rho = 7840$ kg/m^3. (*Cambridge*)

Solution

From equation (22.10), the stress induced by changing the velocity suddenly by an amount v is

$$\sigma = v\sqrt{\rho E}$$

Thus

$$\sigma = 3[(7840)(200 \times 10^9)]^{1/2} = 119 \text{ MN/m}^2$$

22.4 Reflection of the stress wave at the ends of a rod

The rod of Fig. 22.4 is stressed uniformly over the length *ct*, since in the distorted region $\sigma = \sigma_0$. The wave-front, Fig. 22.4(i), travels along the rod with velocity *c*. When the wave-front reaches a free end of the rod, there can be no resultant longitudinal stress in the rod. This can only be achieved if, as in Fig. 22.4(ii), the tensile stress σ_0 is countered at the free end of the rod by a compressive stress σ_0. But the sudden application of a compressive stress at the free end sets up a compressive stress wave travelling from right to left, with velocity *c*, as in Fig. 22.4(iii). We superpose the effects of tensile and compressive stresses, and find that the two waves of Fig. 22.4(iii) are equivalent to a

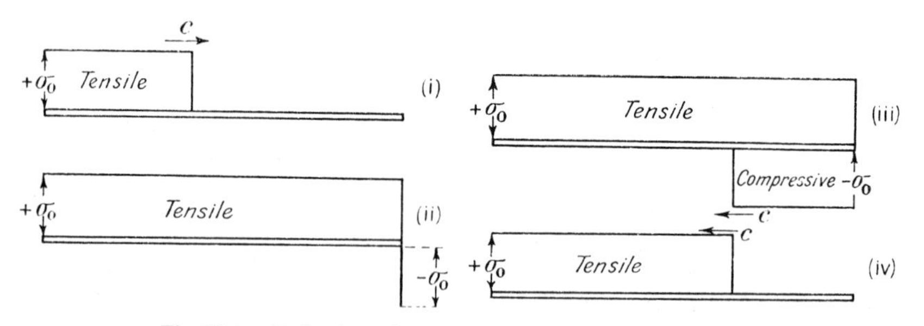

Fig. 22.4 Reflection of a stress wave at the free end of a rod.

single *receding* wave of tensile stress, Fig. 22.4(iv). If the disturbing tensile stress σ_0 be removed as soon as the receding wave reaches the near end of the rod, the rod is left quite unstressed and at rest.

When the remote end of the rod is built-in to a rigid support, at reflection of the tensile stress wave there is no displacement or velocity of the rod at the support. This is achieved by applying an additional tensile stress σ_0 at the built-in end at the instant the stress wave reaches that end, Fig. 22.5(ii). Superposition of the two tensile waves leads to the stress wave of Fig. 22.5(iii); at the built-in end the magnitude of the tensile stress is $2\sigma_0$.

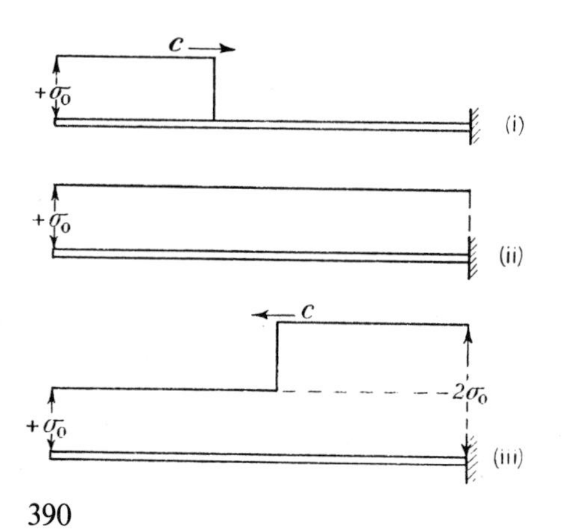

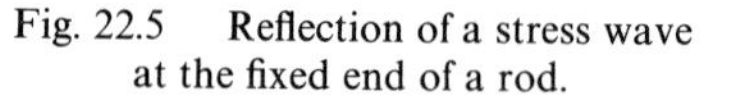
Fig. 22.5 Reflection of a stress wave at the fixed end of a rod.

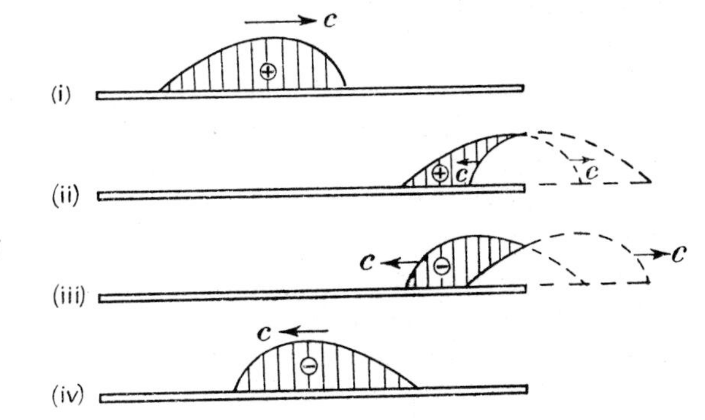

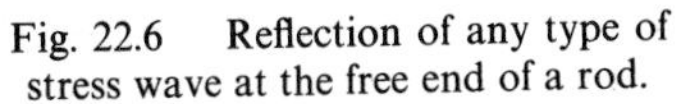
Fig. 22.6 Reflection of any type of stress wave at the free end of a rod.

The reflections of stress waves shown in Figs. 22.4 and 22.5 are for the relatively simple case of uniform tensile stress σ_0 travelling along the rod. In a more general case the stress wave may have any form, Fig. 22.6(i); when a free end is reached a reflected compressive wave is set up, Fig. 22.6(ii) and (iii), and finally the wave is completely reflected as a compressive wave, Fig. 22.6(iv). When the remote end of a rod is fixed rigidly, Fig. 22.7(i), a tensile stress wave on reaching the fixed end induces an equal and opposite tensile stress wave, Fig. 22.7(ii). Momentarily the stresses at the fixed end may attain maximum values twice as great as those induced at any other section of the rod. Ultimately the tensile stress wave is completely reflected, Fig. 22.7(iii).

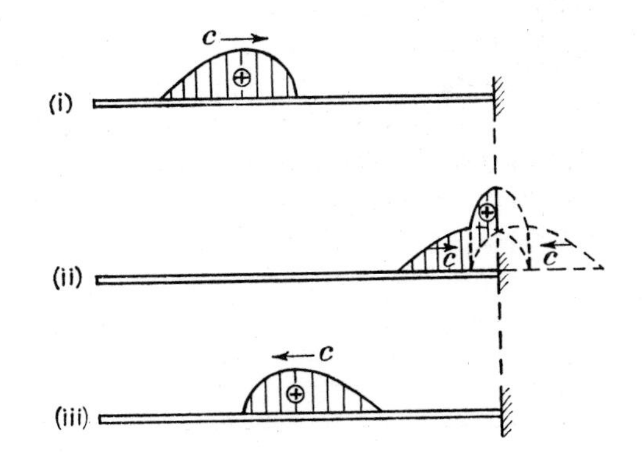

Fig. 22.7 Reflection of any type of stress wave at a fixed end of a rod.

22.5 Longitudinal impact of rods

Consider first the case of two equal rods approaching one another at the same velocity v, Fig. 22.8(i). If contact takes place simultaneously at all points over the sections at the ends of the rods, then by symmetry the plane of contact remains at rest after the initial impact, and equal compressive waves are propagated along the rods with the wave velocity c, Fig. 22.8(ii). The compressed regions of the rods are at rest; but we have seen

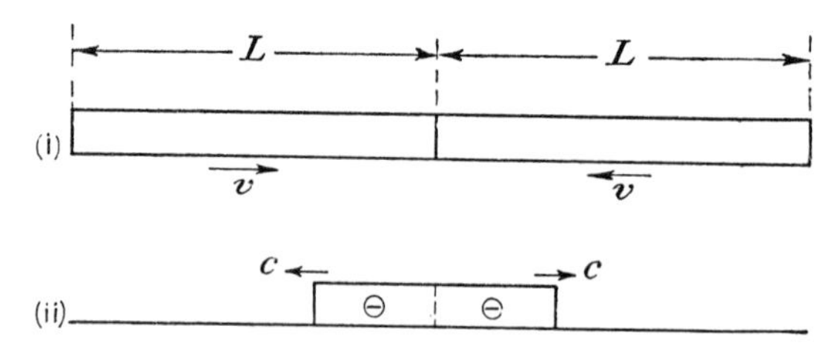

Fig. 22.8 Impact of two equal rods approaching with equal velocities.

already that the particle velocity due to a uniform stress σ_0, equation (22.10), is

$$\dot{w} = -\frac{\sigma_0}{\sqrt{\rho E}}$$

when the rod is initially at rest. When the particles have initially a velocity v, and are then given an additional particle velocity, the final velocity is zero if

$$v = -\frac{\sigma_0}{\sqrt{\rho E}}$$

Thus the intensity of compressive stress induced by the impact is

$$\sigma_0 = -v\sqrt{\rho E} \tag{22.11}$$

When the compressive waves reach the free ends of each rod, equal tensile waves are propagated in the reverse directions, in the manner already outlined in §22.4, and the waves of compression recede to the contact surface. When these receding waves finally reach the contact surface, the two rods separate with velocity v. The duration of the impact is the time taken for the compression waves to advance and recede along the rods at velocity v; this is given by $2L/c$.

Suppose we superpose a constant velocity V, Fig. 22.9, on both rods in Fig. 22.8. Then we have the case of two rods approaching at different velocities. The two rods approach at a *relative* velocity $2v$, and separate, as we have seen already, with an equal *relative* velocity $2v$. When a velocity V is superposed on all velocities of the rods, we see that before the impact the *absolute* velocities of the two rods are

$$V + v, \qquad V - v$$

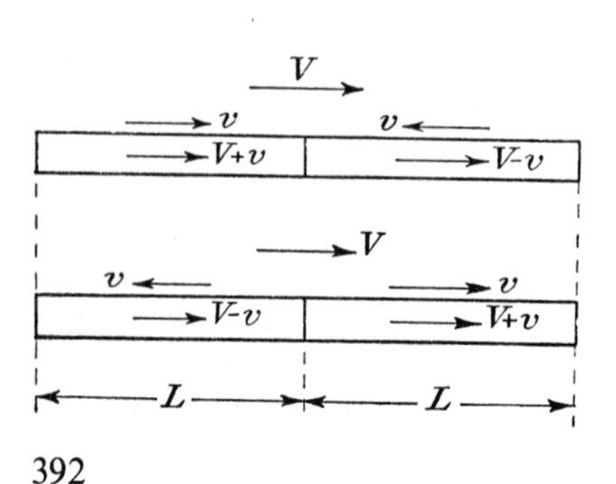

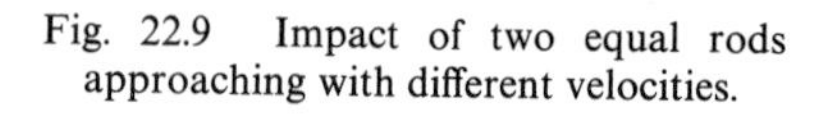
Fig. 22.9 Impact of two equal rods approaching with different velocities.

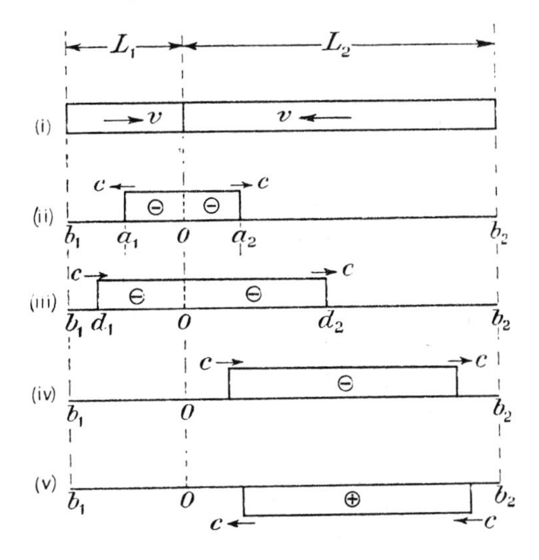

Fig. 22.10 Impact of two rods of different lengths approaching at the same velocities.

and after the impact

$$V - v, \qquad V + v$$

respectively. So that the rods exchange their *absolute* velocities.

Consider now the case of two rods of unequal length, but equal in other respects, approaching at the same velocities v, Fig. 22.10(i). Immediately after the impact, compression waves are set up at the contact surface O, Fig. 22.10(ii), and these are propagated in the two rods at a velocity c; the regions a_1b_1 and a_2b_2, Fig. 22.10(ii), are still moving with velocities v, while the compressed region a_1a_2 is at rest. If $L_1 < L_2$, the wave of compression reaches b_1 before it reaches b_2; the compression wave in Ob_1 then recedes from b_1, after reflection at b_1, so that d_1d_2, Fig. 22.10(iii), progresses as a *standing wave* along the two bars. Since the stresses are compressive at the surface of contact O, this standing wave can be transmitted through the surface of contact, Fig. 22.10(iv). After reflection at b_2, the standing wave is reflected as a tensile stress wave, which however cannot be transmitted across the surface of contact O. Consequently, when the tensile wave reaches the surface of contact, the two rods separate, and the stress waves are afterwards restricted to the rod Ob_2. The rod Ob_1 is travelling ultimately with velocity v, in the reverse direction to that with which it started.

When the rods approach with unequal velocities the behaviour can be deduced by superposing a suitable velocity on all velocities of the rods.

22.6 Rod struck by a moving mass

A rigid body of mass M, Fig. 22.11(i), travelling with velocity V, strikes the free end of a stationary rod of length L. The rod is built-in to a rigid support at the remote end. If the body is absolutely rigid we may assume that the velocity imparted instantaneously

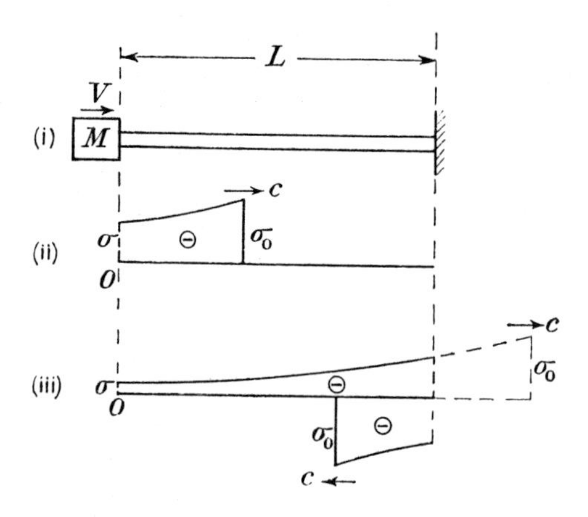

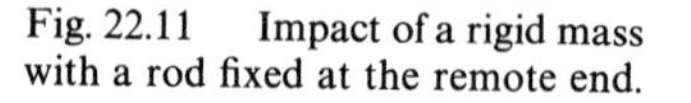
Fig. 22.11 Impact of a rigid mass with a rod fixed at the remote end.

to the particles at the end O of the rod is also V. If the particle velocity is V, we have from equation (22.10) that the induced compressive stress is

$$\sigma_0 = V\sqrt{\rho E} \tag{22.12}$$

But the motion of the mass M is now resisted by the action of this compressive stress at O, and the mass is consequently retarded. Suppose v is the velocity of the mass M at a time t after the impact, and that σ is the compressive stress at the surface of contact at the same instant. Then the equation of motion of the mass is

$$\sigma A = -M\frac{dv}{dt} \tag{22.13}$$

where A is the cross-sectional area of the rod. But v is also the particle velocity in the rod at the surface of contact, and from equation (22.10)

$$v = \frac{\sigma}{\sqrt{\rho E}} \tag{22.14}$$

So that equation (22.13) may be written

$$\frac{M}{\sqrt{\rho E}}\frac{d\sigma}{dt} + A\sigma = 0 \tag{22.15}$$

Hence

$$\sigma = Ce^{-(A\sqrt{\rho E}/M)t} \tag{22.16}$$

where C is a constant; if $\sigma = \sigma_0$ when $t = 0$, we have

$$\sigma = \sigma_0 e^{-(A\sqrt{\rho E}/M)t} \tag{22.17}$$

The intensity of compressive stress at the end O of the rod diminishes exponentially therefore, Fig. 22.11(ii). The compression wave, on reaching the fixed end of the rod, is reflected as a compression wave, and momentarily the compressive stress at the fixed end attains the value $2\sigma_0$. When the reflected wave reaches the end O of the rod, it is again reflected as though O were a fixed end, since the mass M cannot suddenly change its velocity; at the instant of reflection at O the compressive stress in the rod is

$$2\sigma_0 + \sigma_0 e^{-(A\sqrt{\rho E}/M)(L/2c)} \tag{22.18}$$

To find the greatest compressive stress during the complete process of impact we may have to consider a number of reflections of the stress wave.

ANSWERS TO FURTHER PROBLEMS

(Answers are given only for numerical and algebraic parts of problems.)

1.17 31·0 MN/m^2 (compressive); 0·098 cm.
1.18 0·902 cm.
1.19 0·865 cm.
1.21 51·0 kN.

1.22 65·5 MN/m^2 tensile in steel; 41·0 MN/m^2 compressive in copper; increase of length 0·611 cm; force to prevent expansion 135·5 kN.

2.5 $0{\cdot}866 \times 10^{-3}$.
2.6 0·306 cm.
2.7 $2L\theta(\alpha_1 - \alpha_2)$.
2.8 43·3 kN; 50·0 kN; 25·0 kN.
2.9 0·547 W.
2.10 38·1 kN.

3.4 480 MN/m^2.
3.5 521 kW.
3.6 295 m/s^2.
3.7 188·5 Nm.
3.8 d = 6·29 cm; cotter thickness 1·57 cm; mean width of cotter 7·98 cm; distance of cotter hole from end of left-hand rod 2·97 cm; diameter of right-hand rod through cotter pin 8·28 cm; maximum diameter of right-hand rod 12·58 cm; distance of end of right-hand rod from cotter hole 2·97 cm.
3.9 8·93 cycles/sec.
3.10 0·6 MN/m^2.

4.7 376 kN/m.

5.7 50·0 MN/m^2 tensile; 28·9 MN/m^2 shearing.
5.8 greatest tensile stress 86·6 MN/m^2, on plane at 34° 44′ to cross-section; greatest shearing stress 64·0 MN/m^2, on planes at 10° 16′ and 79° 44′ to cross-section.
5.9 30 MN/m^2 tensile; 120 MN/m^2 compressive.
5.10 81·0 MN/m^2 inclined at 23° 27′ to horizontal.
5.11 90 MN/m^2 tensile; 60 MN/m^2 compressive; $5{\cdot}40 \times 10^{-4}$ tensile; $4{\cdot}35 \times 10^{-4}$ compressive.
5.12 7·5 MN/m^2 normal; 51·9 MN/m^2 shearing.

6.8 11·0 MN/m^2.
6.9 1·03 kg/m.
6.10 0·114 per cent.
6.11 (*a*) copper: 38·2 MN/m^2; wire: 83·9 MN/m^2;
(*b*) copper: 28·6 MN/m^2 (compressive); wire: 230 MN/m^2.
6.12 1·19 MN/m^2.
6.13 171 MN/m^2.

7.10 489 kNm.
7.13 238 kNm; 0·75 m from *A*.

8.2 1128 kNm at 10·3 m from one end.
8.6 622 kNm.
8.7 $z(140 - 1{\cdot}5z)$ for $z < (100/3)$ m; $(100 - z)(1{\cdot}5z - 5)$ for $z > (100/3)$ m.

9.12 40·6 kNm.
9.13 69·9 MN/m^2.
9.14 15·9 MN/m^2.
9.15 86·0 MN/m^2.
9.16 6·17 cm.

10.3 1 cm thickness; 5 cm spacing of rivets, assuming one rivet at any cross-section.
10.4 maximum tensile stress of 124 MN/m^2 is greater than the allowable stress; maximum shearing stress of 18 MN/m^2 is less than the allowable stress.
10.5 96 per cent of shearing force carried by web; 88 per cent of bending moment carried by flanges.
10.6 web thickness 0·67 cm; weld throats 0·33 cm.
10.7 $\tau = 2450(RL/t) \sin\theta$, where θ is the angular position of any section from the vertical line through the centre of the tube.
10.8 bending is limiting, and gives an allowable superimposed load of 45 kN/m; required welds 0·26 cm throat thickness.

11.8 $0{\cdot}378 \times 10^{-3}$ m^2; 13·02 kNm.
11.9 114 kN.
11.10 wood 4·56 MN/m^2; steel 52·9 MN/m^2; glue 0·21 MN/m^2.
11.11 (i) 120 MN/m^2.
(ii) 1·00 MN/m^2.
(iii) 100 kN/m.
(iv) 0·75 cm.

12.3 tensile 155 MN/m^2, compressive 147 MN/m^2; neutral axis 0·365 m from outside of box-section.
12.4 17·68 kN; 11·8 MN/m^2 compressive.
12.6 maximum tensile 38·0 MN/m^2; maximum compressive 46·0 MN/m^2.
12.7 161 kN.
12.8 13·8 MN/m^2; 5·94 cm from tip of *T*.

13.12 1·80 cm and 2·48 cm.
13.13 3·06 cm.

14.6 maximum bending moment 105 kNm; points of inflexion at 1·75 m from each end.
14.7 169·7 kNm at left-hand end; 150·0 kNm at right-hand end; 1·52 m from left-hand end; 1·69 m from right-hand end.
14.8 910 kN at end supports; 1370 kN at intermediate supports.
14.9 (*a*) 901 kNm; 211 kN.
(*b*) 1173 kNm; 234 kN.
14.10 central peg 3·28 N, outside pegs 0·74 N, others 2·48 N.

15.2 217 kN.
15.3 62·4 kN/m.
15.4 required elastic section modulus 791 cm^3.
15.5 required elastic section modulus 2030 cm^3.
15.6 84·2 kN/m, with collapse in the end spans.
15.7 3·26 cm.

16.7 38·1 MN/m^2; 1·09°; 39·2 cm.

16.8 40·3 MN/m^2; 3·83°.
16.9 shearing stress 37·7 MN/m^2; maximum tensile stress 37·7 MN/m^2; angle of twist 4·31°.
16.10 0·644:1.
16.11 38·7 Nm.
16.12 147 MN/m^2 (tensile) at 34·8° to axis, 70·5 MN/m^2 (compressive) at 57·2° to axis.

17.12 $4WL/EA$.
17.13 $5WL/3EA$.
17.14 $2W$.
17.15 no horizontal deflection.

18.7 $55WL/6EA$.
18.8 31·6 kN in outside wires, 49·4 kN in central wire.
18.9 609 kNm and 423 kNm.
18.10 $3WR/4\pi$, at the support, where W is the weight of the ring.
18.11 12·45 PR^3/EI.

19.3 44·1 MN/m^2 and 9·75 cm.
19.4 2·41 cm and 5·7 turns.
19.6 (i) 13·0 cm.
(ii) 177 MN/m^2
(iii) 13·0 Nm.
19.7 16 plates; 3·75 cm; 3·34 m.

20.2 970 N.
20.3 0·10 cm.
20.4 1·65 kN.
20.5 24·5 kN.

21.3 7·00 cycles/sec, 28 cycles/sec, etc.; 11·85 cycles/sec, 47·4 cycles/sec, etc.
21.4 4·73 cycles/sec.
21.5 $(3EI/ML^3)^{1/2}/2\pi$.

INDEX

Numbers refer to pages